AA007391

Hydraulics-2019

IOP Conference Series: Materials Science and Engineering
Volume 779

Moscow, Russia
27 November 2019

ISBN: 978-1-7138-3511-0
ISSN: 1757-8981

Printed from e-media with permission by:

Curran Associates, Inc.
57 Morehouse Lane
Red Hook, NY 12571

Some format issues inherent in the e-media version may also appear in this print version.

This work is licensed under a Creative Commons Attribution 3.0 International Licence.
Licence details: http://creativecommons.org/licenses/by/3.0/.

No changes have been made to the content of these proceedings. There may be changes to pagination and minor adjustments for aesthetics.

Printed with permission by Curran Associates, Inc. (2021)

For permission requests, please contact the Institute of Physics at the address below.

Institute of Physics
Dirac House, Temple Back
Bristol BS1 6BE UK

Phone: 44 1 17 929 7481
Fax: 44 1 17 920 0979

techtracking@iop.org

Additional copies of this publication are available from:

Curran Associates, Inc.
57 Morehouse Lane
Red Hook, NY 12571 USA
Phone: 845-758-0400
Fax: 845-758-2633
Email: curran@proceedings.com
Web: www.proceedings.com

TABLE OF CONTENTS

IOP Conf. Series: Materials Science and Engineering **779** (2020) 011001 doi:10.1088/1757-899X/779/1/011001

Hydraulics-2019

The conference "Hydraulics" is held annually in November since 2014at the Bauman Moscow State Technical University. The Department of Fluid Mechanics, Hydraulic Machines and Hydraulic and Pneumatic Automation is the conference organizer. The best conference reports are published as articles in the journal "Hydraulics", indexed in the Russian Science Citation Index. Every year, representatives of various scientific, educational and industrial organizations of Russia and other countries come to us with reports and speeches. The main topics of the reports are hydromechanics, hydraulic machines, and hydraulic drives. A part of the reports is devoted to industrial ecology issues. We are pleased to see at our conference scientists from any countries and from any organizations devoting their work to industrial hydraulics, ecology and solving fundamental issues in these areas.

Sincerely, A. Protopopov

Content from this work may be used under the terms of the Creative Commons Attribution 3.0 licence. Any further distribution of this work must maintain attribution to the author(s) and the title of the work, journal citation and DOI.
Published under licence by IOP Publishing Ltd

IOP Conf. Series: Materials Science and Engineering **779** (2020) 011002 doi:10.1088/1757-899X/779/1/011002

Peer review statement

All papers published in this volume of *IOP Conference Series: Materials Science and Engineering* have been peer reviewed through processes administered by the proceedings Editors. Reviews were conducted by expert referees to the professional and scientific standards expected of a proceedings journal published by IOP Publishing.

Content from this work may be used under the terms of the Creative Commons Attribution 3.0 licence. Any further distribution of this work must maintain attribution to the author(s) and the title of the work, journal citation and DOI.
Published under licence by IOP Publishing Ltd

IOP Conf. Series: Materials Science and Engineering **779** (2020) 012001 doi:10.1088/1757-899X/779/1/012001

The use of various types of circulation in the design of excavator hydraulic drive

A Myrzakhmetov[1] and A Akinfiev[1,2]

[1]Bauman Moscow State Technical University

[2]E-mail: akasimul@yandex.ru

Abstract: This article is devoted to the review, study and analysis of the existing types of circulation used in the design of hydraulic drives of construction machines. Three main types of circulation are distinguished: closed, open, and semi-closed. A comparison is given and the characteristic features of each of them are described. Particular attention is paid to semi-closed circulation because it is under investigated. The sequence of movements of the equipment elements during the work cycle is shown. The results are presented, in the form of a diagram of flow rates in the actuators of an excavator planner, made by semi-closed circulation. This work will be of interest to specialists in the field of designing volumetric hydraulic drives of construction vehicles.

Introduction

Excavators are multi-engine machines with a rigid suspension of working equipment, in which a hydraulic volumetric drive is used to transfer power from the engine to the working mechanisms. Compared to mechanical, hydraulic excavators have a wider range of interchangeable work tools, a greater number of main and auxiliary movements of the working equipment, which significantly expands their technological capabilities.

The hydraulic drive allows [1,2]: to significantly simplify the kinematics of the transmission and working equipment; expand the range of interchangeable work tools; reduce the size of the machine; rationally combine work operations; maximize the power of the power plant; increase the mobility and versatility of machines and improve the quality of work performed; ensure smooth movement and precise orientation of the work tool; realize great digging efforts; improve machine performance.

Fig. 1 shows various types of circulations used to create the hydraulic drive of construction vehicles. When creating hydraulic circuits for hydraulic machines, the first two types of circulation are widely used [3].

A feature of the closed circulation circuit is the work on the hydraulic mechanism, which has the same volume of cavities, in the case of using a hydraulic cylinder with a two-sided rod or a hydraulic motor. The flow rate diagram for a circuit made by closed circulation is shown in Fig. 2a. Make-up is provided by a make-up pump 1, this pump also compensates for leakage in the hydraulic machine and the actuator. When the rod moves, the flow displaced from the cavity of the hydraulic cylinder enters the suction of the pump minus volume losses in the pump [4,5].

Content from this work may be used under the terms of the Creative Commons Attribution 3.0 licence. Any further distribution of this work must maintain attribution to the author(s) and the title of the work, journal citation and DOI.
Published under licence by IOP Publishing Ltd

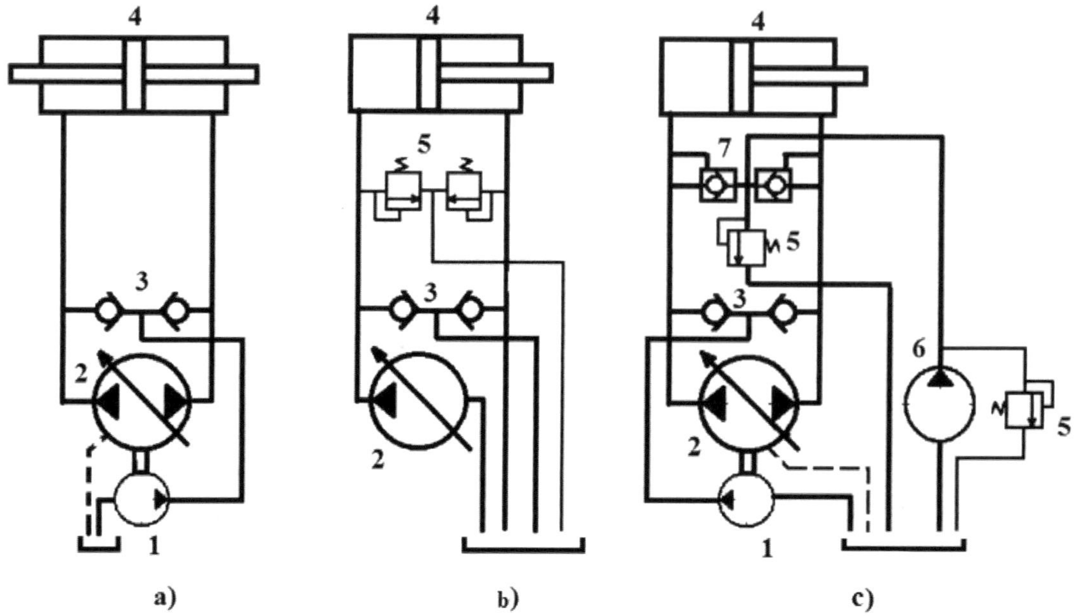

Figure 1. Types of circulation
a) — closed circulation; b) — open circulation; c) — semi-closed circulation;1 — make-up pump; 2 — the main pump; 3 — make-up valves;4 — a hydraulic cylinder; 5 — safety valves; 6 — pump compensating for the lack of working fluid in the cavity of the differential hydraulic cylinder

A feature of the circuit with open circulation is the difference in the volume of the piston and rod cavities of the hydraulic cylinder, which is determined by the coefficient:

$$\varphi = \frac{S_P}{S_R};$$

where, S_P — the area of the piston cavity of the hydraulic cylinder;

S_R — the area of the rod cavity of the hydraulic cylinder.

The pump draws liquid from the tank, and the liquid is also drained into the tank, so this scheme is most often used for machines in the hydraulic drive of which several differential hydraulic cylinders are used. For example, in excavators, where differential hydraulic cylinders are used in the mechanisms of the shovel boom, handle and bucket: in the drive of the mechanism for lifting the shovel boom and telescoping. In the hydraulic drive of excavators, secondary valves (Fig. 1b, pos. 5) are used to protect the drive and metal structures, which protect the drive and the metal structure from overload, in the case when, for example, the excavator handle mechanism loads the locked cavity of the boom or bucket hydraulic cylinders while digging [6, 7]. When the valve is activated in the rod cavity, the piston moves and the rod extends, since the fluid is drained through the valve. Since the hydraulic cylinder cavities are locked, the piston cavity is fed by gravity from the tank through the check valves Fig. 1b pos. 3, due to the difference between the pressure in the tank and the pressure in the cavity.

The flow rates diagram for the scheme made by open circulation is shown in Fig. 2b. When the rod is extended, the pump feed enters the piston cavity, the discharge rate from the rod cavity is less than the pump feed, due to the difference between the areas of the piston and rod cavities [8].

$$Q_D < Q_P = Q_{PUMP};$$

$$Q_D = Q_R = \frac{Q_P}{\varphi};$$

When retracting the rod, the flow rate entering the drain is greater than the flow rate entering the rod cavity.

$$Q_D > Q_R = Q_{PUMP};$$

$$Q_D = Q_R \cdot \varphi;$$

Fig. 1c shows a circuit made by semi-closed circulation. The pump, designed to work in closed circulation, is used to drive a mechanism with a differential hydraulic cylinder. And the difference in the volume of the piston and rod cavities necessitates the introduction of additional devices [9,10].

When the rod is extended, Fig. 2c, a pump feed is supplied to the piston cavity, and a smaller volume of liquid is supplied from the rod cavity to the suction, the difference in these rates is compensated by an additional pump operating in open circulation and performing, in addition to compensation, other functions in the drive. When the rod is pulled the pump feed is supplied into the rod cavity, and a larger volume of liquid is drained from the piston cavity, the difference in these rates is dumped into the tank so that the balance of rates on the pump operating in a closed circulation remains unchanged.

Despite the complication of the hydraulic circuit used for semi-closed circulation, it is used, for example, on Slovakian excavators, which are widely used in the CIS [11,12].

To control the drive speed in open circuits, throttling hydraulic valves are used; in semi-closed circuits, speed is controlled by changing the pump's working volume, which is more energy-efficient than throttling [13,14].

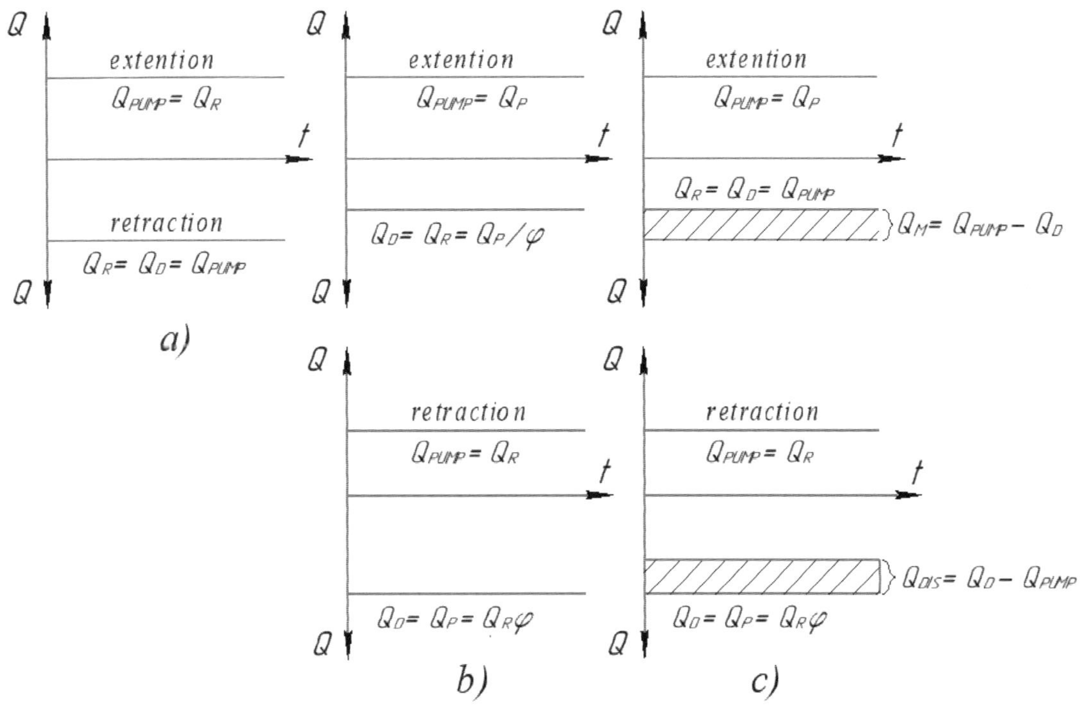

Figure 2. Diagrams of fluid supply to the piston cavity and drain from the rod cavity
a) closed circulation; b) open circulation; c) semi-closed circulation; Q_R— flow rate into the rod cavity, Q_P— flow rate into the piston cavity, Q_{PUMP}— pump supply, Q_D— drainage flow rate, Qm— make-up flow rate, Q_{DIS}— discharge flow.

The sequence of movements of the elements of the working equipment during the work cycle
In the work cycle, the sequence of operations of the excavator planner can be divided into two groups of operations. Digging: by turning the bucket or moving the telescope. Transport operations: lifting the working equipment from the bottom, turning the bucket to keep soil in the bucket, turning from the bottom, unloading, turning into the bottom, lowering the working equipment, installing the bucket at a new digging point [15].

Duty cycle operations can be performed without combining the movements of mechanisms or with combining, for technological reasons, as well as to reduce the duration of the cycle. Some operations, for example, planning slopes or planning the bottom cannot be performed without combinations. Combining operations gives the operator more opportunities, and therefore, when creating hydraulic circuits for excavators, the need to combine the movements of mechanisms is always taken into account.

The table below lists possible combinations of movements of mechanisms of six elements in two, which usually tend to ensure the use of a dual-pump circuit [16].

Table 1. Combination of movements of six mechanisms in two movements

Mechanism	Telescope	Bucket	Boom	Rotation	Movement
Telescope		"+"	"+"	"+"	"+"
Bucket	"+"		"+"	"+"	"+"
Boom	"+"	"+"		"+"	"+"
Rotation	"+"	"+"	"+"		"+"
Movement	"+"	"+"	"+"	"+"	

When digging, the movements of the bucket and telescope do not combine, since before digging, when turning the bucket, the cutting angle is set, which does not change during digging. If digging is done with a bucket, it is also possible to combine with a boom to reduce the thickness of the chips. And it is also possible that, when digging, the movement of the telescope is combined with the turn of the bucket for processing slopes.

The movement of the boom is combined with rotation during transport operations, with the movement when pulling the machine from the ground when working in viscous water-saturated soils.

Excavation of the soil with telescopic working equipment with a backhoe bucket can be done both by turning the bucket and retracting the movable boom sections.

With above information, the retraction mechanism is considered the main, and the bucket rotation mechanism is considered auxiliary. The latter can dig weak soils, cut the bucket when digging the calculated soil, but its main purpose is to fix the bucket relative to the movable section when it is retracted.

All possible combinations of six elements of two can be used in the work cycle. In some cases, it becomes necessary to combine three movements, for example, when planning the bottom of the trench. Any sequence of movements in the work cycle for digging and transport operations can be accepted. Fig. 3 shows a cycle in which there is a combination of three movements when digging and when performing transport operations [17].

When controlling mechanisms in the work cycle, it is desirable that the speeds are not dependent on the load, because operations often must end at the same time. For this, volume control of the speed of movement of the actuators and semi-closed circulation of the flow can be applied.

An example of such a solution is the hydraulic circuit of the UDS-214 excavator. In this hydraulic circuit, some of the mechanisms are powered by a liquid in a closed circulation circuit, and some are separately in an open circulation circuit.

IOP Conf. Series: Materials Science and Engineering **779** (2020) 012001 doi:10.1088/1757-899X/779/1/012001

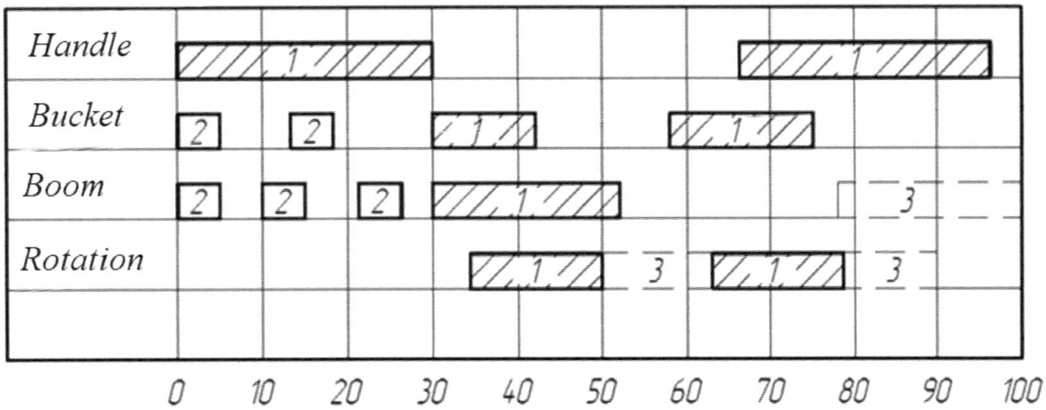

Figure 3. Excavator work cycle with backhoe work equipment
1 — operations performed by energy; 2 — short-term operations; 3 — operations performed without energy consumption (the boom lowers under its own weight, the telescope lowers the bucket without soil into the face)

Having data on the flow of pumps, the parameters of hydraulic cylinders and hydraulic motors, as well as assigning the stroke of the hydraulic cylinders and the angles of platform rotation to unload and to the bottom, you can determine the flow rate flowing into the hydraulic mechanisms and out of them during uncombined and combined operations. Using these data it is possible to determine the required make-up for pumps operating in a closed circulation, as well as the flow rates discharged into the tank when there is an excess of flow to the suction of pumps operating in a closed circulation.

The tables below show the parameters of pumps and hydraulic motors obtained from the UDS-214 operating manual. The parameters of the hydraulic cylinders are also given and the costs in the piston and rod cavities, as well as in the cavity of the platform rotation hydraulic motor, are calculated.

Table 2. Parameters of pumps and hydraulic motors

Excavator Pumps and Hydromotors					
Designation according to manual		Volumetric displacement	Rotation speed	Volumetric Efficiency	Flow rate
		V_0	n	η	Q
		cm^3	rpm	%	l/min
416.3.90	1st pump, twin	90	2000	0,9	162
	Make-up pump	19,8	2000	0,9	35,64
416.3.90	2nd pump, twin	90	2000	0,9	162
	Make-up pump	19,8	2000	0,9	35,64
416.3.71	3rd open circuit pump	71	2000	0,9	128
	Make-up pump	19,8	2000	0,9	35,64
406.0.71	Movement hydraulic motor	71	2000	0,9	128
406.0.71	Rotation hydraulic motor	71	2000	0,9	128
406.0.71	Head rotation hydraulic motor, 2 pcs	71	2000	0,9	128

IOP Conf. Series: Materials Science and Engineering **779** (2020) 012001 doi:10.1088/1757-899X/779/1/012001

Table 3. Parameters of hydraulic cylinders

Pumps, hydraulic cylinders UDS						
	D	d	S	Fp	Fr	φ=Fp/ Fr
	mm			m^2		
Telescope	110	70	4150	0,0095	0,0057	1,66
Boom x2	150	70	1000	0,0177	0,0138	1,28
Bucket	140	70	500	0,0154	0,0115	1,34

Table 4. Flow rates in the piston cavity of hydraulic cylinders

Flow rates in the piston cavity of hydraulic cylinders (excluding efficiency)				
hydraulic cylinders	In the piston cavity	Piston Speed	From the rod cavity	
	Qpump	Vp=Qpump/Fp	Qr=Vp*Fr	
	l/min	m/s	m^3/s	l/min
Telescope	162	0,284	0,0016	97
Boom x2	162	0,152	0,002	126
Bucket	128	0,138	0,0016	95,2

Table 5. Flow rates in the rod cavity of hydraulic cylinders and the hydraulic motor of the platform rotation during the work cycle

Flow rates in the rod cavity of hydraulic cylinders and the hydraulic motor of the platform rotation (excluding efficiency)						
hydraulic cylinders	In the rod cavity	Piston Speed	From the piston cavity		In the hydraulic motor of rotation	Out of the hydraulic motor of rotation
	Qpump	Vp=Qpump/Fr	Qp=Vp*Fp			
	l/min	m/s	m^3/s	l/min	l/min	
Telescope	162	0,474	0,0045	270	128	128
Boom x2	162	0,195	0,0034	207	128	128
Bucket	128	0,185	0,0028	171	128	128

The following is a typical work cycle of an excavator with telescopic equipment for two main operations: during digging and during transport operations. It should also be noted that the work cycle and sequence of actions may be different, and depends on the type of operator control and depending on the operations performed.

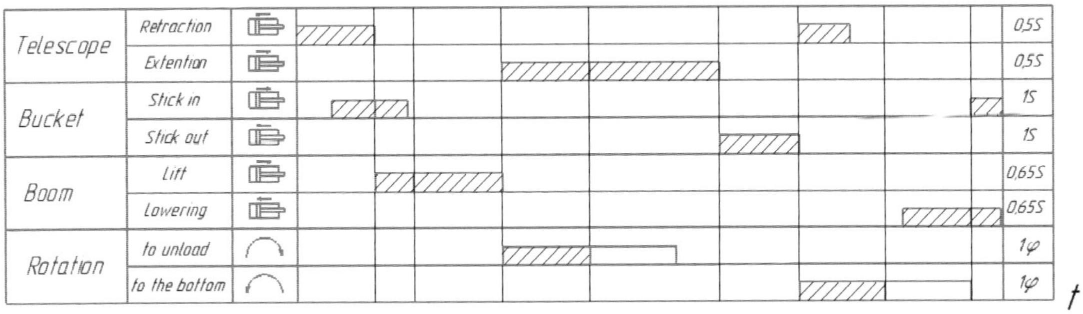

Figure 4. The sequence of operations of the work cycle indicating possible combinations of working bodies

Table 6. Digging suction and discharge flow rates

Delivery to hydromechanisms and discharge and suction flow rates, l / min				
Cycle operations		Digging		
The mechanisms	Direction	telescope, bucket		
Telescope, H1	Retraction	162		
	Extension			
Bucket, H3	Stickin	128	128	
	Stickout			
Boom, H2	Lift		162	162
	Lowering			
Rotation, H3	Tounload			
	Tothebottom			
Suction flow rate H1		270		
Suction flow rate H2			126	126
Drainage flow rate H3		95,2	95,2	
$Q_{dis} - Q_{dr}$ for H1		-108		
$Q_{dis} - Q_{dr}$ for H2			+36	+36

Table 7. Flow rates of suction and discharge during transport operations

Delivery to hydromechanisms and discharge and suction flow rates, l / min							
Cycle operations		transport operations					
The mechanisms	Direction	Boom, rotation, bucket, telescope					
Telescope, H1	Retraction				162		
	Extension	162	162				
Bucket, H3	Stickin						128
	Stickout			128			
Boom, H2	Lift						
	Lowering					162	162
Rotation, H3	Tounload	128	128				
	Tothebottom				128	128	
Suction flow rate H1		97	97		270		
Suction flow rate H2						207	207
Drainage flow rate H3		128	128	171	128	128	95,2
$Q_{dis} - Q_{dr}$ for H1		+65	+65		-108		
$Q_{dis} - Q_{dr}$ for H2						-45	-45

The analysis of suction and discharge flow rates in table 6 and 7 shows that when working in the telescoping cylinder rod cavity during digging there is a need to discharge excess flow rate to 108 l/min, and when working in the telescoping cylinder piston cavity during transport operations, need to be fed with a flow rate of 65 l/min. When the boom is lowered, the discharge flow rate is 45 l/min, and when the boom is raised, it is necessary to feed the suction cavity with a flow rate of 36 l/min. These values are determined by the maximum feeds to the actuating hydromechanisms without taking into account volumetric efficiency (volumetric loss coefficient).

Thus, the control valve and, in general, the hydraulic circuit must be able to provide both make-up and discharge of excess fluid for suction. A semi-closed circuit, consisting of two parts, can provide this possibility: the first part is a closed circulation circuit; the second is an open circulation circuit. In addition, a hydraulic device should be provided that provides make-up and discharge of excess liquid for suction.

Below is a diagram of the flow rates entering and leaving the hydraulic mechanisms during combined and not combined movements of the working bodies. Using these data it is possible to determine the required make-up for pumps operating in a closed circulation, as well as the flow rates discharged into the tank when there is an excess of flow to the suction of pumps operating in a closed circulation.

IOP Conf. Series: Materials Science and Engineering **779** (2020) 012001 doi:10.1088/1757-899X/779/1/012001

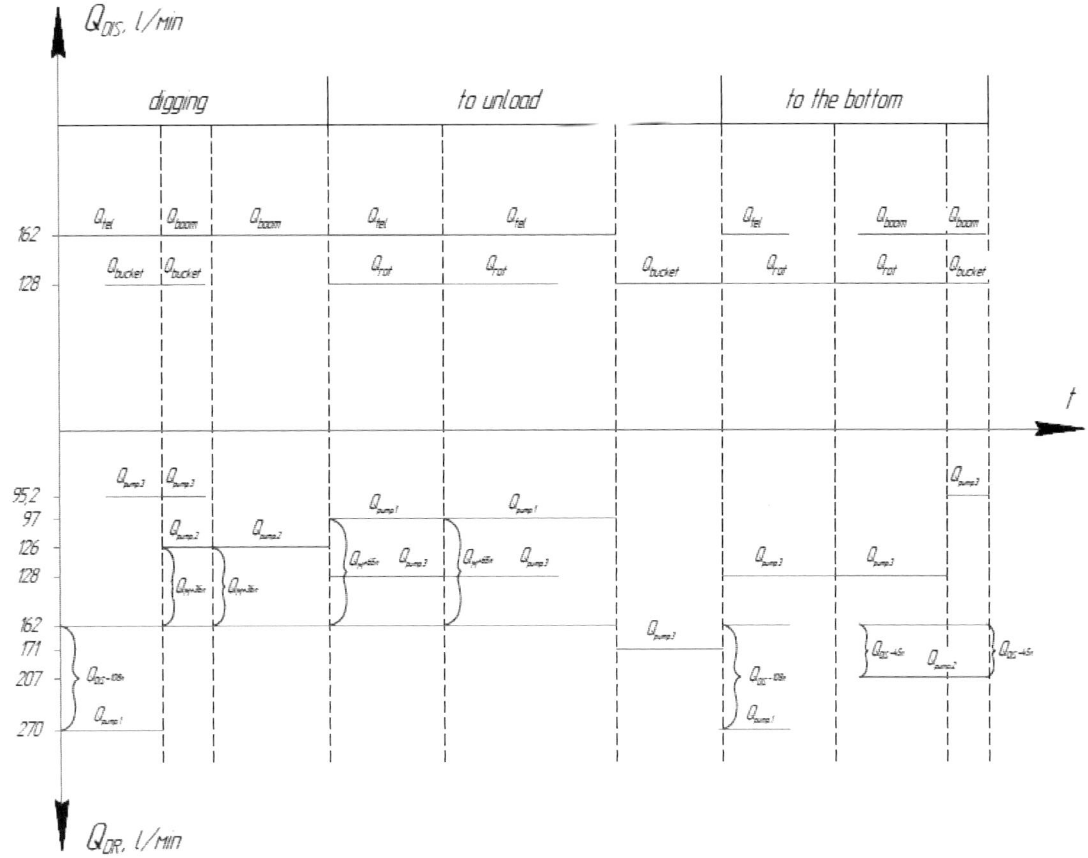

Figure 5. Diagram of flow rates in executive hydraulic mechanisms of an excavator planner

REFERENCES

[1] OS "Pnevmostroimashina" , Prospect PSM on the GST-71, GST-90, Yekaterinburg, 2009.

[2] "Hydraulic and pneumatic power control systems" Under. ed. J.Blackbourne et al .; Per. from English V.M.Dvoretsky and A.M. Plungyan; Under. ed. Cand. tech. Sciences V.A. Khokhlova. — Moscow: Publishing house of foreign countries. lit., 1962 — 614 p.: ill .; 21 cm. Aut. Chapters: Blackburn J.F., Rithof G., Stenning A. etal.Fundamentals of the theory and design of volumetric hydraulic transmissions. Kulagin A.V., DemidovYu.S., Prokofiev V.N., Kondakov L.A. Higher School, 1967, p. 400.

[3] AnatoliyKomissarov, YuliyaLagunova, Viktor Shestakov and Olga Lukashuk2018 IOP Conf. Ser.: Mater. Sci. Eng. 323 012015 .

[4] ATLAS designs of hydraulic machines and hydraulic transmissions. Textbook for students of engineering specialties of universities / B.M. Bim-Bad, M.G. Kabakov, V.N. Prokofiev, etc.— M.: Mechanical Engineering, 1990.—136p.: Ill.

[5] A V Kozhukhova, O N Sakharova and I A Dmitrieva2019 IOP Conf. Ser.: Mater. Sci. Eng. **643** 012098.

[6] Volumetric hydraulic machines: a training manual / B.P. Borisov. — Moscow: Publishing house of MGTU im. N.E. Bauman, 2018.— 237 pp., Ill.

[7] Lewis E., Stern H. Hydraulic control systems., M .: MIR, 1966. Chapter 6, p. 232.

[8] C Y Zhang, G Z Chen, C W Miao, Y F Zhao and S Zhong2018 IOP Conf. Ser.: Mater. Sci. Eng. **392** 062040.

[9] Oliver Unruh, Thomas Haase and Martin Pohl 2016 J. Phys.: Conf. Ser. **744** 012018.

[10] A Protopopov and V Vigovskij 2019 IOP Conf. Ser.: Mater. Sci. Eng. **492** 012003.
[11] A Shablovskiy and E Kutovoy 2019 IOP Conf. Ser.: Mater. Sci. Eng. **492** 012035.
[12] A Gouskov et al 2019 IOP Conf. Ser.: Mater. Sci. Eng. **492** 012013.
[13] V Lomakin et al 2019 IOP Conf. Ser.: Mater. Sci. Eng. **492** 012012.
[14] I Kolodin and M Ryabinin 2019 IOP Conf. Ser.: Mater. Sci. Eng. **589** 012018.
[15] S Barmin and O Nikitin 2019 IOP Conf. Ser.: Mater. Sci. Eng. **589** 012020.
[16] S Korsakova and AProtopopov 2019 IOP Conf. Ser.: Mater. Sci. Eng. **492** 012032.
[17] Construction and road cars and the basics of automation: a training manual / I.F. Dyakov; / Ulyan. state tech. un. — Ulyanovsk: UlSTU, 2017.—516s.

IOP Conf. Series: Materials Science and Engineering **779** (2020) 012002 doi:10.1088/1757-899X/779/1/012002

Traditional methods for the design of radial-axial hydraulic turbines with verification in CFD simulation

I Handal[1,2], V Tkachuk[1], A Petrov[1]and A Protopopov[1]

[1]Bauman Moscow State Technical University

[2]E-mail: handal.ignacio15@gmail.com

Annotation. In the modern world, the use of hydrodynamic power plants is one of the fundamental ways of generating electricity. The main part of such power plants is the hydraulic turbine impeller, therefore, for the developer, the main task is to design it correctly. In this work are used the traditional methods for designing radial-axis hydraulic turbines and test their operability with CAD and CFD programs. For that was created a hydro turbine model that will be installed in El Salvador. Subsequently, in the future, this hydroturbine will be optimized.

Introduction

Hydraulic turbines have been used for over a century and are based on reliable technology. However, growth and competition in the energy market requires an increase in electricity production at lower costs, which leads to the need to improve equipment for hydropower plants. Therefore, it is important to increase the efficiency of hydroelectric power plants, namely the efficiency of turbines.

The efficiency of the turbine can be increased by changing the parameters of the geometry of the turbine. Sequential tests are needed to increase turbine efficiency. Although it is possible to predict the characteristics of the turbine using model tests in the laboratory; time and budget constraints, prototype constraints contributed to the use of CFD tools for turbine optimization.

Each hydropower project requires a different turbine design. Preliminary design is based on experimental methods, traditional design procedures, and the hydrodynamic theory of machines. Preliminary model is subjected to CFD analysis to evaluate hydraulic performance. CFD analysis allows to determine undesirable flow conditions, such as flow separation, cavitation zones, which cannot be predicted in traditional design methods. This is an iterative process leading to optimal turbine performance. It should be noted that the maximum efficiency of the turbine is achieved not under design conditions, but under partial load. The efficiency of the turbine is slightly lower than the maximum efficiency at the design pressure and the estimated flow: this case is called the optimal operating condition of the turbine throughout the entire operation. It is worth noting that approximately 96% of hydraulic losses occur in the blades of the radial axial turbine. Therefore, the main design work is therefore focused on the rotor blades. Minor changes in the solid turbine model, especially in the rotor, result in optimized turbine geometry with optimized efficiency and flow dynamics. Problematic areas can be changed in order to achieve the target efficiency at the design load and design load.Issues related to the design and CFD simulation of hydrodynamic turbines are widely covered in the literature [1–16].

Content from this work may be used under the terms of the Creative Commons Attribution 3.0 licence. Any further distribution of this work must maintain attribution to the author(s) and the title of the work, journal citation and DOI.
Published under licence by IOP Publishing Ltd

A radial-axial turbineis applicable for a wide range of head and flow rates and may be preferable to a rotary-blade hydraulic turbine due to its compactness; or bucket hydroturbine due to its higher efficiency at a better efficiency point.All those factors make the radial-axial turbine profitable, and it is the focus of this study. It has been selected to design a turbine expected to work under the following parametrers.

Table 1. Turbine work parameters

Head pressure, м	Flow Rate, m3 /s	Rotor rotational speed, rpm	Theorical Efficency, %	Specific speed,n_s
72	50.5	300	93	293

The radial-axial turbine design begins with the impeller, and then the guide vanes, fixed vanes and spiral casing are developed. On this article it has been only descrived the process to obtain the impeller geometry, nevertheless the rest of the elements have been calculcated accordingly. After obtaining the theoretical geometry of the turbine, using the CFD software, it is necessary to compare the efficiency obtained at different opening angles of the guide vane.

Method

Hydraulic profile diagram, Bovet method [1]

Theodore Bovet (1964), based on hydraulic turbines already built, determined a number of speed factors to obtain the hydraulic profile of a new turbine.

A specific dimensionless speed is used

$$n_0 = \frac{n_{sync}\left(\frac{Q}{\pi}\right)^{1/2}}{(2gH)^{3/4}} \quad (1)$$

Where n: rad / s, Q: m^3 / s, H: (m), g: (m / s2)

For radial axial turbines the specific number of revolutions is usually within the range of $0.1 < n_0 < 0.8$. Figure 1 shows the variations of the profiles according to their n_0 which serves as a guide for the layout of the hydraulic profiles.

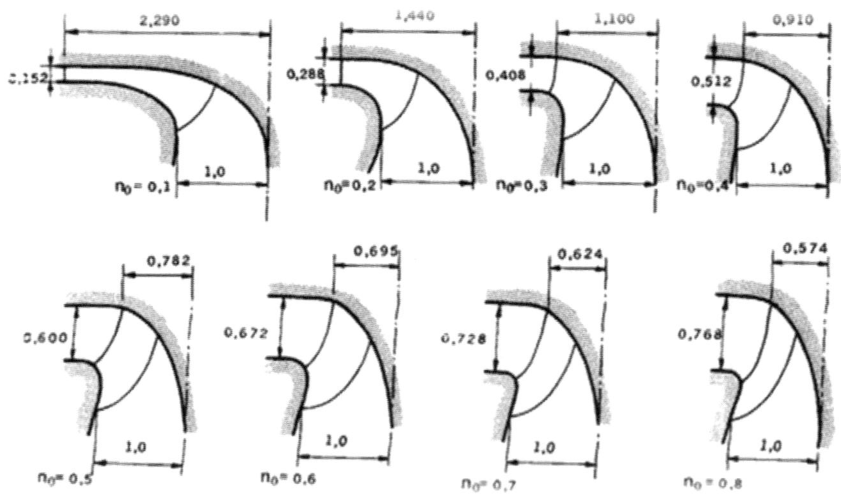

Fig. 1. Variations of the hydraulic profiles.

IOP Conf. Series: Materials Science and Engineering **779** (2020) 012002 doi:10.1088/1757-899X/779/1/012002

To draw the hydraulic profile of the turbine composed of curves *i* and *e*, the following equations are used. It is important to highlight that each of the calculated measures are referred to the basic dimension ρ_{2e} shown in Figure 3, said measurement is taken as a unit equal to the radius of point 2e [2].

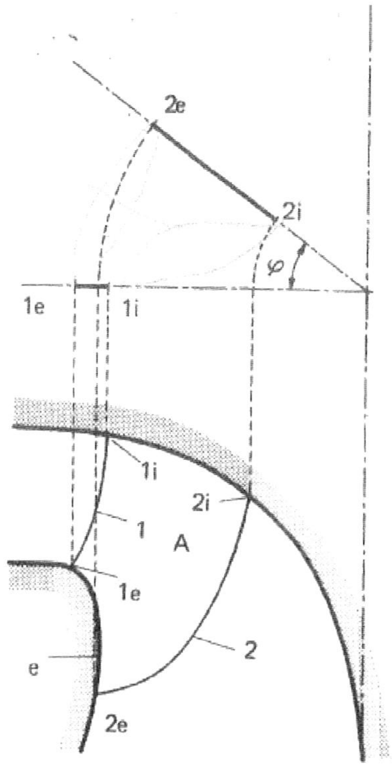

Fig. 2. Limit curves of the volume occupied by the blades.

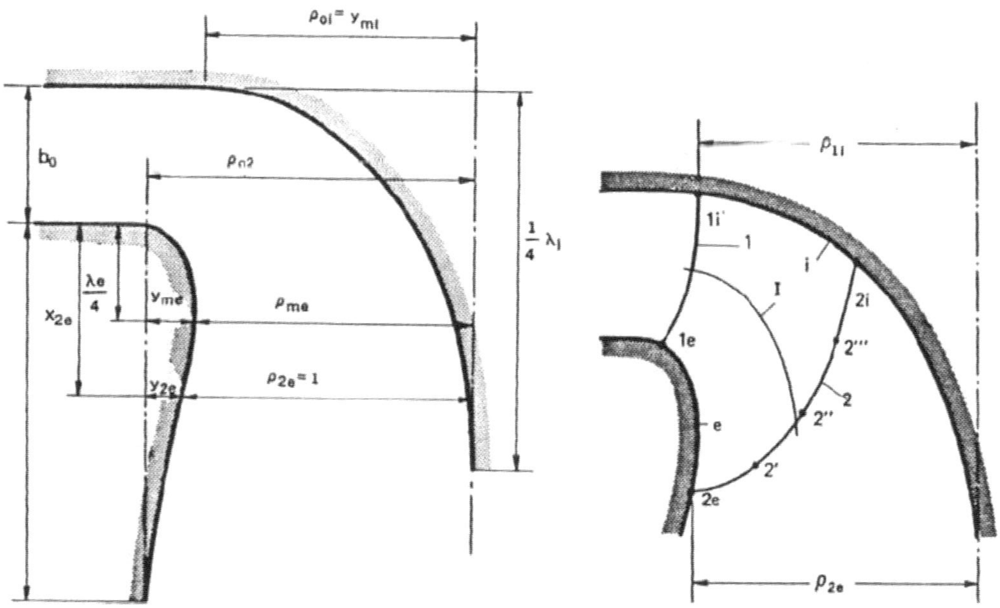

Fig. 3. Typical dimensions of the rotor flow channel

IOP Conf. Series: Materials Science and Engineering **779** (2020) 012002 doi:10.1088/1757-899X/779/1/012002

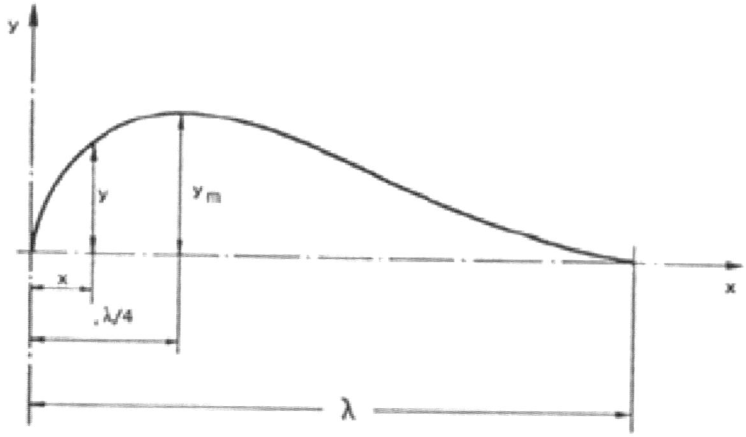

Fig. 4. Boundaries of the curves inside and outside the hydraulic profile

As shown in Figure 4, the internal *i* and external curves *e* of the hydraulic profile are defined by the following equation:

$$\frac{y}{y_m}=3.08*\left(1-\frac{x}{\lambda}\right)\sqrt{\frac{x}{\lambda}\left(1-\frac{x}{\lambda}\right)} \quad (2)$$

Y_m values for the inner curve:

$$y_{mi}=\rho_{oi}=0.7+\frac{0.16}{n_o+0.08} \quad (3)$$

$$\lambda_i=3.2+3.2(2-n_0)n_0 \quad (4)$$

And for the outer curve:

$$\lambda_e=2.4+1.9(2-n_0)n_0 \quad (5)$$

Applying the equation for point *2e*, we have:

$$\frac{y_e}{y_{me}}=3.08*\left(1-\frac{x_{2e}}{\lambda_e}\right)\sqrt{\frac{x_{2e}}{\lambda_e}\left(1-\frac{x_{2e}}{\lambda_e}\right)} \quad (6)$$

Where it is necessary to determine x_{2e} and y_{2e}.

For x_{2e}, a constant value is taken that is independent of n_o:

$$x_{2e}=0.5 \quad (7)$$

For y_{2e}:

$$y_{2e}=\rho_{oe}-1 \quad (8)$$

For ρ_{oe} and other quantities that take into account the geometry of the hydraulic profile, the following values are recommended:

$$\rho_{oe}=\frac{0.493}{n_o^{2/3}}\left(For n_0<0.275\right) \quad (9)$$

$$\rho_{oe} = 1.255 - 0.3n_0 \left(For n_0 > 0.275 \right) \quad (10)$$

$$b_0 = 0.8\left(2 - n_0\right) n_0 \quad (11)$$

Using equation 9 or 10, ρ_{oe} is calculated, this value is taken into equation, and y_{2e} is calculated, and using equation 8, we obtain the relation y_{2e}/y_{me} :

$$y_{me} = \frac{y_{2e}}{y_{2e}/y_{me}} \quad (12)$$

$$\rho_{me} = \rho_{oe} - Y_{me} \quad (13)$$

Determining the actual impeller size [3]

As mentioned above, all dimensions are relativeto ρ_{2e}, this measurement will provide the real size of all measurements of the internal and external curves, and its value is calculated, giving a convenient value for the flow coefficient φ_{2e} expressed by the equation 14.

$$\varphi_{2e} = \frac{Q}{n(\rho_{2e})^2} \cdot \frac{1}{u_{2e}} \quad (14)$$

If you substitute $u_{2e} = 2\pi n \rho_{2e}$, the following equation is obtained

$$\rho_{2e} = \left(\frac{Q}{2\pi^2 n \varphi_{2e}}\right)^{1/3} \quad (15)$$

According to turbines previously built and tested, it is known that the value φ_{2e}, to guarantee optimum performance, it is between 0.26 and 0.28, so the average value is selected $\varphi_{2e} \cong 0,270$.

Similarly, you can determine the true value ρ_{1i}, in this case, the pressure coefficient u_{1i} is used, which is calculated as follows.

$$\psi_{1i} = \frac{H}{(u_{1i})^2 / 2g} \quad (16)$$

If u_{1i} is replaced again and $\psi_{1i} - 1.72$ is provided based on the experience in equation 16, the following equation is obtained for ρ_{1i}

$$\rho_{1i} \cong \frac{60}{2\pi n}\left(\frac{2gH}{1.72}\right)^{1/2} \quad (17)$$

Designing a radial-axial impeller blade in a potential meridional flow (Bowersfeld — Vosnesensky method) [4]

The method of designing the impeller blade in a potential flow was proposed by Bowersfeld in 1913, and then developed by I. N. Voznesensky. This method has been widely used in engineering practice for designing impellers of radial-axial hydraulic turbines, since it allows reasonably and fairly simply to determine the spatial surface of the impeller blade for given conditions.

The method proposes that the flow in the meridionalsection of the rotor is potential; therefore, before calculating the blade, it is essential to construct the potential flow and determine the field of constituent meridional speeds in the rotor flow tract obtained by the Bovet method.

In addition, you need to determine the law of change for the product of the variation of the tangential velocity by the radius $c_u r$ along some of the border flow lines. It is recommended to give the dependency $c_u r$ (s) in the form of a straight line or concave curve. It is necessary to assume a theoretical value of hydraulic efficiency.

The main equation that allows by means of numerical integration to find the shape of the blade is

$$\Delta\theta = \frac{r^2\omega - c_u r}{r^2 c_m}\Delta S \quad (18)$$

Using this equation, the angle ΔΘ in the plane corresponding to the given parameters and to the randomly selected segments ΔS in the current line in the meridional plane of the blade is determined. The sum of the drag angle in the plane Θ=Σ ΔΘ must be within the limits accepted for rotors of similar specific speeds. The procedure is graphical analytical and it's described as follows in order to complete the next table.

Table 2. Tabulated results for drag angle calculation

Point	r (m)	$c_u r$ (m^2/s)	r^2	$r^2\omega$	Cm (m/s)	$r^2 c_m$	$\frac{r^2\omega - c_u r}{r^2 c_m}$	$(8)_{prom}$	Δs (m)	ΔΘ (rad)	Θ (rad)
1	2	3	4	5	6	7	8	9	10	11	12

- Draw the meridional section of the impeller hydraulic tract on a sufficiently large scale. Plot construction flow lines of the potential flow (not less than 5 lines). Draw the entrance contour of the blade (In Figure 5 A_1A_6) and identify the intersection point of the exit contour with the impeller band. The recommendations for placement of the input contour and point B1 can be determined by analogy with other impellers of the same specific speed. Then line A_1B_1 is divided into equal sections of length 8–10 mm.

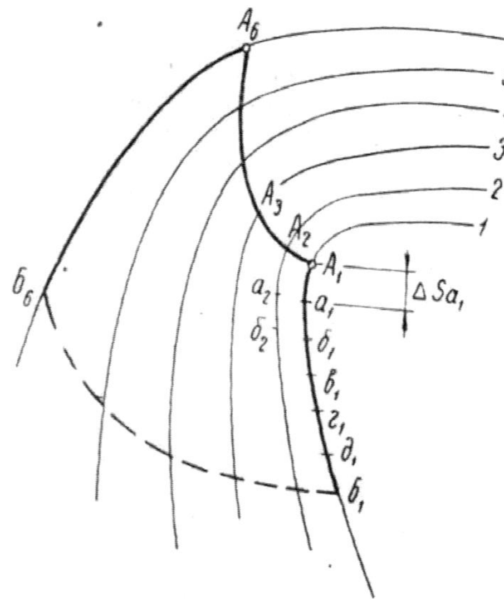

Fig. 5. Hydraulic turbine profile and inlet and outlet contours

- The graph $c_u r$ (s) is constructed for the first current line between limits A_1 and B_1. For this, the values $c_u r$ for the input contour are determined using the main turbine parameters

$$(C_u r)_{in} - (C_u r)_{out} = \frac{\eta g H}{\omega} \quad (19)$$

It can be assumed that $(c_u r)_{out}=0$, which corresponds to a normal speed at the exit, and therefore

$$(C_u r)_{in} = \frac{\eta g H}{\omega} \quad (20)$$

The input and output contours are in meridional planes and are vortex lines along which $c_u r$ = const. Therefore, the result of the previous equation for the input contour will be the same for all its points and for all the points of the output contour it will be 0.

Having chosen a suitable scale, the $c_u r$ (s) curve is arbitrarily determined, as shown in Figure 5. On the S axis the initial coordinate (point 0) corresponds to point A_1 on the input contour. The section Δs of the flow channel drawing is transferred to the graph and consequently for each of them is found the $c_u r$ value.

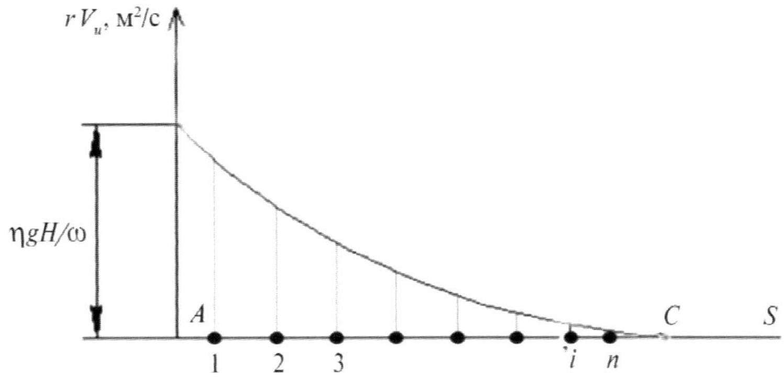

Fig. 6. Variants of dependencies $c_u r = f(s)$

- The c_m (s) graph representing the change in the meridional velocity along the current line, obtained from the construction of the potential flow, is constructed. From the starting point of the coordinates on the s axis for each current line, the point on the input contour is overlaid (A1, A2, ..., A6)

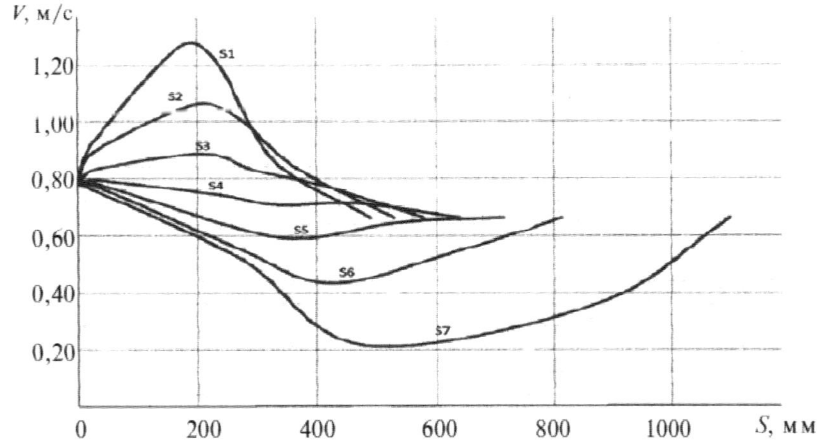

Fig. 7. Variation of meridional velocities along streamlines

- The table previously seen for the first current line is completed. In space 2, the radius values are transferred at points A1, a1, б1 ... Б1, making a direct measurement of the plane of the hydraulic tract. In space 3 the cur values corresponding to these points of the cur (s) graph are placed. The spaces

IOP Conf. Series: Materials Science and Engineering **779** (2020) 012002 doi:10.1088/1757-899X/779/1/012002

4,5,7,8 are completed by means of calculations. In space 6, by the data of the graph cm (s). In space 9, an average is made between the data in space 8 before and after between lines. In space 10 there are the values Δsbetween points A1, a1, б1 ... Б1 at real scale. The values to be placed in space 11 result from multiplication (9) x (10). In spaces 12 and 13, the values Θ of the angles in radians and degrees in summation are transferred, measuring from the input contour (from point A1) using equation 18.

- Line $A_1Б_1$ is constructed in the plane. The horizontal line $0A_1$ corresponds to the meridional section in which the entrance contour is located. The meridional surface on which point a_1 is rotated with respect to point $0A_1$ at an angle $\Delta\Theta_a$, which is taken from the table.With the radius r_a, the point is cut on the line $0a_1$ and the points A_1 and a_1 are joined. Subsequently, the angles $\Delta\Theta_{б1}$, $\Delta\Theta_{в1}$ are placed and the points corresponding to the radius are cut in the lines, the polyline which represents the current line A_1 $Б_1$ in the plane is found. The central angle between the lines $0A_1$ and $0Б_1$ is the drag angle of the blade.

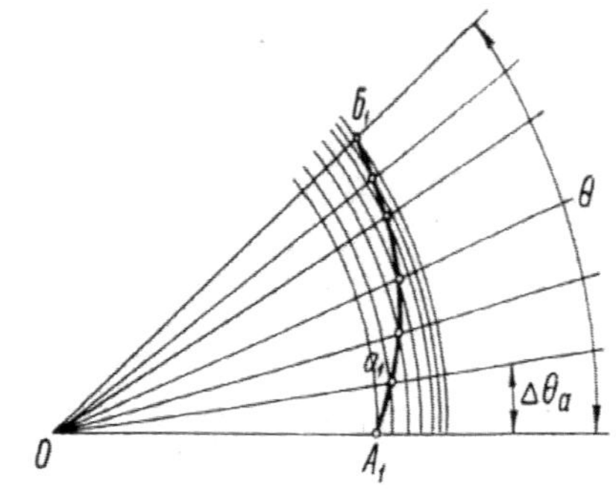

Fig. 8. Planar and meridional coordinates of the surface of the turbine blade

- The calculation is completed for the following current lines. The task is to find sections ΔS_{A2}, $\Delta\Theta_{a2}$that, after carrying out the calculation of the table, must obtain the same angles $\Delta\Theta_{a,}$ $\Delta\Theta_{б,}$ $\Delta\Theta_{в}$as those previously obtained in the first line and are contained within the drag angle.

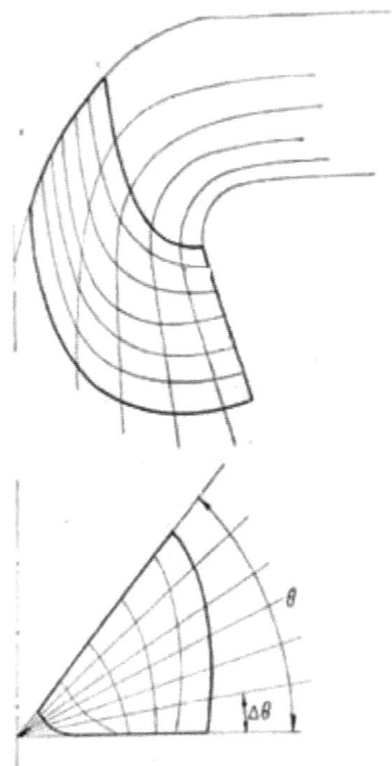

Fig. 9. Planar and meridional coordinates of the surface of the turbine blade

Mathematical model. In order to obtain the hydrodynamic modeling, the following equations were used:

Equation of continuity of the liquid medium:

$$\frac{\partial \overline{u_x}}{\partial x}+\frac{\partial \overline{u_y}}{\partial y}+\frac{\partial \overline{u_z}}{\partial z}=0, \quad (21)$$

Where $\overline{u_i}$ — time-averaged projections of fluid velocities on the corresponding axes;

Equation of the change in the amount of motion averaged over time:

$$\rho\left[\frac{\partial \overline{u_i}}{\partial t}+\overline{u_j}\frac{\partial \overline{u_i}}{\partial x_j}\right]=-\frac{\partial \overline{p}}{\partial x_i}+\frac{\partial}{\partial x_i}\left[T_{ij}^{(v)}-\rho u_i u_j\right], \quad (22)$$

Where $\overline{u_i}\,\overline{p}$ — averaged speed and pressure;

$\tilde{T}_{ij}^{(v)}=2\mu\tilde{s}_{ij}$ — viscous stress tensor for incompressible fluid;

$\tilde{s}_{ij}=\frac{1}{2}\left[\frac{\partial \overline{u_i}}{\partial x_j}+\frac{\partial \overline{u_j}}{\partial x_i}\right]$ — strain rate tensor;

$\rho u_i u_j$ — Reynolds stresses.

The Reynolds system of equations is open due to the presence of unknown Reynolds stresses. The system is closed using the k-ω SST turbulence model.

Terms of computer simulation. In the process of thecomputer simulation, a three-dimensional model of the flow part of the turbine was created, and physical models and parameters of the computational grid were set for it. The density value remained unchanged throughout the experiment. For the boundary conditions, the inlet and outlet pressure values were determined to create a pressure drop equal to the nominal pressure (see Table 3). The rotation speed is preliminarily assigned based on the initial turbine design conditions. Impeller torque and flow rates measured due to pressure drop were measured.

Table 3. Initial conditions

Parameter	Value
Input stagnation	711125,0 Pa
Outlet pressure	0,0 Pa
Rotational speed	300 rpm

To build the grid, the following construction models were selected:

- Generator of polyhedral cells;
- surface mesh generator;
- Prismatic layer generator.

The grid parameters are shown in Table 4:

Table 4. The parameters of the computational grid

Parameter	Value
Base size	50 мм
Prismatic Stretching	1.3
Prismatic layer thickness, relative to the base size	50%
The number of prismatic layers	5
Number of Cells Generated on Impeller	7541332
The number of generated cells on Inlet	5096089
The number of generated cells on Outlet	1024312

IOP Conf. Series: Materials Science and Engineering **779** (2020) 012002 doi:10.1088/1757-899X/779/1/012002

In the impeller, the mesh was thinned 2 times to reduce the calculation error. The resulting mesh in the cross section of the flow part is shown in Fig. 4:

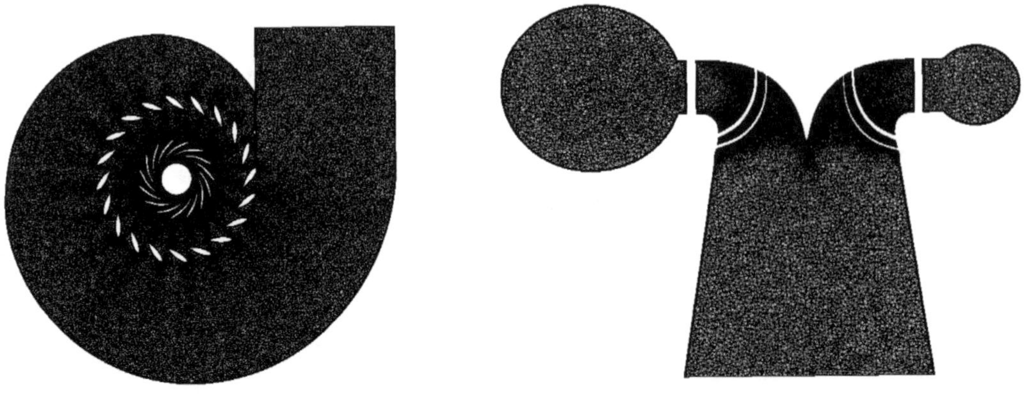

Fig. 10. The resulting mesh in the cross section of the flowing part

Results

Applying the Bovet method the geometry of the hydraulic tract is obtained using the main design parameters of the turbine Q= 50 m3/h; H = 72.5 m; P = 33 MWand applying the Bowersfeld — Vosnesenskymethod, the geometry of the blade is obtained.

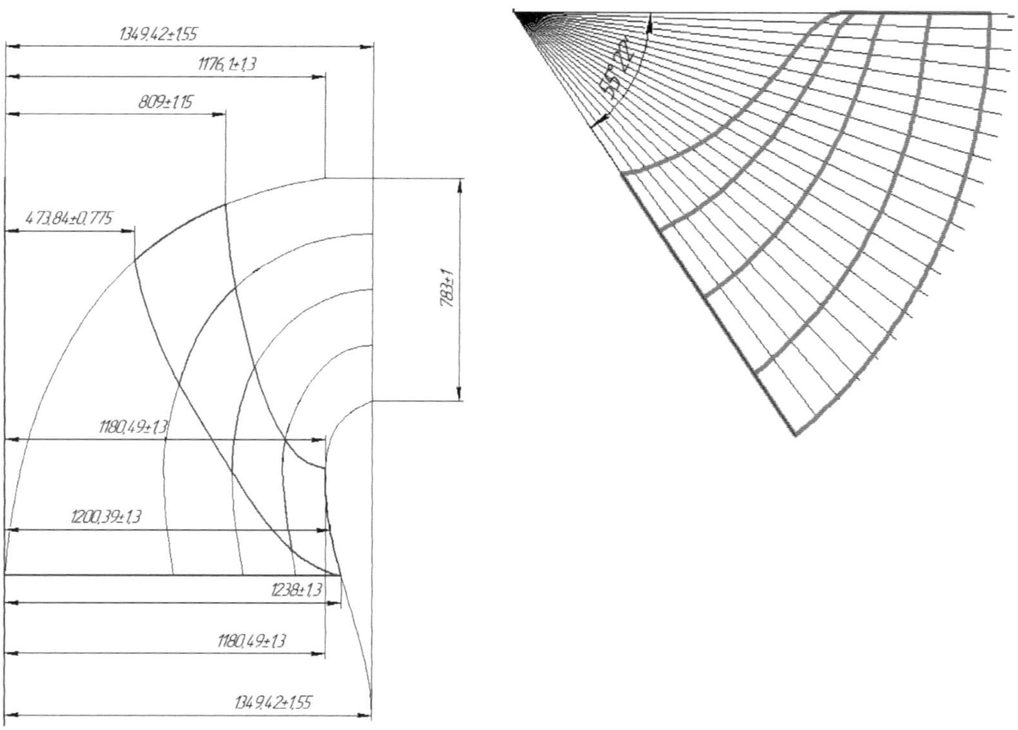

Fig. 11. Dimensions and Coordinates obtained

The spiral chamber, the distributor and the discharge tube take their dimensions based on the nominal diameter of the impeller. The geometry of the simulated turbine is constructed using CAD programs [5]. The CFD program is used to obtain various parameters of the turbine, changing the angle on the distributor. The results calculated are shown in Table 5.

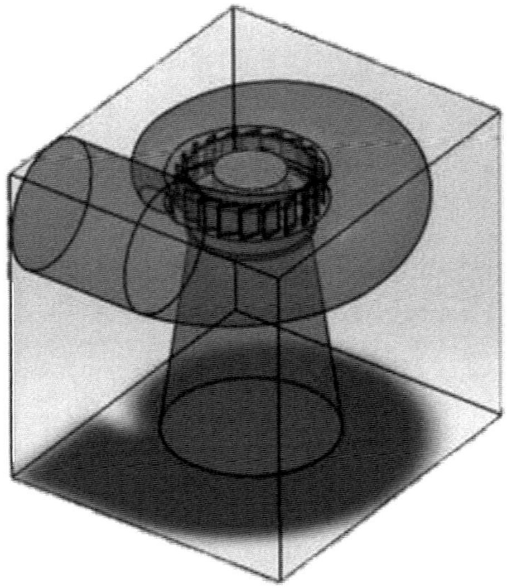

Fig. 12. Hydraulic model of the turbine composition

Table 5. Parameters obtained

Distributor opening (grades)	**Flow Rate** (m^3/s)	**Head** (m)	**Torque** (Nm)	**Power impeller** (MW)	**Power fluid** (MW)	**Efficiency**
26	28.93	71.75	378062.63	11.88	20.36	58.80%
30	43.6	69.66	627163.53	19.70	29.79	64.00%
33	46.3	69.89	693619.86	21.79	31.74	68.00%
35	48.55	69.66	723487.34	22.73	33.18	68.40%
38	51.83	69.26	747819.34	23.49	35.22	66.00%
40	53.77	69	758076.03	23.82	36.40	65.00%
43	57.48	68.4	762232.88	23.95	38.57	62.00%
46	60.36	67.85	744574.59	23.39	40.18	58.00%

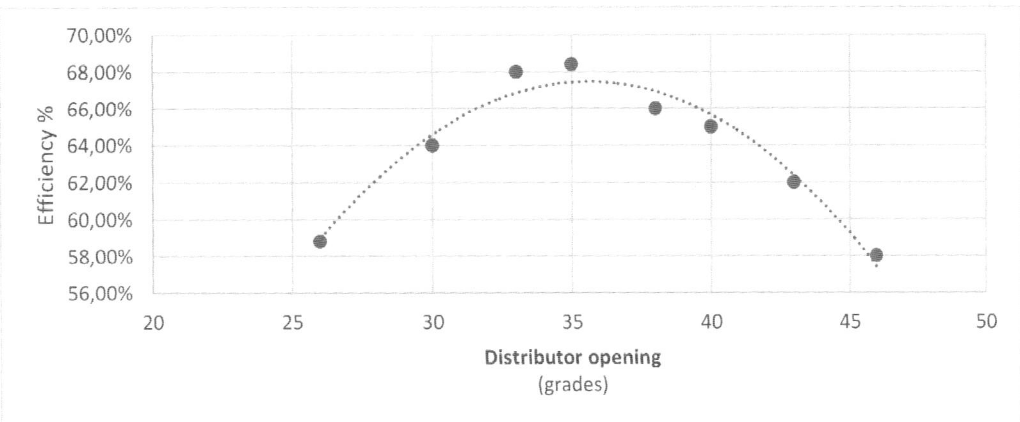

Fig. 13. Efficency graph

IOP Conf. Series: Materials Science and Engineering **779** (2020) 012002 doi:10.1088/1757-899X/779/1/012002

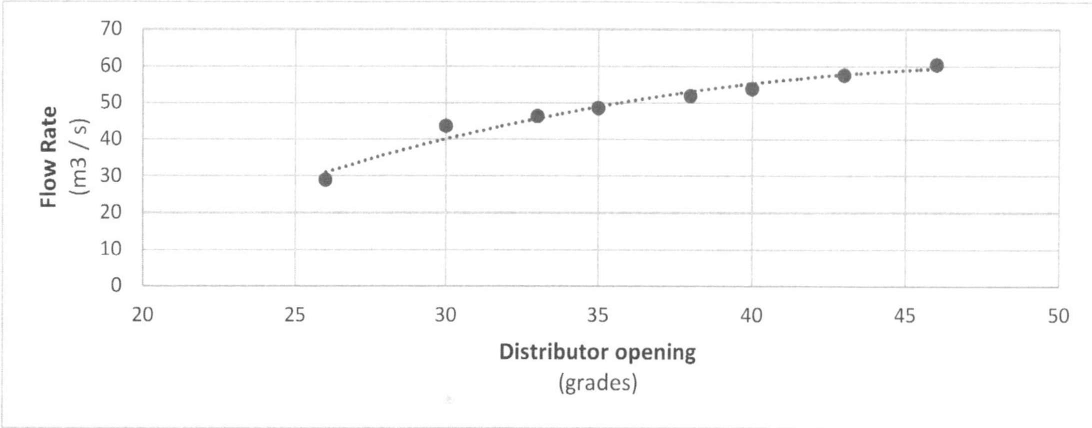

Fig. 14. Flow Rate graph

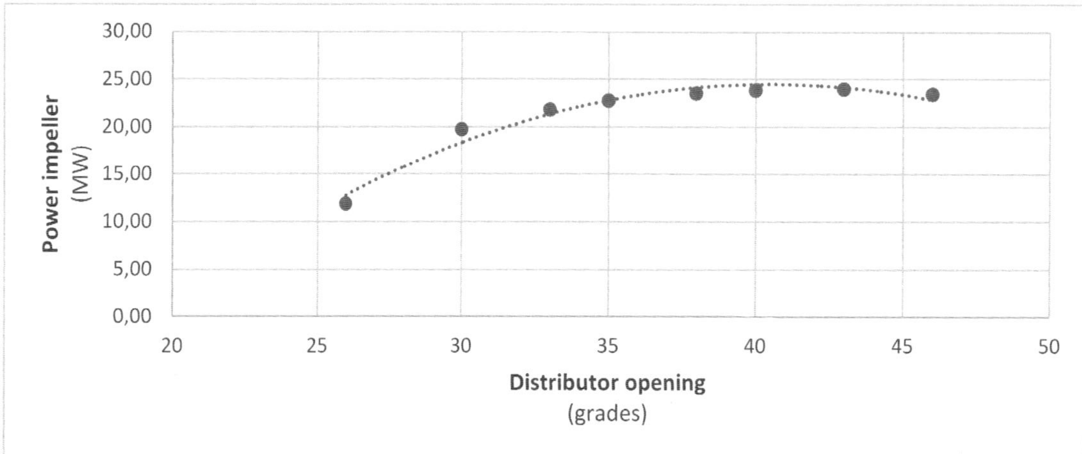

Fig. 15. Power on the Impeller Graph

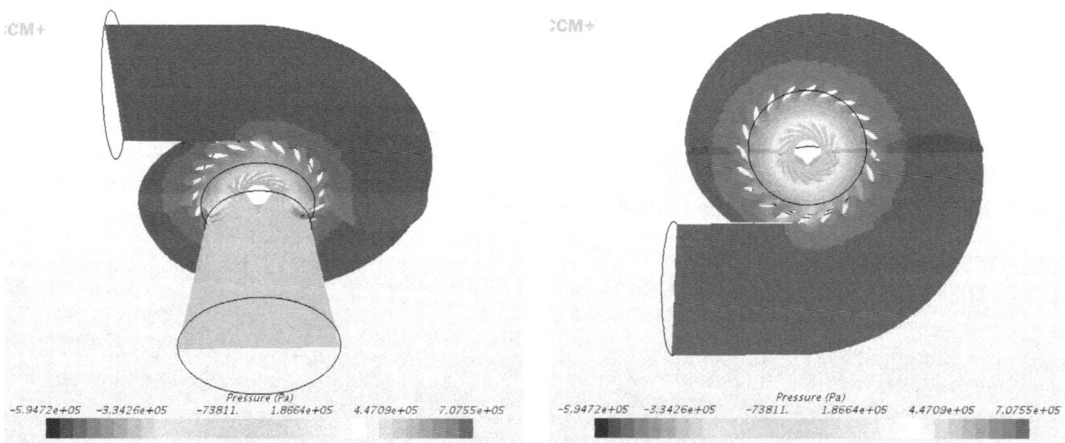

Fig. 16. Pressure distribution in the hydraulic circuit

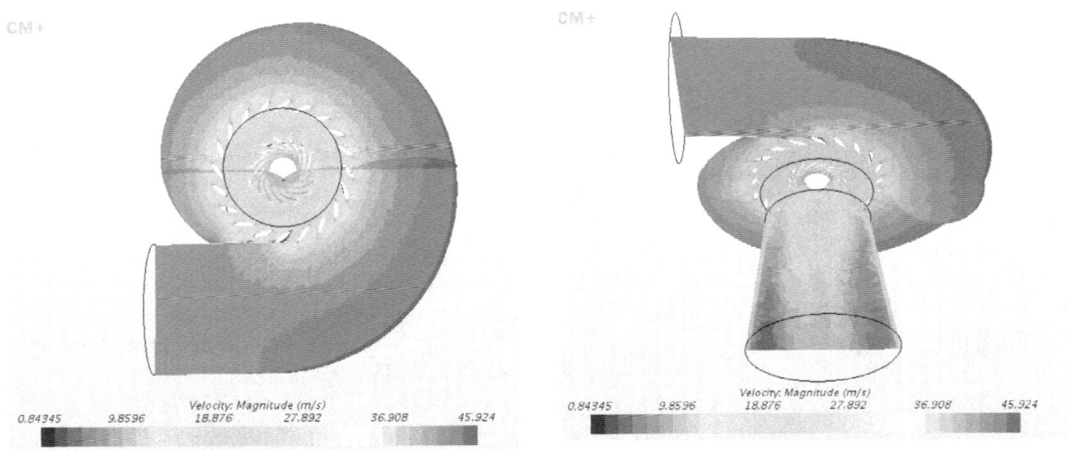

Fig. 17. Speed distribution in the hydraulic path

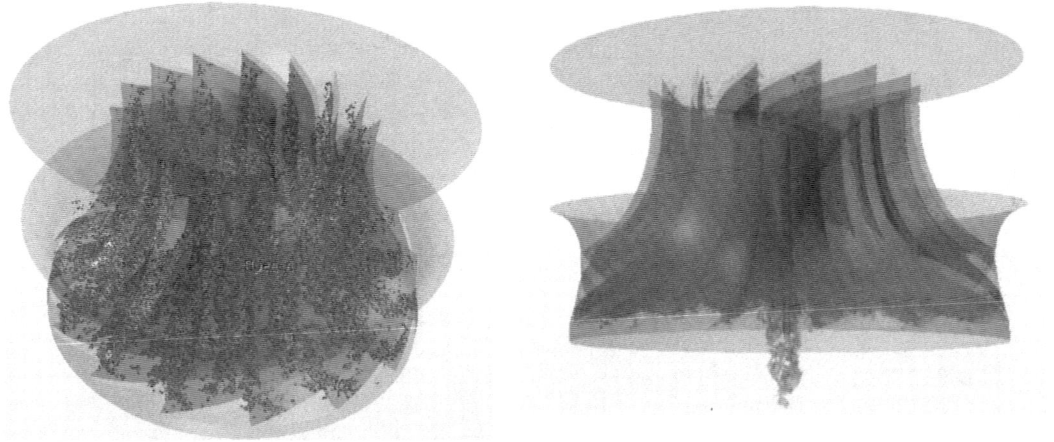

Fig. 18. The presence of cavitation in the impeller

Conclusion

As a result of the experiment, it can be concluded that a Radial Axial Turbine was designed with the following specifications: Net height of 72.5 m, installation flow 50 m3 / s. For this point of operation, an overall efficiency of 93% is projected. The behavior of the parameters is that expected for an axial radial type turbine, in which when the distributor is opened, the flow is increased, having a constant head, reaching an efficiency of 68.4% at an opening of 35 degrees and a flow rate of 48.55 m3 / s

The traditional design for Radial Axial Turbine impllers design, specifically the Bovet and Bowersfeld — Vosnesensky Methods, can be applied in the design of a functional, achieving an acceptable performance according to the efficiency results obtained from the 3D numerical simulations. Nevertheless, considering that the process is iterative, the results cannot be expected to be optimal at the first attempt, so it is necessary to repeat the process until the best results are achieved.In conclusion, we must add that the obtained efficiency has a lower value than that of analogues, since this model has not yet been optimized in CFD packages. However, the power value is acceptable, which means that further modeling should lead to an increase in efficiency based on optimization methods of the flow part.

Literature

[1] Bovet , T. Contribucion of the study of trace of blading of a reaction turbine of the Francis type (1963) New York ASME Winter Meeting 1961, pp. 47–79.

IOP Conf. Series: Materials Science and Engineering **779** (2020) 012002 doi:10.1088/1757-899X/779/1/012002

[2] Mataix, C. “Hyraulic Hydromachines” (2003) Editorial ICAI
[3] Vasiliev Yu. S. Samorukov I. S. Khlebnikov. S. N. Basic energy equipment of hydro power plants. Composition and selection of basic parameters (2002) Saint Petersburg State Polytechnical University
[4] Topazh, G. I. The Basics of the Working Process and Calculation of Hydro Turbines (2011) Saint Petersburg State Polytechnical University, pp. 139–151
[5] De Andrade Correia, J. A. Mechanical and Hydraulic Design for a Tubular Francis Turbine (2006)Simon Bolivar University
[6] Busyrev, A.I. Topazh, G.I. Selection of basic parameters and elements of the flowing part of reactive hydroturbines (2006) Saint Petersburg State Polytechnical University
[7] Laubah D.V. Design of the flow part and the rotation mechanism of the vanes of the guide vane for the Myatlinskaya HPP (2017) Saint Petersburg State Polytechnical University
[8] Barlit V.V. Hydraulic turbines (1977) Editorial Union “Vishcha Shkola”, pp. 360
[9] Kovalev, N. N. “Hydroturbines” (1971), Editorial “Mashinostroienie”, Leningrad.
[10] Morales, M. P. Francis Turbines Design in C++ Builder (2000) Costa Rica University, pp. 125–136
[11] Mathematical modeling of the mechanisms of volumetric hydraulic machines. S. Semenov and D. Kulakov 2019 IOP Conf. Ser.: Mater. Sci. Eng. 492 012042
[12] The effect of design parameters of the closed type regenerative pump the energy characteristics. N Isaev 2019 IOP Conf. Ser.: Mater. Sci. Eng. 492 012026
[13] Hydrodynamic modelling of the impact of viscosity on the characteristics of a centrifugal pump. V Tkachuk 2019 IOP Conf. Ser.: Mater. Sci. Eng. 589 012007
[14] The development of the theory of calculation of the hydrodynamic coupling. V Lomakin 2019 IOP Conf. Ser.: Mater. Sci. Eng. 492 012012
[15] N Egorkina and A Petrov 2019 IOP Conf. Ser.: Mater. Sci. Eng. 492 012015
[16] A Petrov 2019 IOP Conf. Ser.: Mater. Sci. Eng. 492 012036

The implementation of digital program control in hydraulic actuators of hydraulic structures

I Zyubin[1], N Kushnir[1], A Akinfiev[2] and A Sinitsyna[2,3]

[1]Moscow Power Engineering Institute
[2]Bauman Moscow State Technical University

[3]E-mail: sin.anastasia23@gmail.com

Abstract. The problems of controlling the gates of hydraulic structures from the point of view of modernizing current systems and expanding their capabilities are considered. A variant of the modernization of a discrete directional control valve is presented, which ensures the further implementation of a digital system for controlling the speed and position of the hydraulic control regulating element of hydraulic structure.

Introduction

In the territory of the Russian Federation there are an extensive number of hydraulic structures, providing a rated culvert operation through canals and reservoirs. Most of them were built in the period from the early to mid-20th century [1].

At many installations, a hydraulic actuator is used to maneuver the different mechanisms (gates, locks, flood hatches, etc.) [2–4]. However, due to the significant age of the facilities, hydraulic actuator equipment installed at the facilities does not meet the current level of technology development and limits the potential for safe operation of the installation.

The main hydraulic equipment is controlled by discrete control valves [5–6]. Using these valves allows controlling the process only in start-stop mode. The simplicity of the drive design and its relative cheapness in this case lead to a number of operational difficulties, for example, complicates the process of synchronizing the operation of hydraulic motors. At current installations, hydraulic motors are synchronized by a synchronous-tracking mechanical system connected to drain valves [7].

Such systems are difficult for settings and require frequent maintenance, which ultimately leads to an increase in the cost of the lock system as in general, as a single design.

The current trend in the development of hydraulic drive systems is the use of servo drives, which are controlled by electronic modules [8]. The synchronization of the hydraulic motors in this case will be carried out using the proportional control valve. Also, it becomes possible to regulate the speed of movement of the actuating element, as well as to apply the laws of movement selected by the adaptive control system when automatically taking into account external factors (for example, the current operating mode of hydraulic structures and the transition process in the after bay of the hydroelectric power station).

However, despite the advantages of such systems, they have a number of significant disadvantages:

a) High starting cost: caused by the requirements for high-precision manufacturing of hydraulic equipment with proportional control;

Content from this work may be used under the terms of the Creative Commons Attribution 3.0 licence. Any further distribution of this work must maintain attribution to the author(s) and the title of the work, journal citation and DOI.
Published under licence by IOP Publishing Ltd

b) Special requirements for the cleanliness of the internal holes of hydraulic equipment and pressure fluid: caused by small radial clearances in hydraulic equipment.

c) The need for highly qualified specialists involved in the installation and operation of the system. It is determined by the ability of stuff to make settings of drive elements and to detect the malfunctioning on time.

The disadvantages listed above are the main limiting factor for the use of proportional control servo hydraulic systems. According to [9], high cost is the main deterrent to the use of proportional equipment in Russia.

At the same time, special requirements for the pressure fluid and the corresponding regeneration systems, makes the use of valves in the current hydraulic systems complicated. Modernization of the hydraulic system requires significant financial and time costs. One of the options for minimizing the amount of work and improving the quality of control is the use of a numerical control electro hydraulic valve developed at the Department of Hydromechanics and Hydraulic Machines in Moscow Power Engineering Institute as part of the existing hydraulic system.

Methods

When developing the device, one of the ways to reduce the cost of manufacture and operation was considered the reduction of the requirements for the precision manufacturing a pair of spool-sleeve. According to the [10], the transition from a radial clearance of 5 μm (typical for proportional spool pairs) to radial clearance of 20 μm (corresponding to discrete valves) reduces the cost of manufacturing the control valve trim in 2 times. The developed part is based on the trim from a standard traditional discrete control valve. Its conversion to a proportional-control control valve is carried out by modernizing the spool displacement drive. As an electromechanical transducer, the stepper motor paired with a screw-and-nut transmission is used in this control valve. The block diagram of the control valve is shown in Fig. 1.

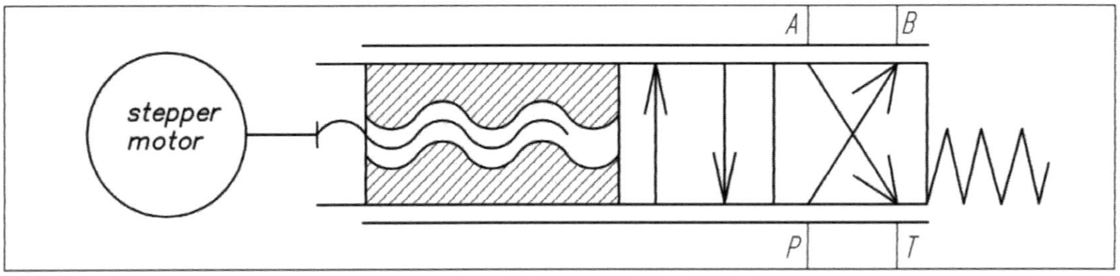

Fig.1. The block diagram of the control valve

Correction of the operating characteristics of the control valve is carried out through the use of a specialized numerical control system. By changing the relationship between the input control signal and the movement of the spool, an automatic linearization of the flow characteristic is performed.

The principle of operation of the developed control valve is described in more detail in [11], [12].

The use will make it possible to integrate the proportional control of the operating mechanisms of hydraulic structures with the ability of adaptive control, or according to specially developed algorithms, as well as with using the modern numerical control systems [13–14]. Due to the use of the trim from the discrete control valve, there is no need for a high quality pressure fluid, since the spool-sleeve pair in such a valve is made with a large radial clearance.

To check the possibility of using control valves as part of hydraulic actuation systems of various hydraulic structures, as well as for the assembly work and testing, a specialized test bench was created at the Department of Hydromechanics and Hydraulic Machines.

The test bench is based on a serial hydrodynamic tray (Fig. 2.). The gate valve is moved due to a double-acting hydraulic cylinder (HC1), which, in one's turn, is controlled by a proportional control

IOP Conf. Series: Materials Science and Engineering **779** (2020) 012003 doi:10.1088/1757-899X/779/1/012003

valve. The position of the gate valve is determined using an encoder. A number of ultrasonic level sensors are located throughout the tray.

When the gate valve is moving in the tray, a non-steady flow mode occurs. The location of the level sensors is selected to provide a record of the parameters of the passage of wave processes along the tray in time at different points in space, and the sensor readings will be used to operate the control system in automatic mode.

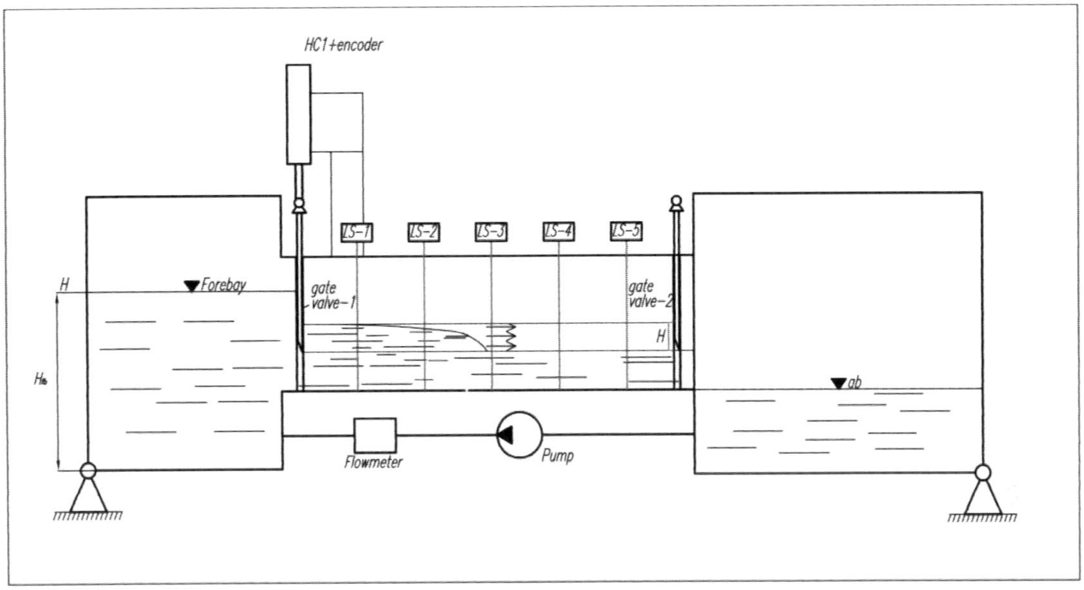

Fig.2. The structural scheme of the test bench

The test bench management system involves working in automatic mode to solve the following application tasks:

a) Maintaining a constant level in the tray when the external factors are changing.

b) Mitigation of transient wave processes in the tray.

c) Deliberate creation of wave processes in order to demonstrate the capabilities of the system.

The work of the test bench in automatic and semi-automatic modes is provided with the help of a specialized control system (Fig. 3). The control system of the test bench consists of two subsystems — a subsystem for collecting and processing data from level sensors and a subsystem for controlling the servo drive and the stepper motor of the control valve, in particular. The examples of using the stepper motors in hydraulic and pneumatic systems are given in [15–16].

In automatic mode, the control system takes data from level sensors, processes them, and determines the direction of the wave process and its amplitude. Then, after analyzing the speed, acceleration and coordinate of the front of the wave, it sends a command for the exact displacement of the gate in relate to the initial state.

All the necessary parameters of the system are recorded on the memory card, and also displayed on the display of a personal computer for the possibility of processing them.

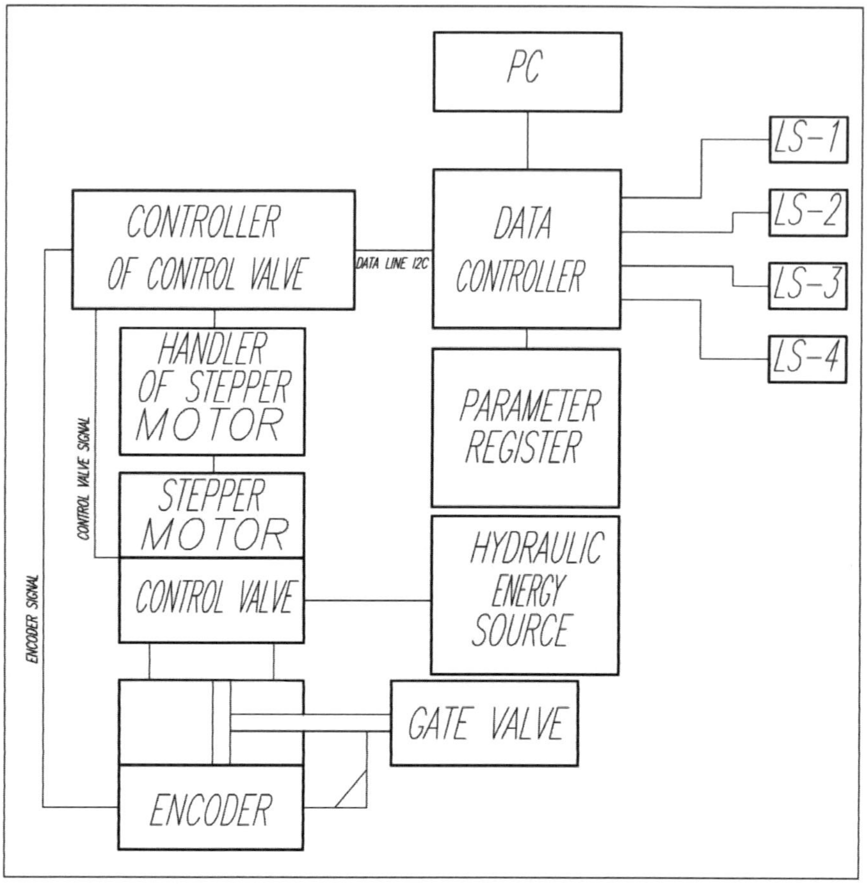

Fig.3. The structural scheme of the digital control system of the test bench

Results

The capabilities of the test bench allow not only to analyze the control modes of hydraulic structures, but also to carry out experiments to study the use of hydraulic actuators with adaptive operation algorithms. By installing the models of hydraulic power system of various structures on the test bench, it becomes possible to emulate their work in laboratory conditions, creating various environmental disturbances. This will allow developing of operation algorithms for the control system, being applied to a specific object.

Separately, one should take into account the possibility of an empirical analysis of the operation of energy-generated devices, in particular, models of wave power stations. The possibility of obtaining regulated wave processes in the channel allows the comparing of the efficiency of various designs of wave power stations as renewable energy sources.

An experimental research of a servo drive with a modified control valve and controller was carried out by the sine sweep method, or by the sine method. Using this method, a Lissajous figure is drawing for which the output signal from the encoder is plotted along the Y axis, which is proportional to the movement of the output rod of the hydraulic actuator, and the displacement along the X axis is proportional to the input signal. The input signal for the actuator in the research is a sinusoidal signal from the master frequency generator. The ellipse for the actuator under study at an input signal frequency of 7 Hz is shown in Fig. 4. The relative coordinate of the displacement of the rod of cylinder of hydraulic actuator is plotted along the X axis, and the relative value of the input signal is plotted along the Y axis. Relative values are obtained as the ratio of the current value to the rated value.

In Fig. 4 it is clearly shown that the practical effect of the extended deadband during control valve modernization is weak and the algorithm used to linearize the control valve characteristics to compensate for large clearance in the sleeve-spool pair works satisfactorily - the figure does not have typical shape distortions in the form of "corners" and "steps".

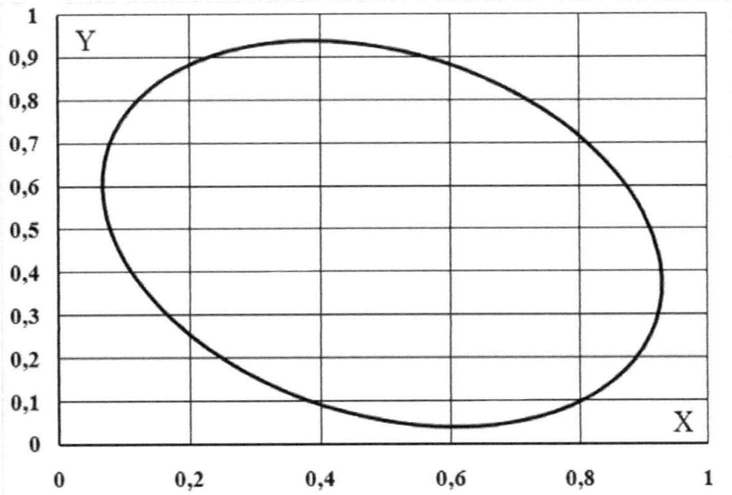

Fig. 4. Graphical view of the sine wave

The cutoff frequency of an actuator with a modified control valve is 10 Hz. The use of the FESTO standard proportional directional control valve, model MPYE-5-010-B within the framework of the servo drive under consideration, allows achieving a cut-off frequency of 25 Hz.

A further increase of the cutoff frequency during conversion of the control valve is connected with an increase in the stepper motor capacity and an increase in its size. In [17], the problem of placing the overall motor was solved due to the the placing of the actuator at the angle of 90 degrees to the axis of movement of the spool, therefore reducing the axial dimension of the control valve, but it does not solve the problem of a general increase in the mass of the device.

Discussions

It should be noted that when using a control valve in control systems of hydraulic structures, the need for good dynamic qualities is not always justified, since the processes take quite long time and relatively high positioning accuracy is not required in comparison with, for example, actuators of metal-cutting machines. Laboratory analyzes did not identify the benefits of a faster hydraulic device. However, for an objective determination of a sufficient response time, an additional analysis of the operation of real hydraulic systems is required.

Thus, it can be noted that the developed control valve will significantly improve the quality of control of the regulating object of the hydraulic structure, and a designated test bench will allow us to accumulate empirical knowledge that will be the basis of accurate calculations of the control system operation algorithms.

List of reference

[1] O.A. Astrakhantseva, V.I. Golubev, V.V. Fedenkov. The use of variable-frequency control of the gate of the water galleries of the Moscow channel gateway and the analysis of dynamic characteristics // Hydraulic machines, hydraulic drives and hydropneumatic automation / XX All-Russian scientific and technical conference of students and postgraduate students. Moscow Power Engineering Institute, December 7, 2016: a collection of materials and reports. — M .: MEI Publishing House, 2016. — 170p.

IOP Conf. Series: Materials Science and Engineering **779** (2020) 012003 doi:10.1088/1757-899X/779/1/012003

[2] Sosnovsky, N., Ganieva, D. Mathematical modeling and analysis of the electrohydraulic actuator with the throttle control (2019) IOP Conference Series: Materials Science and Engineering, 589 (1), article № 012016.
[3] Pilgunov, V., Efremova, K. Ring-shaped fluorescence of mineral oil on the front edge of the choke coil (2019) IOP Conference Series: Materials Science and Engineering, 492 (1), article № 012028.
[4] Kulakov, D., Kulakov, B. Study of the dynamics of an electro-hydraulic servo drive physical model on FESTO learning system (2019) IOP Conference Series: Materials Science and Engineering, 492 (1), article № 012029.
[5] Sosnovskiy, N.G., Popov, D.N., Siukhin, M.V. Sensitivity of dynamic characteristics of electrohydraulic servo drive (2018) Proceedings — 2018 International Conference on Industrial Engineering, Applications and Manufacturing, ICIEAM 2018, article № 8728669.
[6] Popov, D.N., Sosnovskiy, N.G., Siukhin, M.V. Control of synergetic processes to ensure the asymptotic stability of hydraulic systems (2017) Herald of the Bauman Moscow State Technical University, Series Natural Sciences, (3), pp. 37–51.
[7] Mel'nikov E., Danilushkin I. Control system for the upper working gates of the shipping lock. Access mode: https://www.cta.ru/cms/f/454331.pdf (Date of access: 04.09.2019).
[8] Oleg, T. The experience of creating compact energy efficiency hydrostatic servo drive (2015) Proceedings of 2015 International Conference on Fluid Power and Mechatronics, FPM 2015, article № 7337187, pp. 603–609.
[9] Sobolev V. Achilles' heel of proportional hydraulics. Access mode: https://konstruktor.net/podrobnee-hidr/axillesova-pjata-proporcionalnoj-gidravliki.html (Date of access: 21.10.18).
[10] Turning accuracy. Access mode: http://gagar.ru/news/tochnost_tokarnoj_obrabotki/2017-05-09-1131. (Dateofaccess: 10.21.18).
[11] Kushnir N.A., Zyubin I.A., Development of a method for converting a discrete control valve into a proportional one using a stepper motor // Network publication "Hydraulics". — 2019. — Issue № 7. Access mode: http://hydrojournal.ru (Date of access: 2.07.2019).
[12] Kushnir N.A., Zyubin I.A., Russian Federation patent № 2686242. Digital control valve. 2019.04.24.
[13] Pavel, S. Electro-hydraulic drive with separate control of piston groups (2015) Proceedings of 2015 International Conference on Fluid Power and Mechatronics, FPM 2015, pp. 498–500.
[14] Kogler, H., Schöberl, M., Scheidl, R. Passivity-based control of a pulse-width mode operated digital hydraulic drive (2019) Proceedings of the Institution of Mechanical Engineers. Part I: Journal of Systems and Control Engineering, 233 (6), pp. 656–665.
[15] Rybarczyk, D. Investigation of electrohydraulic valve with the stepper motor (2018) AIP Conference Proceedings, 2029, article № 020068.
[16] Vanin, V.A., Kolodin, A.N., Rodina, A.A. Stepper hydraulic motor with pneumatic (jet) control system in machines with hydromechanical forming links (2019) Journal of Physics: Conference Series, 1278 (1), article № 012019.
[17] Ivanov G. M., Sveshnikov V. K., Sazanov I. I. Russian Federation patent №2505716 Digital control valve. 2014.01.27.

IOP Conf. Series: Materials Science and Engineering **779** (2020) 012004 doi:10.1088/1757-899X/779/1/012004

Evaluation of prospective directions of production of fluidics systems

A Davydov[1], I Zyubin[1], O Gavryushina[2,3] and A Sinitsyna[2]

[1]Moscow Power Engineering Institute
[2]Bauman Moscow State Technical University

[3]E-mail: gavros7@rambler.ru

Abstract: the modern possibilities of manufacturing technologies of pneumatic automation systems based on fluidic switching elements are considered. The perceptivity of using the fluidic systems as the alternative to using the relay automation systems is noted.

Introduction

The fluidic systems have proven themselves in various industries where high reliability and explosion and fire safety are needed, for example, in process automation in the food and chemical industries [1–5]. However, a certain technological complexity of the design and manufacture of control units limited the general industrial use of the fluidic.

At the same time, sequencing automation systems based on electromagnetic relays have found quite wide application [6–7]. Relay automation has good qualities in terms of compact installation, maintainability, but have a relatively high cost and short service life. The use of fluidic automation makes it possible to reduce the cost of automation modules and increase their service life.Modern design and manufacturing technologies make it possible to almost completely replace relay unitsto units of fluidic switching elementswith a significant increase in the reliability and service life of sequencing automation systems. Indeed, the service life of typical industrial coupling relaysis about 10^5 switching cycles, while the resource of afluidic control system is limited by the quality of the membrane amplifier and provides about 10^6 cycles. At the same time, the number of critical elements limiting the life of the system is much smaller, since membrane amplifiers are installed only at the output of the control system, and arithmetic logic units are formed on the basis of fluidic elements, with practically unlimited service life due to the absence of moving parts.

The changes in technology make it possible to obtain fluidic elements at a significantly lower cost than coupling relays, even within small batches. Also design techniques for connecting fluidic elements do not differ from the techniques used in the formation of schemesof electronic device in the widely known ECAD systems.

As the advantages of the relay can be specified the load capacity when up to 20 of the same type of relay are connected to one contact block and each such relay provides control of four independent contact blocks. At the same time, as fluidic elements have a relatively low (up to 5 similar connected elements) load capacity. However, a huge margin on the speed of fluidic elements makes it possible to align this disadvantage by using couplingfluidic power amplifiers of the data signal.

A separate question is the power provision of the fluidic automation system with low-pressure air. There are a number of ways to reduce the air consumption by the system as in general, as well as

Content from this work may be used under the terms of the Creative Commons Attribution 3.0 licence. Any further distribution of this work must maintain attribution to the author(s) and the title of the work, journal citation and DOI.
Published under licence by IOP Publishing Ltd

IOP Conf. Series: Materials Science and Engineering **779** (2020) 012004 doi:10.1088/1757-899X/779/1/012004

reduce the cost of its manufacturing as part of a pneumatic actuator, the consideration of which is beyond the scope of this work.

The factors that provide the advantages of the fluidic system over relay automation are a long service life and operation speed. The experimentally proved that the maximum achievable cutoff frequencyin the system on bistable fluidic elements with a blast nozzle width of 0,4 mm is currently from 0,5 to 9 kHz, depending on the supply pressure. At the same time, electromagnetic relays are limited by the actuation timeand relay-releasing time of the lock of 25 ms, which corresponds to frequencies up to 20 Hz.

The factor of increasing actual frequencies and reducing air consumption are associated with the need to reduce the dimensions of the fluidic element, in particular, to reduce the width of the blast nozzle. Taking into account the need of the preparation compressed air (drainage and cleaning) for industrial use, elements with a size of 0,4 to 0,05 mm are especially interesting. In current systems of volumetric pneumatic automation, for example, made byFesto and Camozzi, splits of less than 0,1 mm are allowed, standard filters provide filtering of particles up to 40 microns. A further decrease in the nozzle width allows increasing the actual frequencies of automation systems, but is accompanied by a significant complication of the power fluid preparation system.

Methods

The sizes of the flow part of the fluidic element allow classifying it as a micro product. The main types of technologies for the manufacturing of micro products are discussed in detail in [8]. Over the past period, new types of technological processes did not appear, but theydeveloped qualitatively.

In general, the list of technologies providing the possibility of manufacturing microfluidic elements includes the following types: mechanical (cold stamping and hot plasticstamping), electrophysical methods (electroerosion and laser), chemical methods (chemical milling, including photochemistry,electromachining, galvanoplasty) [9–10].

Stamping, both cold and hot, of miniature products can be effective in mass production. However, the cost of machine-tool attachment and machine accessories manufacturing is very high, and there is no flexibility in production. The requirements for feedstock are also pretty high in terms of stability of stress-related properties. Due to these disadvantages, this method will not be considered in detail in this article.

The most accurate replication of a given contour is possessed by electrodischarge machine with continuous wire feed. Modern types of machines work with a diameter of 20 microns of wire and can ensure the accuracy of the sizes of the flow part (maintaining a certain technology) ± 5 microns. This method has good manufacturing flexibility, but the manufacturing speed of parts is relatively low.Although thin plates are processed in a single batch and one cycle makes it possible to obtain from 10 to 20, and with corresponding technological development a larger number of modules, refixturing the thread into new holes makes it difficult to combine elements into functional blocks. Respectively, such a technology is more suitable for model-experimental work and the production of small series of typical elements.

The laser cutting technology is a pretty interesting area nowadays. Cutting can be carried out both by the material itself and by the easily removed sacrificial layer applied to it. For the manufacture of flow parts according to a variant where the flow part is a through profiled hole, it is possible to use the methods used in the electronic industry in the manufacture of templates for applying solder paste to printed circuit boards. Moreover, solid-state lasers, including fiber ones, are the most effective for cutting metals. For plastics processing it is possible to use cheaper lasers based on gas filled laser tubes.

Most fiber-optical lasers come with lenses with a focal length of 2,85-4,0 inches, which allows to get a minimum spot diameter of 25 microns [11,12]. For a gas filled laser with a focal length of 2 inches, the minimum spot diameter will be much larger — 120 μm, which is associated with a significantly longer radiation wavelength [13]. Respectively, laser cutting can theoretically provide nozzles with a minimum width of 25 to 75 microns in metal when using solid-state cutting heads and

IOP Conf. Series: Materials Science and Engineering 779 (2020) 012004 doi:10.1088/1757-899X/779/1/012004

from 0,12 to 0,4 mm on gas-filled lasers, depending on the quality of realization of the machine's position control system.

The main disadvantage of metal platescutting is the thermal reflow and contamination of the edges of the cutting, which leads to low-precision manufacturing of fluidic products. Moreover, with a decrease in the thickness of the blank and an increase in the width of the blast nozzle, the influence of this factor decreases. According to the results of studies of the processing of stainless steel by an fiber laser inMoscow Power Engineering Institute, the blast nozzle width of 0,2 mm was practically achievable with a ratio of this width to the blank thickness of 1: 1.Moreover, after cutting the dirt on the surface of the plate (along the edges of the cut), it is necessary to remove, for example, by electromachining. This ratio (1: 1) in comparison with the often used ratio 1: 2 reduces the operation speed and load capacity of the fluidic element, but also reduces air consumption and increases the compactability of the stack installation. It should be noted that the potential of the method has not been exhausted due to the limited time of the researches.

An interesting idea was not to carry out cut through by a gas laser, but to carry out the so-called laser engraving. A non-metal layer is applied to the metal surface, for example, epoxy resins or various types of coatings based on polyester or acrylic resins. There are various technological features in this process. Some of them are associated with the features of focusing and the effect of the beam on the polymer. When polymer is burned, combustion products are emitted, which, settling, change the processing space. Also, single holes are made in the form of a crater, tapering to the bottom. Both problems can be solved by carrying out the processing in two steps. As a result, the combustion products are burned out and, with subsequent washing, are carried away from the surface of the base plate, the bottom of the crater is burned up, while the boundaries of the crater are out of the energy flow and saved. It turns out a relatively clean surface with good verticality of walls.

Processing in several steps has a disadvantage - a component of the mechanical positioning of the focusing system is added to the beam focusing error, which additionally limits the possibility of reducing the width of the nozzle. The real result achieved in practice on typical industrial equipment corresponds to a nozzle width of 0,2 mm, which is relatively close to the technological capabilities of CO_2 lasers.

At the same time, laser technologies are more likely methods acceptable for the manufacture of prototypes and small series of devices, forthemassproductionit is economically feasible to produce the products by chemical methods.

It should be immediately noted that the ratio of the nozzle width to theelement height 1: 2 excludes the possibility of effective use the chemical milling,as the speed of the process across the thickness of the material is comparable with the pickling of metalspeedin plan. Experiments with aluminum have confirmed this phenomenon: the contours of the hole are greatly distorted.

Electrochemical processing (when pickling of metal is carried out along the contour of the template) has a greater accuracy of repeating the contour, however, the manufacture of templates is difficult, since their accuracy must be higher than the accuracy of the flow parts of the fluidic elements themselves. Also, maintaining the quality of the electrochemical process, due to the small size of the channels, is difficult. Modern machines provide reproduction accuracy up to 1 μm, however, this applies to the repeatability of the contour while limiting the cutting width to 20 μm. Thus, taking into account the wear of the template, the prospects of the method for both prototype and serial models can be estimated extremely low.

An interesting direction in the production of blocks of fluidic elements in the form of volumetric structures, uniting several elements in layers is photolithography using photosensitive glass-ceramics(glass-crystalline material obtained by crystallization of photosensitive glass). However, the authors do not have data on the practical use of this method in the field of production of fluidic elements. Nevertheless, it was mentioned in [14] that by such methods it is possible to obtain bulk sewers with nozzles width up to 60 μm. However, the depth of the processed material, in which several layers of the hydraulic scheme will be located, does not outnumber 1,5 mm.

A more studied direction in the production of fluidic elements, and not so single ones as blocks, is galvanoplasty. The main idea is that there are technological materials, the so-called photoresists, which have the ability to dissolve in a solvent under the influence of radiation (most often UV radiation).The production scheme consists of several stages: preparing a negative photographic mask, exposure a photoresist, clearing not light-exposedphotoresist and obtaining a protruding profile of the elements system with sections of connecting lines, applying an electroconductivelayer, galvanic metal deposition, filling the obtained profile with a compound, clearing the light-exposedphotoresist and obtaining a final plate with elements and connecting lines.

There are different types of photoresists, providing different thicknesses of the photosensitive layer. In particular, the positive liquid photoresistFP-051Ku (TU F-11833392-0-2006) provides a resolution of 0,8...1,2 μm per layer with a minimum thickness of 2 μm. An increase in the layer thickness due to the application of several layers of a photoresist does not significantly reduce the resolution, so it is possible to obtain elements with a nozzle width of 20...30 μm. The uniformity of the deposition of layers is highly dependent on production technology. Modern machines used in the manufacture of printed circuit boards provide discrepancy in thickness is 5...14% on significant dimension. The image exposure can be carried out both by the photographic mask and by the scanning exposure method directly from the electronic model of the hydraulic system profile.

The application of an electroconductive layer for the galvanoplasty process is possible according to the technologies described in [15–20]. Most often, micronized graphite is used as an electrically conductive material.

The next step is the galvanic deposition of the material (for example, copper) on the prepared surface. Then, the resulting cavities are poured to provide the compoundstrength. The existence of galvanically deposited material reduces the probability of distortion of the contours of the flowing parts captured in the process of gas bubbles filling. After the compound has cured, the light-exposed photoresist is washed out. The place freed from the photoresist is the flow part of the block of fluidic switching devices.

This method can be used for mass production of groups of fluidic elements combined into blocks in the form of automation plates opened on one side with a minimum nozzle width of 30 μm. The production of three-dimensional blocks by this method is difficult due to the complexity of reliable removal of the photoresist. Unlike photosensitive glass-ceramics, the method does not require special materials and has well-proven technologies, which ultimately reduces its total cost in the production of plates with elements of a large surface area.

Results

From the most modern methods, the use of 3D printers for the manufacture of parts of pneumatic automation systems can be noted. However, the method of forming layers of plastic or metal leads to a degeneration of surface quality, that's why this method is not suitable as a method for forming the flowing parts of the fluidic devices.Nevertheless, the combination of electric erosion machining for the manufacture of typical elements and a 3D printer for "printing" base plates with sewers for a specific particular task is commensurate in costs with the manufacture of similar small-lot relay automation units when electromagnetic relays are installed in a printed circuit board. And much cheaper if the volumetric installation is needed.

Discussion

In general, it can be noted that complex and element-filled systems can be performed by chemical methods, however, the specifics of their application will be still limited to special areas where miniaturization and operation speed are more important than the cost of the module.

The possibilities have expanded for the manufacture of dies by the electrochemical method for mass casting of typical active elements layers, included in multilayer assemblies with switching layers, which are plates with holes and can be made on laser machines, or even on milling machines. Templates for electrochemistry can be made by the electroerosion method.

The formation of the block topology can be carried out on the base of existing ECAD software packages.

General changes in manufacture technologies make it possible to rely on the economic efficiency of implementation of fluidic control systems not only in special fields, but also in industry as a whole. At the same time, for small series it the use of electrical discharge machining and laser cutting with a nozzle width limitation of up to 0,4 mm can be recommended. At the same time, it is economically feasible to produce typical fluidic elements and prepare their unifying units using 3D printing for a specific customer.

List of reference

[1] Zverev, A.V., Ivanov, A.I., Pishimova, A.A., Andronik, M., Echeistov, V.V., Mikhailov, S.A., Ryzhikov, I.A., Rodionov, I.A. Optofluidic lab-on-chip platform for realtime sensing applications (2017) Progress in Electromagnetics Research Symposium, pp. 1267–1272.

[2] Agustini, D., Fedalto, L., Agustini, D., de Matos dos Santos, L.G., Banks, C.E., Bergamini, M.F., Marcolino-Junior, L.H. A low cost, versatile and chromatographic device for microfluidic amperometric analyses (2020) Sensors and Actuators, B: Chemical, 304, article № 127117.

[3] Zhang, X., Xia, K., Ji, A. A portable plug-and-play syringe pump using passive valves for microfluidic applications (2020) Sensors and Actuators, B: Chemical, 304, article № 127331.

[4] Jia, Y., Su, C., He, M., Liu, K., Sun, H., Lin, Q. Isothermal titration calorimetry in a 3D-printed microdevice (2019) Biomedical Microdevices, 21 (4), article № 96.

[5] Oleg, T. The experience of creating compact energy efficiency hydrostatic servo drive (2015) Proceedings of 2015 International Conference on Fluid Power and Mechatronics, FPM 2015, article № 7337187, pp. 603–609.

[6] Miyanaga, K., Matsumoto, S., Takahashi, K., Shah, S.K., Takei, F., Nakamura, A., Kurata, Y. A Study on Growth of Dust of Needle-like Crystal from a Flame Retarder of Electromagnetic Relay Housing (2019) Electrical Contacts, Proceedings of the Annual Holm Conference on Electrical Contacts, 2018–October, article № 8611773, pp. 516–520.

[7] Kirschbaum, L., Dinmohammadi, F., Flynn, D., Robu, V., Pecht, M. Failure Analysis Informing Embedded Health Monitoring of Electromagnetic Relays (2019) Proceedings — 2018 3rd International Conference on System Reliability and Safety, ICSRS 2018, article № 8688839, pp. 261–267.

[8] Krasnikov, V.F. Technology of miniature products. - M.: Mechanical Engineering, 1976. — 327 p.

[9] Adamopoulou, T., Nawada, S., Deridder, S., Wouters, B., Desmet, G., Schoenmakers, P.J. Experimental and numerical study of band-broadening effects associated with analyte transfer in microfluidic devices for spatial two-dimensional liquid chromatography created by additive manufacturing (2019) Journal of Chromatography A, 1598, pp. 77–84.

[10] Zhou, Z., He, G., Zhang, K., Zhao, Y., Sun, D. 3D-printed membrane microvalves and microdecoder (2019) Microsystem Technologies, 25 (10), pp. 4019–4025.

[11] Manufacture of stencils for applying solder paste [Electronic resource] / ELINFORM — 2007, August 2. — Access mode: http://www.elinform.ru/articles_13.htm.

[12] Types of laser equipment [Electronic resource] / Company Reklab. — 2013 — Access mode: http://reklab.ru/articles/laser-types.

[13] Spilka S. Digital flexography and types of laser exposure devices. —Flexo Plus magazine: Flexography and special types of printing. — 2004, No. 3. — Access mode: http://www.vipsys.ru/upload/files/00/03/c1.pdf.

[14] Laser-induced structural phase modification of glassy materials / V.P. Veyko, E.I. Ageev, A.V. Kolobov [and others] // Journal of Instrument Engineering. — 2014, b. 57, № 6, pp. 7–31.

[15] Handbook of Thin Film Technology [trans. from English]/ Leon I. Maissel, ReinhardGlang. — M.: Publishing House, 1977, b. 1 — 768 p.

[16] V Cheremushkin and V Lomakin 2019 *IOP Conf. Ser.: Mater. Sci. Eng.***492** 012039

IOP Conf. Series: Materials Science and Engineering **779** (2020) 012004 doi:10.1088/1757-899X/779/1/012004

[17] T Valiev and APetrov 2019 *IOP Conf. Ser.: Mater. Sci. Eng.* **492** 012038
[18] V Lomakin and O Bibik 2019 *IOP Conf. Ser.: Mater. Sci. Eng.* **492** 012037
[19] A Petrov et al 2019 *IOP Conf. Ser.: Mater. Sci. Eng.* **492** 012036
[20] A Shablovskiy and E Kutovoy 2019 *IOP Conf. Ser.: Mater. Sci. Eng.* **492** 012035

Typical cases of destruction of the main hydraulic cylinders, causes of destruction and engineering solutions adopted for their prevention

A Vorob'yev[1], A Gribkov[1], A Akinfiev[2] and A Sinitsyna[2,3]

[1]Moscow Power Engineering Institute
[2]Bauman Moscow State Technical University

[3]E-mail: sin.anastasia23@gmail.com

Abstract. The problems of operation of hydraulic actuators in hydraulic press are considered. Design errors of hydraulic cylinders are identified. As the result of the simulation in the AnsysCFX, a new design with better strength properties is obtained, and the operational life of the hydraulic cylinder is increased. Also the pictures of the stressed-deformed state before and after structural changes are obtained.

Introduction

Powerful hydraulic presses are part of the industries, determining the military capability and industrial potential of Russia. Increasing the quantity and improving the quality of products obtained by hydraulic presses in a substantial way determine the development of the national economy. The modern technological level of present powerful hydraulic presses is provideddue to the modernization of hydraulic actuator, controlling and mechanization systems that is $\approx 20\%$ of mass and cost of the press [1–3]. The force of the hydraulic press, necessary for material processing under pressure, is created and perceived by its basic parts (main hydraulic cylinders, columns, cross-beams, elements of frame racks), that is $\approx 80\%$ of mass and cost of the press. Therefore, theessential condition for maintaining and increasing the technological capabilities of present powerful presses is the further failure-free operation of the present basic parts [4].

The method for engineering analysis of the hydraulic cylinder substantially depends on how it is attached to the press frame. The methods of attaching of the hydraulic cylinder with collars or flanges made in conjunction with the cylinder are widely used. Flanges are attached on the cylinder by threading or welding. Mounting designs are used with a split ring key. The support constructions of the cylinder by split-ring connector are used. The cylinder support on its bottom is used. After choosing the pressure P with which the pressure fluid will be supplied to the cylinder, it is necessary to determine the diameter of the piston or ram due to which the specified force T will be developed. The resulting size D is rounded off to the nearest larger size according to GOST 6540. Then, depending on the accepted design, the inner diameter of the cylinder is determined, which must also be in compliance with the standard above. When determining the wall thickness, the cylinders of hydraulic presses are calculated as thick-walled cylindrical shells. In thick-walled cylinders being under the action of internal pressure P, a complex stress state arises, as the normal stresses act simultaneously in the radial direction at a point remote from the cylinder axis in the tangential and axial directions.

Content from this work may be used under the terms of the Creative Commons Attribution 3.0 licence. Any further distribution of this work must maintain attribution to the author(s) and the title of the work, journal citation and DOI.
Published under licence by IOP Publishing Ltd

Methods

Operating experience shows that the largest number of failures of powerful hydraulic presses with the most serious consequences is primarily associated with destruction and damage of the hydraulic cylinder. With an increase in operating time, the number of failures increases, that's why it is necessary to significantly increase the strength of cylinder to meet the increasing production requirements and to increase its operating time.

The occurrence and development of fatigue cracks in the fillet of the bottoms are one of the main reasons for the destruction of large cylinders of powerful hydraulic presses under the influence of a cyclic load. The operation experience of mold presses has shown that cylinders made of steel 35 (0,32 - 0,4% C, 0,17 - 0,37% Si, 0,5 - 0,8%Mn, ~ 97% Fe) are damaged along cracks in the fillet of the bottom in the same way as cylinders made of steel 25 GS (0.2 - 0.26% C, 0.6 - 0.9% Si, 1 - 1.3% Mn, ~ 96% Fe)that is shown in the Fig. 1.

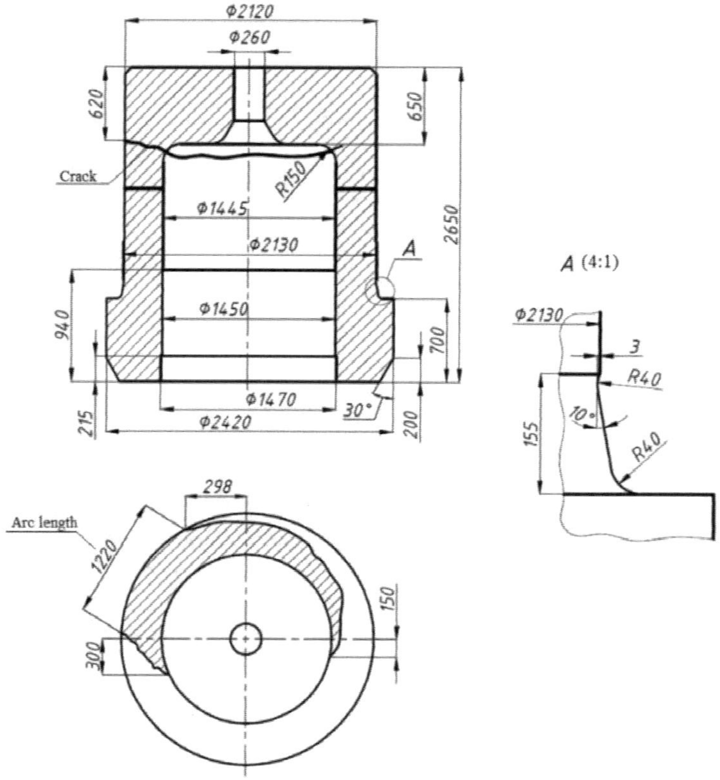

Fig. 1. Scheme of destruction of a cylinder with a force of 50 MN along a crack in fillet of the bottom

High quality structural carbon steel 35 is used for the manufacture of products of low strength, working with low stresses - cylinders, columns of presses, axles, crankshafts, rods, traverses, and other details. Steel 35 has a yield stress $\sigma = 315 MPa$.

Steel 25 GS structural low alloyed steel is used for welded assemblies. Yield stress is $\sigma = 275 MPa$. The properties of 25GS steel depend on the type of heat treatment, tempering temperature and section. An important difference between steel 25GS and steel 35 is good welding properties. Welding can be carried out without preheating, by any methods, without heat treatment of the welded joints. Thereby, the use of steel 25 GS is more economically feasible.

It should be noted separately that after the examination of a number of presses after repair, during which cracks in hydraulic cylinders are eliminated by welding and the yield stress of welding material

IOP Conf. Series: Materials Science and Engineering **779** (2020) 012005 doi:10.1088/1757-899X/779/1/012005

is $\sigma = 450 MPa$, destruction of hydraulic cylinders in the fillet area directly along the welding joint is observed.

To determine the causes of the destruction of the cylinders, their FEM was made. The stress state of the main cylinders of mold presses with a force of 35 MN and 50 MN obtained from Ansys is shown in Fig. 2, the maximum primary stresses σ_I in the fillets of the cylinder bottoms are 140.1 MPa and 130.5 MPa, respectively.

In cylinders made of 25GS steel and 35 steel, fatigue cracks occur in the same zone (fillet of the bottom) at maximum stresses of 130-140 MPa. It means, that the cause of fatigue cracks and of the destruction of the cylinder bottoms is determined not by the grade of steel, but by the technology of manufacturing the bottoms, which was the same with different grades of steel. Different cases of fatigue cracks occurrence are described in [5–7].

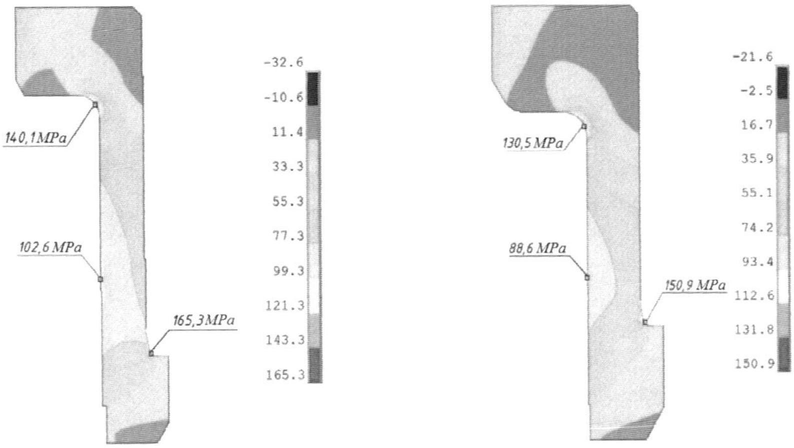

a)b)

Fig. 2. The stress state of cylinders with a force of 35 MN and 50 MN made of steel 35: a) — cylinder with a force of 35 MN; b) — cylinder with a force of 50 MN

The proposition of the connection of the cause of the cylinder bottoms destruction with the adopted technology of blanks forging confirms the results of fatigue tests. The weight of the shell ingot (79 tons) is more in 1,4 times than the mass of the bottom ingot (57 tons), therefore, with the same forging technologies, the working intensity and endurance of the bottom material should be greater than the shell material. But the fatigue limit of the bottom in the fillet zone is $\sigma_{0bottom} \leq 130 MPa$, the fatigue limit of the shell is σ0 $\sigma_{0shell} = 235 - 250 MPa.$

A modern, more intensive technology for forging the bottom blank significantly increases the fatigue limit of the fillet zone. But the maximum stresses in the fillet of the bottom are higher than the stresses on the inner contour of the shell. So, for a cylinder with a force of 35 MN and 50 MN, the stresses on the bottom fillet contour are in 1,4–1,5 times more than the stresses on the inner shell contour (Fig. 2). This, in the general case, significantly reduces the cylinder's factor of safety for fatigue and, due to the corrosive attacks of the pressure fluid, or due to accidental damagesof the fillet surface, leads to fatigue cracks and destruction of the cylinder.

Results

A significant decrease in stresses in the fillet is achieved by its embedding in the cylinder bottom with an increase in the fillet radius [8]. The maximum stresses in the fillet decrease in 1,4 times, from 140,1 to 103,4 MPa for a cylinder with a force of 35 MN and from 130,5 MPa to 95,2 MPa for a cylinder with a force of 50 MN (Fig. 3). These stresses practically coincide with the stresses acting on the inner

IOP Conf. Series: Materials Science and Engineering **779** (2020) 012005 doi:10.1088/1757-899X/779/1/012005

contour of the shell and provide unlimited durability of the inner surface of the cylinder even when using fresh water as a pressure fluid [9].

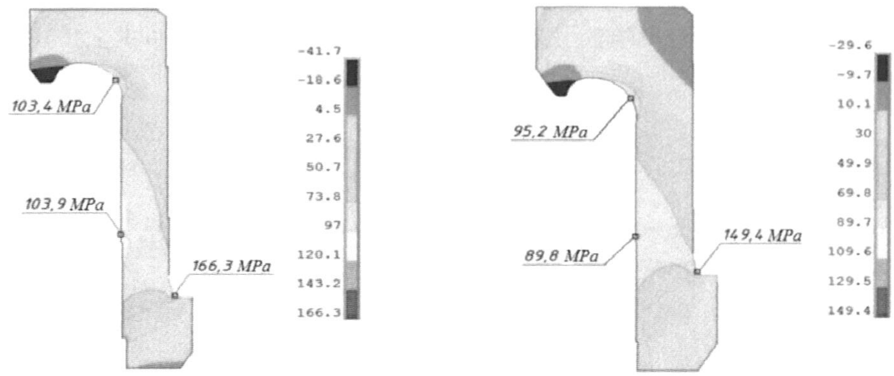

a)b)

Fig. 3. The stress state of the inner surface of the cylinders with a force of 35 MN and 50 MN with the new design of the fillets of the bottom: a — cylinder with a force of 35 MN; b — cylinder with a force of 50 MN

Cylinders, the bottoms of which are manufactured using the new design and technology solutions presented above, have been working in hydraulic presses of the railway wheel production line at OMK since 2005. The line's capacity is 820,000 wheels per year, which is the world record for productivity for powerful hydraulic presses. For ten years of operation, the occurrence of cracks in the fillets of the cylinder bottoms has not been recorded.

Discussions

This optimization with Ansys CFX allows increasing the operation life of the hydraulic actuator through the strength analysis. The destruction of the main cylinders caused by fatigue cracks in the fillet of the bottom, are explainedby structural and technology errors made during the creation of the cylinders.The discussed problem can be solved during a scheduled repair without significant financial costs. Besides the optimization of hydraulic systems, Ansys CFX can solve a number of problems caused in any technical devices. More problems occurring in technic devices solving with AnsysCFX are described in [10–20].

Reference list

[1] Van Nguyen, T., Ha, C. The Actuator and Sensor Fault Estimation Using Robust Observer Based Reconstruction for Mini Motion Package Electro-Hydraulic Actuator (2019) Lecture Notes in Computer Science (including subseries Lecture Notes in Artificial Intelligence and Lecture Notes in Bioinformatics), 11645 LNAI, pp. 244–256.

[2] Aizhambaeva, S.Z., Maximova, A.V. Development of control system of coating of rod hydraulic cylinders (2018) IOP Conference Series: Materials Science and Engineering, 289 (1), article № 012020

[3] Solazzi, L. Design and experimental tests on hydraulic actuator made of composite material (2020) Composite Structures, 232.

[4] Kuznetsov, A.P., Utenkov, V.M., Shirshov, A.G. Thermal assessment of metal-cutting machines (2015) Russian Engineering Research, 35 (1), pp. 8–13.

[5] Ling-Xiao, Q., Yuan, C., Hong-Liang, L., Rui, G., Haixin, G. Fatigue analysis of the cylinder in the axial piston pump (2015) Proceedings of 2015 International Conference on Fluid Power and Mechatronics, FPM 2015, pp. 116–122.

IOP Conf. Series: Materials Science and Engineering **779** (2020) 012005 doi:10.1088/1757-899X/779/1/012005

[6] Huang, B.S., Lu, J. Research on fatigue failure of steel E4330 for fluid cylinder (2015) Materials Science Forum, 814, pp. 241–248.
[7] Yan, Z., Tang, M., Chen, G., Wang, T., Li, X., Zhuang, Z. Development of analogy method for thermal-fatigue crack propagation in pressurized cylinder by using permeation diffusion-fracture model (2019) Engineering Fracture Mechanics, 222.
[8] Birger I.A., Shorr G.F., Iosilevich G.B. Stress analysis of machine parts. Publisher: Machine engineering. 1977, 230 pp.
[9] Kubachek V.V., Fazil'akhmetov R.S., Aleksandrova N.N., KarpovP.P.Improving the manufacturing technology of hydraulic cylinders of presses for wheel rolling shopfloors. Hydropress equipment manufactured by "URALMASH" and the problems of its development. Sverdlovsk. 1988.S. 52–59.
[10] Murashov, M.V., Vlasov, A.V. Three-dimensional Finite Element Modelling of the Cylindrical Specimen Upsetting (2018) MATEC Web of Conferences, 220.
[11] Gavryushin, S.S., Skvortsov, P.A., Skvortsov, A.A. Numerical modeling and optimization of the design of the corrugated pressure sensor membrane (2018) PeriodicoTcheQuimica, 15 (Special Issue 1), pp. 174–181.
[12] Sun, B., Ding, S.-H., Li, S.-X. Failure analysis of cylinder for reciprocating piston compressor based on ANSYS (2017) Petrochemical Equipment, 46 (5), pp. 40–44.
[13] Alshare, A.A., Calzone, F., Muzzupappa, M. Hydraulic manifold design via additive manufacturing optimized with CFD and fluid-structure interaction simulations (2019) Rapid Prototyping Journal, 25 (9), pp. 1516–1524.
[14] Ali, B., Ouahiba, G., Hamid, O., Ahmed, B. Aerodynamic optimization of active flow control over S809 airfoil using synthetic jet (2019) 2018 International Conference on Wind Energy and Applications in Algeria, ICWEAA 2018.
[15] Gunasekharan, S., Raj, K.T. Structural design and modeling of keystone butterfly valve (2019) International Journal of Recent Technology and Engineering, 7 (5), pp. 288–291.
[16] V Cheremushkin and V Lomakin 2019 *IOP Conf. Ser.: Mater. Sci. Eng.***492** 012039
[17] T Valiev and APetrov 2019 *IOP Conf. Ser.: Mater. Sci. Eng.***492** 012038
[18] V Lomakin and O Bibik 2019 *IOP Conf. Ser.: Mater. Sci. Eng.***492** 012037
[19] A Petrov et al 2019 *IOP Conf. Ser.: Mater. Sci. Eng.* **492** 012036
[20] A Shablovskiy and E Kutovoy 2019 *IOP Conf. Ser.: Mater. Sci. Eng.***492** 012035

IOP Conf. Series: Materials Science and Engineering **779** (2020) 012006 doi:10.1088/1757-899X/779/1/012006

Obtaining the maximum permissible gas content at the inlet to the ESP by computational fluid dynamics modeling

A Petrov[1] and A Sinitsyna[1,2]

[1]Bauman Moscow State Technical University

[2]E-mail: sin.anastasia23@gmail.com

Abstract. Methods for increasing the operation stability of the electrical submersible pump (ESP) are listed. Examples of computational fluid dynamics (CFD) applications for designing and studying the flow part of the pump are given. The results of mesh convergence study, Pump performance curve, and dependences of head and hydraulic efficiency on gas content at the inlet, obtained by CFD methods, are presented. The results of CFD modeling are compared with experimental curves.

Introduction

According to statistics, in the present time, the percent of oil wells in Russia still equipped with sucker rod pumps is 34%. Installations of submersible centrifugal pumps (ESPs) account for 63% of oil wells, while 82% of oil in Russia is produced using ESPs, which indicates the effectiveness of this extraction technique [1,2]. At the same time, a significant percent of the ESP works in conditions where the petroleum contains associated gas. With high associated gas contents, reliable operation of the pumps becomes problematic, pressure and energy characteristics deteriorate, and the service life of a pump is significantly reduced. Similar results are observed during the operation of more than half of the oil wells equipped with ESP, therefore, it is necessary to learn as accurately as possible to simulate the pump operation process on a gas-liquid mixture, as well as to competently analyze the results and predict the behavior of the pump in an oil well to make further changes to the geometry of the pump flow part and for its optimization.

The problem of ESP operation on a mixture of oil an dassociated gas has existed for a long time, but in recent times, due to the oil wells depletion, the problem of the operation of ESPs on oil-associated gas mixture is especially topical. When gas enters the pump's flow part, gas pockets are formed, the size of which can be compared with the dimensions of the impeller channels and the pump diffuser. In this case, the energy exchange process between impeller and the liquid is frustrated, the bubbles condense, and the pressure inside the bubbles remains constant and equal to the saturated vapor pressure. The pressure of the liquid increases as the bubbles move along the channel; therefore, the pressure differential between the liquid and the pressure inside the bubble increases. With complete condensation of the bubble, a collision of fluid particles occurs, accompanied by a simultaneous local increase in pressure, reaching hundreds of MPa. This leads to the destruction of the pump working surface, the stability of its operation is frustrated, pump performance curve degradation and, as a result, a significant reduction of service life of ESP installation. To increase the stability, service life and efficiency of the ESP, the following methods are used:

1. Application of a gas disperser

It is installed at the inlet to the pump and is designed to disperse gas bubbles.

Content from this work may be used under the terms of the Creative Commons Attribution 3.0 licence. Any further distribution of this work must maintain attribution to the author(s) and the title of the work, journal citation and DOI.
Published under licence by IOP Publishing Ltd

2. Application of a gas separator

It is installed at the inlet to the pump and is designed to separate the gas-liquid mixture into liquid and gas and to remove the separated gas into the annular space of the oil well.

3. Application of a gas separator-disperser (compression-dispersing modules)

It is installed instead of a gas separator or disperser in oil wells with a particularly high content of gas, where neither a gas separator nor a disperser ensures stable operation of the ESP installation.

4. The use of multiphase pumps stably working in gas-saturated environments.

Multiphase pumps are capable to pump multiphase liquids without separating and / or dispersing modules at the inlet, which allow them to adapt to changing oil extraction conditions.

To increase the stability of the ESP installation operation, other methods describing in detail in [3–5] are also used.

However, it is not always necessary to use one of the above methods to maintain or increase the stability of the pump - with a certain permissible gas content, the pump continues to operate stably and there is no need to use additional modules at the pump inlet, which significantly saves the financial costs of purchasing the ESP installation and reduces its dimensions. Thus, it is very important to know critical or "permissible" gas content for the designing a flow part of the pump, its optimization, the arrangement of the ESP installation and the calculation of production economic performances. It is also necessary to know the operating conditions, operation modes of pump and permissible percentage of gas as accurately as possible in order to improve the design of the flow part, the pump performance curve and increase the service life of installation.

Obtaining gas content at which the pump starvation occurs on a test bench is not a trivial task and causes a number of problems: the need to develop a test bench, and, as a result, the financial costs of its design and manufacture. In addition, significant amount of time is spent on the design and manufacture of the test bench and the manufacture of the flow part of the pump stage under test. Therefore, for obtaining per missibleg as content the computational fluid dynamics method was chosen.

Method

Today the computational fluid dynamics method for modeling flows in CFD software packages is widely used for designing the flow part, calculation the required parameters, for optimization and conducting various researches. CFD methods allow the engineer to significantly save time on calculation, optimization the parameters of the pump flow part, helping to improve the pressure and energy characteristics in the shortest possible time, as well as significantly save financial costs. There are a significant number of software packages for CFD modeling: ANSYS, STAR CCM +, FlowVision, and others. Examples of using the CFD packages for optimization the flow part of pumps and conducting a research are given in [6–11].

For computational fluid dynamics modeling, the STAR CCM + software package was chosen. The flow part of explored submersible pump for obtaining "permissible" gas content was designed in the CF Turbo software package. The flow part contains impeller and diffuser. The designed flow part and the 3-D stage model are shown in Fig. 1, 2, respectively.

Calculation parameters:

• The pumping fluid is water, the recalculation of the flow part parameters and the characteristics of the pump for oil will be carried out according to fundamental work of D.Ya. Sukhanovor other existing methods [12].

• H (Head) = 4.9 m;

• Q (Flow) = 400 m3 / day = 16, 67 m3 / h;

• n (rotationalspeed) = 2900 rpm.

Using the STAR CCM + software package, the performance curve of the explored pump stage at 0% gas content was obtained, shown in Fig. 3.

IOP Conf. Series: Materials Science and Engineering **779** (2020) 012006 doi:10.1088/1757-899X/779/1/012006

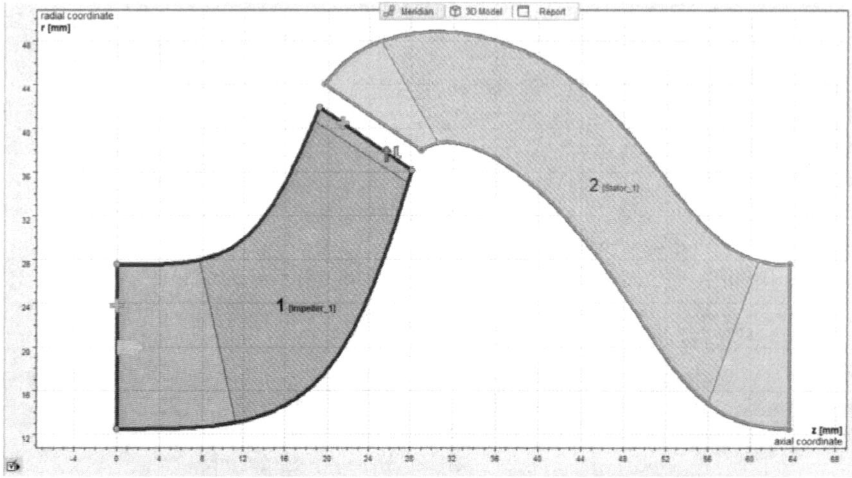

Fig. 1. Explored flow part designed in CF Turbo software package

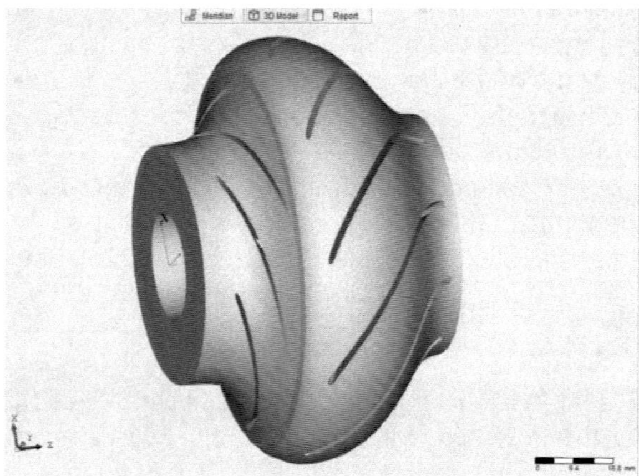

Fig. 2. 3-D model of explored flow part

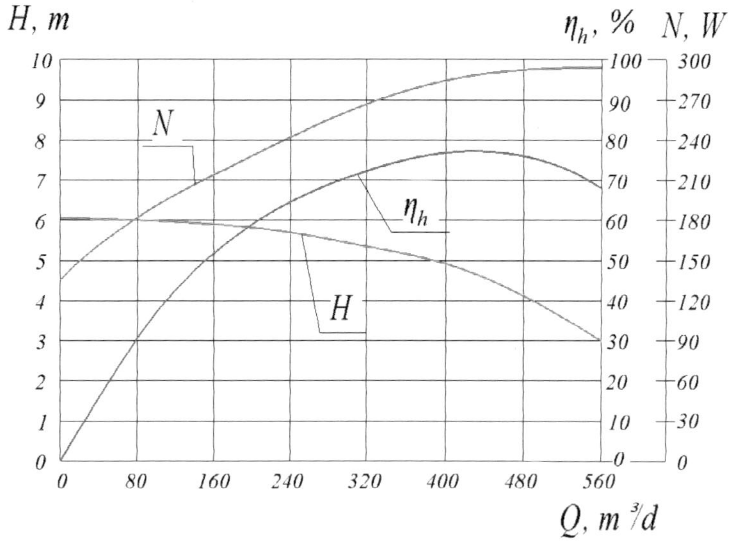

Fig.3. Performance curve, obtained using STAR CCM+ software package

To select the optimal base mesh size, a mesh convergence study was conducted: the head value was determined at 3 values of the base mesh size: 2 mm, 5 mm and 7 mm. The time step was chosen equal to $5 \cdot 10^{-5}$ s; accordingly, ≈ 2000 iterations per 1 impeller rotation. The modeling was conducted at 0% gas, ≈ 6000 iterations were performed for each calculation - 3 full rotations of the impeller.The most accurate result (the one closest to the previously calculated one for performance curve) and the best convergence were expectedly achieved with a mesh base size of 2 mm — the error of calculated value was less than 0.5%, however, the calculation time increases by more than 2 times in comparison with the calculation for mesh with a base size of 5 mm. The calculation of the head on a mesh with a base size of 7 mm the error is about 4%. The error in calculation the head with a base mesh size of 5 mm is about 1%; the time of one calculation is acceptable for further obtaining of dependences of head and hydraulic efficiency on gas content at the inlet. Moreover, to accelerate the calculation, only the H values were compared, assuming that hydraulic efficiency would have approximately the same error. The results of mesh convergence study are shown in Table 1:

Table 1. The results of mesh convergence study

Base size, mm	Number of cells	Time of calculation, min	H, m	H error, %
2	≈ 765 000	≈ 310	4,92	≈ 0,4
5	≈ 330 000	≈ 140	4,95	≈ 1
7	≈ 105 000	≈ 50	4,78	≈ 4

After analyzing the results, a base size of 5 mm was chosen, which is optimal for this flow part in time and accuracy of calculation. For generating the mesh, the following mesh models were selected:

- Polyhedral mesher;
- Surface remesher;
- Prizm layer mesher.

Parameters of generated mesh are presented are shown in Table 2:

Table 2. Parameters of generated mesh

Parameter	Value
Base size	5 mm
Prismatic Streching	1,5
Prismatic layer thickness, relative to the base size	25%
Number of prismatic layers	5

In the impeller and the pump diffuser, the mesh was reduced in 2 times to minimize the calculation error. An outlet pipe was also completed to stabilize the flow at the outlet of the pump diffuser and an inlet pipe to set the initial conditions. The resulting mesh in the cross section of the flow part is shown in Fig. 4.

To calculate the pressure and hydraulic efficiency at different values of gas content were selected physical models and were created 2 phases: gas (air) and liquid (water). Next, the "water-air" phase interaction was created and interaction models were selected.

As the boundary conditions at the inlet boundary, a constant mass flow rate was set corresponding to the flow at the optimal operation mode of the pump, at the outlet boundary constant pressure = 0 Pa was set.

In the initial conditions, only the gas content at the pump inlet was set and changed. The first 2 calculations were carried out at 0%, 1% gas, then the gas content at the inlet increased with step of 2% at first, then as the pressure and efficiency decreased with step of 1% to a permissible gas content.

With permissible gas contentthe head or hydraulic efficiency decreased in 2 times on a test bench or during the exploitation. To compare the obtained dependence of head on gas content the values of head were also obtained at a gas content of 9, 10, and 11%.

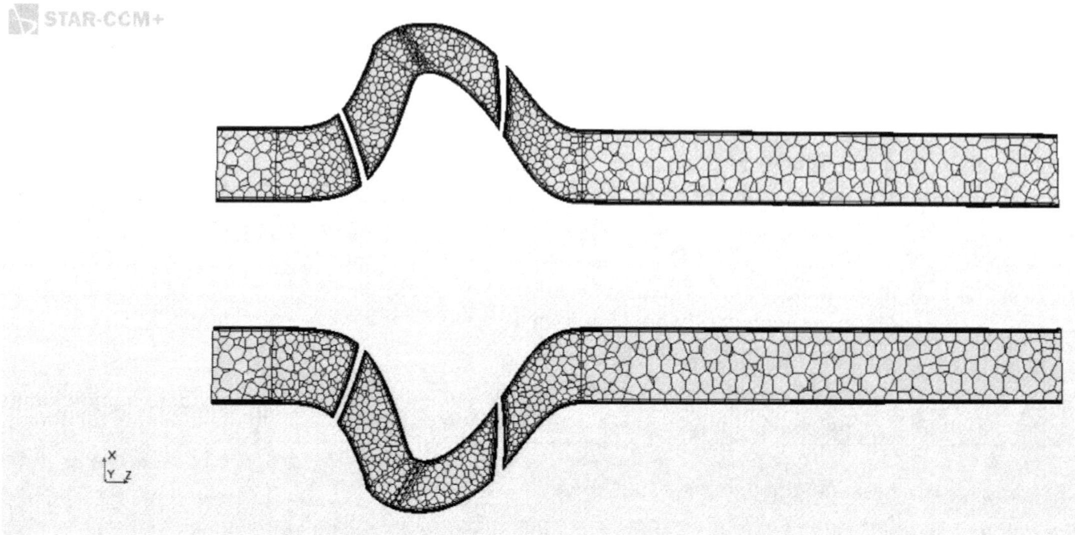

Fig.4. Resulting mesh in the cross section of the flow part

Results

The modeling included ≈20 000 iterations for each gas content for which stabilization of the flow occurred, which made it possible to obtain the correct values of head and hydraulic efficiency. Examples of multiphase flows modeling are given in articles [13-14].The results of computational fluid dynamics modeling are presented in the form of Table 3:

Table 3. The results of computational fluid dynamics modeling

Gas content, %	H, m	η_h, %
0	4,95	76,9
1	4,88	76,6
3	4,76	75,6
5	4,48	71,5
6	3,86	61,7
7	2,45	32,5
8	1,65	5,6

Also, basing on the obtained result, dependences of head and hydraulic efficiency on gas content at the inlet were plotted — Fig. 5, 6, respectively.

To verify the obtained modeling results, the dependence of head on gas content at the inlet was dimensionless and superimposed on the graph of head on gas content obtained experimentally by American scientists from Tulsa University, Oklahoma, USA [15–16]. Superimposed experimental and obtained dependences of head on gas content are presented in Fig. 7.

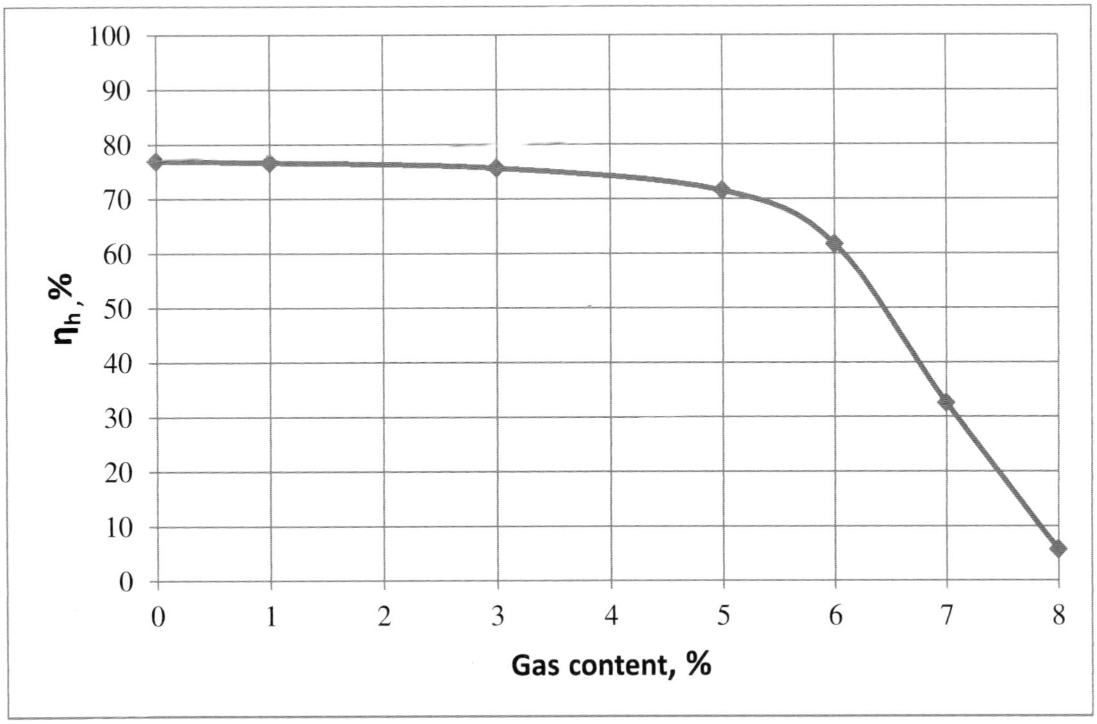

Fig.5. Dependence of hydraulic efficiency on gas content at the inlet

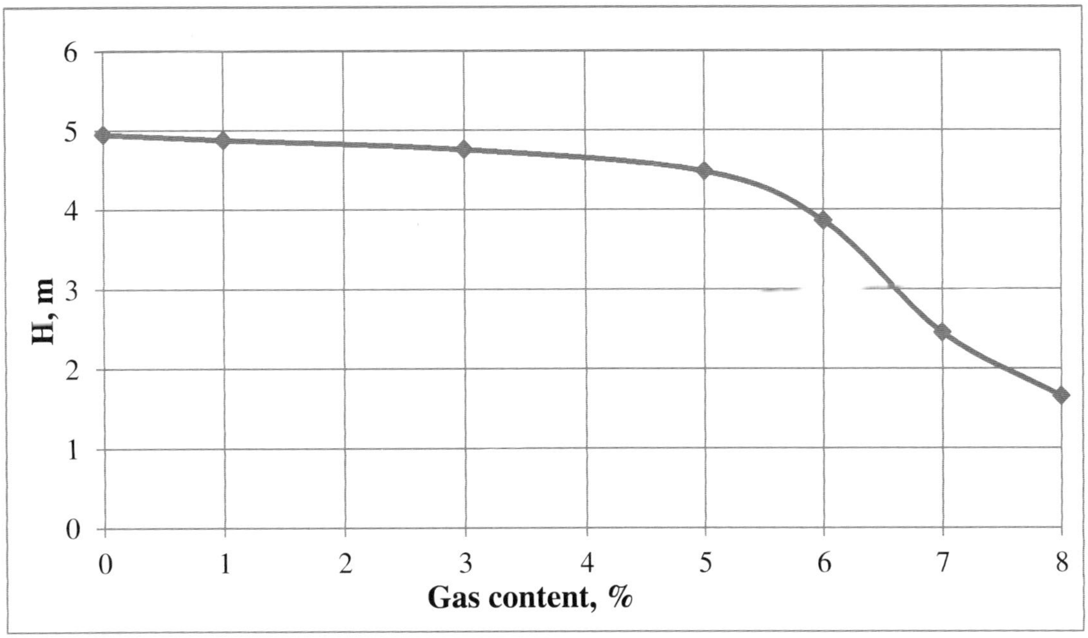

Fig.6. Dependence of head on gas content at the inlet

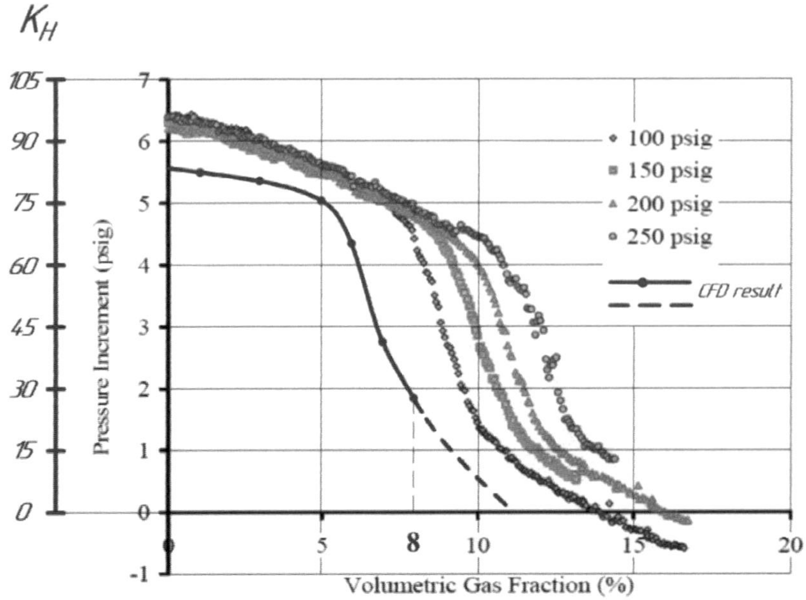

Fig.7. Superimposed experimental and obtained dependences of head on gas content are presented

$K_H = \frac{H}{n^2 \cdot D_2^2} \cdot 10^{-6}$ — dimensionless pressure coefficient, D_2 — is the outlet diameter of the impeller (m), n — is the set pump shaft rotational speed (rpm). The part of the dependency graph after 8% gas content is highlighted separately, since obtaining the average head for a gas content more than permissible there are some difficulties associated with the developed process of cavitation. Some difference in the value of the head coefficient from the experimental data is explained by the difference in the delivery coefficient of pumps.

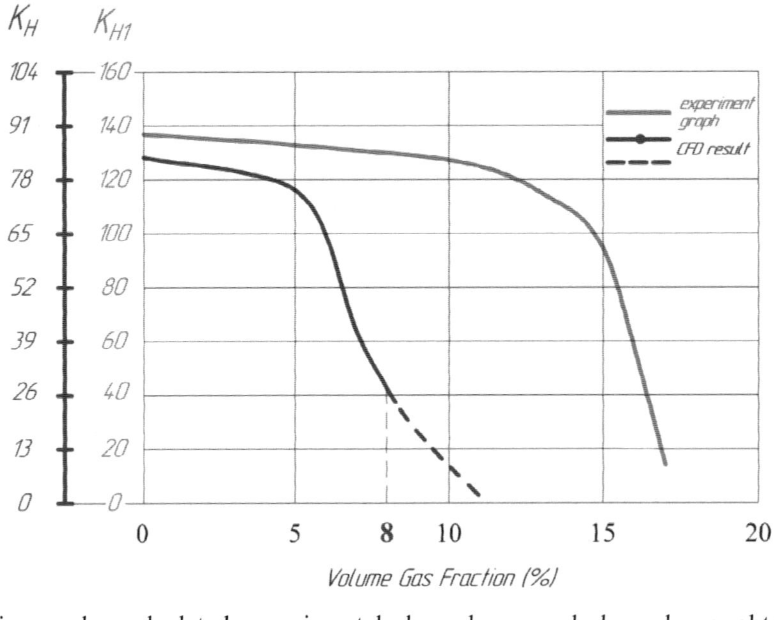

Fig. 8. Superimposed recalculated experimental dependence and dependence obtained by CFD methods

The obtained results can be superimposed on the experimental graphs obtained in the laboratory of the Gubkin Russian State University of Oil and Gas. The characteristic of the pump on the water-gas mixture was recalculated to the optimal operation mode of the pump under test; the head was dimensionless. To verify the results obtained by computational fluid dynamics modeling, Fig. 8 shows superimposed recalculated experimental dependence and dependence obtained by CFD methods.

$K_{H1} = \frac{H}{n^2 \cdot D_2^2} \cdot 10^{-6}$ — dimensionless pressure coefficient, D_2 — is the outlet diameter of the impeller of the pump under test (m), n — is the set pump shaft rotational speed(rpm).

It should be noted that the calculation was carried out at the designed pump stage, which was not optimized according to the criterion of increasing the permissible gas content. Therefore, this pump has a significantly smaller permissible present of gas for stable operation, what can be noticedon the graphs of comparison with existing pump stages. In the future, it is supposed to carry out such optimization by computational fluid dynamics modeling.

Discussion

From Figures 7 and 8 show that the obtained dependence of head on gas content at the inlet is in good agreement with the graphs obtained experimentally, which confirms the adequacy of the mathematical model in the CFD software packages STAR CCM +.

Apparently, for designed pump stage, we can assume that the critical gas content at which the pressure and hydraulic efficiency are reduced by more than 2 times in comparison with the optimal operating mode is 7%. Good convergence of the computational fluid dynamics modeling results with experiment graphs means that to obtain a critical gas content it is not necessary to manufacture a pump or even pump stage, but rather to design a 3-D model of the flow part of one stage and to carry out a virtual experiment using computational fluid dynamics methods, which can significantly save the time on designing, optimization and manufacturing of the pump flow part.

Evaluation of the ability of a pump to operate on a multiphase flowusing computational fluid dynamics method scan be carried out from any sufficiently powerful computer with installed CFD software package (for example, ANSYS or STAR CCM +) in a matter of days. Besides, the hydrodynamic modeling does not require the manufacture of a test bench, which significantly saves time and money on its design and manufacture.

When doing geological and geophysical explorationon a well, the percentage of associated gas in oil is pretty important indicator, which depends on depth of oil occurrence, the configuration and location of the oil formation, and other factors. In wells with associated gas content less than critical, it is possible to install a pump for pumping the oil-associated gas mixture, and then there is no need to install additional modules at the pump inlet, which significantly saves financial costs for the purchase of pump equipment and reduces its dimensions.

Thus, the value of the critical or permissible gas content is a very important indicator during the designing of the flow part of the pump, in calculating the cost of ESP installation for oil production and in calculating the economic indicators of production.

List of reference

[1] Barrios Castellanos, M., Serpa, A.L., Biazussi, J.L., Monte Verde, W., do Socorro Dias ArrifanoSassim, N. Fault identification using a chain of decision trees in an electrical submersible pump operating in a liquid-gas flow (2020) Journal of Petroleum Science and Engineering, 184, article № 106490.

[2] Curkan, B., Klaczek, W., Reede, C. High temperature ESPs for geothermal production: The ideal pump (2018) Transactions — Geothermal Resources Council, 42, pp. 1894–1908.

[3] Wang, Q., Liu, Y., Yang, J., Cui, M., Qi, D. A novel downhole gas separator in ESP systems (2018) Society of Petroleum Engineers — SPE Asia Pacific Oil and Gas Conference and Exhibition 2018, APOGCE 2018.

[4] Camilleri, L., El Gindy, M., Rusakov, A., Ginawi, I., Abdelmotaal, H., Sayed, E., Edris, T., Karam, M. Increasing production with high-frequency and high-resolution flow rate measurements from ESPs (2017) Society of Petroleum Engineers — SPE Electric Submersible Pump Symposium 2017, pp.
[5] Boyarshinova, A., Lomakin, V., Petrov, A. Comparison of various simulation methods of a two-phase flow in a multiphase pump (2019) IOP Conference Series: Materials Science and Engineering, 589 (1), article № 012014.
[6] Lomakin, V.O., Kuleshova, M.S., Kraeva, E.A. Fluid Flow in the Throttle Channel in the Presence of Cavitation (2015) Procedia Engineering, 106, pp. 27–35.
[7] Lomakin, V., Cheremushkin, V., Chaburko, P. Investigation of vortex and hysteresis effects in the inlet device of a centrifugal pump (2018) 2018 Global Fluid Power Society PhD Symposium, GFPS 2018, article № 8472374.
[8] Petrov, A., Isaev, N., Kuleshova, M. Test bench flow straightener design investigation and optimization with computational fluid dynamics methods (2019) IOP Conference Series: Materials Science and Engineering, 492 (1), article № 012036.
[9] Tkachuk, V., Navas, H., Petrov, A., Protopopov, A. Hydrodynamic modelling of the impact of viscosity on the characteristics of a centrifugal pump (2019) IOP Conference Series: Materials Science and Engineering, 589 (1), article № 012007.
[10] Budaev, G., Danilov, D., Kuznechov, A., Lomakin, V., Cheremushkin, V. Research of centrifugal gas-liquid separator (2019) IOP Conference Series: Materials Science and Engineering, 589 (1), article № 012035.
[11] Saprykina, M., Lomakin, V. The evaluation of the effect of gas content on the characteristics of a Centrifugal Pump (2019) IOP Conference Series: Materials Science and Engineering, 589 (1), article № 012017.
[12] Petrov, A., Morozov, A., Kuleshova, M. Investigation of hermetic low-speed pumps performance when operating on highly viscous liquids (2019) IOP Conference Series: Materials Science and Engineering, 589 (1), article № 012030.
[13] Tarodiya, R., Gandhi, B.K. Numerical simulation of a centrifugal slurry pump handling solid-liquid mixture: Effect of solids on flow field and performance (2019) Advanced Powder Technology, 30 (10), pp. 2225–2239.
[14] Olive, N.R., Hong-Quan-Zhang, Redus, C.L., Brill, J.P. Experimental study of low liquid loading gas-liquid flow in near-horizontal pipes (2001) Proceedings of the Engineering Technology Conference on Energy, B, pp. 777–782.
[15] Zhang, H.-Q.;Wang, Q.; Sarica, C.; Brill, J.P. Unified model for gas-liquid pipe flow via slug dynamics—Part 2: Model validation. J. Energy Resour. Technol. 2003, 125, 274–283.
[16] Zhu, J.; Zhang, H.-Q. A Review of Experiments and Modeling of Gas-Liquid Flow in Electrical Submersible Pumps. Energies 2018, 11, 180.

Research of the dependence of the hydraulic friction coefficient of the liquid (λ) on the reinolds number (Re) applicable to the hydraulic drive

V Nikolskiy[1,2] and K Makarov[1]

[1]Bauman Moscow State Technical University

[2]E-mail:vlanikilsk@gmail.com

Annotation. Experimental studies of the dependence of the coefficient of hydraulic friction λ on the Reynolds number Re at the stand "Hydraulics" in relation to the hydraulic drive. The formula for determining the coefficient of hydraulic friction for a turbulent regime of fluid motion is refined. Based on the results of studies, a graph of the dependence of the coefficient of hydraulic friction λ on the Reynolds number Re with an analysis of the four zones of this dependence is constructed.

Introduction

It is known that in the laminar regime of fluid motion, energy (pressure) losses along the flow are explained by the resistance of internal friction forces and, therefore, depend on the viscosity of the fluid. The coefficient of friction resistance along the length λ during laminar motion is a function of only the Reynolds criterion, and the material and the state of the surface of the walls surrounding the flow are independent. A different situation is observed with turbulent motion. As the experiments show, the coefficient of friction resistance λ along the flow length, taking into account the hydraulic flow conditions under the turbulent regime, depends not only on the viscosity of the liquid, but also on the roughness of the walls Δ / d, i.e. Re, $\lambda = f(\mathrm{Re}, \Delta/d)$. Therefore, the reliability of the calculation of energy losses during turbulent motion largely depends on the correct determination of the coefficient of friction resistance along the length λ.

Pipe resistance research has been the subject of many works in Russia and abroad [1–8]. However, until now, due to the complexity of the turbulent flow [9–20], there is no general theoretical method for determining λ for hydraulic pipes.

There are many empirical formulas for determining λ [2, 4]. Each of these formulas is valid only for those conditions for which it is obtained. This practically complicates the choice of the optimal value of λ for each specific system of the pipeline system of hydraulic drives. The complexity and variety of factors that determine the movement of fluid, in most cases, does not allow to be limited to a strictly theoretical solution for a particular case of fluid motion. The obtained solutions need correction coefficients, which are determined as a result of experimental work. This is confirmed by our studies of the dependence of the coefficient of hydraulic friction λ on the Reynolds number at the experimental stand "Hydraulics", created in relation to the hydraulic drive. The stand includes a gear pump, piping systems and instrumentation. The initial data of the hydraulic system: the length of the studied section is ℓ_{ab} = 0.43 m, the internal diameter of the pipeline is d_{in} = 6 mm, the density of

Content from this work may be used under the terms of the Creative Commons Attribution 3.0 licence. Any further distribution of this work must maintain attribution to the author(s) and the title of the work, journal citation and DOI.
Published under licence by IOP Publishing Ltd

mineral oil is ≈900 kg / m3. Measured parameters: pressure at the boundaries of the pipeline section with pressure gauges, the time taken to pass the given volume of fluid is measured using an electronic stopwatch, and the volume of fluid passed through is measured using a flow meter, and the temperature of the working fluid is also measured. The variables of flow rate Q, velocity U, pressure P, kinematic viscosity coefficient ν were varied by changing the temperature during operation of the hydraulic system of the experimental bench [4–5],[9–13].

Methods

Based on the experimental data determined on the bench in the temperature range 18–40 ° C, the hydraulic friction coefficients λ and Reynolds number using previously conducted theoretical and experimental studies. The hydraulic coefficient of friction was determined using the Darcy formula [1–4].

$$h_e = \lambda \frac{l}{d}\frac{U^2}{2g} \quad (1)$$

$$\lambda = \frac{2gd}{l}\frac{h_e}{U^2} = 0.274\frac{h_e}{U^2} \quad (2)$$

$$U = \frac{Q}{S} \quad (3)$$

whereP₁, P₂— liquid pressure according to manometers MH1 and MH2, Pa; ρ— fluid density, 900kg / m3; Q — the volume of fluid traversed during time t. The Reynolds number was determined based on the solution of the system of equations:

- for the laminar regime of fluid motion

$$\begin{cases} \lambda_{on} = 0{,}274\dfrac{h_e}{U^2} \\ \lambda = \dfrac{A}{Re} \end{cases} \quad (4)$$

where A — is a value depending on the state of pipelines, the type of liquid (water, oil), etc.

For oil pipelines and hydraulic pipelines, it is usually taken [4]. ThenRe_{lam}= 75/λ_{op}, and$\lambda_{op.cr}$= = 75/2320 = 0,0323 —friction coefficient for Re_{cr}.

- for the turbulent regime of fluid motion

$$\begin{cases} \lambda_{on} = 0{,}274\dfrac{h_e}{U^2} \\ \lambda = B\left(\dfrac{\Delta_e}{d} + \dfrac{75}{Re}\right)^{0{,}25} \end{cases} \quad (5)$$

Where Δ_e = 0,002 mm —vivalent roughness; B— coefficient determined at Re = Re_{cr}, $\lambda_{op.cr}$ = 0.0323.

$$B = \frac{\lambda_{kr}}{\left(\frac{\Delta_e}{d} + \frac{75}{2320}\right)^{0{,}25}} \quad (6)$$

For the conditions of the experimental stand "Hydraulics" B = 0.0763.

$$Returb = \frac{75}{\left(\left(\frac{\lambda}{B}\right)^4 - \frac{\Delta_e}{d}\right)} \quad (7)$$

It should be noted that formulas (5); (7) are valid for all turbulent motion zones. So, at the boundaries of zone III, this formula is transformed, which corresponds well to the experimental conditions, the dependences for λ: for hydraulically smooth pipes, it turns into the Blasius formula, for a zone of completely rough pipes (self-similar region) into the Shifrinson formula, with only one condition: instead of the coefficient 0.11, the coefficient B is used, determined by the formula (6).

Based on the obtained experimental data by the above-described method, a graph was plotted of the dependence of the Darcy coefficient λ on the Reynolds number Re (figure).

Analyzing this graph, we see that in the laminar mode of motion all the experimental points (λ values) lie on the same line. In logarithmic coordinates, this is line I, described by equation (3)

In turbulent mode of motion and hydraulically smooth pipes, all points are also described on one straight line II.

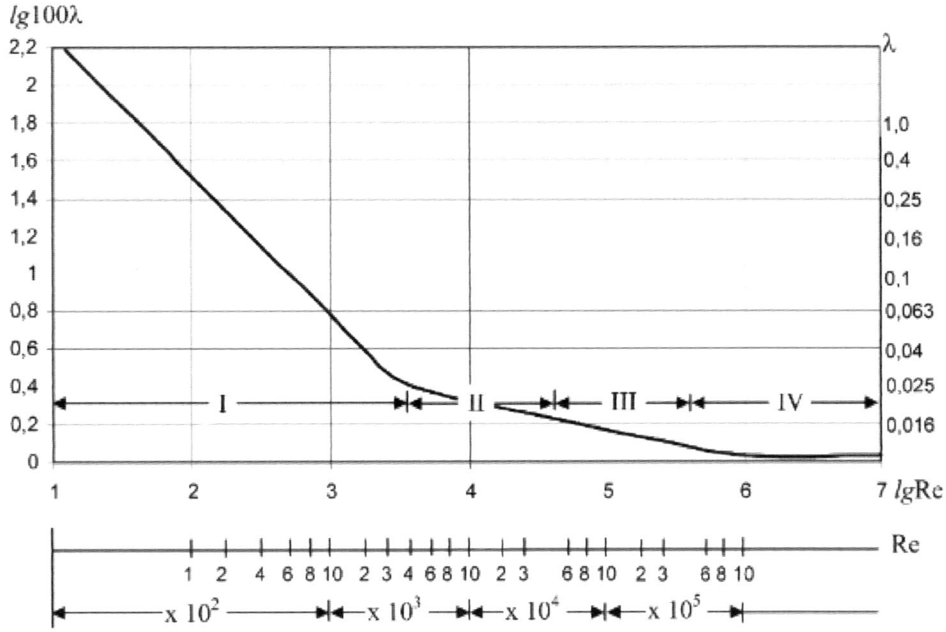

Graph of the Darcy coefficient (λ) versus the Reynolds number (Re)

As can be seen from the graph, in the turbulent regime of motion between the zones of smooth and quite rough pipes (zone IV), there is another (transitional) zone III, in which λ depends both on Re and Δ / d. A smooth decrease in λ with increasing Re in this zone obtained good convergence with the calculated data of the coefficient of hydraulic friction according to the recommended formula.

Results

Conducted experimental studies established the following:

1. Based on the research results (experimental data), we plotted the dependence λ = f (Re) and Δ / d = const, in the temperature range T = 18–40 °, which revealed four zones of dependence λ = f (Re, Δ / d): one for laminar motion and three for turbulent.
2. An improved formula is proposed for determining the coefficient of hydraulic friction for a turbulent regime of fluid motion (3)
3. In the regime of fluid motion between the zones of smooth and completely rough pipes (transition zone III), the curve λ = f (Re, Δ / d) has a smooth decrease in λ with increasing Re and agrees with high accuracy with the calculated data of the hydraulic friction coefficient according to the recommended formula (3)
4. The proposed methodology for determining the coefficient of hydraulic friction is recommended to be used in the calculation of oil pipelines of hydraulic pipelines.

References

[1] M Kraposhin et al 2018*IOP Conf. Ser.: Mater. Sci. Eng.***88**c.79–99
[2] A Kondrat'ev2017*IOP Conf. Ser.: Mater. Sci. Eng.***51**c.107–117
[3] A Leontiev and V Olimpiev 2007 *High Temp.* **45(6)** 844–870
[4] S Isaev et al 2019*Inter.Jorn. Heat and Mass Tran.* **145**118737

[5] AEpikhin 2019*Inter.Jorn. Heat and Mass Tran.***128**10008
[6] A Pugachuk2019 *AIP Conf. Proc.: Mater. Sci. Eng.***2141**030056
[7] K Abramov 2019 *IOP Conf. Ser.: Mater. Sci. Eng.***589** 012013
[8] Y Varaksin 2018*IOP Opt. Eng.***307** 012038
[9] SDergachev et al 2019 *IOP Conf. Ser.: Mater. Sci. Eng.***94**105374
[10] A Leontievet al 2019 *Opt. Eng.***105**206–215
[11] Y Zakharchenko 2020*Opt. Eng.***59(6)** 061604
[12] I Burkov et al 2020*Int.Jor. of Heat trans***147**118946
[13] B Kichatov et al 2020*IOP Conf. Ser.: Mater. Sci. Eng.***198** 106230
[14] A Rudinskii 2019*IOP Conf. Ser.: Mater. Sci. Eng.***55(5)**c. 566–573
[15] A Protopopov et al 2019*IOP Conf. Ser.: Mater. Sci. Eng.***589** 012003
[16] T Lapushkina et al 2019*High Temp.***57(3)**379–387
[17] V Kalugin et al 2018*Rus. Aer.* **61(3)**c. 900–909
[18] S Shchepetov et al 2018PPR **44(6)**c. 539–543
[19] A Gaponov 2017 **52(6)**c.769–776
[20] M Kalugin 2018Her. BMSTU *ser.nat.sci.***(3)**c.32–47

IOP Conf. Series: Materials Science and Engineering **779** (2020) 012008 doi:10.1088/1757-899X/779/1/012008

The Use of Variable Frequency Drives of Centrifugal Pumps atthe Pumping Station

I Zyubin[1], S Pankratov[1], A Petrov[2] and D Babikova[2,3]

[1]National Research University "Moscow Power Engineering Institute", Department of Hydromechanics and Hydraulic Machines named after V.S. Kvyatkovskiy, Krasnokazarmennaya, 14, 111250, Moscow, Russia
[2]Bauman Moscow State Technical University

[3]E-mail: babikova.dina@gmail.com

Abstract: The article presents the experience of operating real pumping stations in urban water supply. The basic principles of practical modeling of the operation of a group of pumps for a common load are considered. The main errors in the modernization of pumping equipment of water supply units are analyzed.

Introduction

Today, many pumping stations (PS) are equipped with a group or individual speed controller of the pump drive motor. In most cases, the rotation speed controller is an electronic device that controls the frequency of the supply voltage of the electric motor of the pump unit (variable frequency drive). Firms distributing variable frequency drives (VFD) have an active policy of introducing their products into the energy product market. The fashion for high technology leads to the fact that sometimes a VFD becomes an end in itself, and not a means of achieving efficient operation of the station. Practical experience of using a VFD shows that there are specific operating features that should be taken into account when a group of controlled pumps as part of a PS operates in conjunction with unregulated pumps.

The task of optimizing energy consumption is to ensure the operation of such a number of pumping units and such a combination of their parameters so that the efficiency of the PS is as high as possible. We will consider the overall efficiency to be a criterion for the quality of the PS operation, although in practice, a part of the reporting documents is drawn up according to such an indicator as the specific energy consumption per cubic meter of water supplied to the consumer — specific energy consumption.

Methods

Satisfaction of both indicators simultaneously is impossible. The desire to reduce the specific energy consumption often leads to the fact that the impeller pumps installed on the PS are operated outside the operating range in the area of high flows, i.e. in overload modes. This is accompanied by a number of negative consequences both for the pumps themselves and for the PS as a whole [1]. Fig. 1 shows the characteristic of a D4000-95 centrifugal pump with an impeller diameter of 825 mm at a reduced rotation frequency (n= 730 rpm) reproduced according tothe catalog [2]; the specific energy

Content from this work may be used under the terms of the Creative Commons Attribution 3.0 licence. Any further distribution of this work must maintain attribution to the author(s) and the title of the work, journal citation and DOI.
Published under licence by IOP Publishing Ltd

consumption $E = N/Q$ is additionally plotted, where N — power on the pump shaft, Q — flow rate. A decrease in this indicator from 0.17 kW·h/m^3 with an optimal Q = 3300 m^3/h to 0.155 kW·h/m^3(only 9%) will be accompanied by an increase in flow rate to 4400 m^3/h and a decrease in pump efficiency by 5%. The characteristic shows that at feeds exceeding the optimum, the NPSH Δh_d sharply increases. By extrapolating the dependence Δh_d to Q = 4400 m^3/h, we obtain Δh_d of 9.5 m. Type D pumps are usually installed with a positive suction lift and a minimum positive suction head, therefore, the available NPSHdoes not exceed 8 ... 9 m. Thus, the pump is guaranteed to fall into the mode of developed cavitation, in which intense erosion of the surfaces of the wet part will occur. An additional decrease in the pump life will occur due to a sharp increase in the radial force of hydraulic origin acting on the shaft and bearings.

In addition, the criterion of specific energy consumption does not allow an objective comparison of PSs with different pressures in the discharge line: for PSs with high pressure, this indicator will always be higher. It is advisable to use this criterion as a reference indicator in relation to the same PS at different periods of its operation. Therefore, from both the technical and economic points of view, when comparing enterprises, one should base on the general efficiency of the station, which is the ratio of the useful hydraulic power to the consumed electric power. The higher this indicator, the higher the technical and economic efficiency of this pumping station.

One can determine the overall efficiency of the pumping station by the ratio

$$\eta_{PS} = \eta_p \eta_m \eta_{vfd}, (1)$$

where η_p is the efficiency of the pump group, η_m is the efficiency of the group of electric motors, η_{vfd} is the efficiency of the power supply, including the frequency control.

PS operating personnel can affect the efficiency of the pumping group only by turning on or off the pumps and setting certain parameters of their work. To solve the task of optimizing energy consumption, we briefly outline the basic principles of hydraulic calculation using simple hydraulic systems as an example, in which all components are connected in series in a single line.

The main purpose of the hydraulic calculation is to determine the parameters of the operating modes, i.e. flow rates and pressures for the hydraulic system as a whole and for its individual elements under typical operating conditions. These parameters can be determined by solving the system of equations:

$$\begin{cases} \Sigma H_s = f(Q), \\ H_r = \varphi(Q) \end{cases} (2)$$

where ΣH_s — total head of elements that are sources of energy, H_r — the required head of the hydraulic system. According to the energy conservation principle at the equilibrium state of the system (i.e., with the steady motion of the fluid flow), these heads must be equal to each other.

The sources of energy can be pumps and the initial elements of the system: various kinds of capacities and collectors. We believe that a pump installed in the considered simple hydraulic system develops a head H, depending on the flow rate Q, and the first element has a head H_f. Then, in the general case, the head of energy sources

$$\Sigma H_s = H(Q) + H_f, (3)$$

The dependence H_r, usually called the hydraulic system characteristic, is written as

$$H_r = H_l + \Sigma h = H_l + kQ^2 \ (4)$$

where H_e is the pressure of the last element, Σh is the pressure loss on the intermediate elements, which for the turbulent regime in the quadratic flow zone [3–20] are proportional to the squared flow rate, k is the proportionality coefficient.

The head of the end elements H_f and H_l in equations (3) and (4) represent the specific potential energy of the liquid, referred to the unit of its weight and having the dimension "meter". This energy consists of two components: potential energy z equal to the geometric (geodesic) height of the liquid level in the tank (for the collector z is the height of its axis) above a certain reference plane, and the pressure energy $p/(\rho g)$, , where ρ is the density of the liquid, g is the acceleration of gravity. When carrying out hydraulic calculations in most cases, pressure is understood not as absolute, but as manometric pressure, and the reference plane is taken in such a way that for the first element of the hydraulic system $z_f = 0$.

The system of nonlinear equations (2) is difficult to solveanalytically, but it is solved quite simply with the use of graphical method. The dependence of the head of the energy sources (3) either coincides with the characteristic of the pump, or differs from it by a constant value of H_i. The characteristic of the hydraulic system (4) is a quadratic parabola with a vertex on the head axis at a point with coordinate H_f. The intersection point of these characteristics on the graph allows you to determine the parameters of the operating mode of the pump (flow and head). Knowing the flow rate in the hydraulic system is equal to the pump flow rate, it is possible to determine the pressure drop on any of its elements.

A similar approach can be implemented when performing a hydraulic calculation for any complex hydraulic system with branches. In this case, one can compose equations similar to equations (3) and (4) for energy sources and for all of the elements of the hydraulic system, and solve the resulting system of two equations graphically or numerically using a computer with appropriate programs.

For some complex hydraulic systems withloop lines, other techniques are used.

Let us consider the operation of the PS in the water supply system, using the simplified model as an example. Water flows from an open or underground reservoir (depending on the type of PS) into the intake manifold. The pumping units installed on the PS take this water from the suction and supply it to the discharge manifold. We suppose that two identical pumping units are installed on the PS. The hydraulic losses in the pipelines connecting the pumps to the manifolds are neglected due to their insignificance compared to losses in the rest of the hydraulic system. Pressure pipelines are connected to the discharge manifold, supplying water to the next stage of the heating system or to the water supply system of a city or industrial enterprise, i.e. to consumers. We believe that the serviced hydraulic system consists of two identical pipelines that connect the pressure manifold of the PS with a reservoir located near consumers. The characteristic of such a hydraulic system containing two parallel pipelines has the form of equation (4), as well as the characteristic of a simple hydraulic system.

Fig. 2 shows the layout of this PS and the hydraulic system it serves.

In the operation of each hydraulic system, it is possible to distinguish characteristic operating modes. Sometimes this is one main mode, but moreoften, there are several. For the water supply system, one can distinguish three regimes, shown in Fig. 3: daytime, nighttime, and peak. In most cases, these modes are ensured by setting the corresponding pressures in the discharge manifold.

The control of the PS operation modes according to a given pressure in the discharge manifold is apparently explained by the relative cheapness and wide distribution of pressure measuring devices (gauges, vacuum gauges, pressure sensors, etc.), as compared to flowmeter devices. Pressure monitoring devices are the basis of measuring systems that allow monitoring and controlling the operation of hydraulic systems and their elements and creating automatic control systems for the operation of the entire station. Often, these tasks are solved without due consideration of factors such as the efficiency and reliability of the machines and other devices that make up the hydraulic system, as well as the PS as a whole. The lack of proper consideration of these factors relates primarily to the operation of pumping units. Let us explain what the reduction in profitability and reliability of their work is.

Let each of the two pumps at the rated speed of rotation have dependences of head, efficiency and NPSH, respectively, H, η and Δh_a, as shown in Fig. 4. Since they are connected to the manifolds in parallel, when the pumps work together with the same pressure, their flow rates add up, and the pressure characteristic of the PS will have the form H_Σ. The intersection point of the characteristics of the hydraulic system H_r and the pump station H_Σ will determine the operating point **A** of the entire system, and will determine the operating point **M** and the operating parameters of each of the pumps, which will correspond to the Q_M flow rate. By the way, this mode is the main or one of the main ones, according to which the selection of pumping equipment for PS is carried out.

We suppose that in a given period in a hydraulic system it is required to have a flow rate less than Q_A, for example, $Q_B = Q_F$. If there is no VFDs, this mode is achieved by partially closing the valve (or several valves) on the pressure part of the hydraulic system, as a result of which the part of the head developed by the pumps, determined by the **BF** section, will be lost on this valve (these valves). The PS will enter the operation mode, determined by point **F**, and the operating mode of the pumps will move to point **N**. The efficiency of the PS will decrease in proportion to the ratio H_B / H_F, which characterizes the decrease in the PSefficiency with this method of regulation. However, the operating modes of pumps will remain within the range of their intended use, limited to lines H_l on the left and H_r on the right(Fig. 4), which will not adversely affect the reliability of their work.

If there is a VFD on the PS, two options of its application are possible with respect to the considered circuit (Fig. 2):

1. Only one pump has VFD.
2. Both pumps have VFDs.

Fig. 4 shows the operation of the PS according to option 1. The dispatching service sets the pressure (head) in the discharge manifold corresponding to mode **B**. The speed of the VFD decreases until the PS characteristic H_Σ'passes through point **B** on the characteristic of the hydraulic system; in this case, the head and efficiency of the pump with VFD will take the position of H'andη'. The parameters of the operating modes of each pump can be determined by drawing a horizontal line through point **B** and finding the points of its intersection with the pressure characteristics of both pumps. These points will be **C** for unregulated and **D** for regulated pump. Both modes are completely unacceptable for operation and are far from the operating range, especially mode **C** for an unregulated pump. The overall efficiency of the PS due to a decrease in the efficiency of each pump will not be higher (rather, lower) than when regulating the valve.The unregulated pump will work in the developed cavitation mode or will not be able to work at all; both pumps will undergo increased wear.

Fig. 5 shows the PS operation in option 2. The speed of both pumps is simultaneously reduced until the PS characteristic H_Σ'passes through the same given point **B** on the characteristic of the hydraulic system, while the dependences of the head and efficiency of the pumps will take the position of H' andη'. The parameters of the operating modes of the pumps are determined by the position of point **C** on the pressure characteristic H', which is in the range of the appropriate use of the pump near the maximum efficiency (or coincides with it). The operation of the pumps in this mode is not associated with any negative consequences, and the efficiency of the PS operation will be maximal. However, for the implementation of this operational option, it is necessary to solve the synchronization problem when changing the speed of the pump units while regulating the PS.

Results

For real PS with a large number of pumping units, the specified mode can be obtained not only with the help of one or several pumps with VFDs, but also by turning on or off unregulated pumping units. The number of possible options for operating conditions and the corresponding design schemes at a

IOP Conf. Series: Materials Science and Engineering **779** (2020) 012008 doi:10.1088/1757-899X/779/1/012008

given pressure (head) in the discharge manifold will be much more thantwo. Moreover, when individual variable frequency drives of pumping units simultaneously operating as part of the PS, inconsistency of their dynamic characteristics is manifested, which is accompanied by a violation of the design operation modes of the PS. Nevertheless, in order to increase the cost-effectiveness and reliability of the operation of pumping equipment at the PS, it is necessary to perform the hydraulic calculations at least for the most characteristic modes of its operation and subsequently use the obtained results in the practical implementation of these modes.

The economic effect of performing hydraulic calculations and subsequent accounting and implementation of their results can be significant at relatively low cost. This is proved by the experience of examining one water supply PS near Moscow. At the examined PS, despite the use of a variable frequency drive for most of the pumping units, the overall efficiency of the station over a significant period of operation turned out to be 52%. There is reason to believe that the quality of the PS operation can be raised to at least 65%, while ensuring both an improvement in the technical performance of the station and significant savings in energy costs.

Discussion

Summing up, we note that the use of a variable frequency drive is an urgent and promising direction in solving the problems of economical operation of such energy facilities as pumping stations. However, in the practice of using such a drive, caution should be taken because a change in the pressure characteristic of a regulated pump (or a group of regulated pumps) leads to a redistribution of the flow rate of the rest of the pumps operating in this hydraulic system. The maximum effect of the variable frequency drive use can be obtained by taking into account the results of hydraulic calculation of the hydraulic system under typical operating conditions.

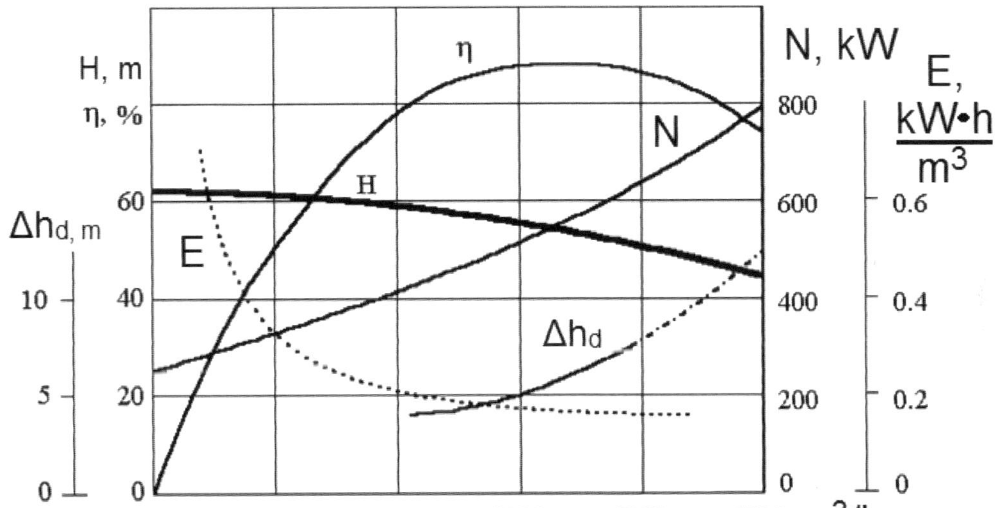

Figure 1. Pump characteristic and specific energy consumption dependence.

IOP Conf. Series: Materials Science and Engineering **779** (2020) 012008 doi:10.1088/1757-899X/779/1/012008

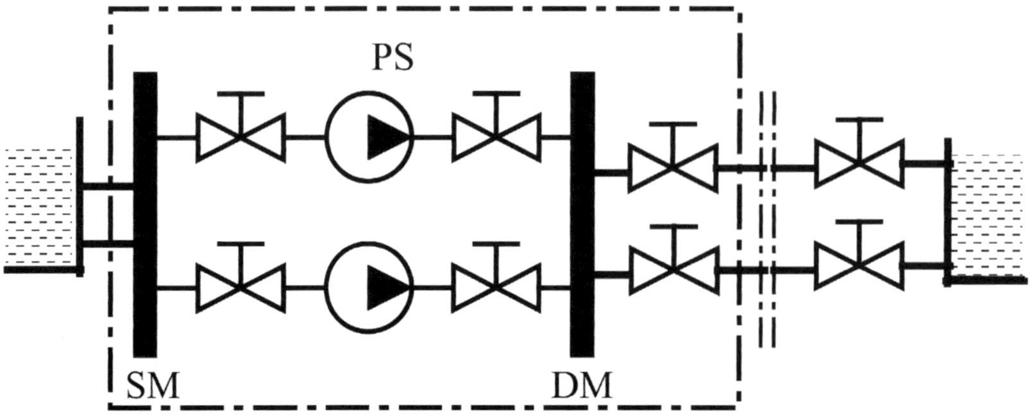

PS — pumping station; SM and DM — suction and discharge manifolds

Figure 2. Diagram of PS and hydraulic system.

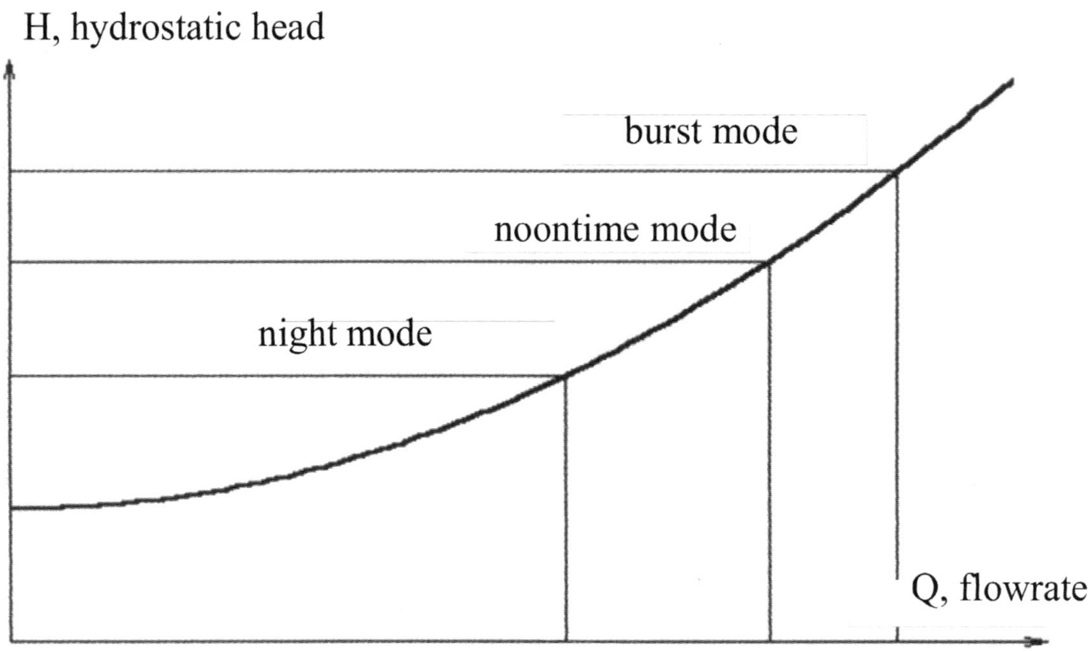

Figure 3. Qualitative head characteristic of the city water supply network and its typical modes of operation.

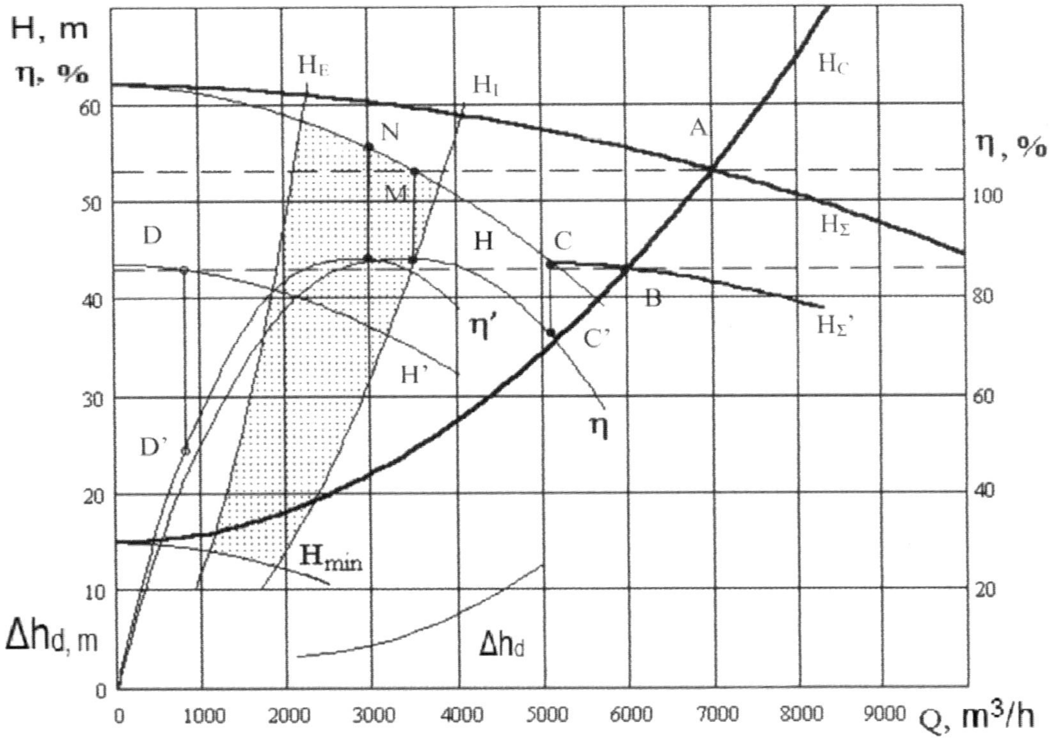

Figure 4.PS with adjustable drive, variant 1.

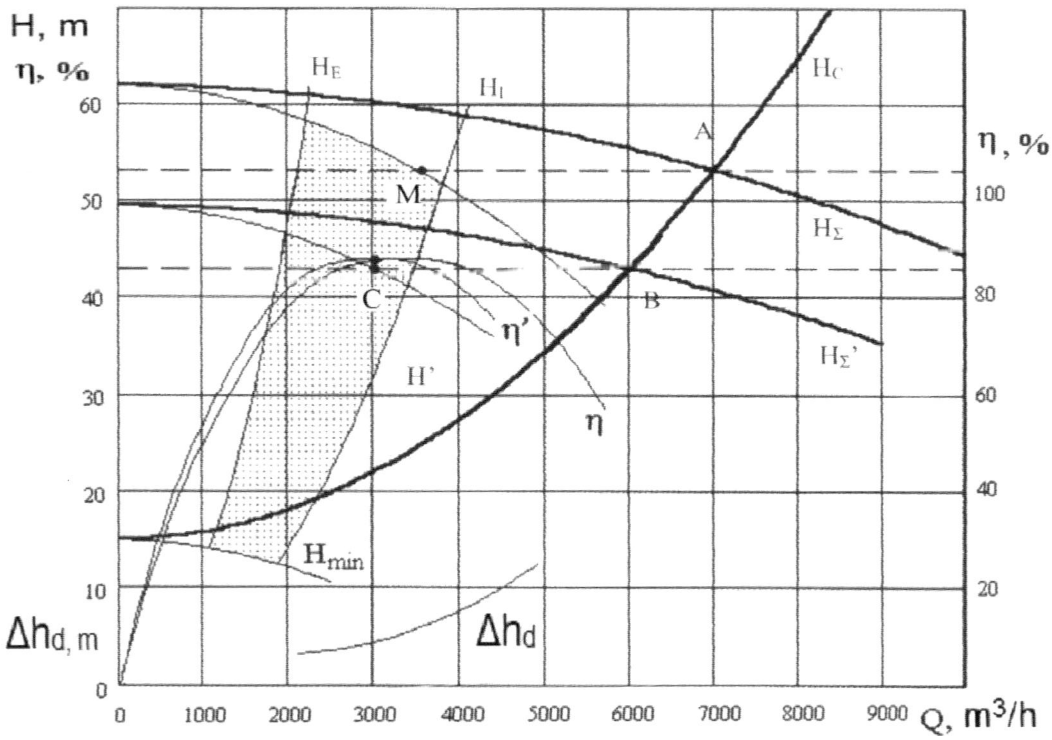

Figure 5.PS with adjustable drive, variant 2.

References

[1] Golubev V.I., Pankratov S.N., Tikhonenkov B.P. Energy saving in water and heat supply systems. *The energy service of the enterprise*, No. 2, 2002. P. 11–13.
[2] Centrifugal pumps of double entry. Catalogue. — Moscow, *TSINTIKhimneftemash*. 1982. 24 p.
[3] Yemtsev B.T. Technical hydromechanics. — Moscow, *Mechanical Engineering*, 1987. 440 p.
[4] Lomakin, V.O., Kuleshovav, M.S. &Bozh'eva, S.M. *Power TechnolEng* (2016) 49: 324. https://doi.org/10.1007/s10749-016-0623-9
[5] A Protopopov and D Bondareva 2019 *IOP Conf. Ser.: Mater. Sci. Eng.***492** 012002
[6] Lomakin V O, Chaburko P S and Kuleshova M S 2017 Multi-criteria Optimization of the Flow of a Centrifugal Pump on Energy and Vibroacoustic Characteristics *Procedia Engineering***176** pp 476–482
[7] A Shablovskiy and E Kutovoy 2019 *IOP Conf. Ser.: Mater. Sci. Eng.***492** 012035
[8] A Protopopov and C Jakovich 2019 *IOP Conf. Ser.: Mater. Sci. Eng.***492** 012034
[9] Lomakin V O and Petrov A I 2012 Verification of the calculation results in the STAR-CCM+ hydrodynamic simulation package of the flow-through part of the centrifugal pump AX 50-32-200 *Proceedings of Higher Educational Institutions. Machine Building* (C) p 6
[10] Petrov A I and Isaev N Y 2017 Hydrodynamic modeling of the centrifugal pump operation in the zone of negative innings *Hydraulics* (3) pp 91–101
[11] V Tkachyk et al 2019 *IOP Conf. Ser.: Mater. Sci. Eng.***589** 012010
[12] K Dobrokhodov and APetrov 2019 *IOP Conf. Ser.: Mater. Sci. Eng.***492** 012016
[13] V Tkachuk et al 2019 *IOP Conf. Ser.: Mater. Sci. Eng.***589** 012007
[14] T Valiev and APetrov 2019 *IOP Conf. Ser.: Mater. Sci. Eng.* **492** 012038
[15] Kraposhin M V, Banholzer M, Pfitzner M and Marchevsky I K 2018 A hybrid pressure-based solver for non-ideal single-phase fluid flows at all speeds *International Journal for Numerical Methods in Fluids*88 (2) pp 79–99
[16] Dazhuan Wu, Shuai Yang, Peng Wu and Leqin Wang 2011 Experimental and numerical study of transient flow in a centrifugal pump during startup *Journal of Mechanical Science and Technology* March
[17] Dazhuan Wu, Shuai Yang and Peng Wu 2014 Transient characterisitcs of a close-loop pipe system during pump stopping periods *Journal of Pressure Vessel Technology, Transactions of the ASME* February
[18] B. J. Chalmers, L. Musaba, D. F. Gosden, "Variable-Frequency synchronous motor drives for electric vehicles", *IEEE Transactions on Industry Applications*, vol. 32, no. 4, pp. 896–903, 1996.
[19] G. Pellegrino, E. Armando, P. Guglielmi, "Direct flux field-oriented control of IPM drives with variable DC link in the field-weakening region", *IEEE Transactions on Industry Applications*, vol. 45, no. 5, pp. 1619–1627, 2009.
[20] T. M. Jahns, G. B. Kliman, T. W. Neumann, "Interior permanent magnet synchronous motors for adjustable-speed drives", *IEEE Transactions on Industrial Applications*, vol. IA-22, pp. 738–747, 1986.

IOP Conf. Series: Materials Science and Engineering **779** (2020) 012009 doi:10.1088/1757-899X/779/1/012009

High-Pressure Positive-Displacement Hydraulic Pumps in Concrete Sample Testing Machines

I Zyubin[1], S Cherkasskikh[1], A Akinfiev[2] and D Babikova[2,3]

[1]National Research University "Moscow Power Engineering Institute", Department of Hydromechanics and Hydraulic Machines named after V.S. Kvyatkovskiy, Krasnokazarmennaya, 14, 111250, Moscow, Russia
[2]Bauman Moscow State Technical University

[3]E-mail: Babikova.dina@gmail.com

Abstract: This article discusses a way to solve the problem of operational quality control of concrete in the construction of critical structures with the use of small-sized testing machines. To reduce the testing machines size, it is proposed to use a high-pressure hydraulic drive. As a promising way to control the machine, the practical experience is described of using a pump with a constant displacement, the drive of which uses an electronic frequency converter.

Introduction

In the construction of hydroelectric power plants (HPP), the quality of concrete is of great importance, since it largely determines the strength of all supporting structures. It is equally important to ensure a short construction time. In this regard, an urgent task is to carry out operational control measures. As practice shows, methods based on the control of the chemical composition of the mixture are not sufficiently effective and reliable. Therefore, methods based on the destruction of control samples remain the most acceptable [1–10]. In this case, verification of compliance of the material with the requirements occurs during testing of the samples for compression, bending, tension and a number of other indicators, on which the durability and bearing capacity of the structures depend.

Currently, these activities are usually carried out in specialized laboratories, since they require special testing equipment. However, this approach leads to a significant increase in the time required for quality control, which is due to the need to deliver samples to the testing laboratory, process delivery, etc. In these conditions, the use of small-sized testing machines that meet modern standards and can be used in temporary laboratories directly at the construction site is promising.

Methods

The specifics of the testing machines allows us to divide them into two main classes: tensile testing machines and compression testing machines. These classes differ significantly in terms of the speed of working body movement, and the requirements for the measured parameters. This article considers only compression testing machines for concrete samples [11–19]. Also promising is the modernization of other types of testing machines and technological devices, for example, machines for pre jacking.

Testing directly at a construction site in a temporary building requires the use of small-sized testing machines. Moreover, such machines must fully comply with the requirements of the standard[2]. In addition, the use of high-strength concrete in construction makes it necessary to test samples with a breaking force of up to 1000 kN inclusive. The need to ensure minimum dimensions with significant

Content from this work may be used under the terms of the Creative Commons Attribution 3.0 licence. Any further distribution of this work must maintain attribution to the author(s) and the title of the work, journal citation and DOI.
Published under licence by IOP Publishing Ltd

IOP Conf. Series: Materials Science and Engineering **779** (2020) 012009 doi:10.1088/1757-899X/779/1/012009

developed forces requires a transition to high working pressures in the hydraulic drive of testing machines. At the same time, a pressure level of 60 MPa and higher is considered promising.

In addition to the above requirements, the design of small-sized testing machines should have a low cost, not require a significant amount of routine maintenance on the hydraulic system, and therefore the level of staff qualification can be lowered.

In the process of testing, continuous loading of the samples should be implemented at a rate that ensures an increase in the design stress in the sample until it is completely destroyed within (0.05±0.02) MPa with automatic measurement of forces [2]. At present, it is not particularly difficult to solve the problem of ensuring the necessary measurement accuracy. At the same time, the need to control the pressure in the hydraulic drive of the loading device for a wide range of efforts faces the problem of the shortage of available high-pressure proportional hydraulic equipment necessary for the implementation of regulation.

In the study of the possibility of creating small-sized testing machines in order to reduce costs, it was considered equipped with typical purchased items. The current situation in the domestic market of hydraulic equipment often stimulates the increased use of foreign equipment, which is due to the lack of domestic analogues [3]. This situation is typical, in particular, for low-flow pumps (less than 5 l/min) and pressures above 60 MPa, which are necessary for the implementation of small-sized testing machines. Therefore, equipment manufactured by domestic firms (jacks, hydraulic tools, etc.) for a given pressure level is usually equipped with hand pumps or foreign pumps (BIERI, HAVE).

Therefore, BIERI radial piston pump HRK was selected as a source of hydraulic energy, according to a combination of technical and economic criteria. Pumps of this type provide pressures up to 70 MPa and are available with capacity range of 0.12...4.52 cm^3 and a drive shaft speed range of 500...3600 rpm.

A feature of the selected pump is the presence of only two plunger pairs, which usually causes an increase in pressure pulsation in the discharge line. The latter circumstance leads to the fact that these pumps are practically not used in a traditional hydraulic drive.

However, they are structurally simple and cheap enough. In addition, if applied in test machines, the resulting pressure pulsation can be leveled due to the large volumetric displacement of the actuating hydraulic cylinder. Further tests showed that the level of the total error in measuring pressure, including the pulsation component, does not exceed 0.3% at an acceptable level of 1% [2].

A double-acting jack DG100G150 manufactured by ZAO NPO Enerprom was used as a loading device, developing a force of up to 100 tf and providing a piston stroke of 150 mm. Its volumetric displacementis 2200 cm^3, and the mass is 48.5 kg.

The operation of the loading device during the compression test involves ensuring the movement of the output link in the following sequence: a relatively fast supply of the loading plate to the sample, testing of the sample at zero speed of movement of the hydraulic cylinder rod and removal of the plate to its original position. To control the pressure level in the actuator, and therefore the loading force, a typical circuit was used with parallel installation of a throttling device. As a throttling device, a high-pressure direct-acting proportional valve type PDV700-P-6-700-2 from BIERI was selected. The electromagnet current was controlled by pulse width modulation with pressure feedback.

Results

In this work, a prototype of a test machine of the proposed design was created and its full-scale tests were performed at the scientific and production base of OOO Firma VNIR. During the tests, the explosive destruction of samples of high-strength concrete (grades above M500) was observed. Analysis of the working process showed that at the time of destruction there is a sharp decrease in the opposing load on the piston rod. As a result, the speed of movement of the output unit increases significantly and the flow rate of fluid entering the actuator hydraulic cylinder increases spasmodically. As tests have shown, this causes erosion of the shutoff-regulating element of the proportional valve and a significant decrease in its resource.

IOP Conf. Series: Materials Science and Engineering **779** (2020) 012009 doi:10.1088/1757-899X/779/1/012009

The traditional way to solve this problem is to limit the flow of fluid at the time of destruction, which can be achieved by installing the appropriate hydraulic equipment. So, for example, in the design of similar stationary-type machines, in particular IP-1000 manufactured by PKC ZIM (Armavir), the use of a three-line flow controller is provided. It is also possible to use a circuit that includes two pumps. However, these approaches lead to construction that is more expensive.

With this background, it was suggested to use an additional speed control of pump shaft to limit the flow rate. Thus, the proposed option has dual control: proportional and precise control of load growth is carried out by a high-pressure proportional valve, and the required flow rate is provided by a variable-speed pumping station with speed control. This pumping station contains a pumping unit of constant supply controlled by a variable frequency electric drive.

The circuit implements a mode of operation with constant power N at which the flow rate Q changes inversely with the pressure downstream the pump p::

$$Q = \frac{N}{p}$$

The use of an electrically controlled pump allows to increase the efficiency of the system, reduce energy consumption and reduce heat generation, which allows to reduce the tank volume and the overall size of the structure.

To implement this solution, a Vesper E2-MINI E2-MINI-SP5L single-phase frequency converter was used, which has a relatively low cost. Tests have shown that to ensure high-quality destruction of the sample and compliance with the required operational characteristics of the testing machine, it is necessary to have a speed control range within 60% ... 200% of the nominal speedwhile maintaining the required moment on the pump shaft.

At present, a design of the HP70 pump has been prepared (Fig. 1) and is undergoing technological development.

Fig. 1. ExterioroftheHP70 pump (Maximum pressure 70 MPa, capacity 0.34 cm^3)

Due to the modular design, this pump can be equipped both with domestic motors with a power of 1.5 kW and a mounting flange with a hole diameter of 165 mm, and with foreign-made engines with power from 1 to 1.5 kW.

The study has also begun of the design of the proportional valve similar in connection size and operating pressure to the PDV700-P-6-700-2 (BIERI) valve. Fig. 2 shows the exterior of a three-dimensional model of a valve with an electromagnet. This valve has longer lifetimethan the analogue and uses the domestic proportional electromagnet EM-25 GOST 19264-82 (NPP S pets electromagnit) as a control.

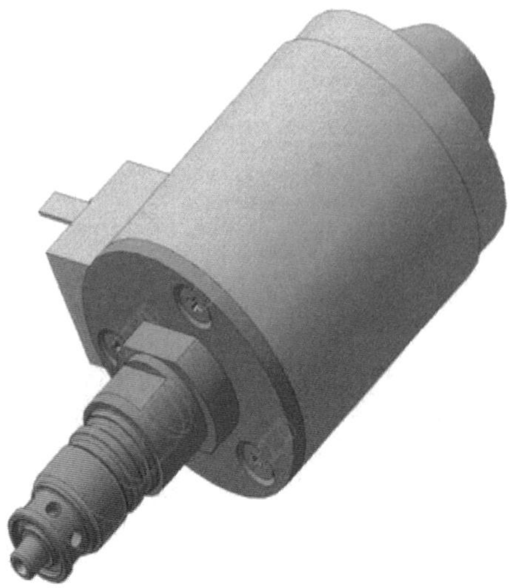

Fig.2. Model of the electromagnetic proportional valve
(Maximum pressure 70 MPa, working flow 10 l/min)

Discussion

The operation of such machines with a high degree of automation in the construction industry confirmed the promise of switching to high pressures.

At the same time, the experience of using BIERI pumps in small-sized testing machines revealed a number of disadvantages. These pumps are poorly aggregated in pump units with domestic electric motors. In this regard, it is required to either equip them with imported analogues, or addto their structure additional elements (coupling, bushing). This leads to an increase in the cost of the aggregate. In addition, the prices of these pumps are subject to changes in exchange rates. Thus, the further direction of work involves the development of a set of high-pressure equipment devoid of these shortcomings.

References

[1] GOST 10180-90. Concrete. Methods for determining the strength of control samples. Moscow,*Standartinform*, 2006. — 30 p.

[2] GOST 28840-90. Machines for testing materials in tension, compression and bending. General technical requirements. Moscow,*Publishing house of Standards*, 2004. — 6 p.

[3] V.K. Sveshnikov, Hydraulics is rapidly developing, *Construction and mechanical engineer*, No. 5, 2010, p. 42–46

[4] A Protopopov et al 2019 *IOP Conf. Ser.: Mater. Sci. Eng.* **589** 012004

[5] N Sosnovsky and D Ganieva 2019 *IOP Conf. Ser.: Mater. Sci. Eng.* **589** 012016

[6] I Kolodin and M Ryabinin 2019 *IOP Conf. Ser.: Mater. Sci. Eng.* **589** 012018

[7] Semenov, S., Kulakov, D. Mathematical modeling of the mechanisms of volumetric hydraulic machines (2019) *IOP Conference Series: Materials Science and Engineering*, 492 (1), article № 012042.

[8] S Barmin and O Nikitin 2019 *IOP Conf. Ser.: Mater. Sci. Eng.* 589 012020
[9] P Shcherbachev and D Vdovin 2019 *IOP Conf. Ser.: Mater. Sci. Eng.* 589 012023
[10] Isaev, A.L., Pushilin, Yu.M.Fracture peculiarities of multilayer structures under local dynamic loading(1994) *ProblemyProchnosti*, (5), pp. 64–68.
[11] Borisov, B.P. Positive-displacement machines. Moscow, *Bauman Moscow State Technical University*, 2018. — p.240.
[12] Tsai, Shir-Tau, "Simulation of the control characteristics of an axial piston pump and aflapper-nozzle valve combination" (1988).*Retrospective Theses and Dissertations*. 8895.
[13] Ivanov, A.S., Fomin, M.V. Classification and Calculation of Drive Loads(2017) *Russian Engineering Research*, 37 (12), pp. 1017–1022.
[14] Manring Noh, D. (2013) Fluid Power Pump and Motors: Analysis, Design and Control, p. 320.
[15] Kim, J.-H., Jeon, C.-S., Hong, Y.-S. Constant pressure control of a swash plate type axial piston pump by varying both volumetric displacement and shaft speed(2015) *International Journal of Precision Engineering and Manufacturing*, 16 (11), pp. 2395–2401.
[16] S V Putintsev et al 2019 *J. Phys.: Conf. Ser.***1177** 012006
[17] Shenoy, P.S., Fatemi, A. Connecting rod optimization for weight and cost reduction(2005) *SAE Technical Papers*.
[18] Nikitin, O.F.: Machine regulation volume hydraulic drive with reciprocating motion of hydraulic motors. In: *International Conference on Fluid Power and Mechatronics (FPM)*, 2015, pp. 593–595
[19] Solodovnikov, V.V., Plotnikov, V.N., Yakovlev, A.V. (1993) The Theory of Auto-matic Control of Technical Systems. Moscow, Bauman MSTU

IOP Conf. Series: Materials Science and Engineering **779** (2020) 012010 doi:10.1088/1757-899X/779/1/012010

Study of Cavitation Characteristics of a Cryogenic Pump by Computational Fluid Dynamic Methods

D Babikova[1,2] and A Petrov[1]

[1]Bauman Moscow State Technical University

[2]E-mail:babikova.dina@gmail.com

Abstract. This article is devoted to the study of cavitation characteristics of a centrifugal cryogenic pump. Hydrodynamic simulation of the fluid flow in the wet part of the pump was carried out in order to determine the output parameters of the pump. A critical increase in the temperature of liquid helium, corresponding to the beginning of cavitation in the pump, was determined. A cavitation performance of the pump was built under given operating conditions and a critical NPSH was determined.

Introduction

In the modern world, cryogenic products are an integral part of many technological processes in the energy sector, oil and gas industry, metallurgy, astronautics, chemical industry, medicine and many other areas of human life [1–4]. Cryogenic products include substances or mixtures of substances at cryogenic temperatures (below 120K). One of the most important cryogenic products is liquid helium. It is used as a refrigerant to obtain and maintain ultra-low temperatures [5–8]. For example, helium is used to cool superconducting magnets in various scientific and technical installations. The low boiling point of helium (4.23K at normal atmospheric pressure) imposes certain restrictions on the design and operation of technical devices used in units with this cryoproduct [9–12]. So, often to maintain the required temperature of helium, additional cooling of the equipment with liquid nitrogen is used. Fig. 1 shows a cryostat for liquid helium, which is a Dewar vessel, the inner vessel of which is filled with liquid helium, and the outer one is cooled with liquid nitrogen.

A pump is located in the inner vessel of the cryostat; it supplies helium directly to the cooling system of superconducting magnets. The pump can be centrifugal or displacement, depending on the required flow and pressure values in the system.

An important problem in the operation of such a pump is the possibility of cavitation in it. Even With the smallest change (increase) in the temperature of liquid helium in a vessel, the pump can switch to super-cavitation mode, which will negatively affect the operation of the entire system. Therefore, it is very important to calculate the operation of the pump on liquid helium in the permissible temperature range, and accurately determine the moment of the beginning of cavitation.

In this work, we studied the cavitation characteristics of a BNHEP-35-000 centrifugal pump manufactured by Barber-Nichols Inc., USA. This is a 240 W submersible pump with overhung impeller. Fig. 2 shows the general view of the pump and its schematic drawing.

Content from this work may be used under the terms of the Creative Commons Attribution 3.0 licence. Any further distribution of this work must maintain attribution to the author(s) and the title of the work, journal citation and DOI.
Published under licence by IOP Publishing Ltd

IOP Conf. Series: Materials Science and Engineering **779** (2020) 012010 doi:10.1088/1757-899X/779/1/012010

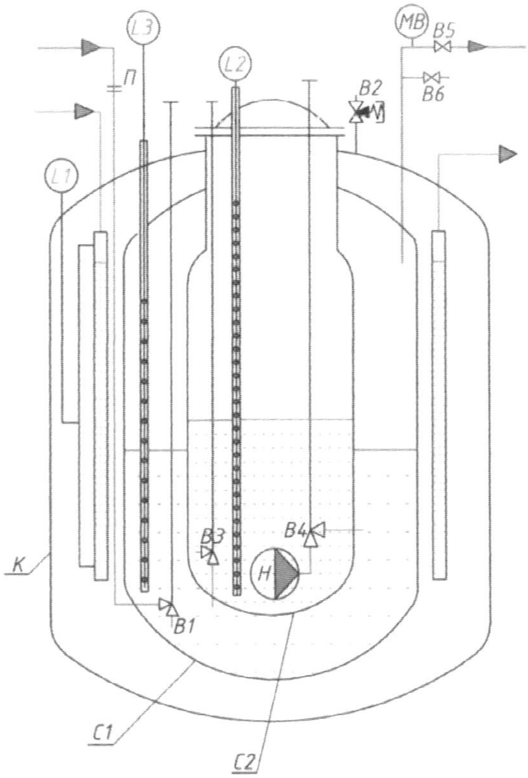

Figure 1. Cryostat for liquid helium. (K — cryostat, C1 — external vessel of the cryostat, C2— internal vessel of the cryostat, L1 — differential pressure gauge, L2 — level sensor in the internal vessel, L3 — level sensor in the external vessel, B1 — valve for filling the external vessel, B2 — valve for evacuation and protection of the insulating chamber from pressure rise, B3 — valve for filling the internal vessel, B4 — valve for regulating the supply of the pump, B5 — valve for the return flow, B6 — valve for evacuating the helium chamber of the cryostat, MB—vacuum pressure gauge, P — plug of the liquid helium filling line)

IOP Conf. Series: Materials Science and Engineering **779** (2020) 012010 doi:10.1088/1757-899X/779/1/012010

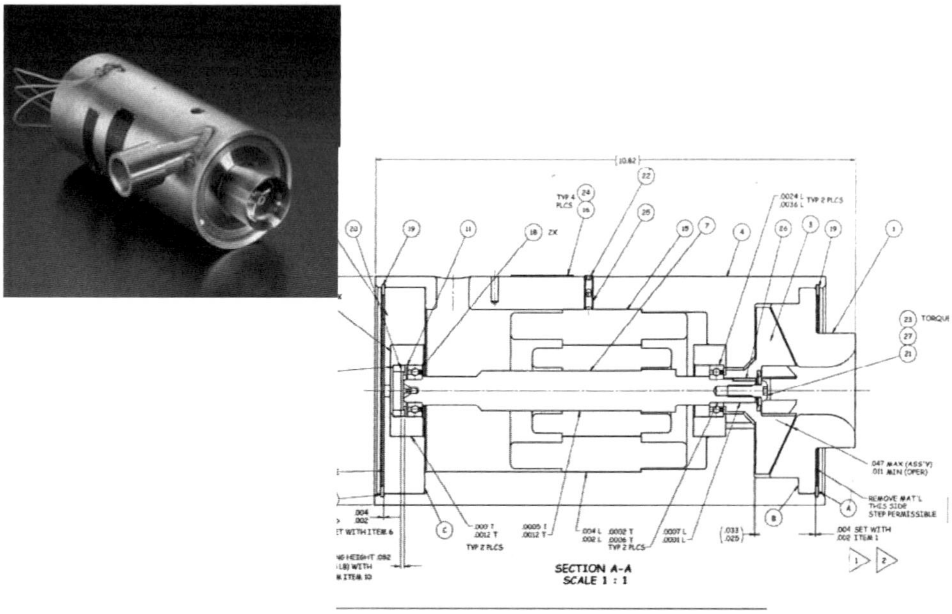

a) b)

Figure 2.BNHEP-35-000 pump. a) general view; b) schematic drawing.

The main parameters of the pump are given in Table 1.

Table1. Parameters of BNHEP-35-000 pump.

Parameter	*Value*
Volumetric flow, l/min	47,67
Shaft rotation speed, rpm	2400
Impeller diameter, mm	100
Inlet diameter, mm	30
Outlet diameter, mm	20
Head, m	10

Methods

The study of the cavitation characteristics of the Barber-Nichols pump was carried out using computational fluid dynamics [13–17]. For this, a three-dimensional model of the pump wet part was created, shown in Fig. 3. The flow part consists of an axial inlet, an impeller with side pockets and a spiral outlet.

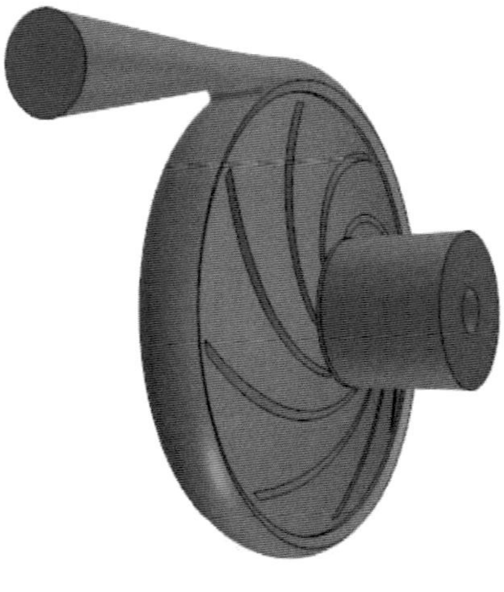

Figure 3. Wet part of the BNHEP-35-000 pump (pockets are not shown).

For this problem, it was decided to simulate a single-phase turbulent flow. To verify the flow regime, the Reynolds number for the flow at the pump inlet was calculated. $Re = \frac{\rho v D}{\eta} = \frac{125.4 \cdot 1.12 \cdot 0.03}{2.94 \cdot 10^{-6}} = 1.4 \cdot 10^6$, which corresponds to turbulent mode. In the simulation, we used the equations of hydrodynamics for an incompressible fluid ($\rho = const$).

The mass conservation equation (continuity equation):

$$\left(\frac{\partial \tilde{u}_i}{\partial t} + \tilde{u}_j \frac{\partial \tilde{u}_i}{\partial x_j}\right) = -\frac{1}{\tilde{p}}\frac{\partial \tilde{p}}{\partial x_j} + \vartheta \frac{\partial^2 \tilde{u}_i}{\partial {x_j}^2},$$

where $\tilde{u}_i = U_i + u_i$ — instantaneous value of velocity (U_i is the average component, u_i is the fluctuation component);

$\tilde{p} = P + p$ —instantaneous value of pressure (P is the average component, p is the fluctuation component);

The Reynolds-averaged Navier–Stokes equations:

$$\rho\left(\frac{\partial U_i}{\partial t} + U_j \frac{\partial U_i}{\partial x_j}\right) = -\frac{\partial P}{\partial x_i} + \frac{\partial}{\partial x_i}\left(T_{ij}^{(v)} - \rho u_i u_j\right),$$

where U, P — averaged values of velocity and pressure, respectively;

$T_{ij}^{(v)} = 2\mu \tilde{s}_{ij}$ — viscous stress tensor for incompressible fluid;

$\tilde{s}_{ij} = \frac{1}{2}\left(\frac{\partial \tilde{u}_i}{\partial x_j} + \frac{\partial \tilde{u}_j}{\partial x_i}\right)$ — instant strain velocity tensor;

$\rho u_i u_j$ — Reynolds stresses.

In addition, a semi-empirical $k - \omega SST$ turbulence model was chosen to close the system of equations. It includes:

kinetic energy of turbulence transfer equation

$$\frac{\partial k}{\partial t} + U_j \frac{\partial k}{\partial x_j} = P_k - \beta k \omega + \frac{\partial}{\partial x_j}\left[\left(v + \sigma_k v_T\right)\frac{\partial k}{\partial x_j}\right];$$

equation for the relative dissipation rate of the kinetic energy of turbulence

$$\frac{\partial\omega}{\partial t}+U_j\frac{\partial\omega}{\partial x_j}=\alpha S^2-\beta\omega^2+\frac{\partial}{\partial x_j}\left[\left(v+\sigma_\omega v_T\right)\frac{\partial\omega}{\partial x_j}\right]+2\left(1-F_1\right)\sigma_{\omega 2}\frac{1}{\omega}\frac{\partial k}{\partial x_i}\frac{\partial\omega}{\partial x_i}.$$

To study the cavitation characteristics of the pump, the task was to determine the critical increase in the temperature of helium at which developed cavitation would be observed in the pump, with the existing overpressure in the inner vessel of the cryostat equal to 0.05 MPa. Several calculations were carried out in a CFD package to simulate different temperatures of the liquid in a cryostat. For each of the points, the pressure of saturated helium vapor and the density of liquid helium were determined from the tables of characteristics of liquefied helium and set in accordance with the temperature.

To visualize cavitation phenomena, the field function "VaporPressure = ($ {AbsolutePressure} <<151025 Pa>)? 1 : 0". In the syntax of the program used, this means that the absolute pressure at each point is compared with the pressure of saturated helium vapor (in the recorded example, the pressure of saturated vapor is 151025 Pa for helium at a temperature of 4.668K). If the absolute pressure at a point is lower than the saturated vapor pressure, then the function takes the value 1 (cavitation is possible), otherwise 0.

Results

In this study, a number of pump operation modes with a temperature of liquid helium from 4.23 to 4.668 K were simulated. The pump head obtained during hydrodynamic simulation, as well as the calculated NPSH (Δh) for each of the points are presented in Table 2.

Table 2. Simulation results.

№	T, K	H, m	Δh, m
1	4,23	10,41	39,220
2	4,3	10,40	33,881
3	4,5	10,39	16,902
4	4,6	10,25	7,315
5	4,65	10,15	2,178
6	4,66	10,13	1,119
7	4,665	10,11	0,586
8	4,668	10,02	0,264

According to the values presented in Table 2, the cavitation performance of the pump was plotted (Fig. 4). The solid line in the graph shows the cavitation performance itself, and the dashed line represents a 3% drop in head relative to the first point. For ease of graph perception, points 1-3 are outside the plot area.

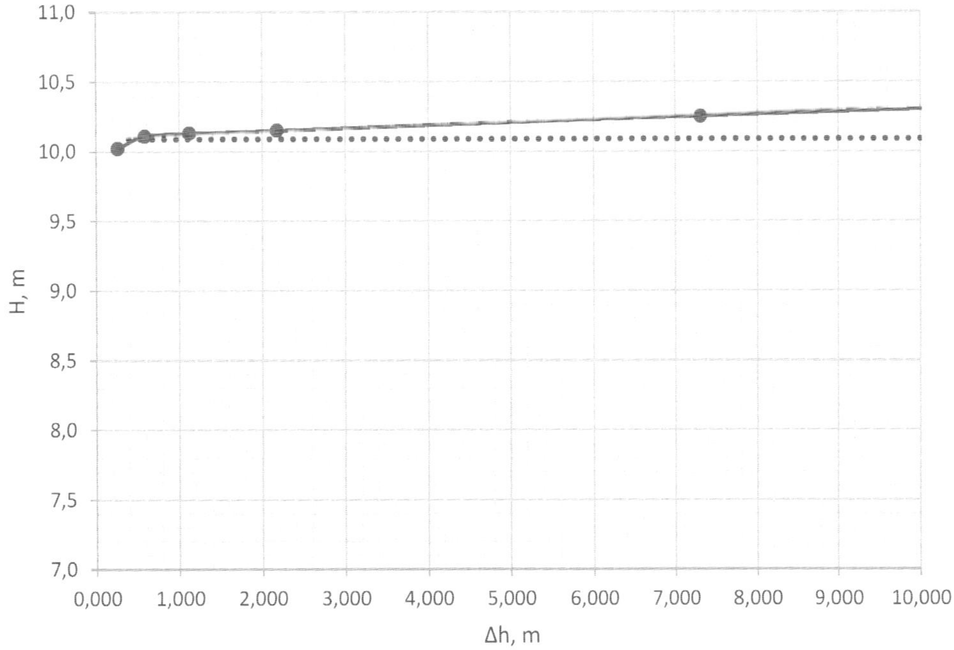

Figure 4. Cavitation performance of the pump.

Using the graph obtained, we determined the critical NPSH of the pump is 0.4 m. Thus, with an increase in the temperature of helium in the cryostat to about 4.667K, cavitation can begin in the pump. Fig. 5 shows the distribution of the created field function in the pump section, in which the zones in which cavitation is possible are red. This scene was obtained at a temperature of 4.668K.

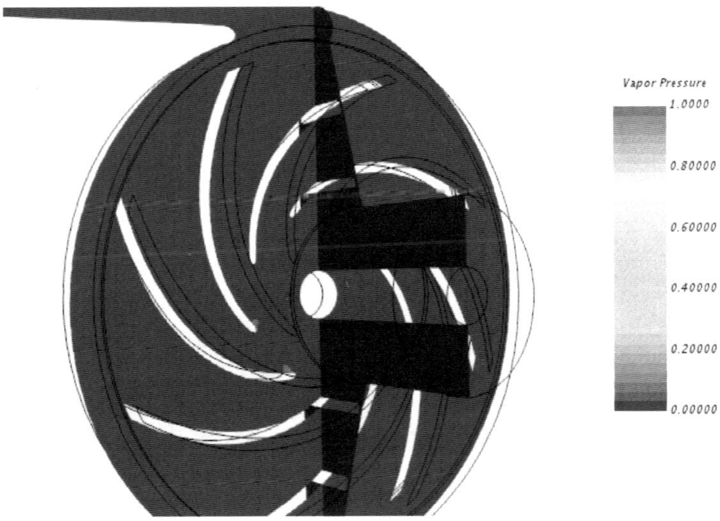

Figure5. Distribution of the field function in the cross section of the pump.

It can be seen that cavitation in the pump is possible only at the inlet tip of the blade. It was not possible to achieve the super-cavitation mode for this pump, since at a temperature greater than 4.668K, the absolute pressure in the inner vessel of the cryostat is less than the pressure of saturated helium vapor, i.e., helium in the cryostat will begin to evaporate. Such pump operating conditions are not permitted.

Conclusion

As a result of hydrodynamic simulation, it was revealed that the BNHEP-35-000 pump does not enter the super-cavitation mode when pumping liquid helium from a cryostat with an overpressure of 0.05 MPa. The beginning of cavitation at the inlet tips of the pump blades can be observed at a temperature of 4.668K. This may be due to the fact that liquid helium has a low density (6–8 times less than the water density), and, accordingly, the pressure drop at the inlet tip of the blade at the pump running on liquid helium is also less significant. Using the plotted cavitation performance of the pump, the critical NPSH for these operating conditions was determined, it amounted to 0.4 m.

References

[1] Vshivkov, A.Y., Delkov, A.V., Kishkin, A.A. et al. *Chem Petrol Eng* (2019) 55: 306. https://doi.org/10.1007/s10556-019-00619-4
[2] Vshivkov, A.Y., Delkov, A.V., Kishkin, A.A. et al. *Chem Petrol Eng* (2019) 54: 746. https://doi.org/10.1007/s10556-019-00543-7
[3] Agapov, N.N., Borzunov, Y.T., Konstantinov, A.V., Klimanskiy, D.I., Arkharov, I.A., Navasardyan, E.S., Arkharov, A.M. Cryogenic targets of the lightest gases (Hydrogen, Deuterium and Helium-4) with GM cryocooler for experiments of high energy physics (2019) *Refrigeration Science and Technology*, Part F147717, pp. 38–44.
[4] Smorodin, A.I., Krasnikova, O.K. &Frolov, I.A. *Chem Petrol Eng* (2017) 53: 44. https://doi.org/10.1007/s10556-017-0292-z
[5] Eselson, B.N., Grigoryev, V.N. (1973) Solutions of Quantum Liquids 3He-4He, p. 423. *Akademizdatcenter "Nauka"*, RAS
[6] Arkharov, A.M., Barishnikov, V.V., Jerdev, A.A., Mikhailov, V.M., Golubev, G.F. Low-Temperature method for enrichment of light isotope 3He concentration in 3He-4He natural concentration mixture (1993) *BMSTU Messenger*, 3, pp. 4–11.
[7] Maksim, K. Low-Temperature natural helium separation processes for obtaining 3helium isotope (2017) *Refrigeration Science and Technology*, 2017-May, pp. 330–335.
[8] Bondarenko, V.L., Grafov, A.P. &Kupriyanov, M.Y. *Chem Petrol Eng* (2016) 51: 697. https://doi.org/10.1007/s10556-016-0107-7
[9] Zolotovskaya, S.V., Molchanov, A.S. (1991) Helium Cryostats for Scientific Research. Basics of Structural Design [in Russian], *MEPhI*, Moscow
[10] Solntsev, Y.P., Stepanov, G.A. (1982) Materials in Cryogenic Technology. [in Russian], *Mashinostroenie*, Leningrad
[11] Malikov, M.P., Danilov, I.B., Zel'Dovich, A.G. (1985) Handbook of Physical and Technical Principles of Cryogenics [In Russian], p. 432. *Energoatomizdat*, Moscow
[12] Belyakov, V.P. (1982) Cryogenic Technology and Engineering [in Russian]. *Energoizdat*, Moscow
[13] S. Patankar, Numerical Heat Transfer and Fluid Flow, *Hemisphere Publishing Corporation*, New York (1980).
[14] V. O. Lomakin and A. I. Petrov, "Verification of analysis results for the flow channel section of an AX-50-32-200 centrifugal pump using the STAR-CCM+ fluid dynamics modeling package",*Izv. VUZov*, 2012, pp. 6–9.
[15] David C. Wilcox, Turbulence Modeling for CFD. 3rd Edition, *DCW Industries* (2006).
[16] Lomakin, V.O., Kuleshovav, M.S. &Bozh'eva, S.M. *Power TechnolEng* (2016) 49: 324. https://doi.org/10.1007/s10749-016-0623-9
[17] Lomakin, V.O., Chaburko, P.S., Kuleshova, M.S. Multi-criteria Optimization of the Flow of a Centrifugal Pump on Energy and Vibroacoustic Characteristics (2017) *Procedia Engineering*, 176, pp. 476–482.

IOP Conf. Series: Materials Science and Engineering **779** (2020) 012011 doi:10.1088/1757-899X/779/1/012011

Methodology for the comprehensive optimization of a centrifugal pump with a magnetic coupling

A Zaytseva[1] and A Protopopov[1,2]

[1]Bauman Moscow State Technical University

[2]E-mail: proforg6@yandex.ru

Abstract: The article considers the design of a leakproof pump with a magnetic coupling operating on a chemical fluid. In this work the optimization of the pump assembly was carried out, including hydrostatic bearings and a magnetic coupling, using the STARCCM + simulation package. The throttle diameters optimal for this design are calculated. The author minimized the overflows in the auxiliary cooling paths of the coupling and the axial force on the pump rotor.

Introduction

Centrifugal pumps with magnetic coupling are designed for use in various industries for pumping clean, viscous and non-viscous fluids, as well as chemically aggressive and toxic fluids. Most often, they are used in pressure circulating systems designed for pumping hazardous substances.

The tightness of the pumping is due to the absence of mechanical seals in the pump. A magnetic drive installed in the pump allows zero leakage of the pumped product, which has a beneficial effect on the environmental safety of the device.

The main problem with centrifugal sealed pumps is the problem of heat balance in a cooled magnetic coupling. A limitation of the use of a pump with a magnetic coupling for pumping liquids having high temperatures is the limit temperature of the coupling magnets [1–7], which begin to lose magnetic properties when the temperature approaches the Curie point. The use of an integrated coupling cooling system is an effective way to solve this problem [8–20].

Another equally important problem is the calculation of the fluid flow in auxiliary paths [14]. The pressure distribution in the auxiliary paths determines the value of the axial forces on the rotor and the heat balance of the entire pump [13]. To solve this problem, it is necessary to have calculation methods that will take into account the hydrodynamic phenomena that occur in the tracts. Modern packages of hydrodynamic modeling are widely used for these calculations [13].

However, now there are no methods for calculating auxiliary paths that can evaluate the working fluid overheating at problem points, estimate the flow rate of the working fluid in the studied channels, evaluate the axial force and cavitation characteristics at the problem points, and optimize the auxiliary paths taking into account all of the above parameters.

This article describes the method for determining leaks in auxiliary cooling paths and the method for determining the axial force acting on the rotor from the magnetic coupling. A method for checking the achievement of the working fluid of the Curie temperature at a minimum flow rate in the auxiliary paths and a method for checking the bearing capacity of sliding bearings, which satisfies the maximum possible axial force, are also considered.

Below there is a diagram of the auxiliary cooling paths of the magnetic coupling.

Content from this work may be used under the terms of the Creative Commons Attribution 3.0 licence. Any further distribution of this work must maintain attribution to the author(s) and the title of the work, journal citation and DOI.
Published under licence by IOP Publishing Ltd

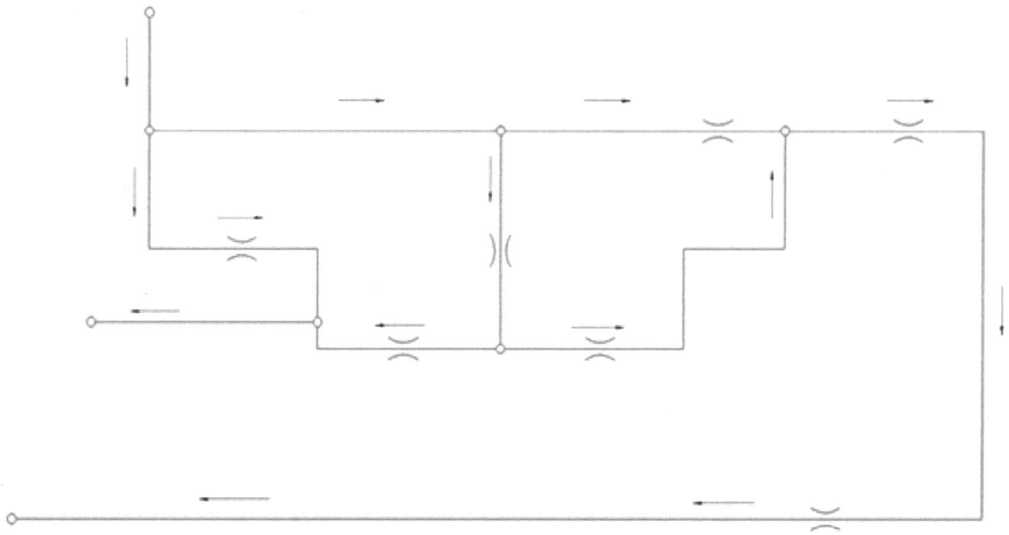

Fig.1. Scheme of auxiliary cooling paths of the magnetic coupling.

The optimization of auxiliary paths is carried out according to two criteria:
• Volumetric efficiency
• Rotor unloading from axial force

Usually, the temperature at the exit from the gap of the magnetic coupling or the maximum value of the axial force on the rotor is set as a limitation, then optimization will be carried out only by the volumetric efficiency.

Plan for calculating the task:

1) Calculate the plain bearings, gap seal and radial clearance of the coupling in the STARCCM + package to obtain the dependence of the flow rate on the differential pressure. As a result, obtain the flow rate characteristic of all small elements of the auxiliary path.

2) Build a picture of the flow of the working fluid in all channels of the auxiliary path, as well as a picture of the distribution of pressures and speeds.

3) Get a thermal picture in all the channels of the auxiliary path by the method of hydrodynamic modeling, introducing the temperature of the working fluid at the entrance to the path and the heat in the magnetic coupling.

4) Estimate the pressure margin to the Curie point.

5) Find the maximum temperature at the problem point and estimate the pressure margin to the boiling point at this temperature using the method of hydrodynamic modeling.

6) Find the axial force acting on the rotor.

7) Calculate the diameters of the two throttles by the Sobol method for the subsequent optimization of auxiliary paths.

8) Build a compromise curve volumetric flow - axial force.

9) Choose the optimal pair of throttles that create a support for the best cooling of the magnetic coupling and the minimum axial force on the rotor.

The object of study is a magnetic coupling with a sealing titanium screen, designed for power transmission of 195 kW at 2900 rpm. The fluid from the discharge pipe enters the cooling circuit, then is discharged to the inlet of the pump unit. When passing through the cooling circuit, the fluid heats up from the titanium screen. The dimensions of the magnetic coupling, the number of magnets, and their arrangement are selected according to the method of S P Subbotin [3]. The fixed titanium screen is in a rotating magnetic field. Losses on screen heating are determined by the formula

$$P_s = 31 \cdot \frac{B_a^2 \cdot D_s^3 \cdot n_{nom}^2 \cdot B \cdot N_{mag} \cdot \delta_s}{\rho},$$

where $B_a = \beta_a * K_i * B_r$ is the value of the current induction in the gap [3];

D_s is the diameter of the screen along the midline;

n_{nom} is the nominal frequency of rotation of the magnetic coupling;

B is the length of the magnet [4];

N_{mag}— the number of rows of magnets in the coupling halves;

δ_s— screen wall thickness;

ρ is the electrical resistivity of titanium [5];

Initial data for the design of the cooling circuit of the magnetic coupling:

V_{in} = 1.003 m/s — pressure at the inlet to the cooling circuit;

p_{out} = 0 Pa — pressure at the outlet of the pump shaft;

t_{out} = 60 °C— inlet temperature;

Methods

The method of numerical simulation is based on solving analogues of the basic equations of hydrodynamics and heat and mass transfer.

Mass conservation equation:

$$\frac{\partial \tilde{u}_j}{\partial x_j} = 0,$$

Where $\tilde{u}_j$— the average value of the fluid velocity in the projection onto the j-th axis (j = 1, 2,3)

The equation of change in momentum in a stationary setting:

$$\rho\left[\tilde{u}_j \frac{\partial \tilde{u}_j}{\partial x_j}\right] = -\frac{\partial \tilde{p}}{\partial x_j} + \frac{\partial}{\partial x_j}\left[T_{ij}^{(v)} - \rho < u_i u_j >\right],$$

where $\tilde{p}$— averaged pressure value;

$\tilde{T}_{ij}^{(v)} = 2\mu \tilde{s}_{ij}$—viscous stress tensor for incompressible fluid;

$$\tilde{s}_{ij} = \frac{1}{2}\left[\frac{\partial \tilde{u}_j}{\partial x_j} + \frac{\partial \tilde{u}_j}{\partial x_j}\right].$$ — strain rate tensor;

$\rho\langle u_i u_j \rangle$— Reynolds stresses, which are modeled on the basis of the k-ωSST – turbulence model;

Optimization of auxiliary pump paths.

The optimization problem is reduced to finding the extremum of the objective function with given constraints. It is necessary to determine the value of the vector of variables $x=(x_1,x_2,\dots,x_m)$ that satisfy the constraints of the form $g_i(x_1,x_2,\dots,x_m) \le b_i$ for all $i=1, \dots, k$, and at which the minimum of the objective function is achieved $f(x_1,x_2,\dots,x_m)$:

$$f(x_1,x_2,\dots,x_m) \to (\max, \min)$$

The final solution to the problem is a pair $(x^*, f^*(x^*))$, consisting of the optimal solution and the optimal value of the objective function.

IOP Conf. Series: Materials Science and Engineering **779** (2020) 012011 doi:10.1088/1757-899X/779/1/012011

- **LP τ search**

To search for the extremum of the function, the LPτ sequence method was chosen, since it has the ability to search for the global extremum of the function.

Throttle optimization was performed using the LPT-search program. With the number of design points of 32, the following results were obtained: d_{thr1} = 2...7 mm, d_{thr2} = 2...7 mm.

- **Calculation of auxiliary cooling paths**

The mathematical apparatus used for research is STARCMM +, Mathcad. The auxiliary path model contains clearances in the plain bearings and clearance in the coupling, which complicates the calculation. To simplify the calculation the model was divided into partsin the Solidworks program: a groove seal, two plain bearings and a radial clearance of the coupling.

Groove seal calculation:

It is necessary to obtain the dependence of the differential pressure on the velocity at the inlet to the groove seal. To do this, you need to configure the calculation model in the STARCCM + program, get a picture of the pressure and velocity distribution in the cross section. Next, use Mathcad to get the dependency function.

Calculation of plain bearings:

The calculation is made similarly to the calculation of the groove seal.

Determination of temperature in a dangerous section of a coupling:

The tuning in STARCCM + of the calculation model is similar to the tuning of the groove seal model. It is necessary to set the equation of fluid energy for temperature and heat gain from the coupling, calculated according to the Subbotin formula.

Results

In the process of modeling the auxiliary cooling paths of the magnetic coupling, a graph of the dependence of the flow velocity on the pressure drop in the gap seal was obtained.

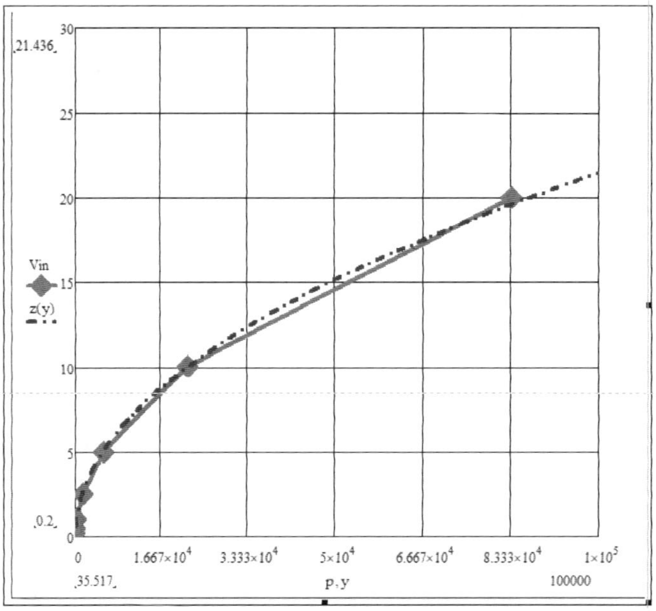

Fig.2. Graph of speed versus differential pressure in a groove seal.

Analyzing the obtained graph, we can conclude that the obtained function, taking into account the errors, fully corresponds to the radical dependence $Q = K * \sqrt{\Delta p}$. Therefore, the desired function is $Q = 0,06779 * \sqrt{\Delta p}$.

During the calculation of plain bearings, the dependence of speed on the differential pressure was obtained, according to which a graph was built in Mathcad. Analyzing the graph, we can conclude that the desired function is $Q = 0,0012 * \sqrt{\Delta p}$.

A set of methods has been obtained that makes it possible to evaluate the energy efficiency of sealed centrifugal pumps with a magnetic coupling. In particular, to evaluate leakage in hydrostatic bearings and fluid flow through auxiliary paths for cooling the magnetic coupling. An assembly optimization method has been developed, which includes hydrostatic bearings and a magnetic coupling.

The ideal sizes of throttles for this task are determined: d_{thr}= 2...7 mm.

Conclusions

1. The design of a leak free pump with a magnetic coupling that operates on a chemical fluid is developed.
2. A 3d-modeling of the design of the pump was carried out.
3. The optimization of the unit, which includes hydrostatic bearings and a magnetic coupling, was carried out using computer simulation methods STARCCM +.
4. The simulation of the magnetic coupling for guaranteed cooling was carried out.

References

[1] Kuznetsov V.S., Yarots V.V. Calculation of the parameters of the fluid flow through the cylindrical throttle channels in the mode of existence of the "locking effect" Engineering and engineering education 2016. —No. 4.— P. 8–14.

[2] Lomakin V.O., Kukushkin P.A., Krylov V.I. Modernization of the auxiliary cooling circuit of the magnetic coupling Territory NEFTEGAZ 2017.—No. 7–8.— P. 84–91.

[3] Lomakin V.O., Kuleshova M.S., Chaburko P.S., Baulin M.N. Comprehensive optimization of the flow part of a sealed pump using the LP-Tau search method Pumps. Turbines. Systems 2016.— No. 1.— S. 55–61.

[4] Makarov K.A. On the physical meaning of the Reynolds number and other criteria for hydrodynamic similarity Engineering Journal: Science and Innovations 2014.— No. 1 (25).

[5] Petrov A.I. Heat balance maintenance systems in modern stands for testing vane pumps Machines and plants: design, development and operation 2015.— No. 5.

[6] Kuznetsov V.S., Shablovsky A.S., Yarots V.V. The effect of backpressure on some hydrodynamic characteristics of fluid flow in valve slots Engineering Journal: Science and Innovation 2013.— No. 4 (16).

[7] Borovin G.K., Protopopov A.A. Calculation of the optimal number of impeller blades for a centrifugal pump Engineering Bulletin of MSTU im. N.E. Bauman, 2014.

[8] Protopopov A. A., Shulzhitsky A. A. Investigation of the influence of supply voltage on the maximum pressure of a centrifugal pump Youth Scientific and Technical Bulletin # 03, March 2016.

[9] Protopopov A.A. Zakharova E.V. Dynamics of small oscillations of a low inertia rotor of a low-flow centrifugal pump with hydrostatic bearings Polytechnic Youth Journal 2017.— No. 5.

[10] Lomakin V.O., Chaburko P.S. The influence of flow swirl on the hydraulic efficiency of the pump Engineering Bulletin. Electronic Journal 2015.— No. 10.

[11] Artyomov A.V., Scherbachev P.V., Tarasov O.I. Application of B-splines for constructing a lateral half-spiral pump inlet Engineering Bulletin. Electronic Journal 2014.— No. 12.

[12] Gavryushina O.S., Mulyarchik I.G. Comparison of flow-differential characteristics of pneumatic drives with executive motors of translational and rotational action Engineering Bulletin. Electronic Journal 2014.— No. 12.
[13] Lomakin V.O., Petrov A. I. Numerical simulation of heat balance in a cooled magnetic coupling of a high-temperature hermetic pump Engineering Bulletin. Electronic Journal 2014.—No. 12.
[14] Cheremushkin V.A., Petrov A.I., Chaburko P.S. Application of stator blades in auxiliary ducts of hermetic pumps Machines and Installations: design, development and operation. MSTU named after N.E. Bauman. Electronic Journal 2017. No. 02. S. 1–12.
[15] Petrov A. I., Poluektov D. A. Development of a new line of centrifugal ship pumps for hot heat carrier Engineering Bulletin. Electronic Journal 2015.— No. 10.
[16] Rudnev S.S., Matveev I.V. The manual on the course design of vane pumps. Editor O. V. Baibakov. Moscow 1974P Chaburko and Z Kossova 2019 *IOP Conf. Ser.: Mater. Sci. Eng.***492** 012011
[17] V Lomakin et al 2019 *IOP Conf. Ser.: Mater. Sci. Eng.***492** 012012
[18] V Cheremushkin and APolyakov 2019 *IOP Conf. Ser.: Mater. Sci. Eng.***589** 012001
[19] V Tkachuk et al 2019 *IOP Conf. Ser.: Mater. Sci. Eng.***589** 012007
[20] N Isaev et al 2019 *IOP Conf. Ser.: Mater. Sci. Eng.***589** 012009

IOP Conf. Series: Materials Science and Engineering **779** (2020) 012012 doi:10.1088/1757-899X/779/1/012012

Improving the performance of a centrifugal vane pump by installing vortex generators on the suction surfaces of blades

O Teplov[1] and V Lomakin[1,2]

[1]Bauman Moscow State Technical University

[2]E-mail: lomakin@bmstu.ru

Abstract. The influence of vortex generators mounting in the wet part of a centrifugal pump on its characteristics was investigated. Dependents of efficiency, cavitation characteristic, head and capacity are given. The vortex generators were mounted on the front edges of suction sides of the blades, where stall occurs, and a vortex is formed. The simulation results were compared with the characteristics obtained for the usual optimized flow part of the pump.

Introduction

With the increase of modern computers power, multicriterial CFD optimization[1–4] has become commonplace in the design of centrifugal pumps. The resulting models of wet parts have almost the maximum possible characteristics for their design. Therefore, it is necessary to look for new ways to improve the quality of wet parts, based on the including new elements to the existing design concept.

One of the problem areas in the design of a vane pump is the zone of stalling on the suction surface of the blade. A following the blade vortex is forming there and occurs an energy dissipation [5]. Its negative impaction increases with increasing of angle of attack [6] and leads to a decrease in efficiency when the nominal flow is exceeded. Also, increased vortex formation leads to a deterioration of the cavitation characteristics and a decrease of the pump's suction capacity.

Figure 1. Location of vortex generators on the suction surface of the blade

Content from this work may be used under the terms of the Creative Commons Attribution 3.0 licence. Any further distribution of this work must maintain attribution to the author(s) and the title of the work, journal citation and DOI.
Published under licence by IOP Publishing Ltd

This problem can be solved by the use of vortex generators, installed on the leading edge of the stall surface (Fig.1). The generators create small eddies, destroying the main stall eddy. It could decrease energy loss due to vortex formation and increase the cavitation characteristics of the flow part by reducing vortex speed.

The existing articles about vortex generators are consider their use on airfoils of helicopter's blades [7], and also on hydrofoils of turbines [8]. There are given detailed description of the nature of processes occurring in the flow after generators. Particular attention should be paid to the article [9], devoted to the influence of vortex generators on the occurrence of cavitation on the hydrofoil NACA0015.

Mathematical model and methods

As an initial model should be chosen the flow part obtained after optimization since it is necessary to exclude the possibility of improving characteristics due to the correction of errors in the construction of geometry. For this role was chosen the model of the wet part of the vane pump with the flow $600\ {}^{m^3}\!/_{hr}$ and head $65\ m$ with a tangential dual volute and increased cavitation resistance (Fig.2). Shaft rotation frequency $1450\ rpm$. Optimization was carried out using LP_τ search on 260 models[10].

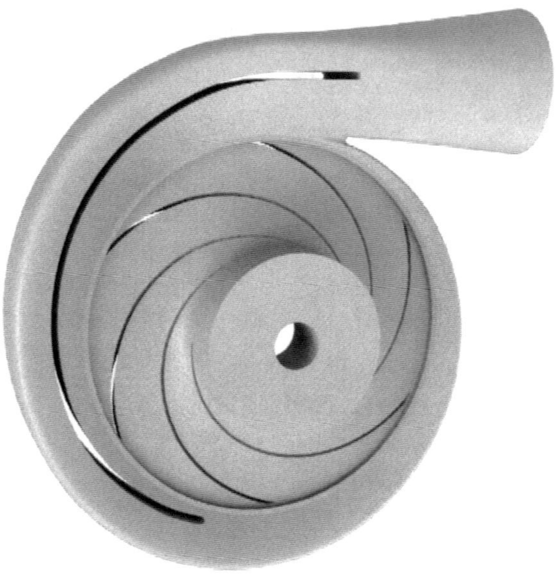

Figure 2. Model of optimized pump flow

Modeling was carried out by methods of computational fluid dynamics based on solving the basic hydrodynamic equations discrete analogs. In case of an incompressible fluid model, when the density is constant, these are:

Continuity equation

$$\frac{\partial \tilde{u}_j}{\partial x_j} = 0,$$

where $\tilde{u}_j$ — j axis projection of an averaged value of the fluid velocity (j=1,2,3);

The momentum conservation equation (Reynolds averaged):

$$\rho\left[\frac{\partial U_i}{\partial t} + U_j \frac{\partial U_i}{\partial x_j}\right] = -\frac{\partial P}{\partial x_i} + \frac{\partial}{\partial x_i}\left[T_{ij}^{(v)} - \rho u_i u_j\right];$$

where U, P — averaged velocity and pressure;

$\tilde{T}_{ij}^{(v)} = 2\mu\tilde{s}_{ij}$ —viscous stress tensor for incompressible fluid;

$\tilde{s}_{ij} = \frac{1}{2}\left[\frac{\partial \tilde{u}_i}{\partial x_j} + \frac{\partial \tilde{u}_j}{\partial x_i}\right]$ —instant strain rate tensor;

$\rho u_i u_j$ — Reynolds stresses.

The introduction of the Reynolds averaged Navier-Stokes equation makes the equations system not closed, as additional unknowns, Reynolds stresses, appear.

To solve this system of equation, the k-ω SST turbulence model was used. It introduces the necessary additional equations: the transport equations for the turbulent kinetic energy and the relative dissipation rate of this energy:

$$\frac{\partial k}{\partial t} + U_j \frac{\partial k}{\partial x_j} = P_k - \beta^* k\omega + \frac{\partial}{\partial x_j}\left[\left(\nu + \sigma_k \nu_T\right)\cdot \frac{\partial k}{\partial x_j}\right]$$

$$\frac{\partial \omega}{\partial t} + U_j \frac{\partial \omega}{\partial x_j} = \alpha \cdot S^2 - \beta \cdot \omega^2 + \frac{\partial}{\partial x_j}\left[\left(\nu + \sigma_\omega \nu_T\right)\cdot \frac{\partial \omega}{\partial x_j}\right] + 2\cdot\left(1 - F_1\right)\cdot \sigma_{\omega 2}\cdot \frac{1}{\omega}\cdot \frac{\partial k}{\partial x_i}\cdot \frac{\partial \omega}{\partial x_i}$$

Boundary conditions are pressure value on the inlet and speed value on the output. The resulting best model was additionally edited manually.

Comparison of the characteristics of the wet parts was made according to the efficiency (η_1, η_2), head (H_1, H_2) and capacity (N_1, N_2) in the value range from $300\, {m^3}/{hr}$ to $900\, {m^3}/{hr}$ in increments of $100\, {m^3}/{hr}$. To determine the cavitation values dependencies of heads on the net positive suction heads which characterize the suction capacity of the pump were obtained. Schnerr-Sauer Cavitation Model was used which is based on a reduced Rayleigh-Plesset equation and neglects the influence of bubble growth acceleration, viscous effects, and surface tension effects[11, 12].

Discussion and results

Table 1. Results obtained during CFD modeling (*1* — model without vortex generators, 2 — model with vortex generators)

Q, ${m^3}/{hr}$	η_1, %	η_2, %	H_1, м	H_2,м	N_1,kW	N_2,kW
300	77,5	76,3	81	81,2	85,6	87,42
400	83	82,3	79,6	79,4	104,11	105,48
500	87,1	86,2	77,13	76,6	119,29	120,05
600	88,8	88,8	73,8	73	132,63	132,95
700	88,2	88,6	67,5	67,6	143,87	143,87
800	84,5	85	60,5	60,9	152,98	152,68
900	79,6	80,8	52,8	53,3	160,27	159,96

As expected the obtained results show an increase in efficiency in the area of increased flow, however with a decrease of the flow below optimal point the inverse effect of a decrease in efficiency is manifested. Values of pressure and capacity practically did not change.

IOP Conf. Series: Materials Science and Engineering **779** (2020) 012012 doi:10.1088/1757-899X/779/1/012012

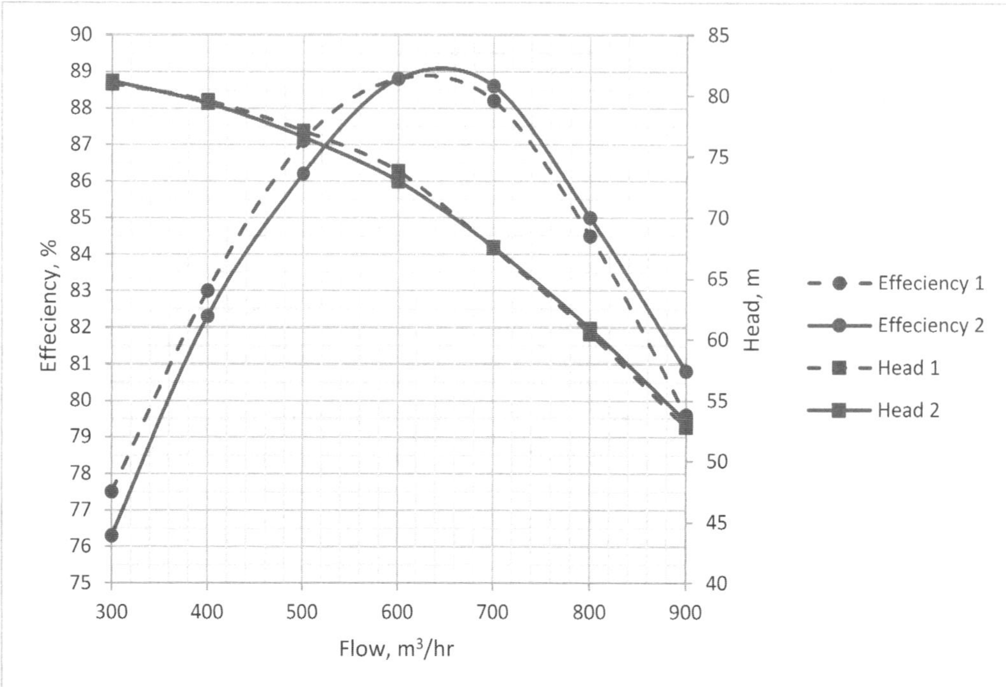

Figure 3. Graphs of efficiency and pressure (*1* — model without vortex generators, *2* — model with vortex generators)

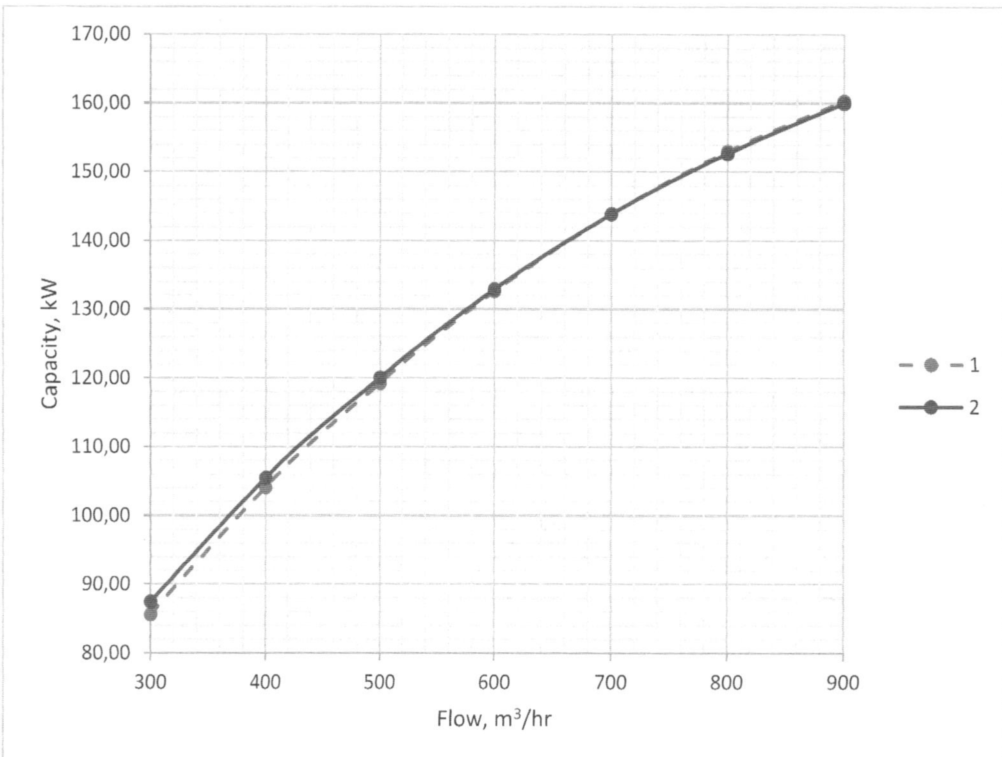

Figure 4. Capacity graphs (*1* — model without vortex generators, *2* — model with vortex generators)

Table 2. Results of cavitation calculation (*1* — model without vortex generators, *2* — model with vortex generators)

Δh_1, m	H_1, m	Δh_2, m	H_2, m
1	8	1	15,3
1,5	19,49	1,25	31,89
2	32,83	1,5	47,6
2,5	47,12	1,75	61,49
2,75	53,29	2	72,13
3,5	72	2,15	73

Cavitation characteristic shows significant reduction of required net positive suction head more than 1.5 meters.

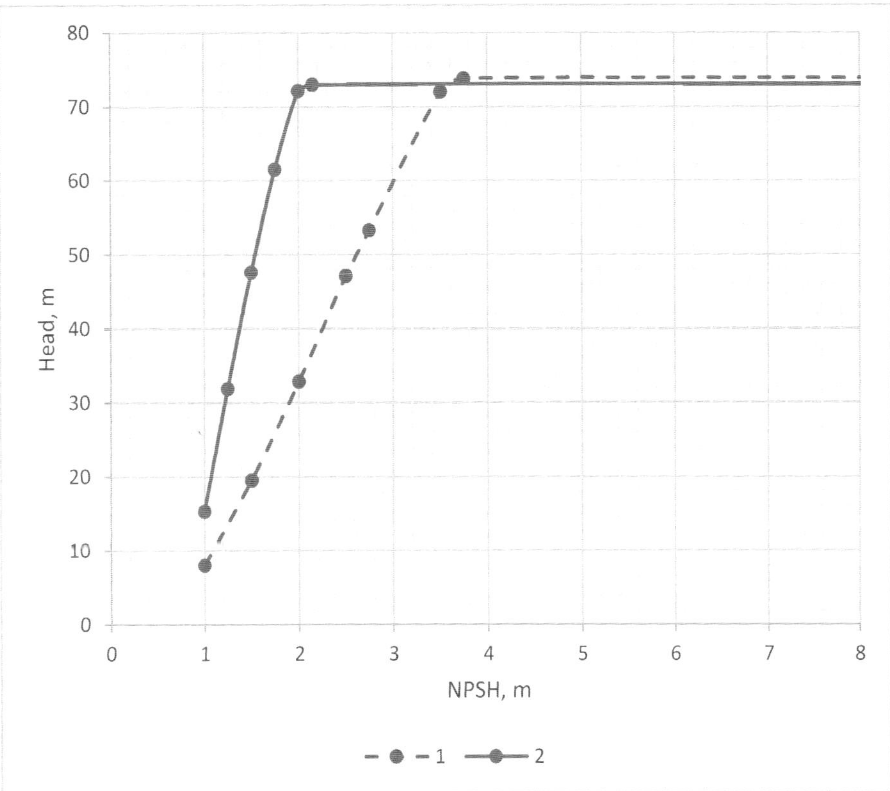

Figure 5. Cavitational characteristics (*1* — model without vortex generators, *2* — model with vortex generators)

Conclusions

1. Application of vortex generators increases efficiency of the pump for exceeding the optimum flow (in the presented model to 0.8%) but decreases for values of flow less then optimum.

2. This solution reduces required net positive suction head (in the considered model it was possible to reduce the $NPSH_R$ almost twice), which in creases pump resistance to cavitation.

IOP Conf. Series: Materials Science and Engineering **779** (2020) 012012 doi:10.1088/1757-899X/779/1/012012

3. Vortex generators can be easily implemented technologically, which economically fully justifies the expediency of their widespread use in industry.

References

[1] Lomakin, V.O., Chaburko, P.S., Kuleshova, M.S. Multi-criteria Optimization of the Flow of a Centrifugal Pump on Energy and Vibroacoustic Characteristics (2017) Procedia Engineering, 176, pp. 476–482.

[2] Gouskov, A., Lomakin, V., Banin, E., Kuleshova, M., Chaburko, P. Investigation of the influence of centrifugal pump wet part geometry on hemolysis index (2019) IOP Conference Series: Materials Science and Engineering, 492 (1), № 012013

[3] Gouskov, A.M., Lomakin, V.O., Banin, E.P., Kuleshova, M.S. Minimization of Hemolysis and Improvement of the Hydrodynamic Efficiency of a Circulatory Support Pump by Optimizing the Pump Flowpath (2017) Biomedical Engineering, 51 (4), pp. 229–233.

[4] Chaburko, P., Kossova, Z. Wet part hybrid optimization method of hermetic pump (2019) IOP Conference Series: Materials Science and Engineering, 492 (1), № 012011.

[5] Tao, R., Zhao, X., Wang, Z. Evaluating the transient energy dissipation in a centrifugal impeller under rotor-stator interaction (2019) Entropy, 21 (3), № 271.

[6] V. B. Maleev, V. A. Prishchenko The influence of the angle of attack in the impeller of the pump for drainage on its suction capacity (2004) Scientific works of Donetsk national technical University ("Mining and electromechanical" series), 83 pp. 208–214.

[7] Mai, H., Dietz, G., Geißler, W., Richter, K., Bosbach, J., Richard, H., de Groot, K. Dynamic stall control by leading edge vortex generators (2008) Journal of the American Helicopter Society, 53 (1), pp. 26–36.

[8] Kundu, P., Sarkar, A., Nagarajan, V. Improvement of performance of S1210 hydrofoil with vortex generators and modified trailing edge (2019) Renewable Energy, 142, pp. 643–657.

[9] Che, B., Chu, N., Schmidt, S.J., Cao, L., Likhachev, D., Wu, D. Control effect of micro vortex generators on leading edge of attached cavitation (2019) Physics of Fluids, 31 (4), № 044102.

[10] Sobol I.M. Numerical Monte Carlo methods. M. Nauka, 1973. — 311p.

[11] Sauer, J., InstationaerkavitierendeStroemungen—Einneues Modell, basierend auf Front Capturing VOF und Blasendynamik, Dissertation, UniversitaetKalrsruhe, 2000.

[12] Lomakin, V., Bibik, O. Numerical prediction of the gas content effect on the cavitation characteristics of the pump using the simplified Rayleigh-Plesset equation (2019) IOP Conference Series: Materials Science and Engineering, 492 (1), № 012037.

IOP Conf. Series: Materials Science and Engineering **779** (2020) 012013 doi:10.1088/1757-899X/779/1/012013

The effect of the throttles diameter installed on the auxiliary channel on the flow rate of the fluid flowing through them and the heating of the fluid in the magnetic field

D Pavlovskiy[1,2], A Protopopov[1] and N Isaev[1]

[1]Bauman Moscow State Technical University

[2]E-mail:pavlovskiy.dmb@mail.ru

Abstract: This article discusses the effect of the diameters of the throttles installed in the auxiliary paths on the flow rate of the fluid flowing through them and the heating of the fluid in the area of the magnetic coupling. The simulation was carried out in the STAR-CCM+ package.

Introduction

Leak-free pumps are used for pumping liquids and do not leak [1–5]. Usually they pump caustic liquids that have a negative impact on human life and health, as well as on the environment.

Leak-free pumps do not require frequent maintenance [6–12], such as cantilever pumps with oil seals or mechanical seals. They are more reliable.

The basic principle of a leak-free pump is to transmit torque from the engine to the pump rotor through a fixed diaphragm.

According to a method of transferring energy from an engine to a liquid, a leak-free pump can be:

1. With electromagnetic coupling,
2. With a leak-free motor.

The leak-free pump with an electromagnetic coupling is divided into a leak-free pump with a synchronizing coupling and a leak-free pump with an asynchronous coupling.

Pumps with hermetic motors are divided into pumps with "wet" rotors and "wet" rotors and stators.

The following advantages and disadvantages of a leak-free pump can be distinguished from conventional pumps.

Benefits:

1. No leak at all
2. No routine maintenance is required during the entire service life.
3. The operating time before the first repair is usually at least 50,000 hours.
4. Reduce noise and vibration (especially in sealed pumps with sealed motors),
5. High fire and explosion safety.

Disadvantages:

1. The cost of the pump is higher
2. Low efficiency
3. Maintenance and repair of the operating company is extremely complex.
4. The difficulty of operating the pump on contaminated and hot liquids.

Content from this work may be used under the terms of the Creative Commons Attribution 3.0 licence. Any further distribution of this work must maintain attribution to the author(s) and the title of the work, journal citation and DOI.
Published under licence by IOP Publishing Ltd

5. Heavy weight and size
6. Power limitation.

There are some difficulties in developing a leak-free pump.

In most cases, most leak-free pumps are equipped with plain bearings that operate on the pumped liquid. If the pumped liquid has abrasive inclusions, this can cause problems.

When a leak-free pump with a magnetic coupling is used on a hot fluid, the magnets heat up, and as a result, they lose their magnetism until the magnetic coupling fails completely.

In addition, since the operation of a leak-free pump at high pressure in the system is very difficult, an increase in the thickness of the sealing separator leads to an increase in losses in the sealing separator.

In addition, hermetic pumps require maximum radial and axial unloading due to the low bearing capacity of sliding bearings, and due to the design characteristics of such pumps, the proximity of the bearings used to diagnose their condition is very limited.

The aim of the project is to develop a sealed pump with a magnetic coupling for the chemical industry.

This article attempts to simulate the fluid flow in the wet part of a leak-free pump with a magnetic coupling using computational fluid dynamics methods in order to study the effect of the diameters of the throttles installed in auxiliary paths on the flow rate of the fluid flowing through them and the heating of the fluid in the magnetic coupling region.

The object of study in this work is the area of auxiliary paths of a sealed pump, limited by a sealing cup.

Figure 1 shows a 3D model of the flow part of the pump of the selected area.

Figure 1. Design geometric model

Methods

The method of numerical simulation is based on the solution of discrete analogues of the basic equations of hydrodynamics [4–8]. For an incompressible fluid $(\rho = const)$ it is:

Mass conservation equation (continuity equation)

$$\frac{\partial \overline{u_j}}{\partial x_j} = 0,$$

where $\overline{u_j}$ is the projection of the averaged fluid velocity on the j-th axis (j=1,2,3);

Equation of conservation of momentum (averaging by Reynolds):

$$\rho\left[\frac{\partial U_i}{\partial t}+U_i\frac{\partial U_i}{\partial x_j}\right]=-\frac{\partial P}{\partial x_i}+\frac{\partial}{\partial x_i}\left[\overline{T_{ij}{}^{(v)}}-\rho u_i u_j\right], \qquad where$$

U, P — averaged speed and pressure;

$\overline{T_{ij}^{(v)}}=2\mu\overline{S_{ij}}$ —viscous stress tensor for incompressible fluid;

$\overline{S_{ij}}=\frac{1}{2}\left[\frac{\partial u_i}{\partial x_j}+\frac{\partial u_j}{\partial x_i}\right]$ —instant strain rate tensor;

$\rho\langle u_i u_j\rangle$ — Reynolds stresses.

The Reynolds system of equations is open due to the presence of unknown Reynolds stresses. The system can be closed using the k-w SST turbulence model. This model combines the advantages of both k-w and k-models. In the parietal region, the k-w model is used, and in the core of the flow, k-e.

For the thermal calculation, the following hydrodynamic equation, the equation of fluid energy for temperature, was used:

$$\frac{\partial}{\partial t}\int_V \rho E dV+\oint_A \rho H v\cdot da=-\oint_A q''\cdot da+\oint_A T\cdot v da+\int_V f_b\cdot v dV+\int_V S_u dV$$

where E is full energy

H — total enthalpy

$\dot{\mathbf{q}}''$ — heat flux vector

$\mathbf{T}$ —viscous stress tensor

$\mathbf{v}$ — speed vector

$\mathbf{f}_b$ —body force vector representing the combined rotational forces of the body, gravity and fan mode

S_u — contributes to terms of energy sources, such as radiation sources, interfacial energy sources, or energy sources due to chemical reactions.

The total energy is related to the total enthalpy H:

$$E=H-p/\rho,$$

where:

$$H=h+|\mathbf{v}|^2/2$$

h is a static enthalpy.

The computational grid constructed for the area of the pump seal with the pump magnetic coupling consisting of about 1 mln. polyhedral cells with a base size of 3 mm, shown in the figures 2,3.

To determine the effect of the diameters of the throttles installed in the auxiliary paths of the hermetic pump on the flow rate of the fluid flowing through them and the temperature of the fluid in the area of the magnetic coupling, several 3D models of the flow part with different diameters of the throttles were simulated. To simplify the hydrodynamic modeling, the throttle diameter in the shaft was changed, since it most affects the temperature of the pumped fluid in the region of the magnetic coupling.

IOP Conf. Series: Materials Science and Engineering **779** (2020) 012013 doi:10.1088/1757-899X/779/1/012013

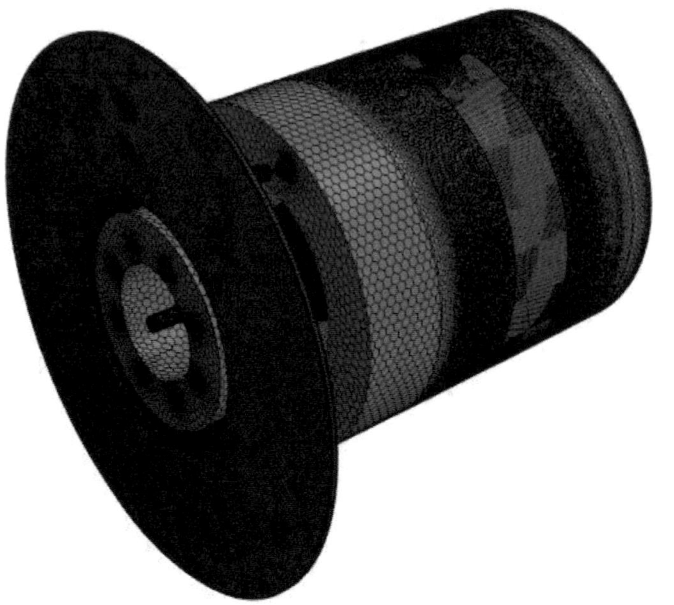

Figure 2. Design grid model

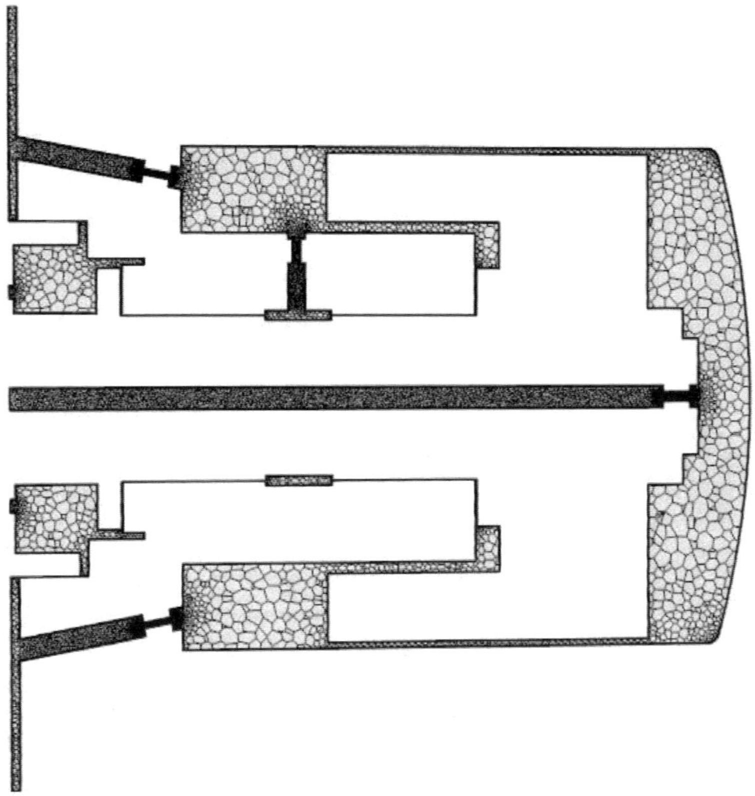

Figure 3. Estimated meridional section

IOP Conf. Series: Materials Science and Engineering **779** (2020) 012013 doi:10.1088/1757-899X/779/1/012013

In order to minimize errors associated with numerical calculation, all models were calculated with the same parameters of the computational grid and with the same boundary conditions.

Heat is generated inside the magnetic coupling due to eddy currents generated on the protective shell in the areas of the ends of the magnets. Heating is set on the surface of the sealed cup in the area of the magnets as 5% of the total transmitted power through a magnetic coupling from the motor shaft to the pump shaft (100 W). Since the working fluid lubricates the plain bearings, heating also occurs in them.

Heating is set on the rotating part of the bearing (sleeve).

The evaluation criteria (quality) for the selection of throttles in the auxiliary paths of the hermetic pump are: the temperature of the fluid in the area of the magnetic coupling, the flow rate flowing through the auxiliary paths, as well as the volumetric efficiency of the pump.

Results

In the process of simulating the flow of liquids in the magnetic coupling region of an leak-free pump, patterns of the distribution of liquid temperatures were obtained.

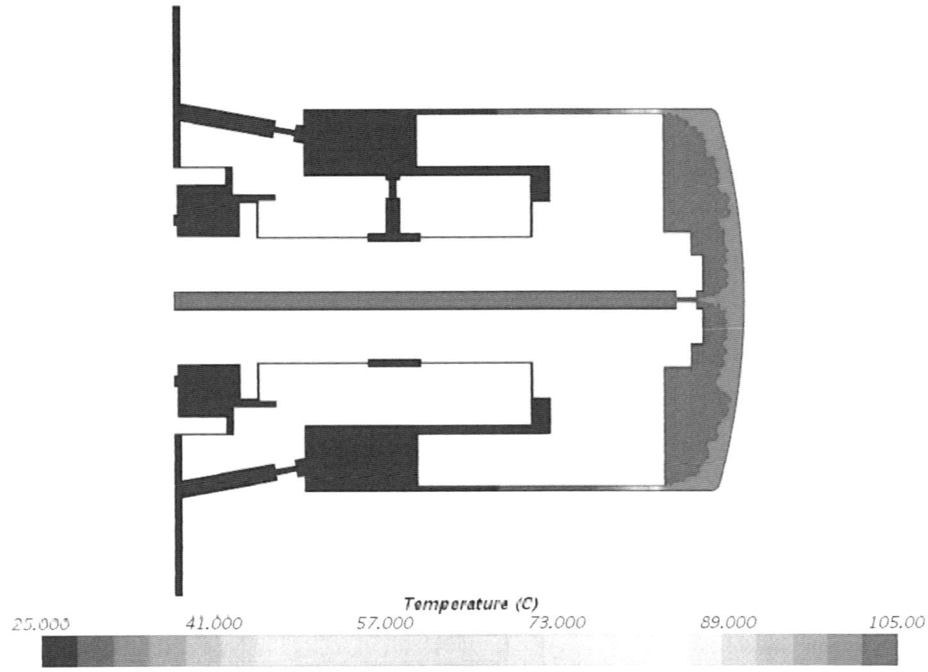

Figure 4. Fluid temperature distribution field (throttle diameter 1mm)

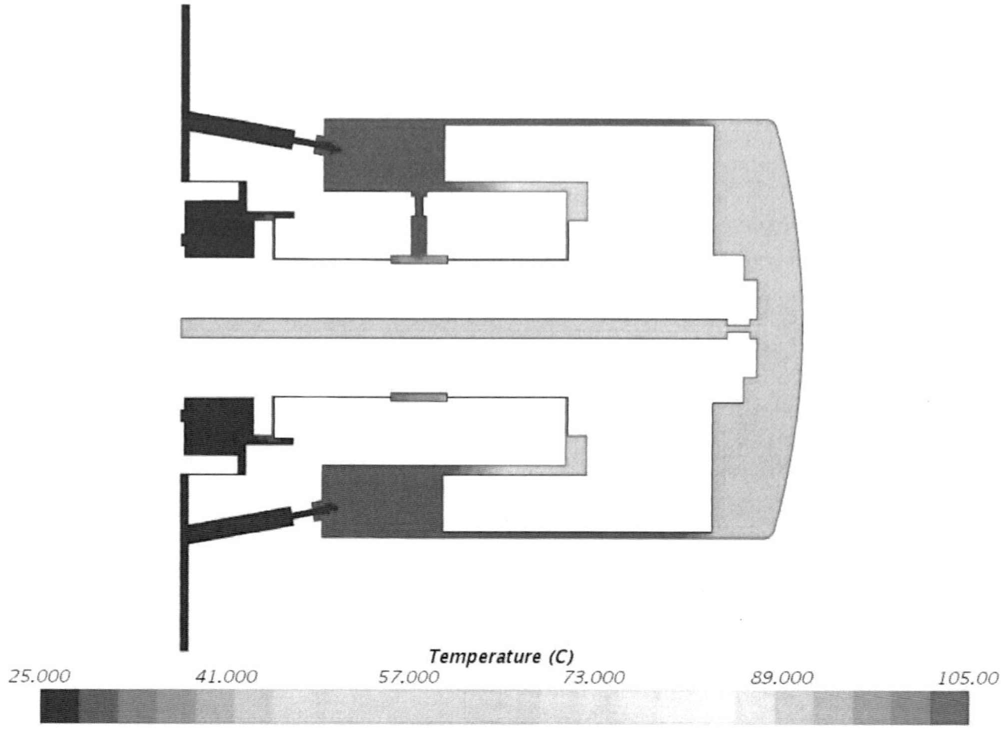

Figure 5. Fluid temperature distribution field (1.8mm throttle diameter)

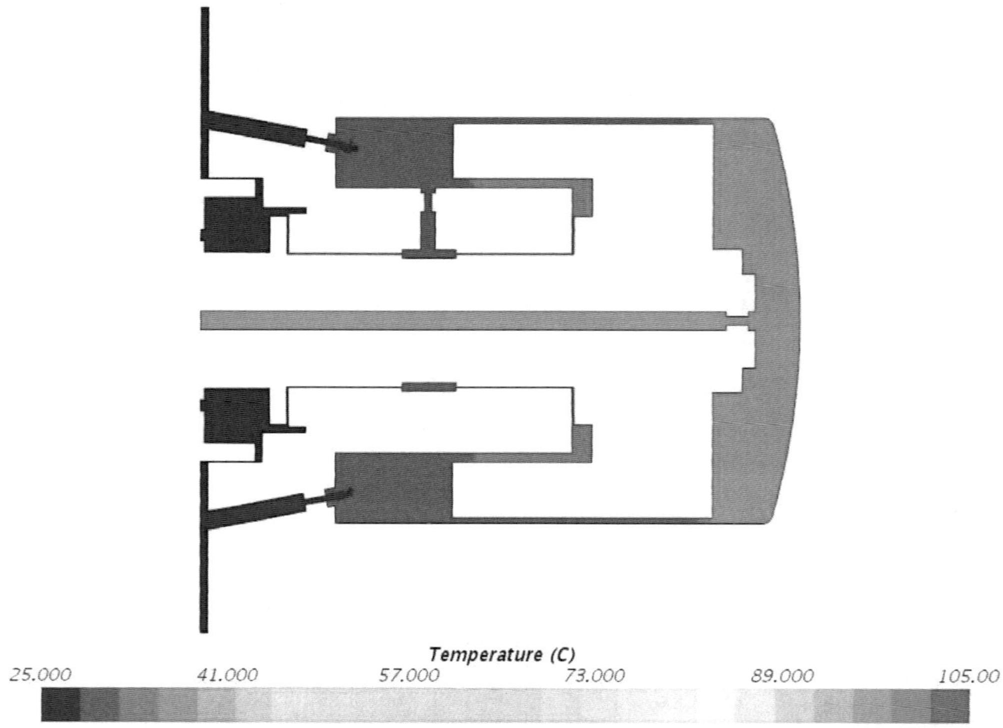

Figure 6. Fluid temperature distribution field (2.3 mm throttle diameter)

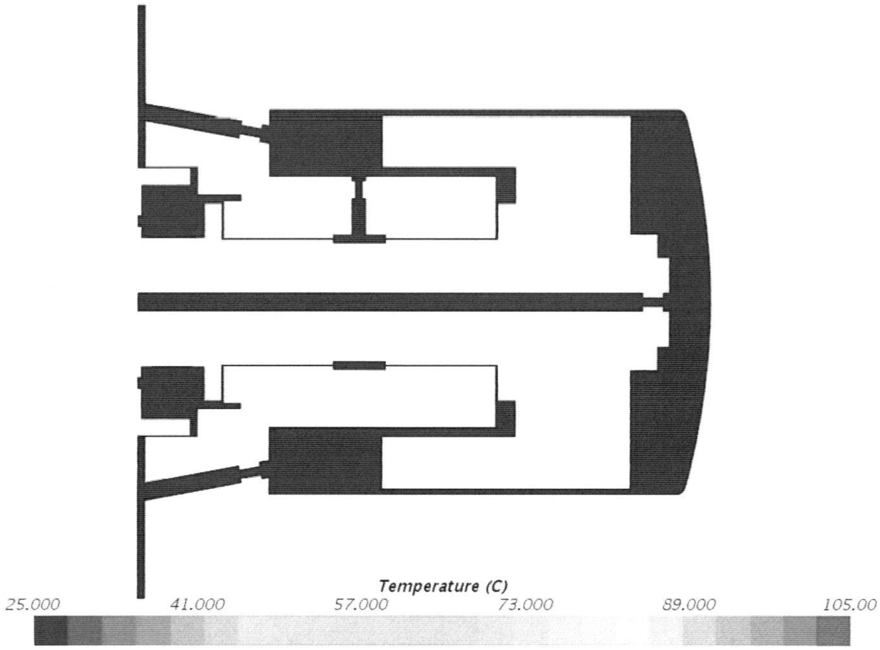

Figure 7. Fluid temperature distribution field (2.5 mm throttle diameter)

Conclusions

In the process of hydrodynamic simulation of the wet part of the pump, the values of the temperature of the pumped fluid in the region of the magnetic coupling and the flow rate flowing through the auxiliary paths were obtained. According to these results, a graph of the temperature versus flow rate is constructed.

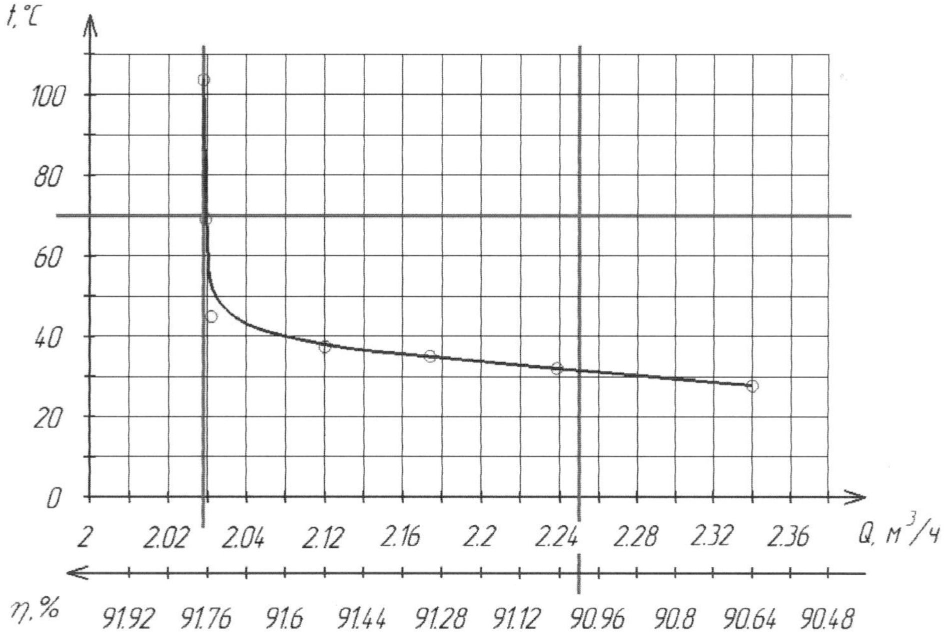

Figure 8. The dependence of temperature on fluid flow in auxiliary paths

IOP Conf. Series: Materials Science and Engineering **779** (2020) 012013 doi:10.1088/1757-899X/779/1/012013

Limitations are imposed on temperature (70 °C), volumetric efficiency (91%) and minimum throttle diameter (1 mm). As a result of which, conclusions were drawn that it is advisable to use a throttle with a diameter of 1.8-2.36 mm.

References

[1] V Cheremushkin and APolyakov 2019 *IOP Conf. Ser.: Mater. Sci. Eng.***589** 012001
[2] K Abramov 2019 *IOP Conf. Ser.: Mater. Sci. Eng.***589** 012013
[3] P Chaburko and Z Kossova 2019 *IOP Conf. Ser.: Mater. Sci. Eng.***492** 012011
[4] V Lomakin et al 2019 *IOP Conf. Ser.: Mater. Sci. Eng.***492** 012012
[5] A Gouskov et al 2019 *IOP Conf. Ser.: Mater. Sci. Eng.***492** 012013
[6] N Egorkina and APetrov 2019 *IOP Conf. Ser.: Mater. Sci. Eng.***492** 012015
[7] K Dobrokhodov and APetrov 2019 *IOP Conf. Ser.: Mater. Sci. Eng.***492** 012016
[8] A Boyarshinova et al 2019 *IOP Conf. Ser.: Mater. Sci. Eng.***589** 012014
[9] M Saprykina and V Lomakin 2019 *IOP Conf. Ser.: Mater. Sci. Eng.***589** 012017
[10] A Martynyuk et al 2019 *IOP Conf. Ser.: Mater. Sci. Eng.***589** 012029
[11] A Shablovskiy and E Kutovoy 2019 *IOP Conf. Ser.: Mater. Sci. Eng.***492** 012035
[12] A Petrov et al 2019 *IOP Conf. Ser.: Mater. Sci. Eng.* **492** 012036

IOP Conf. Series: Materials Science and Engineering **779** (2020) 012014 doi:10.1088/1757-899X/779/1/012014

Investigation of the influence of the front end clearance on the parameters of a centrifugal pump with an open type impeller

S Kalinkin[1] and A Petrov[1,2]

[1]Bauman Moscow State Technical University

[2]E-mail: alexeypetrov@bmstu.ru

Abstract: The paper considers the influence of the size of the front end gap between the impeller and the housing on the energy performance of the pump, selects the parameters, compiles a mathematical model used in hydrodynamic modeling, presents the results of the study, draws conclusions about the influence of the size of the front gap between impeller and housing on efficiency and pump head. A comparison is made of the dimensionless pump characteristics obtained by the methods of hydrodynamic modeling with experimental dependences.

Introduction

In the case of the development of centrifugal pumps with a low speed coefficient, the question often arises of choosing the type of impeller [1]. The impeller can be closed, half open and open type, each of which has its advantages and disadvantages. So the open type impeller consumes much less power when working at high viscosity [2] relative to other types, does not have a front gap seal, which allows pumping contaminated liquid [3–9]. However, due to the lack of a driven disk, it is impossible to accurately calculate volumetric losses, since the fluid flow in the gaps between the impeller and the walls of the casing is part of the working process of the impeller [10–12]. It is also worth noting that currently there is not enough literature for the design of impellers with low and ultra-low speed coefficient, and theoretical formulas give large errors. Therefore, at this stage, it is necessary to make sure that the calculations of the complex flow in the open impeller, including the end gap region, can be performed by numerical simulation methods and to verify the obtained calculated dependences by comparing them with experimental data.

Methods

The method of numerical simulation is based on the solution of discrete analogues of the basic equations of hydrodynamics. In the case of the selected model of an incompressible fluid $(\rho = const)$, this is:

The continuity equation of a liquid medium [1]:

$$\frac{\partial \overline{u_x}}{\partial x} + \frac{\partial \overline{u_y}}{\partial y} + \frac{\partial \overline{u_z}}{\partial z} = 0,$$

where $\overline{u_i}$ — time-averaged projections of fluid velocities on the corresponding axes;

Content from this work may be used under the terms of the Creative Commons Attribution 3.0 licence. Any further distribution of this work must maintain attribution to the author(s) and the title of the work, journal citation and DOI.
Published under licence by IOP Publishing Ltd

The equation of change in momentum [7]

In such a simulation, there is a flow region that rotates with the rotor - this is the flowing part of the impeller. The flow equation for a given region is written in terms of relative motion [4], so the equation of change in momentum is as follows:

$$\rho\left(\frac{\partial \overline{W_i}}{\partial t}+u_j\frac{\partial \overline{W_i}}{\partial x_j}\right)=-\frac{\partial \overline{p}}{\partial x_i}+\frac{\partial}{\partial x_i}\left(T_{ij}^{(\nu)}-\rho W_i W_j\right)+F_i+F_\kappa,$$

where $\overline{W_i}$ — average speed: $\bar{W}=\bar{V}-\bar{u}$ — ratio of relative, absolute and transfer speeds;

$\bar{p}$ — averaged pressure;

$T_{ij}^{(\nu)}=2\mu s_{ij}$—viscous stress tensor for incompressible fluid;

$s_{ij}=\frac{1}{2}\left(\frac{\partial \overline{W_i}}{\partial x_j}+\frac{\partial \overline{W_j}}{\partial x_i}\right)$ — strain rate tensor;

$\rho W_i W_j$ — Reynolds stresses;

$F_i=\rho\omega^2 r$ —inertia force;

$F_\kappa=2\omega W_{xy}$ —cariolis stress.

The Reynolds system of equations [1] is open due to the presence of unknown Reynolds stresses. System closure is performed using the k-ωSST turbulence model [5]. This model combines k-ω and k-ε models: the first is used in the near-wall region, the second in the central part of the flow.

This model includes two equations of transport of turbulence parameters [12]:

1. The equation of transfer of kinetic energy of turbulence

$$\frac{\partial k}{\partial t}+\overline{u_j}\frac{\partial k}{\partial x_j}=P_k-\beta\cdot k\omega+\frac{\partial}{\partial x_i}\left[\left(\nu+\sigma_k\nu_T\right)\frac{\partial k}{\partial x_j}\right],$$

where $k=\frac{1}{2}\left(u_x'^2+u_y'^2+u_z'^2\right)$ — kinetic energy of turbulence;

u_i' —pulsation of speed;

P_k —turbulence energy generation;

ω —relative turbulence dissipation rate;

ν_T —turbulent viscosity.

2. The equation of transport of the relative velocity of dissipation of turbulence energy:

$$\frac{\partial \omega}{\partial t}+\overline{u_j}\frac{\partial \omega}{\partial x_j}=\alpha S^2-\beta\omega^2+\frac{\partial}{\partial x_i}\left[\left(\nu+\sigma_\omega\nu_T\right)\frac{\partial \omega}{\partial x_j}\right]+2\left(1-F_j\right)\sigma_{\omega 2}\frac{1}{\omega}\frac{\partial k}{\partial x_j}\frac{\partial \omega}{\partial x_j}$$

Reynolds stresses in the equations of dynamics are based on the Boussinesq hypothesis:

$$\rho u_i u_j=2\mu_T\left[\frac{1}{2}\left(\frac{\partial u_i}{\partial x_j}+\frac{\partial u_j}{\partial x_i}\right)-\frac{1}{3}\frac{\partial u_k}{\partial x_k}\delta_{ij}\right]-\frac{2}{3}\rho k\delta_{ij},$$

where δ_{ij} —Kronecker symbol.

Using empirical closure coefficients of these equations [11], one can obtain a numerical solution of the turbulent fluid flow in the computational domain.

The impeller has been designed for the following parameters.:

$H=25m$ — pressure head,

IOP Conf. Series: Materials Science and Engineering **779** (2020) 012014 doi:10.1088/1757-899X/779/1/012014

$Q = 25\,m^3 / h$ — flow rate,

$n = 2900\,rpm$ —rotation speed.

The drawing of the impeller under study is shown in Figure 1. The drawing shows the size δ_1 of the front axial clearance between the wheel and the casing wall, as well as the width b_2 of the impeller at the exit, along which dimensioning was carried out. The size δ_2 of the rear axial clearance between the wheel and the housing wall is fixed and equal to half the width b_2.

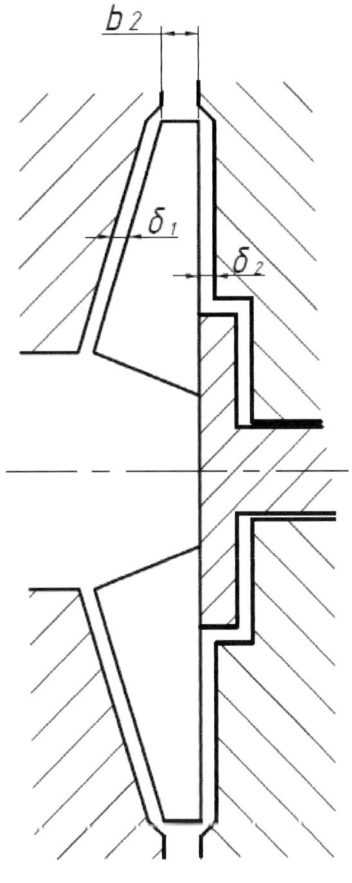

Figure 1. Impeller drawing

When modeling the fluid flow, it is necessary to break the flow geometry into cells, for which the equations given above [5] will be applied. For this, a model was chosen that generated polyhedral cells, as well as a model for constructing prismatic cells for modeling fluid flow near the wall [9]. A value of 5 mm was taken as the base size. The maximum and minimum cell sizes and the thickness of the boundary layer were specified as a percentage of the base size — 80%, 40%, and 24%, respectively. For the mesh in the impeller, the cell sizes are different from those specified for the entire model: the maximum cell size is 40%, the minimum cell size is 20%. The stretching of the prismatic layer occurs with a coefficient of 1.3. 5 layers have been created. At the exit from the spiral outlet, an extruder was created with a constant speed parameter in the normal direction [10]. The length of the extruder is 1 m, the number of layers is 500. It is necessary to equalize the flow at the output boundary [12] in order to reduce the influence of reverse currents, which significantly reduces the calculation accuracy near the boundary. An example of the generated grid is shown in Figure 2. To check the accuracy of the results, a calculation was performed with increased accuracy with the following values of the main parameters: the base size is 2.5 mm, the maximum cell size is 50%, the minimum cell size is 20%, boundary layer thickness — 48%; for the impeller: the maximum cell size is 20%, the minimum cell size is 10%. The discrepancy between the results of a more accurate and simplified modeling is insignificant (less than 0.5%). Thus, all calculations were performed on simplified models to reduce the calculation time.

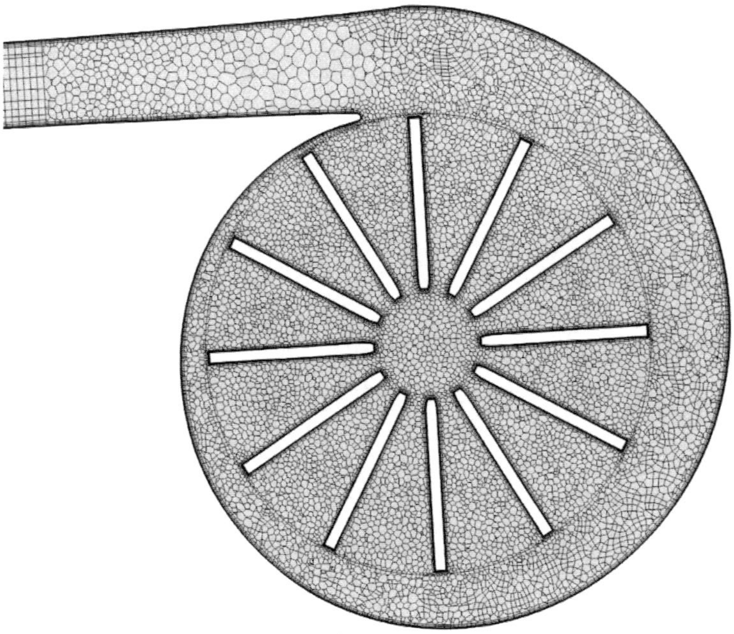

Figure 2. Impeller cross section

In this study, the value of only the front gap was changed, since the influence of the dimensions of the rear gap does not strongly affect the power characteristic [2]. The relative gap value is calculated by the formula

$$\Delta = \frac{\delta_1}{b_2}.$$

For the calculation, models with a relative gap Δ from 0.1 to 1.0 in increments of 0.1 were used. The correspondence between the numbers of the studied models and the values of the relative gap Δ are given in table 1.

Table 1

Model no.	1	2	3	4	5	6	7	8	9	10
Δ	0,1	0,2	0,3	0,4	0,5	0,6	0,7	0,8	0,9	1,0

Results

When simulating the flow, the following were determined: pressure at the inlet to the inlet and outlet of the spiral outlet, the moment at the impeller. After processing the results, the following values were obtained: the pressure created by the impeller, the moment on the impeller, and efficiency. The results are shown in table 2.

Table 2

Model no.	H, m	M, N·m	η
1	25,79	2,679	68,95
2	25,97	2,740	67,87
3	25,55	2,723	67,18
4	25,26	2,710	66,73
5	25,09	2,700	66,55
6	24,87	2,680	66,44
7	24,71	2,668	66,33
8	24,58	2,665	66,06
9	24,38	2,655	65,76
10	24,27	2,660	65,32

In [2], experimental results are presented. For comparison, the pressure was reduced to a dimensionless form by the formula

$$\bar{H} = \frac{H}{(D_2 \cdot n)^2},$$

and for efficiency the coefficient is used

$$k_\eta = \frac{\eta}{\eta_{max}}.$$

The obtained values are displayed on the graphs (Figure 4, 5) of the dependence of the dimensionless pressure and efficiency coefficient on the dimensionless axial clearance — $\bar{H}(\Delta)$ and $k_\eta(\Delta)$.

The dependences obtained by calculation were compared with the experimental data given in [2]. The results, recalculated in the same dimensionless quantities, are shown in Figures 4 and 5. Good convergence of the type of experimental and theoretical characteristics is visible. A certain deviation can be explained by a limited number of experimental points and a certain difference in the speed coefficients of the impeller, on which the experiments were performed, and the impeller investigated by hydrodynamic modeling methods, as well as by a small (for such pump parameters) absolute value of efficiency, which increases the "price" of its minimum deviation in absolute terms.

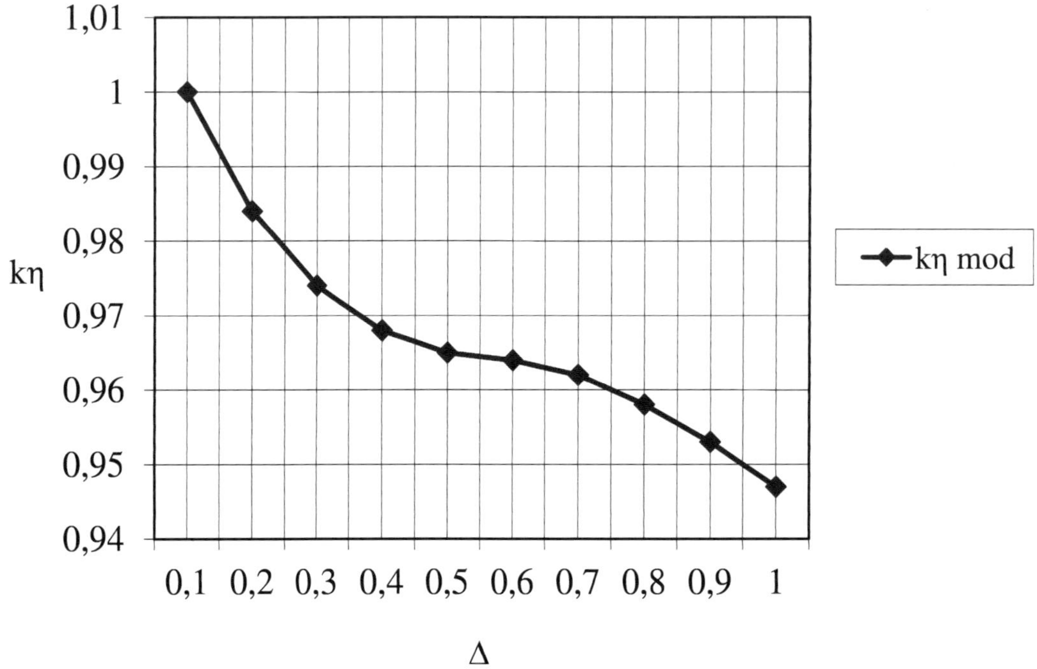

Figure 3. Dependence graph $k_\eta(\Delta)$ obtained by hydrodynamic modeling

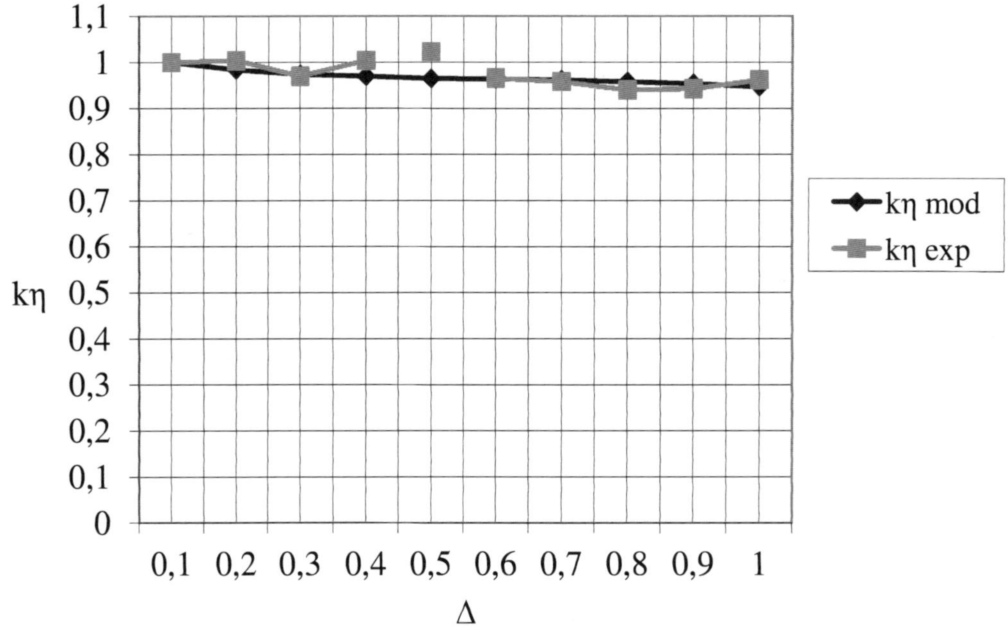

Figure 4. Dependency graph $k_\eta(\Delta)$

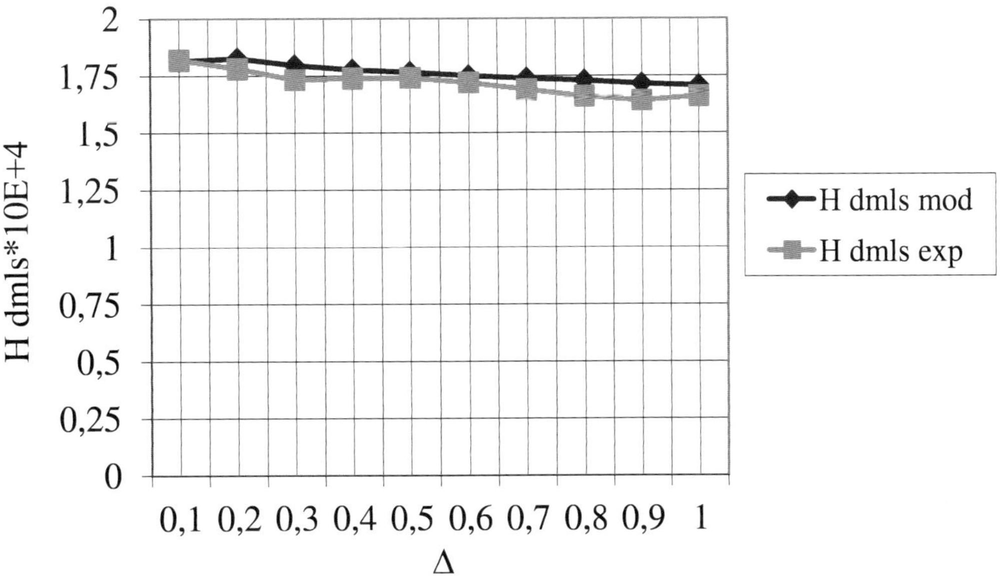

Figure 5. Dependency graph $\bar{H}(\Delta)$

Conclusion

1. Good convergence of the efficiency coefficient graphs was obtained, with the exception of one experimental value at $\Delta = 0,5$, which is explained by small values of efficiency, and therefore, a large influence of the error.

2. The graph of the dimensionless pressure repeats the shape of the experimental curve, however, it has slightly larger values and does not completely coincide. Such a difference could arise due to a different slope of the pressure characteristic due to a different speed coefficient. This issue requires for there study.

3. On the obtained efficiency graph, the area is clearly visible on which the change in the values of the function of the end gap is small. The presence of such a zone can save on ensuring the accuracy of manufacturing parts included in the dimensional chain, while maintaining a high value of efficiency.

References

[1] Loitsyansky L.G. Fluid and Gas Mechanics — 4th ed. — M.: Publishing House "Science", 1973.
[2] Matveev I.V., Petrov A.I. The study of the influence on the pressure characteristic of the pump of the front clearance between the open impeller and the housing. // 50 years kaf. GM: Thesis. Doc. MNTK — M., MPEI, 1996.— P. 87.
[3] Petrov A.I. The effect of high viscosity of the working fluid on the parameters of ultra-low speed centrifugal pumps with open impellers. // Hydraulic Engineering. Present and future .: Abstract. Doc. MNTK — M., MSTU, 2004.— S. 50.A Petrov et al 2019 *IOP Conf. Ser.: Mater. Sci. Eng.* 492 012036
[4] A Protopopov and D Bondareva 2019 *IOP Conf. Ser.: Mater. Sci. Eng.* 492 012002
[5] A Protopopov and V Vigovskij 2019 *IOP Conf. Ser.: Mater. Sci. Eng.* 492 012003
[6] K Dobrokhodov and APetrov 2019 *IOP Conf. Ser.: Mater. Sci. Eng.* 492 012016
[7] N Egorkina and APetrov 2019 *IOP Conf. Ser.: Mater. Sci. Eng.* 492 012015
[8] N Isaev 2019 *IOP Conf. Ser.: Mater. Sci. Eng.* 492 012026

[9] T Valiev and APetrov 2019 *IOP Conf. Ser.: Mater. Sci. Eng.* 492 012038
[10] V Cheremushkin and V Lomakin 2019 *IOP Conf. Ser.: Mater. Sci. Eng.* 492 012039
[11] V Lomakin and O Bibik 2019 *IOP Conf. Ser.: Mater. Sci. Eng.* 492 012037
[12] V Lomakin et al 2019 *IOP Conf. Ser.: Mater. Sci. Eng.* 492 012012

IOP Conf. Series: Materials Science and Engineering **779** (2020) 012015 doi:10.1088/1757-899X/779/1/012015

Obtaining of the drying time of a vertical cylindrical tank

A Shablovskiy[1,2], I Bilenko[1], V Chadin[1], I Budarin[1], A Makhlaeva[1] and A Sivova[1]
[1]Bauman Moscow State Technical University

[2]E-mail: shabl@bmstu.ru

Abstract: The work is devoted to the finding of the formula for calculating the drying time of a vertical cylindrical tank. The structure, applications and main types of this vertical tank and its advantages in comparison to the horizontal tank are described. The figuring out of the formula is based on the method of solving a differential equation of first order for pump capacity (performance), a time-dependent operation of the pump and liquid volume, depending on the height of the decreasing liquid.

Introduction

Currently, steel tanks are widely used for oil depots and gas stations, namely for long-term storage of oil, petroleum products and other liquids. Most tanks for oil depots consist of several parts: a cylindrical enclosure made of metal (horizontal or vertical type) with additional components: flanges, rings and technological man ways.

Many modern publications [1–5] are enlightened in the calculation of various hydraulic systems. In particular, some of them [6–10] are devoted to numerical modeling of hydraulic processes. However, the issues of integrating the drying time of various vessels are not sufficiently covered.

We consider a detailed structure of a vertical one on the example of a steel cylindrical tank with a volume of 1000 m^3 Figure 1, which consists of a enclosure, outer and inner tank shells, a spacer angel, has a ladder and a base plates. And in order to keep the substances stored in the tanks from freezing at low ambient temperature, the tanks are equipped with a heater and thermal insulation. In addition, the outer surface of the walls is treated with soil, covered with anti-corrosion paint and sheathed with galvanized (or aluminum) steel sheets to protect the tanks from the environment.

There are ground and underground types of tanks (according to technological features) that are installed vertically or horizontally. However, it is worth considering that the vertical canister will take several times less area than the horizontal.

The most common types of vertical (metal) tanks are:

- The tank of VSR (vertical standard reservoir). It's vertical steel canister with a cone-shaped bottom, designed for filling various types of liquids, including water, aggressive acids and alkalis.
- Underground gas station tanks. Such tanks are manufactured for storage of dark and light oil products in oil storage facilities, gas stations and gas stations. Moreover, modern tanks for gas stations can have two-layer tanks, which are two metal cylinders, one of which is installed inside the other, and the inter-wall space is filled with a liquid that is more dense than the density of stored fuel. This construction of the tank allows you to prevent leakage of fuel into the environment in the event of damage to its enclosure.
- Fire metal tanks. Fire containers are filled with drinking, technical or waste water with a volume of 5 to 60 cubic meters. They can be installed both vertically and horizontally, but above the ground.

Content from this work may be used under the terms of the Creative Commons Attribution 3.0 licence. Any further distribution of this work must maintain attribution to the author(s) and the title of the work, journal citation and DOI.
Published under licence by IOP Publishing Ltd

IOP Conf. Series: Materials Science and Engineering **779** (2020) 012015 doi:10.1088/1757-899X/779/1/012015

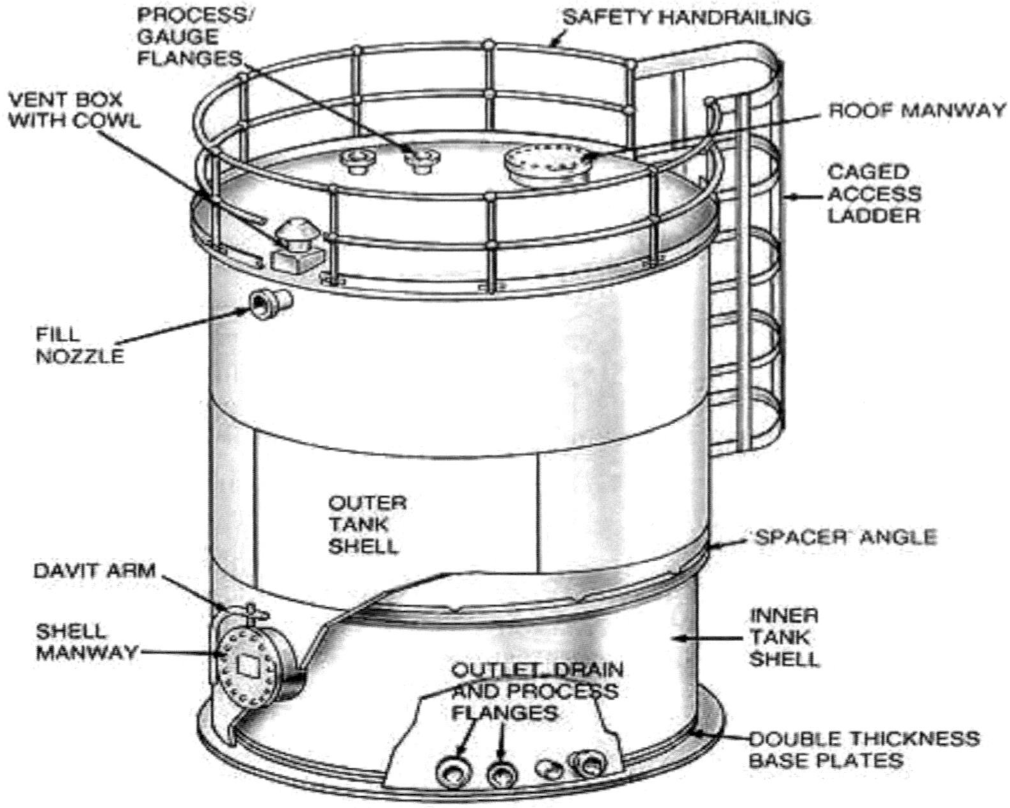

Figure 1. Structure of vertical tank by the vertical cylindrical tank VCR-1000 m^3 example

In this article we attempt to use mathematical analysis methods to derive a general formula for determining the drying time of vertical cylindrical tanks.

Method

The method of mathematical analysis is based on differentiation and integration. In our model $y(0)=0, y(T)=h$. In figure Figure 2we show a scheme of the drainage tank.

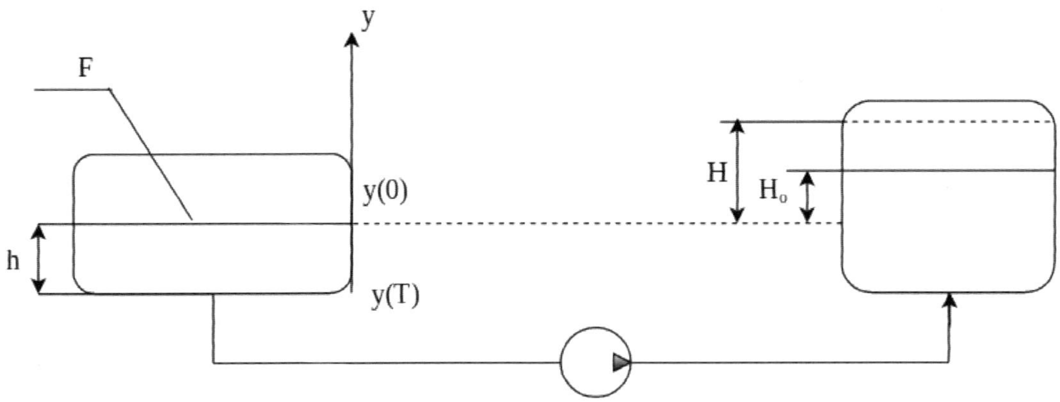

Figure 2. High-quality scheme of the drainage tank

A differential equation of first order for volume flow rate Q and area F:

$$-Q(t)dt = F(y)dy$$

where dt — pump operating time;

dy —height of the liquid in the draining tank.

The height of the pump, depending on flow rate:

$$H_{pump}(Q) = A + B\cdot Q + C\cdot Q^2$$

where A, B, C — pump constants.

So the height is equal to:

$$H = H_o + y + \alpha\cdot Q^2$$

where $\alpha = 0{,}0827\dfrac{x\cdot l}{d^5}$ — the constant.

If we equate the above heights and express flow rate Q, we get:

$$Q(y) = \frac{-B \pm \sqrt{B^2 - 4(\alpha - C)(H_o + y - A)}}{2(\alpha - C)}$$

Analysis of the obtained formula for flow rate Q to the sign before the square root:

$$\alpha > 0,\ C < 0$$

It means that:

$$\alpha - C > 0$$

$$H_{static}^{friction} = H_o + y$$

$$H_{pump}^{o} = A$$

$$H_{static}^{friction} < H_{pump}^{o}$$

Also:

$$H_o + y - A < 0$$

Then:

$$\sqrt{B^2 - 4(\alpha - C)(H_o + y - A)} = \sqrt{B^2 + 4|(\alpha - C)(H_o + y - A)|} > |B|$$

$$Q(y) = \frac{-B + \sqrt{B^2 - 4(\alpha - C)(H_o + y - A)}}{2(\alpha - C)}$$

We substitute the final formula for the volume flow rate Q(y) in the differential equation $-Q(t)dt = F(y)dy$ under the condition that $F = const$:

$$-dt = F\frac{-B + \sqrt{B^2 - 4(\alpha - C)(H_o + y - A)}}{2(\alpha - C)}dy$$

IOP Conf. Series: Materials Science and Engineering **779** (2020) 012015 doi:10.1088/1757-899X/779/1/012015

We get the desired formula for the time of the draining tank, having integrated:

$$T=-F\int_{0}^{h}\frac{2(\alpha-C)dy}{\sqrt{B^{2}-4(\alpha-C)(H_{o}+y-A)}-B}$$

Results

During the study working formula was obtained for determining the drying time of the tank, which depends only on the pump constants and initial heights in the tanks. The ultimate formula is as follows:

$$T=-F\int_{0}^{h}\frac{2(\alpha-C)dy}{\sqrt{B^{2}-4(\alpha-C)(H_{o}+y-A)}-B}$$

where A,B,C — pump constants;

F — the area of the tank, from which the liquid is pumped;

α — the constant;

$H_{o}, y=h$ — the initial height of the fill and drained the tank, respectively.

Conclusion

The obtained method allows us to determine the time of drying of the tank under the condition of the cylindrical shape of its walls, which narrows down its capabilities, but nevertheless it can find its application in various fields of hydraulic engineering.

References

[1] V Cheremushkin and APolyakov 2019 *IOP Conf. Ser.: Mater. Sci. Eng.***589** 012001
[2] V Tkachyk et al 2019 *IOP Conf. Ser.: Mater. Sci. Eng.***589** 012010
[3] V Tkachyk et al 2019 *IOP Conf. Ser.: Mater. Sci. Eng.* **589** 012011
[4] E Morozova et al 2019 *IOP Conf. Ser.: Mater. Sci. Eng.***589** 012012
[5] K Abramov 2019 *IOP Conf. Ser.: Mater. Sci. Eng.***589** 012013
[6] A Boyarshinova et al 2019 *IOP Conf. Ser.: Mater. Sci. Eng.***589** 012014
[7] N Sosnovsky and D Ganieva 2019 *IOP Conf. Ser.: Mater. Sci. Eng.***589** 012016
[8] M Saprykina and V Lomakin 2019 *IOP Conf. Ser.: Mater. Sci. Eng.***589** 012017
[9] A Boyarshinova et al 2019 *IOP Conf. Ser.: Mater. Sci. Eng.***589** 012028
[10] A Martynyuk et al 2019 *IOP Conf. Ser.: Mater. Sci. Eng.***589** 012029

IOP Conf. Series: Materials Science and Engineering **779** (2020) 012016 doi:10.1088/1757-899X/779/1/012016

Predicted resource of the most loaded bearing of a horizontal centrifugal pump

A Shablovskiy[1,2], N Menagarishvili[1], A Makhlaeva[1] and E Emshina[1]

[1]Bauman Moscow State Technical University

[2]E-mail:shabl@bmstu.ru

Abstract: The paper gives a derivation of the formula for calculating the predicted resource of the most loaded bearing of a horizontal centrifugal pump. A model of a horizontal centrifugal pump, as well as the action of axial forces and radial reactions, are considered.

Introduction

Currently, centrifugal pumps [1–9] are widely used in various fields of industry and national economy, such as agriculture. The choice of this particular type of pump is due to the fact that it has a simple design, is easy to operate, and can use water, fuel and chemically hazardous compounds as a working fluid.

Consider the structure of the horizontal centrifugal pump in Figure Figure 1. On the shaft of the electric motor 1, the impeller 2 and two bearings 3. The working fluid is captured by the blades of the wheel 2, is thrown to the walls of the chamber, creating excess pressure. Water is pushed out under pressure. During the discharge of liquid from the working chamber, a vacuum is created, which ensures the suction of a new portion of the liquid.

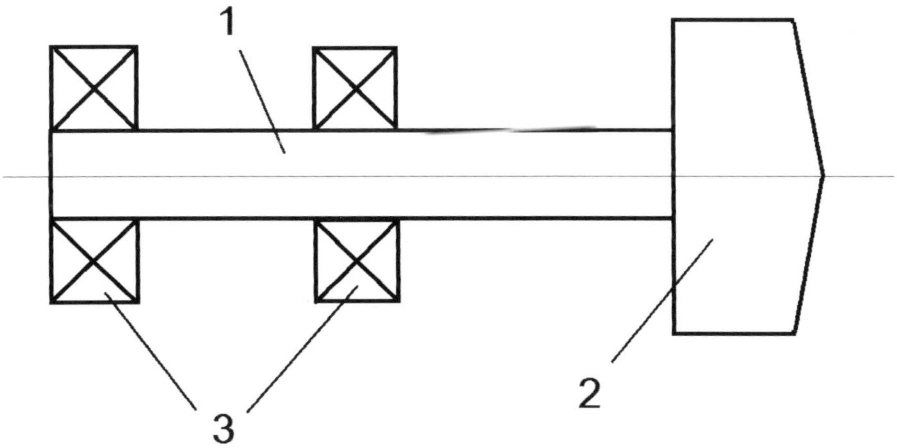

Figure 1. The structure of a horizontal centrifugal pump

Content from this work may be used under the terms of the Creative Commons Attribution 3.0 licence. Any further distribution of this work must maintain attribution to the author(s) and the title of the work, journal citation and DOI.
Published under licence by IOP Publishing Ltd

In this article, an attempt was made using the method of mathematical analysis to derive a formula for calculating the predicted resource of the most loaded bearing.

Methods

Consider the figure of Fig. 2. The impeller 2 creates a radial force P_{r1}, as a result of which a radial reaction P_{r2} occurs on the first bearing, and P_{r3} on the second bearing. P_{o2} axial force occurs on the front impeller disk, and P_{o3} on the rear impeller disk. Also, its weight *G* acts on the rotor. The bearing arrangement is such that the total axial force from the impeller and the weight of the rotor is compensated by the axial reaction in the second bearing P_{o3}.

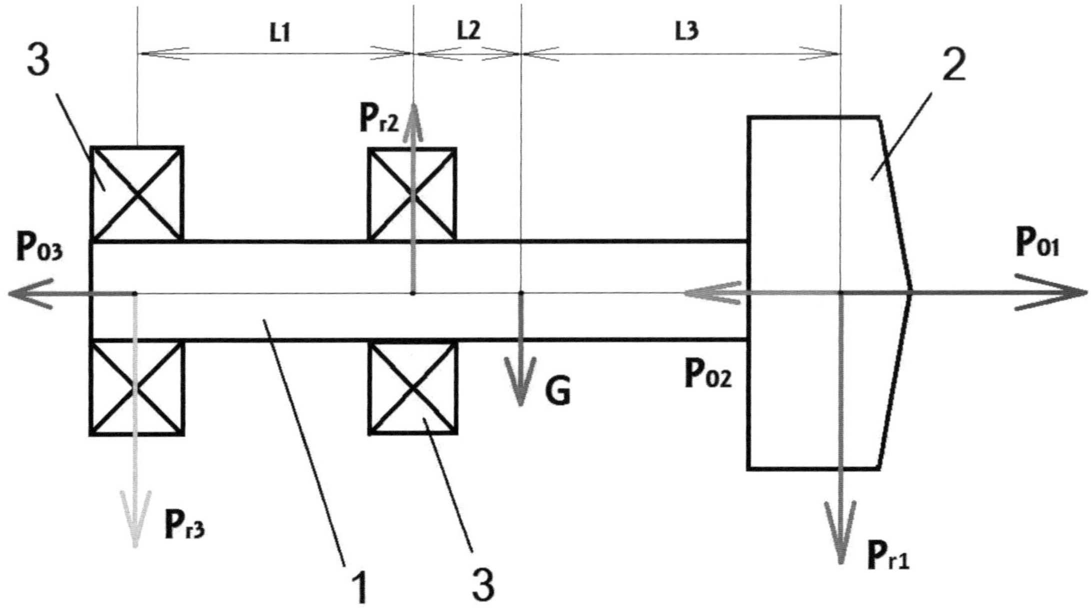

Figure 2. Reactions of a horizontal centrifugal pump

From the equations of the equilibrium condition (the sum of the acting moments and forces on the structure is zero $M_0 = 0$ $\sum_i P_i = 0$) we get the following system:

$$P_{a3} = P_{a1} - P_{a2} = P_{a\kappa} \quad (1.1)$$

$$P_{r3} \cdot L_1 = G \cdot L_2 + (L_2 + L_3) \cdot P_{r1} \quad (1.2)$$

$$G = M_p \cdot g \quad (1.3)$$

where G is the weight of the rotor,

M_p is the mass of the rotor, L_1 is the distance between the bearings, L_2 is the distance between the first bearing and the center of mass of the shaft, L_3 is the distance between the shaft and the impeller (Figure 2).

The total axial force created by the impeller is:

$$P_a = \pi\rho g\left(R_y^2 - r_s^2\right)H -$$
$$-\frac{\pi^3 \cdot \rho \cdot n^2}{8 \cdot 30^2}\left(R_y^2 - r_s^2\right)\left[R_2^2 - 0{,}5\cdot\left(R_g^2 - r_s^2\right)\right] \quad (1.4)$$

where ρ is the density of the working fluid,
r_s is the radius of the sleeve,
andR_g is the radius of the gap seal.

In our case, the design of the impeller is such that $r_{вт}$= 0, $R_g \approx R_1$. As will be shown below, the optimal value of R_1 will be equal to:

$$R_1 = 1{,}125 \cdot \sqrt[3]{\frac{Q \cdot 30}{\pi^2 \cdot n}} \quad (1.5)$$

Thus, the formula (1.4) takes the form:

$$P_a = \pi \rho g R_1^2 H - \frac{\pi^3 \cdot \rho \cdot n^2}{8 \cdot 30^2} R_1^2 \left[R_2^2 - 0{,}5 \cdot R_1^2 \right] \quad (1.6.)$$

The radial force for spiral bends is:

$$P_{r1} = 0{,}4 \left(1 - \frac{Q}{Q_{опт}} \right) \rho \cdot g \cdot H \cdot D_2 \cdot b_2 \quad (1.7)$$

whereQ_{opt} is the optimum pump flow.

As will be shown below, the optimal value of the impeller width at the output is:

$$b_2 = \frac{\pi \cdot n \cdot 0{,}044 \cdot \left(\frac{Q}{n} \right)^{\frac{2}{3}}}{\sqrt{\frac{H \cdot g}{y \cdot \eta_г}}} \quad (1.8)$$

Thus, in the worst case, the magnitude of the radial force is determined by:

$$P_{r1} = 0{,}4 \cdot \rho \cdot g \cdot H \cdot 2 \cdot R_2 \cdot \frac{\pi \cdot n \cdot 0{,}044 \cdot \left(\frac{Q}{n} \right)^{\frac{2}{3}}}{\sqrt{\frac{H \cdot g}{y \cdot \eta_г}}} \quad (1.9)$$

The resulting P_{r1} value corresponds to a zero pump flow.

Substituting formulas (1.3), (1.4) and (1.9) into the system of equations (1.1)–(1.2), we obtain the reaction values in the rotor bearings. After calculating the axial and radial reaction forces in the bearings of both bearings, we determine the radial forces equivalent to the reaction forces obtained. The worst situation will be observed in the second bearing, as both radial and axial forces act on it. Equivalent radial force is defined as:

$$P_{eq} = \left(V \cdot X \cdot P_{r3} + Y \cdot P_{o3} \right) \cdot K_б \cdot K_m \quad (1.10)$$

where$V$, X, K_b and K_t are coefficients depending on various conditions,

Y is a functional dependence characterizing the fraction of axial forces in the equivalent radial force.

We take the indicated coefficients in accordance with [10] according to the given working conditions equal to the following values: $V = 1$; $K_t = 1$; $K_b = 1.2$; $X = 0{,}56$. Function Y is calculated by the formula:

$$Y = \frac{0{,}44}{e} \quad (1.11)$$

where the coefficient e is:

$$e = 0,28 \cdot \left(\frac{f_0 \cdot P_{ок}}{C_{or}} \right)^{0,23} \quad (1.12)$$

where $f_0 = 15$, (determined depending on the design parameters of the bearing),

C_{or} is the basic static radial bearing capacity.

Next, we determine the value of the resource of this bearing for various pump operation modes.

The estimated bearing life is determined by the formula:

$$T = a_1 \cdot a_{23} \cdot \left(\frac{C}{P_{rэкв}} \right)^k \cdot \frac{10^6}{60 \cdot n} \quad (1.13)$$

where a_1, a_{23} are coefficients depending on the special operating conditions of the bearing and the probability of failure-free operation.

The coefficient k in the formula (1.13) depends on the type of bearing. In this case, it is a ball bearing and coefficient $k = 3$.

Substituting (1.1)–(1.12) into (1.13), we obtain the dependence of the resource on the frequency

$$T = 0,7 \cdot \left(\frac{C}{\left(0,56 \cdot P_{r3}(n) + \frac{0,44}{e(n)} \cdot P_{ок}(n) \right) \cdot 1,2} \right)^3 \cdot \frac{10^6}{60 \cdot n} \quad (1.14)$$

Results

During the study, a working formula was obtained to determine the predicted resource of the most loaded bearing of a horizontal centrifugal pump. As a result, the formula has the form:

$$T = 0,7 \cdot \left(\frac{C}{\left(0,56 \cdot P_{r3}(n) + \frac{0,44}{e(n)} \cdot P_{o3}(n) \right) \cdot 1,2} \right)^3 \cdot \frac{10^6}{60 \cdot n}$$

where C is the basic static radial bearing capacity, P_{r3} - radial reaction on the second bearing, P_{o3}- axial force on the rear disk of the impeller, n is the rotation frequency.

Conclusion

The obtained method allows us to predict the life of a centrifugal pump under the assumption that its operating mode is constant. The disadvantages of the obtained method include the assumption that failure is possible only in the most loaded bearing, which is not entirely true, however, if the pump is operated on a clean, uncontaminated and chemically inactive liquid, the described model can be considered suitable for use in the calculation of centrifugal pumps.

Reference

[1] A Protopopov and D Bondareva 2019 *IOP Conf. Ser.: Mater. Sci. Eng.* **492** 012002
[2] A Protopopov and V Vigovskij 2019 *IOP Conf. Ser.: Mater. Sci. Eng.* **492** 012003
[3] P Chaburko and Z Kossova 2019 *IOP Conf. Ser.: Mater. Sci. Eng.* **492** 012011
[4] V Lomakin et al 2019 *IOP Conf. Ser.: Mater. Sci. Eng.* **492** 012012

[5] A Gouskov et al 2019 *IOP Conf. Ser.: Mater. Sci. Eng.* **492** 012013
[6] N Egorkina and APetrov 2019 *IOP Conf. Ser.: Mater. Sci. Eng.* **492** 012015
[7] K Dobrokhodov and APetrov 2019 *IOP Conf. Ser.: Mater. Sci. Eng.* **492** 012016
[8] N Isaev 2019 *IOP Conf. Ser.: Mater. Sci. Eng.* **492** 012026
[9] S Korsakova and AProtopopov 2019 *IOP Conf. Ser.: Mater. Sci. Eng.* **492** 012032
[10] A Protopopov and C Jakovich 2019 *IOP Conf. Ser.: Mater. Sci. Eng.* **492** 012034

IOP Conf. Series: Materials Science and Engineering **779** (2020) 012017 doi:10.1088/1757-899X/779/1/012017

Predicted resource of the most loaded centrifugal pump bearing

A Protopopov[1], A Makhlaeva[1] and P Kaplenkova[1,2]
[1]Bauman Moscow State Technical University

[2]E-mail:kaplenkova2014@gmail.com

Abstract: The paper gives a derivation of the formula for calculating the most loaded bearing of a centrifugal pump. The structure, applications, the main types of this centrifugal pump and the comparison of these bearings with hydrostatic are described. The resulting model is well applicable to pumps running on clean, chemically inactive liquids.

Introduction

Currently, centrifugal pumps [1–12] are widely used in various fields of industry and national economy, such as agriculture. The choice of this particular type of pump is due to the fact that it has a simple design, is easy to operate, and can use water, fuel and chemically hazardous compounds as a working fluid.

Unlike hydrostatic bearings, which have a significant resource, the life of pumps with ball bearings is very limited. The life of a ball bearing is largely dependent on the choice of rotational speed of the pump rotor.

In this article, an attempt was made using the method of mathematical analysis to derive a formula for calculating the predicted resource of the most loaded bearing.

Methods

The method is based on solving the equation for an incompressible fluid (ρ=const)

We determine the effect of the shaft speed on the pump resource. To do this, we first determine the value of the equivalent radial force acting on the most loaded bearing. The design diagram of the radial and axial forces acting on the rotor is shown in the figure.

An impeller and two bearings are mounted on the motor shaft. Impeller creates radial force Pr_1, as a result, a radial reaction occurs on the first bearing Pr_2, and on the second bearing, a radial reaction Pr_3. Axial force occurs on the front of the impeller P_{a2}, and axial force on the back of the impeller P_{a1}. The rotor also affects its weight. P_{a4}. The arrangement of the bearings is such that the total axial force from the impeller and the weight of the rotor is compensated by the axial reaction in the second bearing P_{a3}.

From the equations of the equilibrium condition (the sum of the acting moments and forces on the structure is zero $M_o = 0 \sum_i P_i = 0$ we get the following system of equations:

Content from this work may be used under the terms of the Creative Commons Attribution 3.0 licence. Any further distribution of this work must maintain attribution to the author(s) and the title of the work, journal citation and DOI.
Published under licence by IOP Publishing Ltd

IOP Conf. Series: Materials Science and Engineering **779** (2020) 012017 doi:10.1088/1757-899X/779/1/012017

$$P_{r1} \cdot (L_2 + L_1) = P_{r2} \cdot L_2 \quad (1)$$

$$P_{r1} + P_{r3} = P_{r2} \quad (2)$$

$$P_{a4} + P_{a1} - P_{a2} = P_{a3} \quad (3)$$

$$P_{r2} = P_{r1} \cdot \frac{(L_2 + L_1)}{L_2} \quad (4)$$

$$P_{r3} = P_{r1} \cdot \frac{L_1}{L_2} \quad (5)$$

$$P_{a3} = P_{a4} + P_{a1} - P_{a2} \quad (6)$$

whereL_2 is the distance between the bearings,

L_1 is the distance between the front bearing and the impeller (see figure).

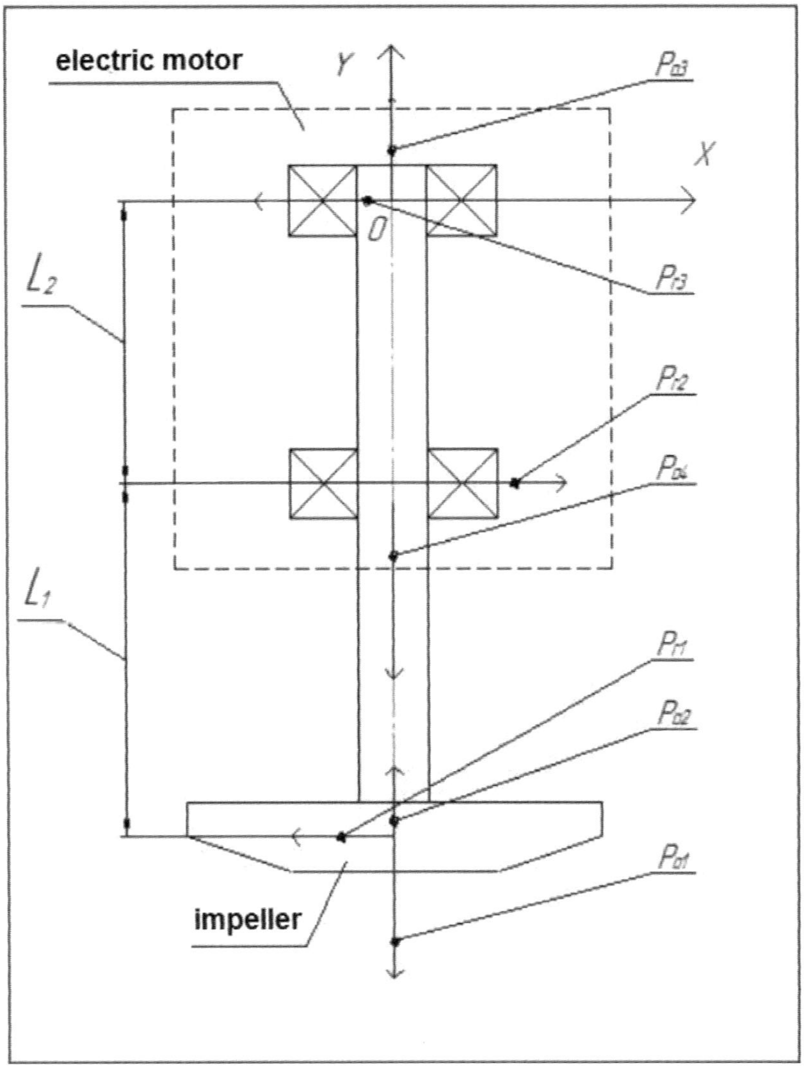

Figure. Design diagram of radial and axial forces, rotor acting

These values are known to us from the design of the electric motor.

Dynamic and static reactions in bearings were not included in the system of equations (4)–(6) due to their smallness for a given pump, while eliminating them from this system allows us to significantly simplify the resulting mathematical model. In order to evaluate their contribution, the magnitude of the static and dynamic reactions can be compared, for example, with the weight of the rotor, which will be calculated below.

Thus, to determine the reaction forces in the bearings, it is necessary to determine the weight of the rotor and the dependence of the axial and radial forces on the shaft rotation frequency. The weight of the rotor is equal to:

$$P_{a4} = M_p \cdot g \quad (7)$$

whereM_r is the mass of the rotor.

The total axial force created by the impeller is:

$$P_{a1} - P_{a2} = \pi \rho g \left(R_s^2 - r_s^2\right) H - \frac{\pi^3 \cdot \rho \cdot n^2}{8 \cdot 30^2}\left(R_y^2 - r_s^2\right)\left[R_2^2 - 0{,}5 \cdot \left(R_s^2 - r_s^2\right)\right] \quad (8)$$

where ρ is the density of the working fluid,

r_s is the radius of the sleeve,

andR_s is the radius of the gap seal. In our case, the design of the impeller is such that $r_{вт} = 0$, $R_s \approx R_1$. As will be shown below, the optimal value of R_1 will be equal to:

$$R_1 = 1{,}125 \cdot \sqrt[3]{\frac{Q \cdot 30}{\pi^2 \cdot n}} \quad (9)$$

Thus, the formula (8) will take the form:

$$P_{a1} - P_{a2} = \pi \rho g R_1^2 H - \frac{\pi^3 \cdot \rho \cdot n^2}{8 \cdot 30^2} R_1^2 \left[R_2^2 - 0{,}5 \cdot R_1^2\right] \quad (10)$$

The radial force for spiral bends is:

$$P_{r1} = 0{,}4\left(1 - \frac{Q}{Q_{opt}}\right)\rho \cdot g \cdot H \cdot D_2 \cdot b_2 \quad (11)$$

whereQ_{opt} is optimum pump flow.

As will be shown below, the optimal value of the impeller width at the output is:

$$b_2 = \frac{\pi \cdot n \cdot 0{,}044 \cdot \left(\frac{Q}{n}\right)^{\frac{2}{3}}}{\sqrt{\frac{H \cdot g}{y \cdot \eta_г}}} \quad (12)$$

Thus, in the worst case, the magnitude of the radial force is determined by:

$$P_{r1} = 0{,}4 \cdot \rho \cdot g \cdot H \cdot 2 \cdot R_2 \cdot \frac{\pi \cdot n \cdot 0{,}044 \cdot \left(\frac{Q}{n}\right)^{\frac{2}{3}}}{\sqrt{\frac{H \cdot g}{y \cdot \eta_г}}} \quad (13)$$

IOP Conf. Series: Materials Science and Engineering **779** (2020) 012017 doi:10.1088/1757-899X/779/1/012017

The resulting P_{r1} value corresponds to a zero pump flow.

Substituting formulas (7), (8) and (13) into the system of equations (4)–(6), we obtain the reaction values in the rotor bearings. After calculating the axial and radial reaction forces in the bearings of both bearings, we determine the radial forces equivalent to the reaction forces obtained. The worst situation will be observed in the second bearing, as both radial and axial forces act on it. Equivalent radial force is defined as:

$$P_{eq} = \left(V \cdot X \cdot P_{r3} + Y \cdot P_{a3}\right) \cdot K_b \cdot K_t \quad (14)$$

where V, X, K_b and K_t are coefficients depending on various conditions, Y is a functional dependence characterizing the fraction of axial forces in the equivalent radial force (see below).

We take the specified coefficients according to the given working conditions equal to the following values: $V = 1$; $K_t = 1$; $K_b = 1{,}2$; $X = 0.56$ and the function Y is calculated by the formula:

$$Y = \frac{0{,}44}{e} \quad (15)$$

where the coefficient e is :

$$e = 0{,}28 \cdot \left(\frac{f_0 \cdot P_{o3}}{C_{or}}\right)^{0{,}23} \quad (16)$$

where $f_0 = 15$, (determined depending on the design parameters of the bearing,

C_{or}—basic static radial bearing capacity.

Next, we determine the value of the resource of this bearing for various pump operation modes.

The estimated bearing life is determined by the formula:

$$T = aa_1 \cdot aa_{23} \cdot \left(\frac{C}{P_{eq}}\right)^k \cdot \frac{10^6}{60 \cdot n} \quad (17)$$

where aa_1, aa_{23} are coefficients depending on the special operating conditions of the bearing and the probability of failure-free operation.

The coefficient k in formula (17) depends on the type of bearing. In this case, it is a ball bearing and coefficient $k = 3$.

Substituting (4)–(16) in (17) we obtain the dependence of the resource on the frequency:

$$T = 0{,}7 \cdot \left(\frac{C}{\left(0{,}56 \cdot P_{r3}(n) + \frac{0{,}44}{e(n)} \cdot P_{a3}(n)\right) \cdot 1{,}2}\right)^3 \cdot \frac{10^6}{60 \cdot n} \quad (18)$$

where C is the basic static radial load bearing capacity,

P_{r3} is the radial reaction on the second bearing,

P_{a3} is the axial force on the rear wheel of the impeller.

Results

Thus, the obtained technique allows us to obtain a working formula for determining the predicted resource of the most loaded bearing of a horizontal centrifugal pump.

IOP Conf. Series: Materials Science and Engineering **779** (2020) 012017 doi:10.1088/1757-899X/779/1/012017

Conclusion

The obtained method allows us to predict the life of a centrifugal pump under the assumption that its operating mode is constant. The disadvantages of the obtained method include the assumption that failure is possible only in the most loaded bearing, which is not entirely true, however, if the pump is operated on a clean, uncontaminated and chemically inactive liquid, the described model can be considered suitable for use in the calculation of centrifugal pumps.

Reference

[1] A Protopopov and D Bondareva 2019 *IOP Conf. Ser.: Mater. Sci. Eng.* **492** 012002
[2] A Protopopov and V Vigovskij 2019 *IOP Conf. Ser.: Mater. Sci. Eng.* **492** 012003
[3] P Chaburko and Z Kossova 2019 *IOP Conf. Ser.: Mater. Sci. Eng.* **492** 012011
[4] V Lomakin et al 2019 *IOP Conf. Ser.: Mater. Sci. Eng.* **492** 012012
[5] A Gouskov et al 2019 *IOP Conf. Ser.: Mater. Sci. Eng.* **492** 012013
[6] N Egorkina and APetrov 2019 *IOP Conf. Ser.: Mater. Sci. Eng.* **492** 012015
[7] K Dobrokhodov and APetrov 2019 *IOP Conf. Ser.: Mater. Sci. Eng.* **492** 012016
[8] N Isaev 2019 *IOP Conf. Ser.: Mater. Sci. Eng.* **492** 012026
[9] S Korsakova and AProtopopov 2019 *IOP Conf. Ser.: Mater. Sci. Eng.* **492** 012032
[10] A Protopopov and C Jakovich 2019 *IOP Conf. Ser.: Mater. Sci. Eng.* **492** 012034
[11] N Isaev et al 2019 *IOP Conf. Ser.: Mater. Sci. Eng.* **589** 012009
[12] V Tkachyk et al 2019 *IOP Conf. Ser.: Mater. Sci. Eng.* **589** 012010

IOP Conf. Series: Materials Science and Engineering **779** (2020) 012018 doi:10.1088/1757-899X/779/1/012018

Determination of the coefficient of economic efficiency of frequency regulator

A Protopopov[1,2], A Makhlaeva[1] and A Panya[1]

[1]Bauman Moscow State Technical University

[2]E-mail:proforg6@yandex.ru

Abstract: The work is devoted to the finding of the formula for calculating economic efficiency of the frequency control method. This method is considered on the example of vertical tank drainage. Also in this work the principle of operation of the frequency controller and its application are described. The figuring out of the formula is based on the expression for calculating efficiency and this system capacity, depending on pump volume flow rate.

Introduction

Centrifugal pumps attract great interest from researchers [1–8]. At the same time, modern methods of calculating such pumps are of considerable interest [9–12]. However, questions of economic efficiency often stay without explanation.

Nowadays, the automation of pumping stations is increasingly using schemes based on frequency regulators. Such automated circuits with a frequency converter provide:

–Protection of drive motors from short circuits, phase breakage, overheating, overloads, voltage drops.

–Smooth change in performance when pressure decreases or increases. The frequency Converter can also operate in the supply control mode for several parameters of the water supply or heating system.

–Fault alarm.

It's also a great way to save energy. This is important for large liquid pumping systems.

The essence of the frequency control method is that by changing the frequency of the pump that pumps the liquid at the right time, it is possible to save energy spent on the system. This method is used in various environments where the main thing is pumping liquids. For example, such as oil pipelines, water pipelines, etc. it is also worth noting that frequency converters are used for automation of simple Autonomous water supply systems, as well as powerful stations with a large number of pumps.

In this article, an attempt was made to use the methods of mathematical analysis to derive a General formula for determining the coefficient of economic efficiency of frequency control.

This formula is remarkable because it can in practice help companies determine individually whether it is profitable for them to save on electricity by spending money on a frequency regulator, or conversely.

Content from this work may be used under the terms of the Creative Commons Attribution 3.0 licence. Any further distribution of this work must maintain attribution to the author(s) and the title of the work, journal citation and DOI.
Published under licence by IOP Publishing Ltd

Method

The method of mathematical analysis is based on expression for calculating efficiency and this system capacity, depending on pump volume flow rate. In our model $y(0)=0, y(T)=h$. We show a scheme of the drainage tank in figure Figure 1.

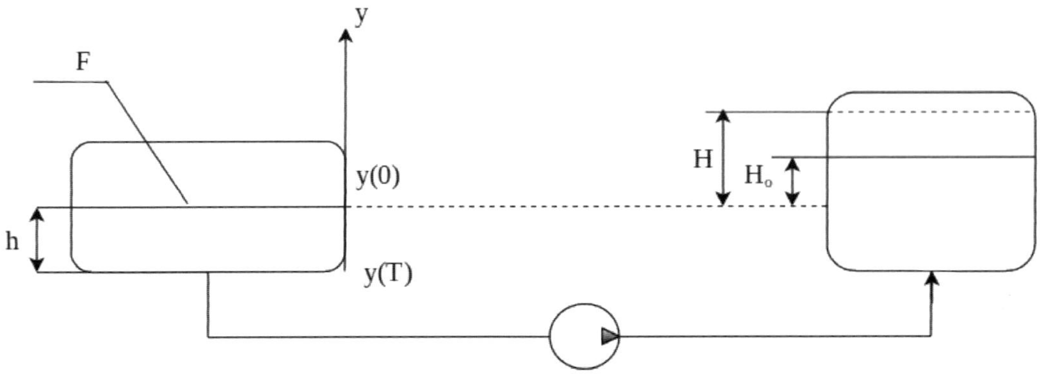

Figure 1. High-quality scheme of the drainage tank

The height of the pump, depending on flow rate:

$$H_{pump}(Q) = A + B \cdot Q + C \cdot Q^2$$

where A, B, C — pump constants.

So the height is equal to:

$$H = H_o + y + \alpha \cdot Q^2$$

where $\alpha = 0{,}0827\dfrac{x \cdot l}{d^5}$ — the constant.

If we equate the above heights and express flow rate Q, we get:

$$Q(y) = \frac{-B \pm \sqrt{B^2 - 4(\alpha - C)(H_o + y - A)}}{2(\alpha - C)}$$

Analysis of the obtained formula for flow rate Q to the sign before the square root:

$$\alpha > 0,\ C < 0$$

It means that:

$$\alpha - C > 0$$

$$H_{static}^{friction} = H_o + y$$

$$H_{pump}^{o} = A$$

$$H_{static}^{friction} < H_{pump}^{o}$$

Also:

$$H_o + y - A < 0$$

Then:

$$\sqrt{B^2-4(\alpha-C)(H_o+y-A)}=\sqrt{B^2+4|(\alpha-C)(H_o+y-A)|}>|B|$$

$$Q(y)=\frac{-B+\sqrt{B^2-4(\alpha-C)(H_o+y-A)}}{2(\alpha-C)}$$

Now we find the maximum efficiency of the system with a frequency regulator.

The efficiency of a system with a frequency regulator frequency n_1 can be found from the equation:

$$\eta(Q)=A_1+B_1*Q+C_1*Q^2$$

A_1, B_1, C_1 —constants of pumps with frequency regulator.

We differentiate this equation. In this case the system flow rate Q is converted to the maximum system flow rate Q_{max}

$$\eta'=2*C_1*Q_{max}+B_1=0$$

We express it from this equation:

$$Q_{max}=\frac{-B_1}{2*C_1}$$

In this case, the formula for finding the maximum efficiency of the system with a frequency regulator is:

$$\eta_{max}=A_1+B_1*\left(\frac{-B_1}{2*C_1}\right)+C_1*\left(\frac{-B_1}{2*C_1}\right)^2$$

We obtain the final formula, using simple algebraic transformations :

$$\eta_{max}=A_1-\frac{B_1^2}{4*C_1}$$

The coefficient of economic efficiency of frequency control can be found by the formula:

$$K_{ef}=\frac{T*N*W_1}{T*N_1*W_1+W}$$

where T —the time resource of the pump, [T]=[hour];

W_1 —the kW electricity per hour cost;

W — the frequency regulator cost;

N_1 —the pump capacity with a frequency regulator;

N — the pump capacity without a frequency regulator;

We write down the formula for finding the power of the pump without a regulator:

$$N=\rho*g*H_H*Q$$

ρ —the density of the fluid in the system

g —acceleration of gravity;

Substitute the expression for H_H then the formula will take the form:

$$N=\rho*g*(A+B\cdot Q+C\cdot Q^2)*Q$$

IOP Conf. Series: Materials Science and Engineering **779** (2020) 012018 doi:10.1088/1757-899X/779/1/012018

Now find the power of the pump using the frequency regulator. It can be calculated as the product of power without a regulator on the ratio of the maximum possible efficiency to the efficiency of the frequency regulator:

$$N_1 = N * \left(\frac{\eta_{max}}{\eta(Q)} \right)$$

Substitute already have found expressions for N, $\eta_{max} and \eta(Q)$:

$$N_1 = \rho * g * \left(A + B \cdot Q + C \cdot Q^2\right) * Q * \left(\frac{A_1 - \dfrac{B_1^2}{4 * C_1}}{A_1 + B_1 * Q + C_1 * Q^2} \right)$$

Substitute all the values for the capacities in the formula for calculating the efficiency coefficient K_{ef}:

$$K_{ef} = \frac{T * \rho * g * \left(A + B \cdot Q + C \cdot Q^2\right) * Q * W_1}{T * \rho * g * \left(A + B \cdot Q + C \cdot Q^2\right) * Q * \left(\frac{A_1 - \dfrac{B_1^2}{4 * C_1}}{A_1 + B_1 * Q + C_1 * Q^2} \right) * W_1 + W}$$

We can substitute the previously calculated pump volume flow rate Q into the formula for the final result.

The economic efficiency coefficient may be less than, equal to, or greater than one. We can conclude whether it is profitable to use a frequency controller for the pump for this operating system, depending on the coefficient value.

If $K_{ef} < 1$, when the using of a frequency regulator is not profitable, because the cost of it and its operation will be more than the system without regulation.

Else if $K_{ef} > 1$, when the using of the frequency control method is advantageous for this operating system.

Else $K_{ef} = 1$, when using the regulator doesn't matter.

Results

During the study working formula was obtained for determining the coefficient of economic efficiency of frequency regulator, which depends on many parameters, including the pump flow rate, the cost of kW, the cost of the regulator, the time resource of the pump, its constants, as well as the density of the liquid in the system. The ultimate formula is as follows:

$$K_{ef} = \frac{T * \rho * g * \left(A + B \cdot Q + C \cdot Q^2\right) * Q * W_1}{T * \rho * g * \left(A + B \cdot Q + C \cdot Q^2\right) * Q * \left(\frac{A_1 - \dfrac{B_1^2}{4 * C_1}}{A_1 + B_1 * Q + C_1 * Q^2} \right) * W_1 + W}$$

where A, B, C — constants of pumps without frequency regulator;

A_1, B_1, C_1 —constants of pumps with frequency regulator;

α —the constant;
T — the time resource of the pump, [T]=[hour];
W_1 —the kW electricity per hour cost;
W — the frequency regulator cost;
ρ —the density of the fluid in the system
g —acceleration of gravity;

We can substitute the previously calculated pump feed value into the formula for the final result. We can draw a conclusion about the economic efficiency of using this method, depending on the value of the efficiency.

Reference

[1] A Gouskov et al 2019 *IOP Conf. Ser.: Mater. Sci. Eng.***492** 012013
[2] N Egorkina and APetrov 2019 *IOP Conf. Ser.: Mater. Sci. Eng.***492** 012015
[3] K Dobrokhodov and APetrov 2019 *IOP Conf. Ser.: Mater. Sci. Eng.***492** 012016
[4] N Isaev 2019 *IOP Conf. Ser.: Mater. Sci. Eng.***492** 012026
[5] S Korsakova and AProtopopov 2019 *IOP Conf. Ser.: Mater. Sci. Eng.***492** 012032
[6] A Protopopov and C Jakovich 2019 *IOP Conf. Ser.: Mater. Sci. Eng.***492** 012034
[7] A Shablovskiy and E Kutovoy 2019 *IOP Conf. Ser.: Mater. Sci. Eng.***492** 012035
[8] A Petrov et al 2019 *IOP Conf. Ser.: Mater. Sci. Eng.* **492** 012036
[9] V Cheremushkin and APolyakov 2019 *IOP Conf. Ser.: Mater. Sci. Eng.***589** 012001
[10] E Morozova et al 2019 *IOP Conf. Ser.: Mater. Sci. Eng.* **589** 012008
[11] N Isaev et al 2019 *IOP Conf. Ser.: Mater. Sci. Eng.***589** 012009
[12] A Boyarshinova et al 2019 *IOP Conf. Ser.: Mater. Sci. Eng.***589** 012028

IOP Conf. Series: Materials Science and Engineering **779** (2020) 012019 doi:10.1088/1757-899X/779/1/012019

Obtaining of the optimal speed of the centrifugal pump

A Shablovskiy[1,2], A Makhlaeva[1] and O Shikova[1]
[1]Bauman Moscow State Technical University

[2]E-mail: shabl@bmstu.ru

Abstract: The work is devoted to the finding of the formula for calculating the optimal speed of the centrifugal pump. The structure, applications and main types of this vertical tank and its advantages in comparison to the horizontal tank are described. The figuring out of the formula is based on the conditions of maximum pump efficiency and it allows us to quickly assess the degree of non optimality of the pump when its operating frequency differs from the optimal frequency.

Introduction
The concept of "house" efficiency as a characteristic of the most optimal operation of the centrifugal pump can often be found in the literature. However, how can you ensure optimal operation of the centrifugal pump [1–5] if it is already selected in a particular hydraulic system and does not fall into the necessary range? This can be achieved mainly by frequency regulator [6–12]. In this work the formula for calculating the optimal rotation frequency of the rotor of a centrifugal pump is obtained.

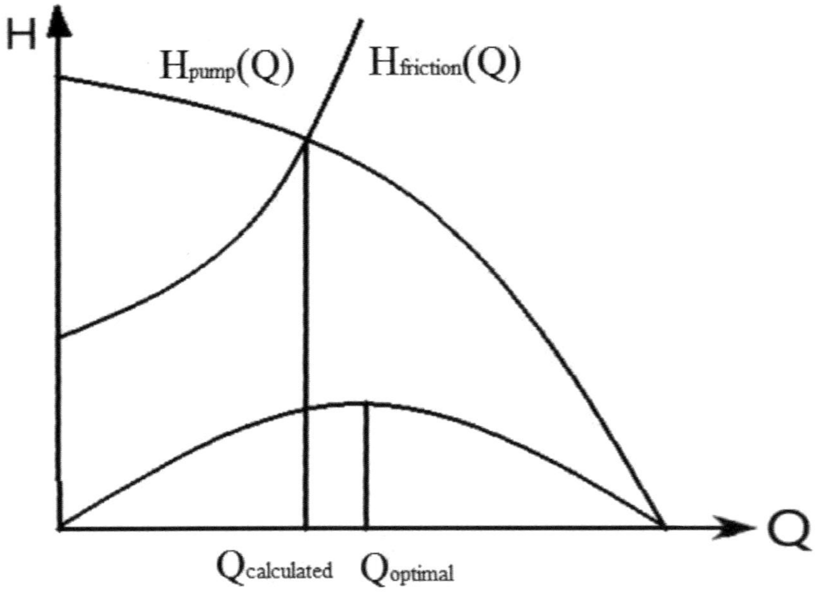

Figure 1. The "house" efficiency

Content from this work may be used under the terms of the Creative Commons Attribution 3.0 licence. Any further distribution of this work must maintain attribution to the author(s) and the title of the work, journal citation and DOI.
Published under licence by IOP Publishing Ltd

IOP Conf. Series: Materials Science and Engineering **779** (2020) 012019 doi:10.1088/1757-899X/779/1/012019

Mathematical model

$$H_{pump}(Q) = A + B \cdot Q + C \cdot Q^2 \tag{1}$$

where A, B, C — pump constants, obtained from the approximation of its hydraulic head characteristic.

Also the hydraulic head at the working point is equal to:

$$H_{friction} = H_{static} + \alpha \cdot Q^2 \tag{2}$$

where $\alpha = 0{,}0827 \frac{x \cdot l}{d^5}$ — the constant.

If we equate the above hydraulic heads and express flow rate Q, we get:

$$Q_{calculated} = \frac{-B \pm \sqrt{B^2 - 4(\alpha - C)(H_{static} - A)}}{2(\alpha - C)} \tag{3}$$

Analysis of the obtained formula for flow rate Q to the sign before the square root:

$$\alpha > 0,\ C < 0 \tag{4}$$

It means:

$$\alpha - C > 0 \tag{5}$$

$$H_{pump}(0) = A \tag{6}$$

$$H_{static} < A \tag{7}$$

Or:

$$H_{static} - A < 0 \tag{8}$$

Then:

$$\sqrt{B^2 - 4(\alpha - C)(H_o + y - A)} = \sqrt{B^2 + 4\left|(\alpha - C)(H_o + y - A)\right|} > |B| \tag{9}$$

$$Q(y) = \frac{-B + \sqrt{B^2 - 4(\alpha - C)(H_o + y - A)}}{2(\alpha - C)} \tag{10}$$

Similarly for the efficiency curve:

$$\eta(Q) = A_1 + B_1 \cdot Q + C_1 \cdot Q^2 \tag{11}$$

where A_1, B_1, C_1 — the constants of the approximately efficiency curve. Differentiate this expression:

$$\eta'(Q) = 2 \cdot C_1 \cdot Q + B_1 \tag{12}$$

Equate equation (12) to 0 to find the maximum efficiency, and we get the following formula:

$$Q_{optimal} = -\frac{B_1}{2 \cdot C_1} \tag{13}$$

We get following expression from the theory of similarity of centrifugal pumps:

$$\frac{n_2}{n_1} = \frac{Q_{optimal}}{Q_{calculated}} \tag{14}$$

where n_1 — frequency at optimal efficiency;

n_2 — initial frequency;
$Q_{calculated}$ — calculated volume flow rate.
Then it turns out, taking into account the formulas (14), (10) and (13):

$$n_2 = n_1 \cdot \frac{-B_1 \cdot 2 \cdot (\alpha - C)}{2 \cdot C_1 \cdot \left(\sqrt{B^2 - 4(\alpha - C)(H_o + y - A)} - B\right)} \tag{15}$$

Conclusion

If the centrifugal pump is selected in a particular hydraulic system so that it is not in the range of the "house" of efficiency, but there is a possibility of frequency control, when the above formula (15) allows us to determine the optimal speed of the pump rotor, providing maximum efficiency.

Reference

[1] A Protopopov and D Bondareva 2019 *IOP Conf. Ser.: Mater. Sci. Eng.* **492** 012002
[2] A Protopopov and V Vigovskij 2019 *IOP Conf. Ser.: Mater. Sci. Eng.* **492** 012003
[3] P Chaburko and Z Kossova 2019 *IOP Conf. Ser.: Mater. Sci. Eng.* **492** 012011
[4] V Lomakin et al 2019 *IOP Conf. Ser.: Mater. Sci. Eng.* **492** 012012
[5] A Gouskov et al 2019 *IOP Conf. Ser.: Mater. Sci. Eng.* **492** 012013
[6] N Egorkina and APetrov 2019 *IOP Conf. Ser.: Mater. Sci. Eng.* **492** 012015
[7] K Dobrokhodov and APetrov 2019 *IOP Conf. Ser.: Mater. Sci. Eng.* **492** 012016
[8] N Isaev 2019 *IOP Conf. Ser.: Mater. Sci. Eng.* **492** 012026
[9] S Korsakova and AProtopopov 2019 *IOP Conf. Ser.: Mater. Sci. Eng.* **492** 012032
[10] A Protopopov and C Jakovich 2019 *IOP Conf. Ser.: Mater. Sci. Eng.* **492** 012034
[11] A Shablovskiy and E Kutovoy 2019 *IOP Conf. Ser.: Mater. Sci. Eng.* **492** 012035
[12] A Petrov et al 2019 *IOP Conf. Ser.: Mater. Sci. Eng.* **492** 012036

Development of hydraulic transmission for implementing an active drive of the landing gear wheels of modern transport aircraft

V Brusov[1,2,3], A Menshikov[1,2] and Y Merzlikin[2]

[1]Bauman Moscow State Technical University
[2]The Central AeroHydrodynamic Institute named after Prof. N.E. Zhukovsky

[3]E-mail: vasyab2@rambler.ru

Annotation An active drive of the wheels of the chassis is necessary on the take-off and run with reduced coupling characteristics of the airplane on an unpaved or icy runway in cross-wind conditions to reduce lateral drift from the axis of the strip. The hydraulic transmission has been developed for the active drive of the wheels by aircraft chassis. At using a control system, that regulates the moment wheels characteristics and includes an adaptive circuit with neural network, it is possible to reduce the lateral drift of the aircraft about 30–40 %. The output parameters of the hydraulic motors intended for the landing gear wheels of aircraft weighting 21 tons are determined.

Introduction

With the development of sparsely populated and inaccessible regions of the Russian Federation, including the Arctic zone, the problem of passenger, transport and special tasks arises. This is explained by a number of reasons, among which the low density of aerodrome and road networks, the lack of economic opportunities and practical feasibility of constructing airfields with concrete coverings of runways, and also the climatic features are of decisive importance.

In many ways, the solution to this problem is associated with the development of aviation communications. For these regions, the development of aviation is directly related to the increased cross-country ability of aircraft on elementary prepared unpaved runways with irregularities at the stages of take-off and run. Under the cross-country ability of aircraft refers to the ability of the aircraft to operate safely on this type of runway airfield. Aircraft cross-country ability is limited by three factors: excessive contact loads on the landing gear support elements and flight strips, unacceptable glider overloads and loss of controllability of the aircraft due to the absence of contact connections between the landing gear and the supporting surface on take-off and landing modes [1, 2]. In this regard, the landing gear is a unit that sets the limiting indicators of the cross-country ability of any aircraft.

At the same time, the quality of cross-country ability is closely related to the takeoff and landing capabilities of the aircraft as a whole, due to the perfection of its aerodynamic layout, mass characteristics, the thrust-weight ratio of the power plant, and the effectiveness of aerodynamic controls. A significant effect on the performance of cross-country ability can have the chosen method of piloting the aircraft. It is especially noticeable on low-strength unpaved runways, when complex

Content from this work may be used under the terms of the Creative Commons Attribution 3.0 licence. Any further distribution of this work must maintain attribution to the author(s) and the title of the work, journal citation and DOI.
Published under licence by IOP Publishing Ltd

IOP Conf. Series: Materials Science and Engineering **779** (2020) 012020 doi:10.1088/1757-899X/779/1/012020

interaction processes, including slippage and slipping, are manifested in contact between the chassis and the deformable soil medium.

One of the ways to increase aircraft flotation is to use an active wheel drive in conjunction with adaptive landing gear moment control algorithms.

The most optimal option for the implementation of an active drive of the chassis wheels is the use of a hydraulic transmission with proportionally controlled elements that control the flow and pressure in the hydraulic lines. Using these elements, it is possible to control the torque of each wheel of the aircraft chassis [3]–[5].

The torque control algorithm of hydraulic motors is carried out based on the following parameters:

1) Specific power spent on movement;

2) The ratio of the useful work of the traction force on the wheel to the perfect work of the torque supplied to the wheels;

3) Wheel slipping;

4) Specific traction and specific energy losses during rolling of the front and main landing gear wheels.

An active drive of the wheels of the chassis is necessary on the take-off and run with reduced coupling characteristics of the airplane on an unpaved or icy runway in cross-wind conditions to reduce lateral drift from the axis of the strip.

The article discusses the prospect of using a hydraulic drive as a transmission element for aircraft chassis wheels (Fig. 1).

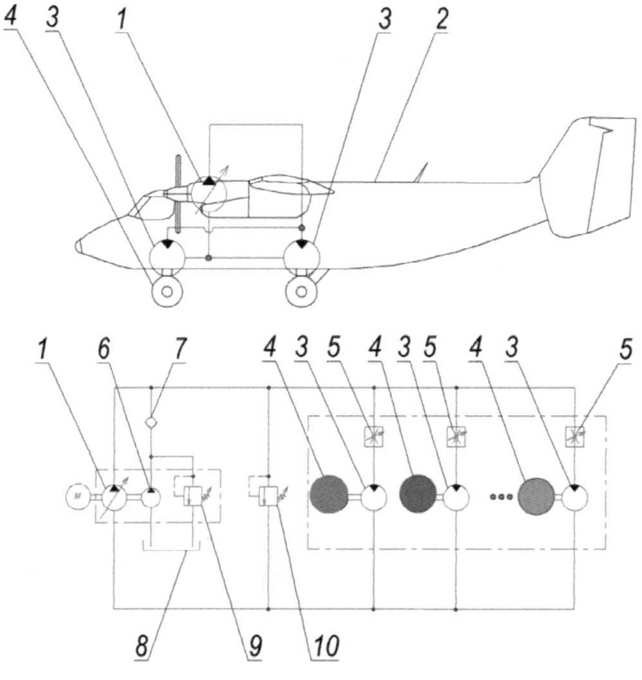

Fig. 1. Variant of adjustable hydraulic transmission of the aircraft for driving of the wheels of the landing gear. Designated: 1 — adjustable pump; 2 — airplane glider; 3 — hydraulic motors; 4 — landing gear wheels; 5 — flow controllers; 6 — make-up pump; 7 — check valve; 8 — tank; 9 — safety valve; 10 — overflow valve.

Using a single hydraulic system to create an effective power transmission of an aircraft provides the ability to control its parameters and characteristics. At the same time, the developers of the hydraulic system are faced with the task of optimal control moving along various types of surfaces with a wide range of speeds.

In transport equipment with a large number of uncertainties and random disturbances, it is advisable to use control systems with several adaptive circuits operating at different speeds to control the object [6]–[9].

The proposed methods for controlling the moment characteristics of the wheels of the chassis can create a theoretical and methodological basis for a new generation of ground-based transport aircraft with increased cross-country ability.

Methods

The study of controlled aircraft movement is carried out on the basis of a mathematical model that is usually used to assess the characteristics of static stability and controllability of an aircraft with wheeled landing gears during take-off (run) along a horizontal unpaved runway. The equations of the model reflect only three degrees of freedom of movement of the aircraft: movement in the longitudinal and transverse directions and angular motion relative to its vertical axis (Fig. 2). The equations use well-known forms of representing the aerodynamic forces and moments from the airframe of the aircraft.

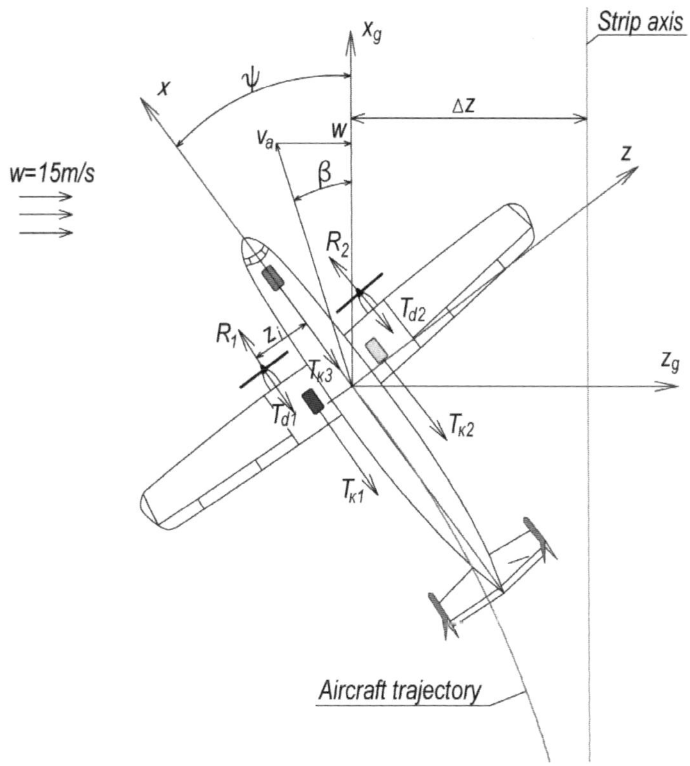

Fig. 2. Diagram of power actions during the take-off of the aircraft in crosswind conditions.

The ability to limit only three degrees of freedom of its movement for describing the lateral movement of an aircraft is due to the fact that in the case of cross-connections introduced into the lateral movement by the aerodynamic rudders and brake elements of the chassis with their real force effects, they lead only to negligible changes in the angle of heel and pitch, not significantly affecting the accuracy of calculations.

The main assumptions and provisions adopted in the construction of the mathematical model:

1) Consider the rectilinear motion of an aircraft on a flat soil base;

2) The relationship of the wheels with the aircraft fuselage in the vertical plane is considered taking into account the elastic and damping properties of the shock absorption system and the longitudinal compliance of the guide elements of the landing gear is taken into account;

3) The fuselage is undeformable;

4) The friction in the joints, bearings is negligible;

5) The engine torque acts directly on the wheel, the elastic-damping properties of the transmission elements are not taken into account;

6) The magnitude of the torque is determined by the fraction of the change in position of the control;

7) Take into account the possibility of separation of the wheels from the ground.

If you use the trajectory $Ox_k y_k z_k$, associated $Oxyz$ and the normal earth $O_0 x_g y_g z_g$ coordinate systems when compiling the differential equations of lateral motion of the aircraft, the differential equations can be reduced to the following form:

$$\dot{v}_n = \frac{1}{m}\cdot\left(X_{\sum}\cdot\cos\beta_a + Z_{\sum}\cdot\sin\beta_a\right)$$

$$\dot{\beta} = \frac{1}{m\cdot v_a}\cdot\left(Z_{\sum}\cdot\cos\beta - X_{\sum}\cdot\sin\beta\right) + \omega_y$$

$$\dot{\omega}_y = \frac{M_{y\sum}}{I_y}$$

$$\dot{\psi} = \omega_y$$

$$\dot{x}_g = v_n\cdot cos\psi_n$$

$$\dot{z}_g = v_n\cdot\sin\psi_n,$$

where: v_n — ground speed of the aircraft; v_a — airspeed of the aircraft; β — slip angle; β_a — slip angle in absolute motion; m — mass of the aircraft; I_y — moment of inertia of the aircraft about the axis Oy; ω_y — angular velocity of yaw; ψ — yaw angle; ψ_n — air speed yaw angle.

Between kinematic parameters there are the following relationships:

$$\beta_a = \psi - \psi_n;$$

$$v_a = \sqrt{\dot{x}_g^2 + \left(\dot{z}_g + \omega\right)^2};$$

$$\psi_n = \arcsin\left(\frac{\dot{z}_g}{\sqrt{\dot{x}_g^2 - \dot{z}_g^2}}\right).$$

In the general case, for an aircraft with N_d marching engines, N_m brake elements (wheeled landing gear), the forces and moments in the equations are presented in the form of the following expressions:

–projection of the resultant force on the axis associated with the aircraft Ox:

$$X_{\sum} = \left(\sum_{i=1}^{N_d} T_{dixg} + \sum_{k=1}^{N_m} T_{mkxg}\right)\cdot\cos\psi - \left(\sum_{i=1}^{N_d} T_{dizg} + \sum_{k=1}^{N_m} T_{mkzg}\right)\cdot\sin\psi;$$

–projection of the resultant force on the axis associated with the aircraft Oz:

$$X_{\sum} = Z + \left(\sum_{i=1}^{N_d} T_{dizg} + \sum_{k=1}^{N_m} T_{mkzg}\right)\cdot\sin\psi + \left(\sum_{i=1}^{N_d} T_{dixg} + \sum_{k=1}^{N_m} T_{mkxg}\right)\cdot\cos\psi;$$

–total moment relative to the vertical axis Oy:

$$M_{y\sum} = M_m + M_d + M_y.$$

Here: $Z = c_z^{\beta} \cdot \sin\beta \cdot \rho \cdot \frac{v_a^2}{2} \cdot s + c_z^{\delta_n} \cdot \delta_n \cdot \rho \cdot \frac{v_a^2}{2} \cdot s$ — transverse aerodynamic force of the aircraft; $X = c_x \cdot \rho \cdot \frac{v_a^2}{2} \cdot s$ — longitudinal aerodynamic force of the aircraft; T_{mkzg}, T_{mkxg} — projection on the $O_0 z_g$ axis and $O_0 x_g$ axis the friction forces of the brake elements, calculated using the ratios:

$$T_{mkxg} = -T_{mk} \cdot \frac{\dot{x}_{mkg}}{v_{mk}};$$

$$T_{mkzg} = -T_{mk} \cdot \frac{\dot{z}_{mkg}}{v_{mk}};$$

$$v_{mk} = \sqrt{\dot{x}_{mkg}^2 + \dot{z}_{mkg}^2};$$

$$\dot{x}_{mkg} = \dot{x}_g + z_{mkg} \cdot \omega_y;$$

$$\dot{z}_{mkg} = \dot{z}_g - x_{mkg} \cdot \omega_y;$$

$$x_{mkg} = \sqrt{x_{mk}^2 + z_{mk}^2} \cdot \sin(\xi_{mk} + \psi);$$

$$z_{mkg} = \sqrt{x_{mk}^2 + z_{mk}^2} \cdot \cos(\xi_{mk} + \psi);$$

moreover, in the last two formulas — for wheeled landing gear on the left side ($k < \frac{N_т}{2}$): $\xi_{mk} = \pi - \arcsin\left(\frac{x_{mk}}{\sqrt{x_{mk}^2 + z_{mk}^2}}\right)$; and for the landing gear located on the right side($k > \frac{N_т}{2}$): $\xi_{mk} = \arcsin\left(\frac{x_{mk}}{\sqrt{x_{mk}^2 + z_{mk}^2}}\right)$; $T_{mk} = \frac{m \cdot g - Y}{N_m} \cdot f_m$,

where: f_m —coefficient of friction of the wheeled landing gear on the surface of the runway; T_{dizg}, T_{dixg} — projection on the $O_0 z_g$ and $O_0 x_g$ axis the forces of the impulse resistance of the engines.

The moments due to the resistance of the brake elements in contact with the surface of the runway, as well as the moments of the impulse resistance of the engines are:

$$M_m = \sum_{i=1}^{N_m} T_{ixg} \cdot z_{ig} - \sum_{i=1}^{N_m} T_{ixg} \cdot x_{ig}.$$

The aerodynamic moment of the aircraft is determined by the ratio:

$$M_y = m_y \cdot \frac{\rho \cdot v_a^2}{2} \cdot s \cdot l + m_y^{\beta} \cdot \sin\beta \cdot \frac{\rho \cdot v_a^2}{2} \cdot s \cdot l + m_y^{\dot{\beta}} \cdot \dot{\beta} \cdot \rho \cdot v_a \cdot s \cdot \frac{l^2}{4} + m_y^{\omega_y} \cdot \omega_y \cdot \rho \cdot v_a \cdot s \cdot \frac{l^2}{4} + m_y^{\delta_n} \cdot \delta_n \cdot \frac{\rho \cdot v_a^2}{2} \cdot s \cdot l$$

For the appearance of various torques on the wheels of the aircraft landing gear in order to reduce lateral drift during a take-off run with a crosswind, it is necessary to apply a signal to the pressure regulators (item 5, Fig. 1). Then the braking forces can be written in the form of the following relations [10]–[13]:

$$T_{\kappa i} = T_{\kappa 0} + sign(\Delta z) \cdot (z - z_0) \cdot k_{zi} \cdot th\left[\frac{z}{z - z_0} \cdot \frac{t}{t_{\max}}\right] + sign(\Delta \psi) \cdot (\psi - \psi_0) \cdot k_{\psi i} \cdot th\left[\frac{\psi}{\psi - \psi_0} \cdot \frac{t}{t_{\max}}\right],$$

where: k_{zi}, $k_{\psi i}$ — the feedback coefficients adopted in the control of the aircraft during take-off when controlled by a wheeled landing gear.

These coefficients can be varied using the adaptive neural circuit implemented in the control system [14]–[21].

Results

Using the described equations, a mathematical model of the movement of the aircraft on an unpaved runway has been developed. Design studies of the aircraft on take-off in lateral wind conditions were carried out. The lateral drift values of the aircraft were obtained at maximum thrust for different wind speeds (Fig. 3). The studies were conducted for type Il-112 aircraft with a take-off weight of 21 tons.

Using an additional controller with elements of a neural network to control the active drives of the wheels of the landing gears allows to reduce lateral drift of the aircraft from the runway:

–at a cross-wind speed of 5 m / s — an average of about 40%;

–at a crosswind speed of 10 m / s — an average of about 36%;

–at a crosswind speed of 15 m / s — an average of about 30%.

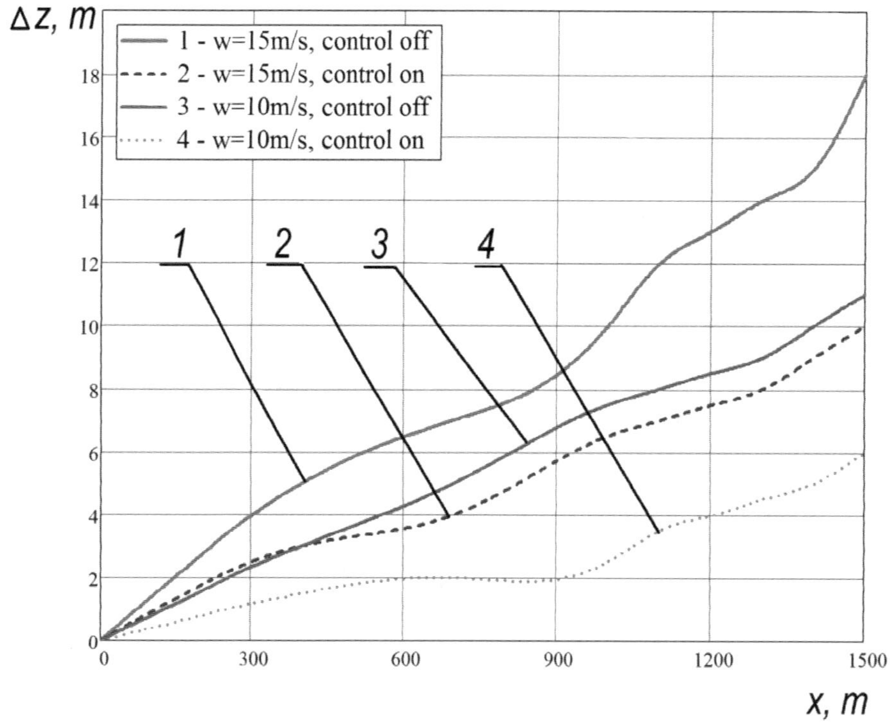

Fig. 3. Computational studies of the takeoff run of an Il-112 type aircraft. Lateral drift at maximum traction for different wind speeds.

Conclusions

The urgent tasks of expanding the conditions for the deployment of promising aircraft, increasing the cross-country ability and safety of the grounding of new aircraft can be successfully solved only on the basis of a sufficiently detailed study of the characteristics of the "elastic aircraft - unprepared runway" system. At the same time, the results of comprehensive studies on the mechanics of loading the chassis on deformable and uneven runways, on the aerodynamics of aircraft in the take-off and landing configuration and the dynamics of their take-off and landing using special cross-country ability criteria should be consistent even at the early stages of aircraft design.

The developed design conditions for the flotation of airplanes, including external ground-based conditions, are subject to standardization and refinement as experience accumulates in the creation and operation of appropriate aircraft.

Mathematical modeling of the movement of an aircraft equipped with active drives of the wheels of the landing gears on an icy and / or snowy runway with randomly changing characteristics in the conditions of side wind at the stages of take-off and run allowed optimization of the developed algorithms to counteract the drift of the aircraft away from the runway axis.

The developed torque control algorithm on hydraulic motors significantly reduced lateral drift of the aircraft from the strip.

The required ranges of changes in the torque of the drive (hydraulic motor) of the landing gear wheel are determined. In particular, for an aircraft of the Il-112 type:

–range of moments - (0 ... 450) Nxm;

– range of rotation frequencies - (100 ... 500) rpm (up to 3000 rpm - on the take-off run to take-off speed).

To increase the safety of take-off and landing of the aircraft on unprepared runways and to further reduce the offset from the axis of the runway in conditions of strong crosswind, especially at high speed (at the end of the take-off phase and at the beginning of the run), it is advisable to introduce a control loop using aerodynamic surfaces of aircraft.

References

[1] A. Bondarets, O. Kreerenko. Using of Artificial Neural Networks (ANN) for Aircraft Motion Math Model Parameters Identification. EANN 2009, CCIS 43, 2009. pp. 246–256.

[2] O.D. Kreerenko. Adaptive control of the nonlinear dynamic object at the stage of breaking under indefinite contact surface conditions. Book of Abstracts CHAOS 2010, Chania, Creete (Greece), 2010. p. 43.

[3] O.F. Nikitin, "Machine regulation volume hydraulic drive with reciprocating motion of hydraulic motors" *2015 International Conference on Fluid Power and Mechatronics (FPM)*, Harbin, 2015, pp. 593-595. doi: 10.1109/FPM.2015.7337185.

[4] D.N Popov, N.G. Sosnovskiy. Automatic pressure regulation at the input of the main line pump during emergency cutout of an electric motor. Science and Education: Scientific Publication of BMSTU, 2013, no. 1, pp. 29-38 (in Russ.). doi: 10.7463/0113.0518041.

[5] Jong-Hycok Kim, Chag-SooJeon, Yeh-Sun Hong. "Constant pressure control of a swash plate type axial piston pump by varying both volumetric displacement and shaft speed", International Journal of Precision Engineering and Manufacturing, October 2015, Volume 16, Issue 11, pp. 2395–2401.

[6] O.I. Chudakov, V.A. Gorelov, E.B. Sarach. Improving traction and active safety of the wheeled vehicle by the distribution of the driving torque between its axles.IOP conference series: materials science and engineering 2019. pp. 72–80.

[7] G.O. Kotiev, A.A. Stadukhin, B.B. Kositsyn, A.V. Miroshnichenko A.V. Determination of mechanical characteristics of high-speed tracked vehicles traction motor with individual drive wheels. Journal of Physics: Conference Series 2019. p. 012058.

[8] N.V. Buzunov, G.O. Kotiev, V.A. Gorelov. Implementation of the interaction of the steering wheel loader control system of the remote-controlled wheeled vehicle operator interface with

a real-time simulation modEL. IOP Conference Series: Materials Science and Engineering 2019. pp. 101–113.

[9] N.V. Buzunov, G.O. Kotiev, B.V Padalkin. Wheel vehicle dynamics real-time simulation for on-board stand-alone moving control system realization. IOP Conference Series: Materials Science and Engineering 2019. pp. 204–214.

[10] A.S. Diakov, G.O. Kotiev. Establishment of production of special wheel and track technology for extreme natural-climate conditions of the arctic. MATEC Web of Conferences 2018. p. 02096.

[11] E. Sarach, G. Kotiev, S. Beketov. Methods for road microprofile statistical data transformation. MATEC Web of Conferences 2018. p. 04009.

[12] M.M. Zhileykin, G.O. Kotiev, M.V. Nagatsev. Comparative analysis of the operation efficiency of the continuous and relay control systems of a multi-axle wheeled vehicle suspension. IOP Conference Series: Materials Science and Engineering FSUE "NAMI". 2018.

[13] V.E. Klubnichkin, E.E. Klubnichkin, G.O. Kotiev, S.A. Beketov, V.S. Makarov. Interaction between elements of the track ground contacting area with the soil at curvilinear motion of the timber harvesting machine. IOP Conference Series: Materials Science and Engineering 2018. p. 012016.

[14] S. Chen, S. Billings. Neural networks for non-linear dynamic system modeling and identification. Int. J. Control. 1992. Vol. 56.no. 2. pp. 319–349.

[15] S. Geman, E. Bienenstock, R. Doursat. Neural networks and the bias/variance dilemma. Neural Computation. 1992. Vol. 4. pp. 1–58.

[16] J. Larsen, L. Hansen. Generalization performance of regularized neural network models. Neural Networks for Signal Processing: Proc. of the IEEE Workshop IV. Brussel (Belgium). 1994. Vol. 66. pp.2396–2399.

[17] M.W. Pederson, L.K. Hansen. Recurrent networks: second order properties and pruning. Neural Information Processing Systems: Proc. of the 7-th Conference. Vienna (Austria), 1994. pp. 673–680.

[18] J. Sjoberg, L. Ljung. Overtraining, regularization and searching for the minimum in neural networks. Adaptive Systems in Control and Signal Processing: Preprint IFAC Symposium. Grenoble (France). 1992. pp. 669–674.

[19] B. Widrow. The original adaptive neural net broom — balancer. Proc. IEEE Int. Symp. Circuits and Sysrems. Phil (USA). 1987. pp. 351–357.

[20] R.V. Monopoli. Model reference adaptive control with an augmented error signal. IEEE Trans. On Automat. Control. 1975. V.25. no. 3. pp.474–484.

[21] K.S. Narendra, L.S. Valavani. A comparison of Lyapunov's and hyperstability approaches to adaptive control of continues systems. IEEE Trans. Automat. Confr. 1980. AC-25. no. 2. pp.243–247.

IOP Conf. Series: Materials Science and Engineering **779** (2020) 012021 doi:10.1088/1757-899X/779/1/012021

The Dynamic Characteristics of Hydraulic-Mechanical Constant Speed Drive

Y Vul[1], R Kalyakin[1,2] and A Shablovskiy [1]

[1]Bauman Moscow State Technical University

[2]E-mail: defendly@gmail.com

Annotation: This paper discusses the development of a mathematical model for the constant speed drive, comprising the power supply of an aircraft power grid. The model is developed in the form of equations of dynamics of the drive main components, such as differential, volumetric hydraulic transmission, centrifugal regulator, auxiliary pumps, as well as using the AMESim calculation package. The model has been verified using experimental data of the AC frequency responses of the existing unit to various influences. The models obtained allow to conclude about the stability of the system, and to optimize the control system parameters.

Introduction

Three-phase AC voltage with frequency of 400 Hz is required to power on-board electrical system of most aircrafts. The most commonly used power supply of this type of electricity is a bunch of alternator and a constant speed drive called 'integrated generator drive' (GD). Due to the fact that aircraft developers need to increase the power of the power source and strict requirements to its weight and size properties [1], the best solutions in the design of the unit are required. To select the optimal parameters of the unit and calculate its stability, its mathematical model is necessary. Mathematical modeling of hydraulic-mechanical devices is an important problem [2]–[6].

Research on the choice of an optimal power source for the power supply systems of aircraft is still underway. Papers [7] conclude that there is no single solution for the choice of aircraft power generation systems. For each individual case, a structural and functional method of analysis and synthesis of electric power systems is required.

The study of the constant speed drive operation in the range of zero revolutions of a hydraulic transmission is demonstrated in the paper [8]. In the paper [9], generator model is used to detect and prevent its faults during the flight.

The GD control system ensures the frequency constancy of current with a deviation from the rated value of no more than ± 2 % at a change of loading in the grid, ambient temperature, or rotation velocity of the aircraft engine to which the input shaft of GD is connected via an aircraft accessory box.

The GD principle of operation is as follows. Mechanical energy at the input is divided into two streams, one of which goes through a mechanical transmission with a constant transfer coefficient, while the other goes through a hydrostatic drive with a variable transfer coefficient. Then, both streams are summed up using a differential gear, which results in the maintenance of a constant rotation frequency of a generator shaft within the tolerance regardless of the engine shaft speed, load in the grid, or ambient temperature. Control of the hydrostatic drive is performed by a centrifugal

Content from this work may be used under the terms of the Creative Commons Attribution 3.0 licence. Any further distribution of this work must maintain attribution to the author(s) and the title of the work, journal citation and DOI.
Published under licence by IOP Publishing Ltd

IOP Conf. Series: Materials Science and Engineering **779** (2020) 012021 doi:10.1088/1757-899X/779/1/012021

regulator and an associated servo piston that moves the inclined washer of the adjustable hydraulic machine. For more details about the operation of the constant speed drive, refer to paper [10].

Method

We will study the dynamic characteristics of GD using its mathematical model. This requires to compile the dynamics equations for the main assemblies of the unit and utilize AMESim application package which allows to perform one-dimensional multi-physical modeling [11–13]. The main assemblies of the unit include a generator, differential gear, hydraulic machine block, centrifugal regulator, low-pressure hydraulic system with booster and exhaust pipes.

The object of regulation in this problem is the frequency of alternating current, but taking the response speed of the electric part much more than the one of the constant speed drive, one can assume that the current frequency and generator rotor speed are connected by a constant coefficient. Therefore, let us write down the rotor motion equation:

$$J\cdot\frac{d\omega}{dt}=M_d-M_g-M_p-M_r\ ,where$$

J is the moment of inertia of the moments of inertia of the moving elements of the drive brought to the generator shaft; M_d is the driving torque; M_e is the torque on the generator rotor generated by the electric field; M_p is the torque on the crown gear being generated by the auxiliary pumps; M_r is the moment of mechanical loss brought to the output shaft of the drive.

$$M_g=\frac{N\cdot cos\varphi}{\eta_g\cdot\omega},where$$

N is the load in power grid; ω is the rotor speed; η_g is the efficiency factor of generator.

The differential gear is a planetary gear which input link (pinion carrier — *n*) is connected to the input shaft of the drive. Output link - crown gear (*b*) transmits rotation to the generator shaft, sun gear (*a*) rotation from the hydraulic machine block. Pinion mates (*g*) and (*f*) being engaged with each other are connected by internal engagement to the crown and sun gears respectively. Axes of pinion mates are fixed in the pinion carrier. Figure 1 illustrates the kinematic scheme of the differential gear.

The relationship between the torques on the crown and sun gears of the differential gear is as follows:

$$M_b=M_a\cdot\left|i_{ab}^H\right|-M_{loss}\cdot\left(-1\right)^{\alpha},where$$

i_{ab}^H is the gear ratio between the sun and crown gears when the pinion carrier is stopped; α is the coefficient characterizing the mode of operation of the generator drive; M_{loss} is the moment of mechanical loss in the differential gear.

Kinematic relation for the planetary gear:

$$i_{ab}^{н}=\frac{\omega_b-\omega_н}{\omega_a-\omega_н},where$$

$\omega_н,\omega_b,\omega_a$ are the rotation speeds of the pinion carrier, the crown, and the sun gears respectively

The hydraulic machine block consists of two axial-plunger hydraulic machines of variable specific supply (HM_1) and constant specific supply (HM_2) connected by a hydraulic flow through a common distribution spool with two cavities of high and low pressures. The HM1 hydraulic machine receives rotation from the pinion carrier, and its inclined washer can turn from an angle of $+\gamma$ to $-\gamma$. The gear of the HM_2 hydraulic machine shaft is engaged to the sun gear of the differential gear.

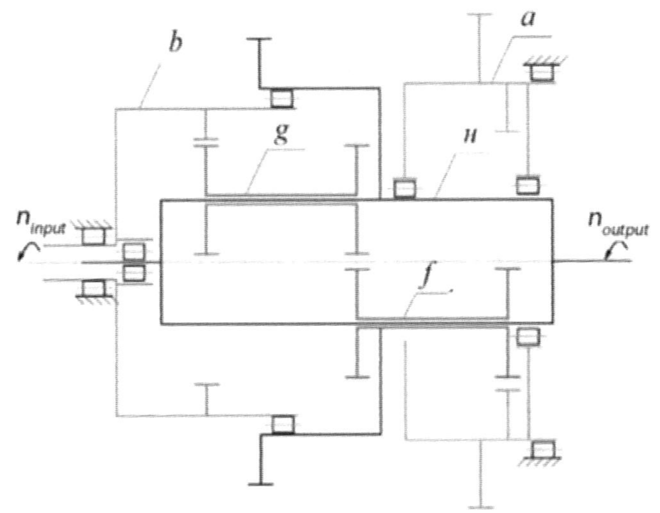

Fig.1. Kinematic scheme of the differential gear

If we do not take into account the compressibility of a liquid in the distribution spool, because of its small capacity, then the relationship between the movements of the hydraulic machines HM_1 and HM_2 will be written in the form as follows:

$$\frac{q_1 \cdot \gamma_1}{2\pi} \cdot \omega_1 = \frac{q_2 \cdot \gamma_2}{2\pi} \cdot \omega_2 + Q_l, where$$

q_1, q_2 is the supply of the HM1 and HM2 hydraulic machines per one shaft revolution, and unit of angles γ_1, γ_2 of inclined washers respectively; ω_1, ω_2 are the rotation velocities of the HM_1 and HM_2 hydraulic machines.

The leakage rate is determined by a linear dependence with coefficient κ_l on the delivery pressure p_p:

$$Q_l = \kappa_l \cdot p_p$$

A detailed model of the applied hydrostatic drive in AMESim considering the processes in each working chamber separately, taking into account the kinematics of an axial-piston hydraulic machine, the size of the dead volume, and resistance in the hydraulic lines is demonstrated in paper [14–15].

The low pressure (1.5...2 MPa) hydraulic system supplies a working fluid to the hydraulic machine block, compensating for leaks to the regulator spool and servo cylinder, as well as to the sliding bearings of the pinion mates and the generator cooling line.

The torque on the crown gear from the auxiliary pumps can be found from the power consumption expression. For the booster pump:

$$N_n = \frac{q_n \cdot \Delta p_v \cdot n_b}{\eta_v}, where$$

q_n is the displacement of the booster pump; Δp_v is the drop of the generated pressure; η_v is the efficiency factor of the booster pump.

The regulator consists of a centrifugal mechanism that receives rotation from the crown gear, spool, and tuning spring. The weights of the centrifugal mechanism rotate and are able to rotate freely relative to the points of its suspension affected by the centrifugal forces when changing the rotation speed of a sleeve while moving the spool (Figure 2). The regulator spool controls a servo piston moving the inclined washer of the GM_1 hydraulic machine.

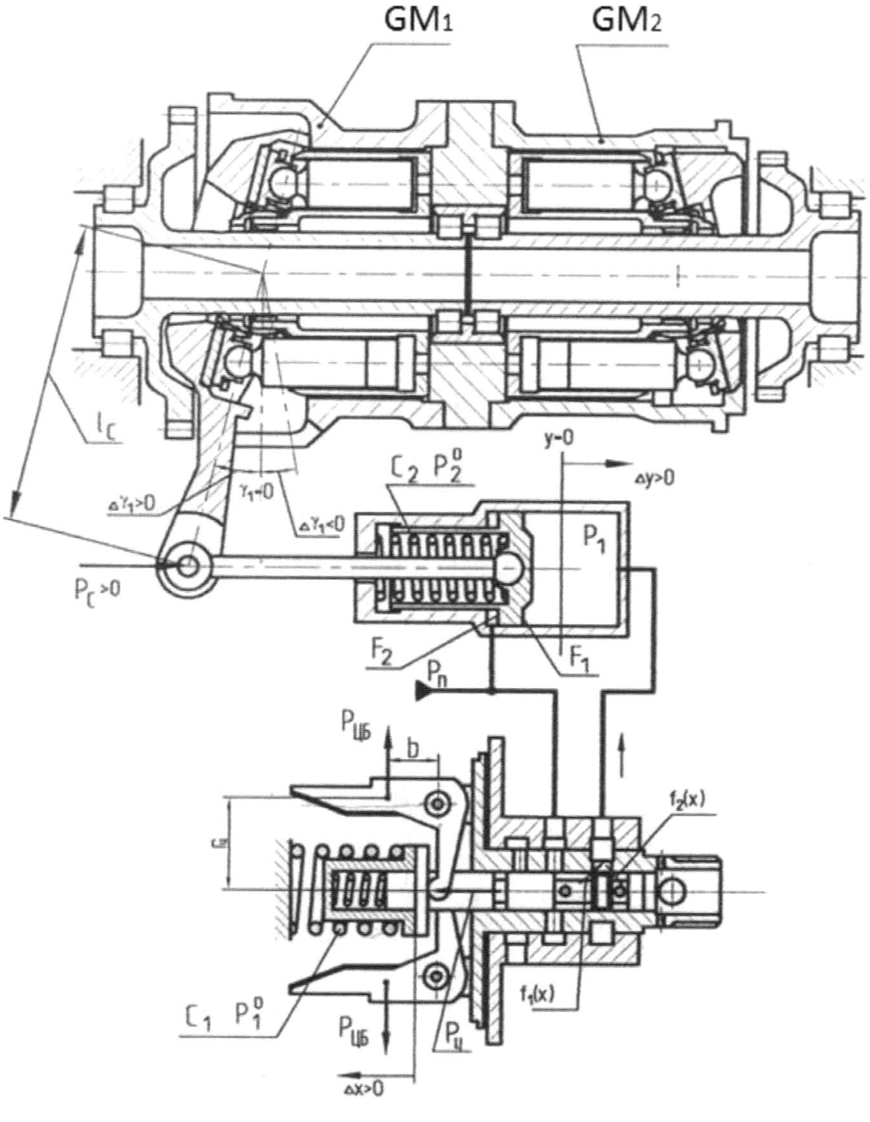

Fig.2. The diagram of the generator drive control system

The spool motion equation without taking into account the weight of spool and friction force:

$$P_1^0 + c_1 \cdot x = 2 \cdot m \cdot r_u \cdot \frac{b}{a} \cdot \omega_p^2, where$$

P_1^0 is the spring force in position of tuning of a regulator to a rated frequency (400 Hz); c_1 is the spring stiffness; x is the spool travel; r_u is the distance between the centroid of the weights and spool axis; b is the distance between the centroid of the weights and rotation axis of the weights; a is the distance between the rotation axis of the weights to the centrifugal force application point; ω_p is the rotational frequency of the weights relative to the spool axis.

The servo piston motion equation:

$$P_2^0 + c_2 \cdot y + P_c = F_1 \cdot p_1 - F_2 \cdot p_n \ , where$$

P_2^0 is the initial spring force; c_2 is the spring stiffness; F_1, F_2 is the area of the servo piston from the control pressure p_1 and booster pressure p_n; P_c is the force acting on the servo piston from the inclined washer.

The equation of the relationship between the movements of the regulator spool and the servo piston:

$$F_1 \cdot \frac{dy}{dt} = \mu \cdot \sqrt{\frac{2}{\rho}} \cdot f_1(x) \cdot \sqrt{p_n - p_1} - \mu \cdot \sqrt{\frac{2}{\rho}} \cdot f_2(x) \cdot \sqrt{p_1 - p_k}\text{, } where$$

$f_1(x), f_2(x)$ are the areas of clear openings of slots in the regulator sleeve on the pressure and drain edges when moving the spool; μ is the discharge ratio; ρ is the density of working fluid.

The relationship between the movement of the servo piston and the inclined washer with sufficient accuracy:

$$y = -\ell_c \cdot \gamma\text{, } where$$

ℓ_c is the length of the lever of the inclined washer

The presented equations can be combined into one system, forming a mathematical model of GD:

$$\begin{cases} J\dfrac{d\omega}{dt} + \kappa_3\omega = (\kappa_1 + 2\kappa_2)\kappa_2\kappa_l\omega_n - \kappa_2\kappa_l p_n - \dfrac{N cos\varphi}{\eta_g \omega} - M_n - M_{loss}(-1)^\alpha \\ P_1^0 + c_1 \cdot x = 2 \cdot m \cdot r_u \cdot i_p^2 \cdot \dfrac{b}{a} \cdot \omega^2 \\ P_2^0 - c_2 \cdot \ell_c \cdot \gamma + P_c = F_2 \cdot p_1 - F_1 \cdot p_n \\ \dfrac{d\gamma_1}{dt} = \dfrac{-1}{\ell_c \cdot F_2} \cdot \left(\left(f_1(x) \cdot \sqrt{p_n - p_1} \right) - \sqrt{p_1 - p_k} \cdot \mu \cdot \sqrt{\dfrac{2}{\rho}} \right) \end{cases} \text{, } where$$

$$\kappa_1 = \frac{q_1 \cdot \gamma_1}{2\pi \cdot \kappa_l} \cdot i_{21}; \quad \kappa_2 = \frac{q_2 \cdot \gamma_2}{2\pi \cdot \kappa_l} \cdot i_{34}; \quad \kappa_3 = \left(\frac{q_2 \cdot \gamma_2}{2\pi\kappa_l} \cdot i_{34} \right)^2 \cdot \kappa_l = \kappa_2^2 \cdot \kappa_l;$$

The analysis of the above model is conducted by its linear model, which is obtained by linearization of the system equations in the vicinity of a certain steady-state regime. By replacing variables with their increments, and derivatives with Laplace operators, and omitting intermediate transformations, we will obtain:

$$\begin{cases} (S + \kappa_{31}) \cdot \Delta\omega = \kappa_4 \cdot \Delta\omega_n - \kappa_N \cdot \Delta N - \kappa_\gamma \cdot \Delta\gamma_1 \\ \Delta x = \kappa_x \Delta\omega \\ S\Delta\gamma_1 = D\Delta x \end{cases} \text{, } where$$

$$\kappa_{31} = \frac{\kappa_3}{J}; \quad \kappa_4 = (\kappa_1 + 2\kappa_2) \cdot \kappa_2 \cdot \kappa_l \cdot \frac{1}{J}; \quad \kappa_N = \frac{cos\varphi}{\eta_r \omega_0 J};$$

$$\kappa_\gamma = \frac{q_1 \cdot i_{21}}{2 \cdot \pi} \cdot \frac{\kappa_2}{J} \cdot \omega_0; \quad \kappa_x = \frac{4 \cdot m \cdot i_p^2 \cdot \dfrac{b}{a} \cdot \omega_0 \cdot r_u^0}{c_{1-} 2m \cdot i_p^2 \cdot \left(\dfrac{b}{a}\right)^2 \cdot \omega_0^2}; \quad D = \frac{\sigma}{\ell_c F_1};$$

$$\sigma = \mu \cdot \sqrt{\frac{2}{\rho}} \cdot \left(\frac{df_1^0(x)}{dx} \cdot \sqrt{p_n - p_1^0} - \frac{df_2^0(x)}{dx} \cdot \sqrt{p_1^0 - p_k^0} \right)$$

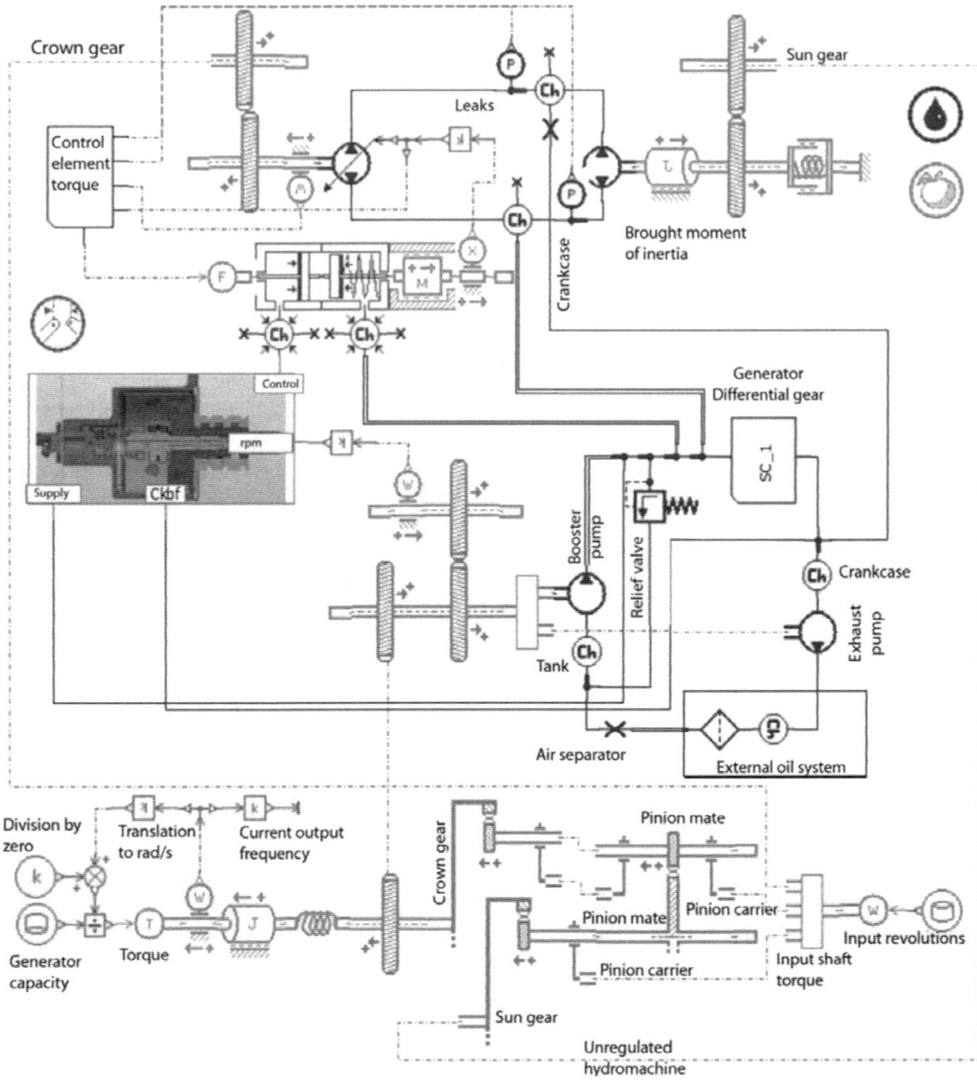

Fig.3. The generator drive model in AMESim

A single equation of this system, which relates the change in the rotation speed of the generator rotor with changes in speed of aircraft engine shaft and the load in the power grid, is an equation of the 2nd order:

$$\left(S^2 + \kappa_{31} S + \beta_c^2\right) \cdot \Delta\omega = \kappa_{н} \cdot S \cdot \Delta\omega_{н} - \kappa_N \cdot S \cdot \Delta N, where$$

$\beta_c = \sqrt{\kappa_\gamma \cdot \kappa_x \cdot D}$ is the frequency of undamped natural oscillations.

The attenuation coefficient is determined by the equality:

$$\xi = \frac{\kappa_{31}}{2\beta_c}$$

In the expression of interest, the attenuation coefficient ξ is always greater than zero. Therefore, according to the Nyquist stability criterion, analytical model of GD is stable under any parameters of the unit.

IOP Conf. Series: Materials Science and Engineering **779** (2020) 012021 doi:10.1088/1757-899X/779/1/012021

The oscillability characterized by the attenuation coefficient largely depends on the magnitude of leaks from the power circuit of the hydraulic machines, i.e. the coefficient κ_{ym}.. It is possible to obtain the following dependence:

$$\xi = f\left(\frac{1}{\sqrt{\kappa_l}}\right)$$

Since the friction pairs in the hydraulic machine wear out during operation, the leakage coefficient κ_l will increase over time, and at the same time the oscillability of the system will increase.

When generating a system model in the AMESim environment, numerous libraries of pre-tested components from various physical domains can be used. Thus, the GD model represents the blocks of gearing, hydrolines, inertial components, etc. connected to each other in a similar way to a real configuration of the unit. In contrast to the analytical GD model, in this case, the forces of dry and viscous friction, weights of the spool and servo piston, compressibility of liquid, and nonlinearity of the system in the form of the restriction of the rotation angle of the inclined washer, etc. are taken in account. Figure 3 illustrates the final model.

Results

To verify the models obtained, test data of the existing unit with rated power of 120 kW for a heavy cargo aircraft were applied. Figures 4 and 5 illustrate the responses of the AC frequency to the changes in the load in the test bench power grid and the input shaft rotation velocity. For the analytical model, the absence of a static error is true, while the value of the initial frequency jump and the nature of the transition process when the load changes in the power grid is close to the experimental data.

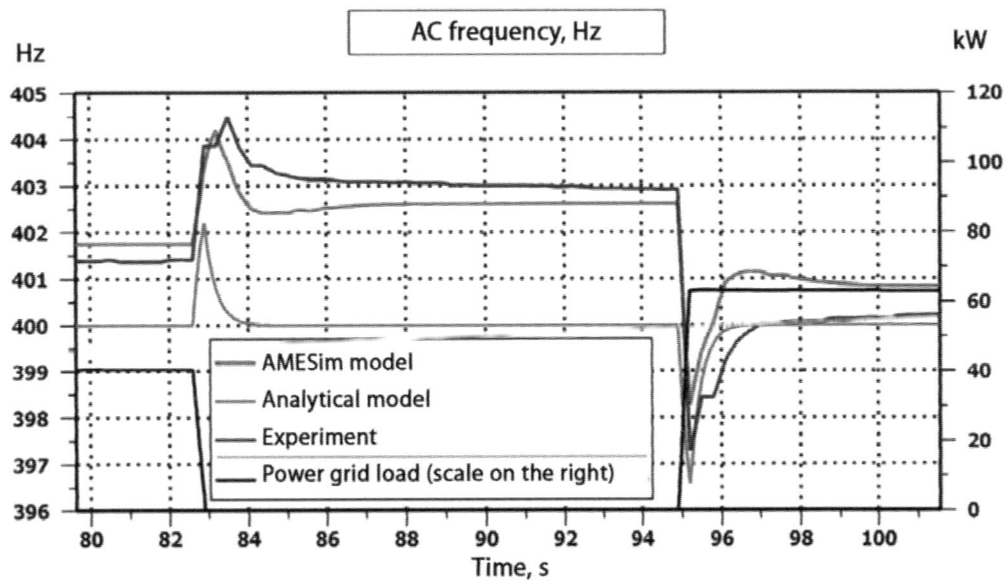

Fig.4. The response of AC frequency to a change of the load in power grid

The model prepared in AMESim has good convergence with the test data for both impacts. The value of the static error strongly depends on the size of gaps appeared during manufacture, level of filling, quality of execution of a particular unit, thus it is inexpediently to perform a comparison on steady-state modes.

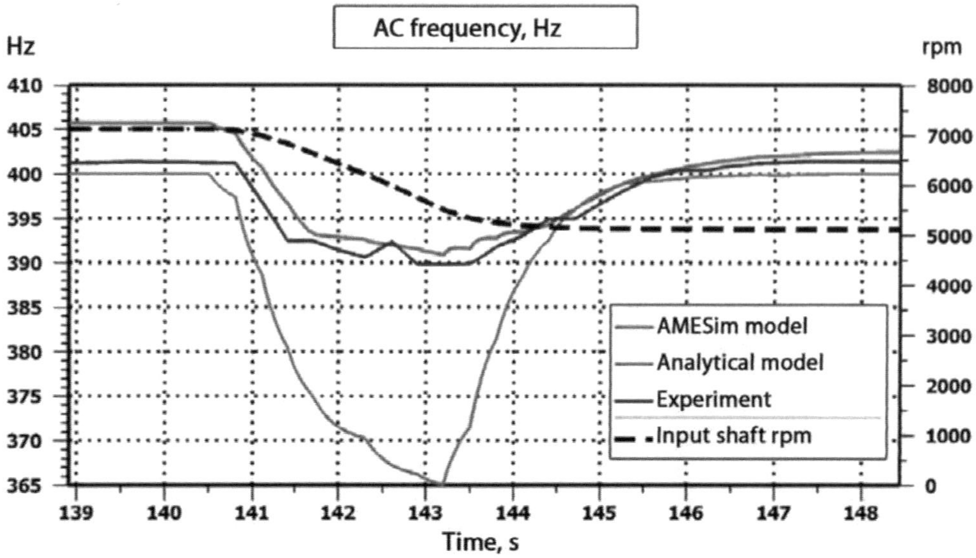

Fig. 5. The AC frequency response to a change inaircraft engine rpm

Conclusion

The analysis of the linear mathematical model has shown unconditional stability of the unit that is confirmed by the long-term experience of operation of the generator drives. The second conclusion is that the oscillability of the transient processes depends on the leakage coefficient and increases throughout the service life. However, cases of frequency mismatch due to severe wear and tear are extremely rare.

Using a model that takes into account the forces of dry and viscous friction, weight of the spool and servo piston, liquid compressibility, as well as the nonlinearity of the system allows to obtain results close to experimental data. Thus, the resulting model can be used as a tool for observing variables that are not available for direct measurement and used to synthesize the parameters of the regulator in order to obtain optimal transients.

Add in gelements to the model that take into account thermal processes, as well as application of the CFD approach to calculate the internal flow of the working fluid is the next stage of the research.

References

[1] S. Guo, J. Chen, Y. Lu, Y. Wang, H. Dong. Hydraulic piston pump in civil aircraft: Current status, future directions and critical technologies, Chinese Journal of Aeronautics, 2019, https://doi.org/10.1016/j.cja.2019.01.013.

[2] I Kolodin and M Ryabinin 2019 IOP Conf. Ser.: Mater. Sci. Eng. 589 012018

[3] Semenov, S., Kulakov, D. Mathematical modeling of the mechanisms of volumetrichydraulic machines (2019) IOP Conference Series: Materials Science and Engineering,492 (1), article № 012042, .

[4] Semenov, S.E. Mathematical modeling of the electro-hydraulic actuation systems of themachines with tree-like kinematic structure (2015) Proceedings of 2015 InternationalConference on Fluid Power and Mechatronics, FPM 2015, article № 7337184, pp. 583–592.

[5] D Kulakov and B Kulakov 2019 IOP Conf. Ser. Mater. Sci. Eng. 492 012029

[6] N Sosnovsky and D Ganieva 2019 IOP Conf. Ser.: Mater. Sci. Eng. 589 012016

[7] A. Eid, H. El-Kishky, M. Abdel-Salam, T. El-Mohandes. Power Quality Investigations of VSCF Aircraft Electric Power Systems. 42nd South Eastern Symposium on System Theory, USA, March 7–9, 2010.

[8] M. Mikhaylov, S. Stazhkov, V. Cvetkov.The Method of Disturbances Analysis on the Hydromechanical Constant Speed Drive. Procedia Engineering. 69.2014, 199–202. https://doi.org/10.1016/j.proeng.2014.02.221.
[9] T. Jing , Y. Chengyu ,Y. Yaowen, S.Xudong. Simulation and Fault Detection for Aircraft IDG System. Procedia Engineering. 15. 2533–2537. (2011) https://doi.org/10.1016/j.proeng.2011.08.476.
[10] Babaev, O.M.; Ignatov, L.N.; Kistochkin, E.S. &Cvetkov, V.A. (2000). Hydromechanical power transmissions, Mechanical engineering, St.-Petersburg, Russia.
[11] Fornarelli F., LippolisA.,OrestaP., PosaA., A computational model of axial piston swashplate pumps, Energy Procedia, Vol. 126, 2017, pp.1147–1154.
[12] Junjun Z., Lingling J. Optimal Design of Hydraulic Position Servo-system Based on AMESim and Genetic Algrhythm. Machine Tool and Hydraulic,2008, p.86–89.
[13] Yongling F., XiaoyeQ.AMESim System Modeling and Simulation,BeiJing:Bei Jing University of Aeronautics Press, 2005.
[14] Kalykin R., Perfil'evA.,DruzhkovD.,Elimination Of Erosion Phenomena Of A Cylinder Block Of Aircraft Axial Piston Hydraulic Machine, «Nasosy. Turbina. Sistemy»№4(29), 2018
[15] Alessandro C., Massimo R., Comparison of 0D and 3D Hydraulic Models for Axial Piston Pumps, Energy Procedia, Volume 148, 2018, Pages 114–121, https://doi.org/10.1016/j.egypro.2018.08.038.

IOP Conf. Series: Materials Science and Engineering **779** (2020) 012022 doi:10.1088/1757-899X/779/1/012022

Review of two-line hydraulic drive research

O Nikitin[1,2] and Zhang He Xuan[1]

[1]Bauman Moscow State Technical University

[2]E-mail: nof1936@yandex.ru

Abstract. The article provides a brief overview of literary sources from the PRC on the study of a hydraulic drive with two-line control. This hydraulic drive is a new electro-hydraulic system for regulating the flow of a positive displacement pump, developed on the basis of control systems for the electric motor and the volumetric displacement of an adjustable positive displacement pump. The hydraulic drive has two lines of control variables: the rotational speed of the drive motor shaft and the adjustable displacement of the pump. The composite function of the two control lines allows you to create a hydraulic actuator with high efficiency and operating speed. An advanced hydraulic drive with two-line control increases the overall natural frequency of the entire system.

Introduction

The proposed review is based on 10 scientific and technical articles of 3 scientific research organizations and 3 Chinese universities, published over the past 10 years.

The hydraulic drive has a number of advantages such as high energy intensity, short response time, high positioning rigidity and convenient adjustment. These advantages of the hydraulic drive are confirmed by the widespread use in world technology for over a hundred years. However, relatively large energy losses, especially in the range of low speeds (low efficiency), limit the further use of hydraulic systems including at high power. The application experience shows that the useful output power of the hydraulic system, especially the throttle control system, maximally amounts to no more than 20 ... 30% of the total input power of the system, and the rest of the energy is dissipated in the form of heat.

In the second half of the twentieth century, a hydraulic drive scheme was widely used, which included an asynchronous electric motor, a positive displacement pump with variable displacement using an electro-hydraulic mechanism that make up one control line, and a hydraulic motor. This hydraulic circuit is relatively simple in structure, relatively energy-saving and efficient, has a low cost of use and maintenance.

The flow rate of the positive displacement pump is determined by the product of the pump displacement $V_{0н}$ and the rotational speed n of the pump shaft $Q = V_{0н}n$. [1] It can be seen that a displacement hydraulic drive in combination with an electric motor and a hydraulic pump with variable displacement can have two control lines with two variable parameters — the output motor shaft rotation speed and the pump displacement. It allows you to obtain an adjustable output (speed) characteristic of the hydraulic actuator according to two control variables simultaneously and separately, while the control actions of the two variables can cancel each other out. [2, 3] The proposed scheme involves increasing efficiency, higher positioning rigidity and other dynamic characteristics.

Content from this work may be used under the terms of the Creative Commons Attribution 3.0 licence. Any further distribution of this work must maintain attribution to the author(s) and the title of the work, journal citation and DOI.
Published under licence by IOP Publishing Ltd

IOP Conf. Series: Materials Science and Engineering **779** (2020) 012022 doi:10.1088/1757-899X/779/1/012022

However, even at present, the second double-sided control line is most often used, which provides control of the speed of the output shaft of the hydraulic motor by changing the flow rate of the pump by changing the volumetric displacement.

Two-line hydraulic drive

A hydraulic drive with two-line control is a new type of hydraulic drive developed on the basis of a traditional line (first) for regulating the displacement of the pump and a line (second) for regulating the frequency of rotation of the motor shaft. [4, 5]

The principle of a hydraulic drive with two-line control is shown in Fig. 1. The electric motor drives a hydraulic adjustable pump. The controller in real time controls the output speed of the motor shaft (power unit) and the pump displacement, which ultimately controls the movement of the hydraulic motor. Changing the pump flow is carried out using a pump displacement control mechanism.

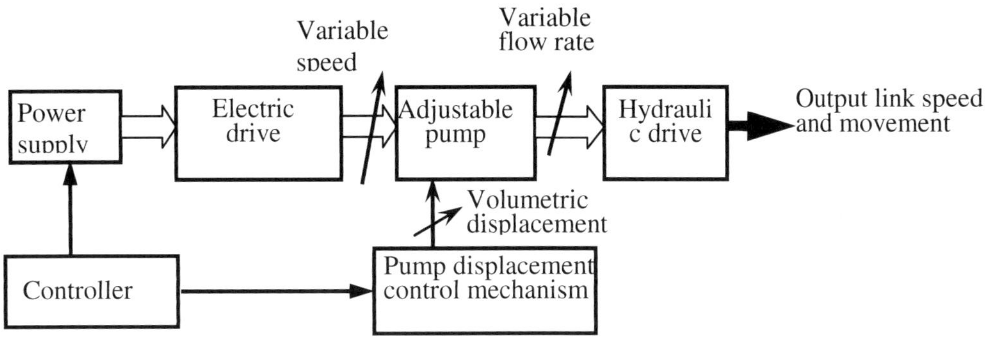

Fig.1. Schematic diagram of a hydraulic actuator with two-line control and two variable parameters

Schematic design of a hydraulic actuator with two-line control

In a hydraulic drive with two-line control (with two variable parameters), an axial piston pump with an adjustable displacement is used (Fig. 2). This type of hydraulic pump has a compact design, low moment of inertia, high working pressure and high volumetric efficiency. [6]

To ensure the accuracy of the pump supply, the vast majority of adjustable pumps use an electro-hydraulic mechanism with a proven circuit, for which the working fluid comes from the discharge line of a controlled pump or a specially equipped auxiliary pump. The block diagram of the electro-hydraulic mechanism is relatively complex. In addition, all this not only reduces the reliability of the entire control system, but also brings large energy losses during throttling of the flow of the working fluid in the electro-hydraulic mechanism. [7]

In order to improve the characteristics of the pump flow control mechanism, an electric drive mechanism was developed, shown in Fig. 2, which includes an electric motor and mechanical gear links.

The proposed scheme of the electric drive for regulating the pump displacement (the first control line) not only increases the performance parameters of the entire hydraulic actuator, but also facilitates integration and increases the level of integration of the control system with a computer system. For such a non-linear system, the difficult task arises - to conduct a relatively perfect simulation, analysis of modeling and research of the control strategy in full. The use of an electric drive eliminates energy losses during throttling of the working fluid flow in the electro-hydraulic mechanism, providing dynamic characteristics of the adjustable mechanism, which increases the energy efficiency and reliability of the executive system.

The use of a ball screw in an electric drive allows you to convert the rotational movement of the shaft into reciprocating motion of a nut that changes the angle θ_p of inclination of the support disk.

Since the dead zone of a ball screw is usually small, this suggests that the process of reversing movement will occur mainly without play (clearance), eliminates non-linearity in the dead zone during switching of the suction and discharge channels of the pump.

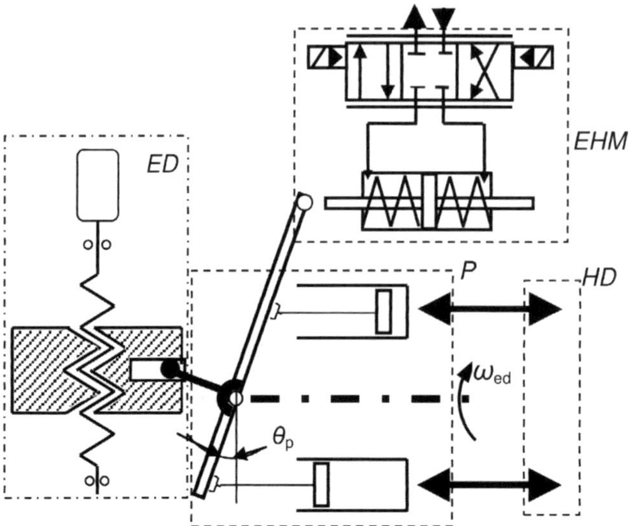

Fig. 2. Schematic diagram of a hydraulic drive with two-line control with two variable parameters; P — adjustable pump; EHM — electro-hydraulic mechanism; ED — electric drive; GD — hydraulic drive

The hydraulic drive with two-line control with two variable parameters there is no need for any switching components. [8] It is necessary not only to change the flow rate of the regulated pump, but also to achieve continuous adjustment of the operating system's gear ratio. The unique composite - compensation function - adjusting the shaft speed of the drive motor of this hydraulic drive and the control strategy of the pump displacement control mechanism can compensate for the defects and shortcomings of the hydraulic drive with a single source variable (regulating the pump flow rate only by changing the pump displacement), and it is expected that it will mainly solve the problem of reducing the energy consumption of the hydraulic drive. Solving the problem of reducing energy consumption and heating the working environment of the hydraulic drive significantly improves the dynamic characteristics of the executive system and further expands the bandwidth. It should be noted that the value of the product of the motor shaft rotation frequency and the volumetric displacement of the adjustable hydraulic pump determines the flow rate of the output flow entering the hydraulic motor. The same product allows you to determine the amount of power. The control effects of two variables can still compensate for each other.

It should be noted that a single-line hydraulic drive has typical non-linear characteristics. The dual control system has many non-linearities of its own, which complicates the nonlinear characteristics of a hydraulic drive with two-line control. For such a non-linear system, it is a difficult task to perform an ideal simulation, analyze the simulation, and study the control strategy.

Although the hydraulic drive with two-line control with two variable parameters has its advantages, consisting in convenient power transmission and simple control, it still has some differences from the traditional control system (with an electro-hydraulic mechanism) in the speed of operation and control accuracy. The proposed scheme of the electric drive for regulating the working volume of the pump (the first control line) not only increases the performance parameters of the entire hydraulic drive, but also facilitates integration and increases the level of integration of the control system with the computer system.

IOP Conf. Series: Materials Science and Engineering **779** (2020) 012022 doi:10.1088/1757-899X/779/1/012022

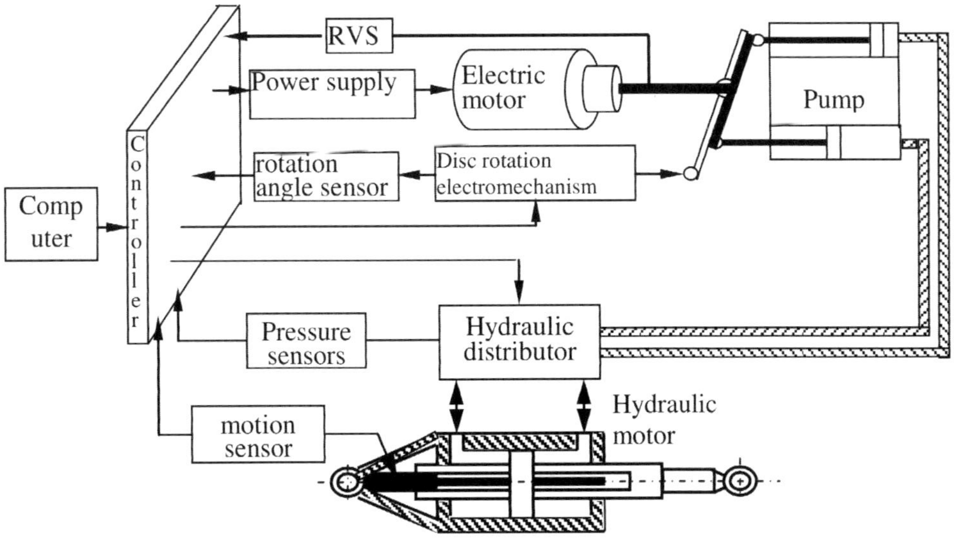

Fig.3. Schematic diagram of an improved hydraulic drive with two-line control

When introducing the second control line into the hydraulic drive (changing the motor shaft speed), it was found that in order to increase the controllability and the values of output characteristics, the pressure in the power discharge line must be increased, and the sum of the pressures in the high and low pressure hydraulic lines of the executive hydraulic motor must be maintained at a constant $p_{dis} + p_{suct} = 2p_{up}$. To ensure this mode of operation of the hydraulic drive, a hydraulic distributor was introduced into its structure, and an improved scheme is shown in Fig. 3. This brings the hydraulic drive with two-line control with two variable parameters in the operating state with a hydraulic drive with a single control line (electrohydraulic control), effectively forms the mechanical characteristic, increases the stiffness and improves the natural frequency and dynamic characteristics of the hydraulic drive. [8,9] The same effect is noted in the Russian sources [10, 11, 12].

During the research, it was noticed that when the feed values and pressures change in the hydraulic lines of the regulated pump, there is an improvement in the continuous adjustment of the transfer functions of the operating system. The unique composite (compensation function) - adjusting the shaft speed of the drive motor of this hydraulic drive and the control strategy — of the pump displacement control mechanism allows you to compensate for the shortcomings of the hydraulic drive with a single source variable (regulating the pump supply by changing the pump volume), and expect that these measures mainly solve the problem of reducing the energy consumption of the hydraulic drive. As a result, solving the problem of reducing energy consumption and heating the working environment of the hydraulic drive significantly improves the dynamic characteristics of the executive system and further expands the bandwidth. [13,14]

Thus, the existing literature mainly indicates the main direction of efforts to improve the structural scheme of the hydraulic drive with two-line control, including a new design of the electric drive mechanism that regulates the pump flow, and the improvement of the two-line control circuit.

Choosing a method for modeling the drive system

In a hydraulic drive with two-line control with two variables (the speed of the motor shaft and the working volume of the hydraulic pump), the parameters are multiplied in the working process and the value of the supply of the regulated pump is determined. At the same time, the hydraulic drive is affected by factors such as changes in the parameters of the hydraulic system, external load disturbances, and temperature changes. All this significantly complicates mathematical modeling and simulation analysis of a two-line hydraulic drive with two variable parameters. [15, 16] Currently, the commonly used methods for modeling a hydraulic drive are the transfer function, state space, and connection graph methods. Among these methods, the transfer function method is the most common method of modeling in theoretical analysis. This method - often called the local linearization method - is used to work with internal change parameters in the simulation process and ignores some complex or difficult to describe model dynamics. The mathematical model obtained by this method is a model of linearization of the object of study.

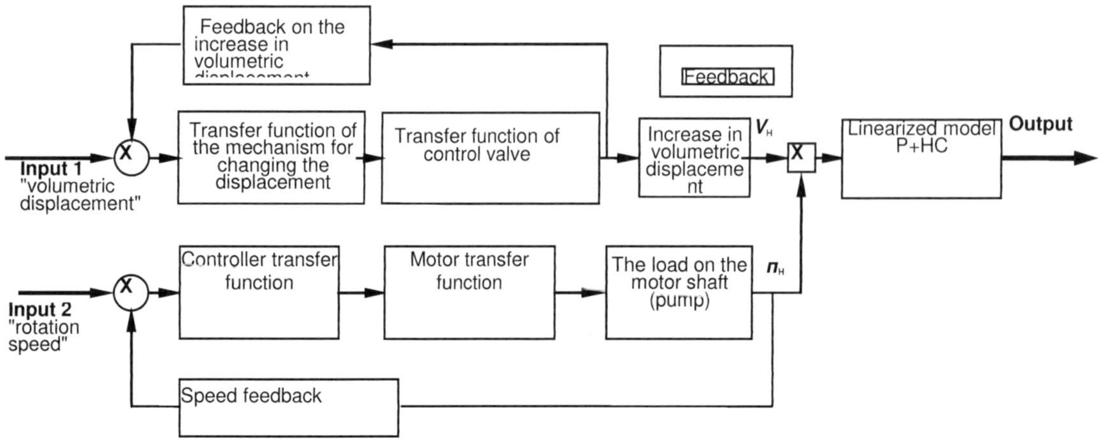

Fig. 4. Two-line hydraulic drive transfer function model

Based on the established linear mathematical model, it is convenient to use the classical control theory to analyze and design the executive system to achieve the required performance. The obtained model of the linearized transfer function of a hydraulic actuator with two-line control, shown in Fig. 4, is used to describe the operating characteristics of the control system.

Studies have established that the transfer function method is not completely suitable for a hydraulic actuator with two-line control with strong nonlinear connection.

The state space method is a widely used modeling method in modern control theory and is suitable not only for traditional systems with a linear constant with one input and one output, but also for non-stationary systems, nonlinear systems, and systems with several inputs and several outputs. When the state space method is used to describe the mathematical model of the object of study, it is no longer necessary to consider in detail non-linear characteristics, such as non-linearity, high order and their change in time, so this can effectively avoid the disadvantages of the modeling process of the transfer function method.

Using the idea of block modeling in the state space method in accordance with the working mechanism of each module, the corresponding differential equations are established and the differential equations of each module are integrated, a model of the state space of the hydraulic drive with two-system control is formed (Fig. 5) with a clear expression of the relationships between different state variables inside the system.

In fig. 5, the input quantities ***u1, u2*** represent two control variables of a hydraulic actuator with two-line control with two variables, respectively. The state variables ***x1, x2, x3, ..., x11*** represent the rotational speed of the motor shaft at the input of the hydraulic pump shaft; rotation angle and angular speed of rotation of the motor shaft at the input of the displacement control mechanism (speed and displacement of the moving nut); angular velocity and angle of rotation of the inclined disk; acceleration, speed and movement of the hydraulic cylinder. Output ***y*** represents the movement of the output link of the hydraulic cylinder. The values ***a1, a2,..., a13, b1, b2*** and ***d1, d2*** are the matrix of coefficients, the input matrix, and the parameters of the output matrix of the state space model, respectively.

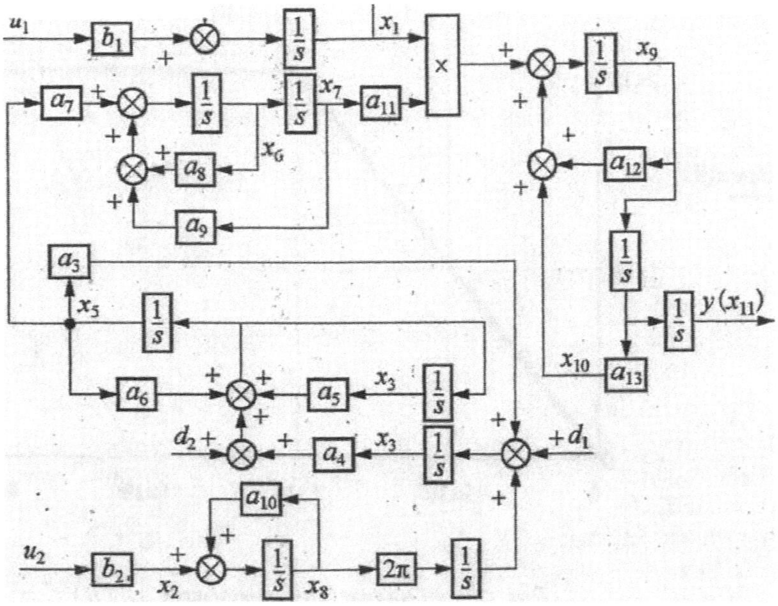

Fig. 5. Spatial model of the state of a hydraulic actuator with two-line control

The Bond graph method and co-simulation technology can take advantage of the simulation of various software platforms, simplify the modeling work, significantly increase the modeling efficiency and more accurately describe the performance of an electrostatic activation system with two variable parameters. This is a method of modeling and simulation with great development prospects. [17, 18]

Results and discussion

Thus, the reviewed literature gives a complete picture:

–of the determining directions of efforts of researchers of a hydraulic actuator with two-line control with two parameters:

– of improvement of the structural diagram of the hydraulic drive, including the new scheme and design of the electric drive mechanism that regulates the pump flow, and, accordingly, the improvement of the two-line control circuit;

–of the analysis of the main problems of improving the dynamic characteristics, energy efficiency and service life.

Conclusion

This hydraulic actuator is a new electro-hydraulic system developed on the basis of drive electric motor control systems and pump displacement control.

An advanced hydraulic actuator with two-line control increases the overall natural frequency of the entire system.

The composite function of the two control lines allows you to create a hydraulic drive that has the advantages of high energy efficiency and operational speed.

References

[1] **O.F. Nikitin O.F., V.V. Yarots**Fundamentals of hydraulics and hydropneumatic drive, Moscow: Publishing house of BMSTU, 2019. — 490 p.

[2] Review of research on dual-variable electro hydrostatic actuation system. ***Feng Lijun, Yan Hao, Dong Lijing, Ren Yukai.*** *Joumal Beijing Jiatong University.* 2018, 42. №4, 104–112. (in Chinese)

[3] ***Krasovskii, A.B.*** An analysis of the conditions for obtaining constant output power in a switched reluctance electric drive (2002) *Electrical Technology Russia,* 2002 (1)

[4] Scheme analysis of power-by-wire airborne actuation systems. ***QI Xiaoye, FU Yongling, WANG Zhanlin.****Joumal of Beijing University of Aeronautics and Astronautics,* 1999, 25(4):426–430. (in Chinese)

[5] Research on dual-variable integrated electro hydrostatic actuator [J]. ***GAO B, FU Y L, PEI Z C, et al.*** *Chinese Journal of Aeronautics,* 2006, 19(1):77–82.

[6] Modeling and AMESim simulation of electrohydraulic compound regulating integrated actuator [J]. ***LANGYan, LI Yunhua.****ActaAeronauticaEtAstronauticaSinica,* 2007, 28(S):152–157. (in Chinese)

[7] The researches of state space modeling method and dynamic properties for double-variable electro hydraulic servo control system [D]. ***LIU Erdong.****Xi'an:Xi'an University of Technology,* 2015. (in Chinese)

[8] Performance analysis based on AMESim and MATLAB co-simulation for dual-variable electro-hydrostatic actuator [J]. ***JI Tieling, QI Haitao, LIU Zilong.*** *Chinese Hydraulics & Pneumatics,* 2015(5):51–57. (in Chinese)

[9] **Engineering. Encyclopedia.** T. IV-2. Electric drive. Hydro and vibro drives. In 2 vol. Vol. 2. Hydro and vibro drives. M .: Engineering. 2012.304 p .;

[10] Analysis and modeling of the multi-disciplinary coupling characteristics in EHA[C]//***LI K, WANG SP.*** *International Conference on Fluid Power and Mechatronics. Beijing,* 2011:602–608.

[11] ***Vorontsov, A.L.*** Theoretical aspects of technological mechanics: 1. Fundamentals of continuum mechanics (2013) *Russian Engineering Research*, 33 (4), pp. 206–210. DOI: 10.3103/S1068798X13040205

[12] ***Barmin, S, Nikitin O.F.***Difinition of positive displacement output regulation mechanism parameters with providing of constant flow capacity in output. *IOP Conference Series: Materials Science and Engineering* 2019. — Vol. 589, Issue 1. — Art.no 012020 https://iopscience.iop.or...57-899X/589/1/012020 DOI: 10.1088/1757-899X/589/1/012020

[13] Motor speed robust control with multiplying nonlinear property [J]. ***BAI Guochang, QI Xiaoye, WANG Zhanlin.****Joumal of Beijing University of Aeronautics and Astronautics,* 2008, 34(7):803-806. (in Chinese)

[14] ***Myagkov, L.L., Strizhov, E.E., Malastovskii, N.S.*** Modeling of the thermal state of the diesel cylinder cover with allowance for liquid flow in the cooling cavity (2008) *Heat Transfer Research*, 39 (8), pp. 685–694. DOI: 10.1615/HeatTransRes.v39.i8.40

[15] Multidisciplinary modeling method and simulation for electro-hydrostatic actuator [C]//***LI K, WANG SP.****5th IEEE Conference on Industrial Electronics and Applications. Taichung,* 2010:544–548.

[16] ***Krut'ko, P.D.*** Robustly stable structures of control systems of high dynamic precision: Algorithms and dynamics of control of model objects (2005) *Journal of Computer and Systems Sciences International,* 44 (2), pp. 273–292.

[17] ***Medvedev, V.S., Romanova, T.A.*** Synthesis of control algorithms providing independence of multidimensional plant subsystems (1995) IzvestiyaAkademiiNauk. TeoriyaiSistemyUpravleniya, (1), pp. 54–71.

[18] ***Bychkov, M., Krasovsky, A.*** Taking into Account Nonlinear Properties of Switched Reluctance Machines in Electric Drive Control Algorithms (2018) 2018 10th *International Conference on Electrical Power Drive Systems,* ICEPDS 2018 — Conference Proceedings, № 8571868. DOI: 10.1109/ICEPDS.2018.8571868

IOP Conf. Series: Materials Science and Engineering **779** (2020) 012023 doi:10.1088/1757-899X/779/1/012023

Profile optimization of the impeller blade of a low-speed centrifugal pump using surrogate modeling

S Valyukhov[1], D Galdin[1], V Korotov[1,3], V Rusin[1] and A Shablovskiy[2]

[1]Voronezh State Technical University, 14 Moscow Avenue, Voronezh, 394026, Russian Federation
[2]Bauman Moscow State Technical University

[3]E-mail: vvkorotov@ya.ru

Abstract. This article presents an alternative approach to optimizing the profile of the impeller blade of a centrifugal pump. The researches were carried out using modern methods of computational fluid dynamics and the ANSYS CFX software package. The target of optimization is to maximize hydraulic efficiency and head. The calculated variables were selected using the Bezier biquadratic curve, and calculation experiments were used to select reference points from the projected space. Particular attention is paid to the construction of a surrogate RSA model, which is used in numerical simulation. To study the optimal points, the NSGA-II genetic algorithm was used. In a process of processing the results, a standard response surface with a second-order polynomial was formed to predict integral parameters. It is concluded that the main problem associated with the optimization of design based on a surrogate model is the accuracy of modeling the approximation functions. Also, the use of several surrogates can increase the reliability of optimization with minimal computational costs.

Introduction

Centrifugal pumps are widely used in engineering and are used in almost all industries. The problem of increasing the efficiency of pumping equipment is relevant, since pumps consume a large amount of energy, so one of the main criteria when choosing a pump is its efficiency [1]–[9].

Today, hydrodynamic modeling methods are widely used as an effective tool to increase hydraulic efficiency in the design and optimization of flow parts of vane pumps. The main element of centrifugal pump optimization is the impeller. Changing its geometric parameters has the greatest impact on the performance of the entire pump. The flow phenomenon inside the impeller is complex. Most CFD-based impellers are trial and error, in which the design process is based on the experience of a design engineer [10]–[13]. However, in the design process, the use of numerical optimization methods makes it possible to perform a rational and systematic search in order to find a profile with improved performance.

To carry out optimization, surrogate models, also called surrogates, metamodels, or low-precision models, are very often used. The surrogate approximation function (approximation function) imitates a real phenomenon or high-precision model that is used for engineering analysis. High-precision model estimation, for example, the CFD method, requires longer computational time, while surrogates can be estimated hundreds of times faster in a limited period of time. It is important to understand that a universal surrogate that could be used to solve all problems does not exist. The main problem of

Content from this work may be used under the terms of the Creative Commons Attribution 3.0 licence. Any further distribution of this work must maintain attribution to the author(s) and the title of the work, journal citation and DOI.
Published under licence by IOP Publishing Ltd

surrogate modeling is the complexity of modeling high-precision simulations, i.e. surrogates depend on a specific problem, and one surrogate cannot be used to solve all problems, so designers need to choose a suitable surrogate that can give sufficient accuracy and will have a more reliable result.

To optimize the profile of the impeller blade, two objective functions were selected: hydraulic efficiency and pressure. Hydraulic efficiency (η) is given by formula:

$$F_1 = \eta = \frac{H}{H_m},$$

where H and H_t are the actual and theoretical pressure, respectively.

According to the Euler equation, the head (H) generated by the impeller blade is given by the formula:

$$F_2 = H = \frac{c_{u2}u_2 - c_{u1}u_1}{g},$$

where c_{u1} and c_{u2} are the input and output components of the peripheral speed; u1 and u2 are the peripheral speeds.

The algorithm of the optimization process based on surrogate modeling is presented in Figure 1. Initially, the task was formulated and a space for design was created. It is limited by the lower and upper limits of the variables. The next step is to build a surrogate and search for optimal points. Optimal points are determined using a genetic algorithm.

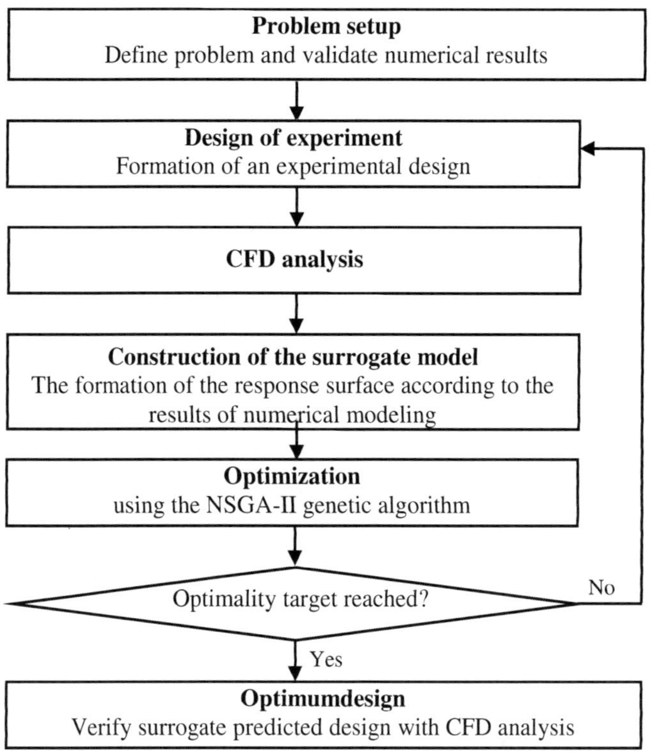

Figure 1. Flowchart of the optimization procedure

Genetic algorithms belong to the class of evolutionary algorithms. These algorithms search for the optimal solutions of various target functions, based on the principles inherent in the processes of natural evolution and the genetic mechanisms used in it. The use of genetic algorithms due to the nonlinearity of the target function is a reasonable choice.

Among the various modifications of genetic algorithms, the most common Non-Dominated Sorting Genetic Algorithm (NSGA-II), which was used in this work. Under the conditions, the value of the initial population was set to 1000, the subsequent ones received a similar restriction. Their maximum number was 20. The purpose of the genetic algorithm is to maximize hydraulic efficiency (η) and head (H).

The accuracy of the surrogate depends on the number and location of design points in the design space. Surrogates also depend on the task, which makes it difficult to choose any surrogate for a particular problem. In this work, the response surface approximation (RSA-model) was used.

The RSA model as a surrogate model is used to estimate the values of the target function. It is a technique of polynomial function selection for discrete responses obtained from numerical calculations. In this work, the second order polynomial is used as a response function:

$$F(x) = \psi_0 + \sum_{j=1}^{n} \psi_j x_j + \sum_{j=1}^{n} \psi_{jj} x_j^2 + \Sigma \sum_{j \neq j}^{n} \psi_{ij} x_i x_j,$$

where n is the number of variables, x and ψ are the regression coefficients.

Real engineering problems are solved using one- or multi-criteria optimization methodologies, where the latter option is preferred [14]–[18]. Multi-objective optimization considers several conflicting criteria that are optimized simultaneously. This approach includes the search for a variety of Pareto optimal solutions and the construction of a relationship between the criteria, the combination of criteria, etc. The multi-objective problem can be solved by evaluating weight factors for each individual criterion, namely:

$$F_C = \Sigma w_i F_i,$$

where i=1, 2, ... n, and n is the number of criteria.

Mathematical model

The ANSYS CFX computational fluid dynamics software module was used to conduct a numerical experiment.

The geometry of the impeller of the centrifugal pump is shown in Figure 2.

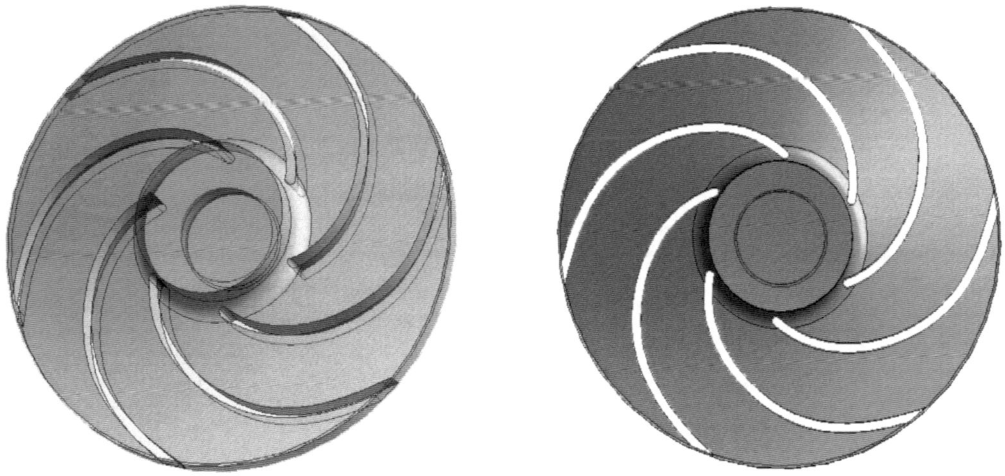

Figure 2. Impeller geometry

The main parameters of the calculation model are presented in table 1.

IOP Conf. Series: Materials Science and Engineering **779** (2020) 012023 doi:10.1088/1757-899X/779/1/012023

Table 1. Model parameters

Parameter	Description
Flow domain	Single impeller
Number of blades	6
Mass flow rate	25 m^3/h
Head	50 m
Rotation frequency	2950 rpm
Mesh model	Unstructured hexahedral
Number of elements	4,7 million
Fluid	Water, 20°C
Turbulence model	*k-ε*
Inlet/Outlet	Inlet — mass flow rate, outlet — pressure

Initially, a parametric model of the impeller was created, in which the meridional section remained unchanged, and the profile of the midline of the blade was determined using parametric dependencies. To define the midline profile, we used the Bezier curve with five control points, shown in Figure 3. Such a function can be fully defined by nine parameters: the angles φ1, φ2, φ3, φ4 and the radiuses r_{in}, r_1, r_2, r_3, r_{out}. It was also decided to reduce the number of variables to 4 — φ1, φ2, φ3 and φ4 to reduce the size of the search space. During the optimization search, the radiuses r_{in}, r_1, r_2, r_3, and r_{out} were constants.

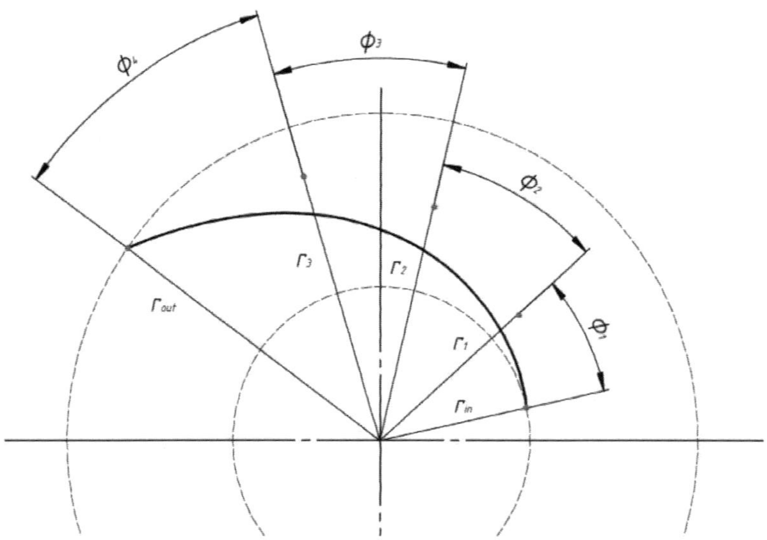

Figure 3. The profile of the blade profile

The angles φ1, φ2, φ3, φ4 were chosen as the calculated variables for constructing the profile of the middle line of the blade. The ranges of variables shown in the table 2, were selected based on preliminary calculations and search.

Various researchers have carried out a lot of work, which considered all kinds of surrogate models, as well as performed a comparative analysis [19] - [20]. In further optimization, one of the varieties of surrogate models, highly dependent on the response received from high-precision models, was used to obtain functional dependencies of the criteria change. It is important to choose the best option, otherwise you can get erroneous results. We used the response surface approximation (RSA-model).

Table 2. Variable Ranges

Parameter	Range of parameter values, deg
φ_1	20 … 40
φ_2	20 … 35
φ_3	15 … 35
φ_4	5 … 20

Figure 4 shows the response surfaces obtained using the RSA model, based on the results of numerical simulation according to the experimental plan of 25 points.

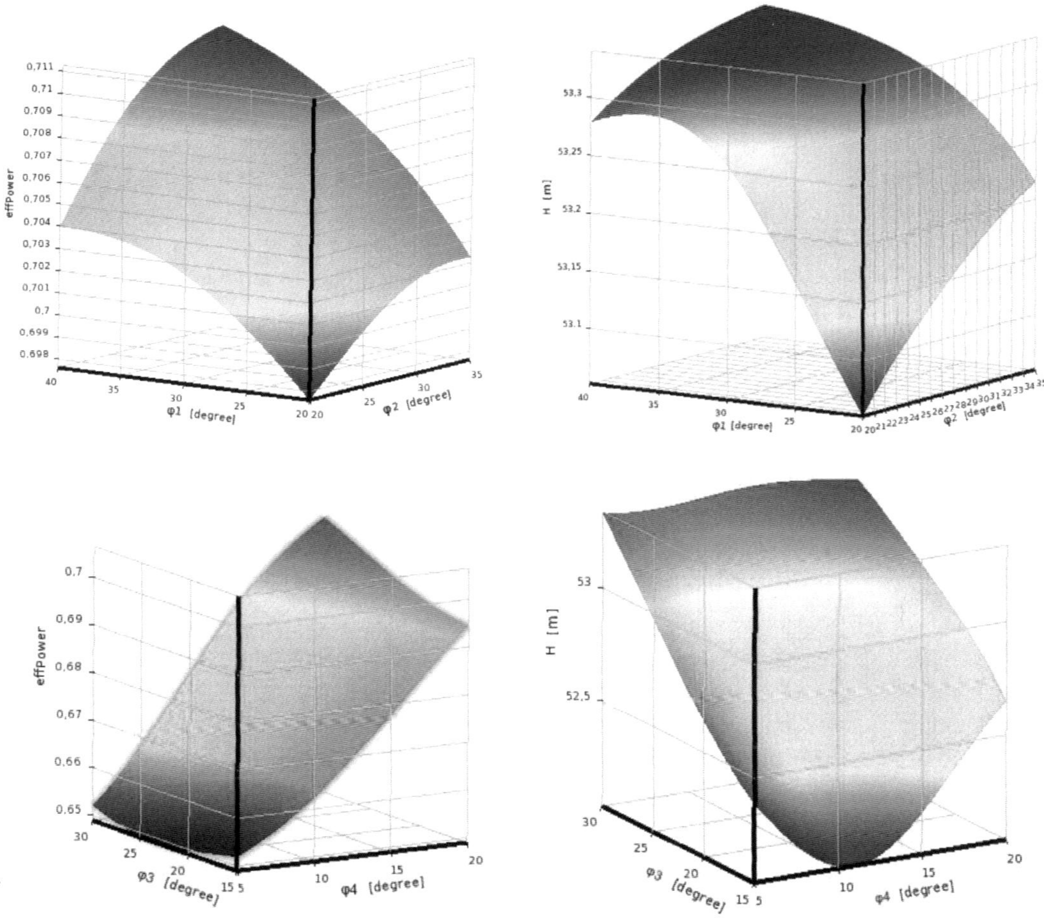

Figure 4. Response surfaces

For this type of surrogate model, a set of Pareto optimal solutions were obtained, from which three optimal points were selected for subsequent verification by numerical simulation. The results are presented in table 3.

Of these three points, one with the highest head and efficiency values was selected for verification.

The results of the calculation of the optimized impeller are presented in table 4.

Table 3. Relative deviations of simulation results

№	Deviation from the predicted value		Deviation of the optimized version from the original	
	H, %	η, %	H, %	η, %
1	1,86	0,22	1,37	6,15
2	1,73	0,19	1,43	6,23
3	1,40	0,03	1,57	6,57

Table 4.Optimization result

Geometry	Variableparameter				Predictedvalue	
	φ_1	φ_2	φ_3	φ_4	H, m	η, %
Original	26,7	30,82	17,55	9,83	52,18	65,86
Optimized	25,4	28,74	34,86	20	53,77	70,47

Conclusion

This article successfully evaluated the design capabilities of a surrogate RSA model. As an example, the impeller of a low-speed centrifugal pump was selected. When optimizing the profile of the midline of the blade, the increase in hydraulic efficiency and pressure amounted to 4.61% with a simultaneous increase in pressure by 1.59 m compared to the initial form. The increase in efficiency occurs due to the reduction of flow losses. Since surrogates do not work well in all cases, it is suggested that they be reused in the calculations. These optimization results indicate the effectiveness of this technique.

References

[1] M Saprykina and V Lomakin 2019 *IOP Conf. Ser.: Mater. Sci. Eng.* 589 012017
[2] V Tkachuk *et al* 2019 *IOP Conf. Ser.: Mater. Sci. Eng.* 589 012007
[3] A Boyarshinova *et al* 2019 *IOP Conf. Ser.: Mater. Sci. Eng.* 589 012014
[4] E Morozova *et al* 2019 *IOP Conf. Ser.: Mater. Sci. Eng.* 589 012012
[5] V Tkachyk *et al* 2019 *IOP Conf. Ser.: Mater. Sci. Eng.* 589 012011
[6] Lobanoff V. S. Ross R. R. Centrifugal Pumps: Design and Application (2013) Gulf Professional Publishing, pp. 305–310
[7] Hydraulic Institute (U.S.) Hydraulic Institute standards for centrifugal, rotary & reciprocating pumps (1983) Michigan University, pp.111
[8] Gülich J.F. Centrifugal Pumps. Springer-Verlag Berlin Heidelberg, 2010, 964 p.
[9] Yousefi H., Noorollahi Y., Tahani M., Fahimi R. and Saremian S. 2019 Numerical simulation for obtaining optimal impeller's blade parameters of a centrifugal pump for high-viscosity fluid pumping *Sustainable Energy Technologies and Assessments* 34 16–26
[10] Zharkovsky A.,2003. Mathematical modeling of working processes in centrifugal pumps of low and medium speed for the solution of problems of the automated designing. Dissertation and the abstract on HAC RF 05.04.13.
[11] Lomakin V.O. Numerical simulation of flow parts of pump models and verification of simulation results by comparison of obtained values with experimental data. Naukaiobrazovanie MGTU im. N.E. Baumana = Science and Education of the Bauman MSTU, 2012, no. 5, pp. 52–62. DOI: 10.7463/0512.0356070 (in Russian).
[12] Valyukhov, S.G., Kretinin, A.V., Galdin, D.N. et al. Chem Petrol Eng (2018) 53: 658. DOI: 10.1007/s10556-018-0398-y

[13] Ding H., Visser F.C., Jiang Y., Furmanczyk M. Demonstration and Validation of a 3D CFD Simulation Tool Predicting Pump Performance and Cavitation for Industrial Applications // Journal of Fluids Engineering. 2011. Vol. 133, no. 1. P. 011101-011101-1 DOI: 10.1115/1.4003196

[14] Lomakin V O 2015 Investigation of two-phase flow in axial-centrifugal impeller by hydrodynamic modeling methods Proceedings of 2015 International Conference on Fluid Power and Mechatronics, FPM 2015 no 7337302 pp 1204–6

[15] Lomakin, V.O., Kuleshovav, M.S., Bozh'eva, S.M. Numerical Modeling of Liquid Flow in a Pump Station (2016) Power Technology and Engineering, 49 (5), pp. 324-327. DOI: 10.1007/s10749-016-0623-9

[16] Lomakin V O, Chaburko P S and Kuleshova M S 2017 Multi-criteria Optimization of the Flow of a Centrifugal Pump on Energy and Vibroacoustic Characteristics Procedia Engineering176 pp 476–82 DOI: 10.1016/j.proeng 2017.02.347

[17] Lomakin V.O. Investigation of two-phase flow in axial-centrifugal impeller by hydrodynamic modeling methods Proceedings of 2015 International Conference on Fluid Power and Mechatronics, FPM 2015 no 7337302 pp 1204–6 DOI: 10.1109/FPM.2015.7337302

[18] Zhu J, Zhu H, Zhang J and Zhang H-Q 2019 A numerical study on flow patterns inside an electrical submersible pump (ESP) and comparison with visualization experiments Journal of Petroleum Science and Engineering 173 pp 339–350 DOI: 10.1016/j.petrol.2018.10.038

[19] Bellary, Sayed Ahmed Imran & Husain, Afzal &Samad, Abdus& Kanai, R. A. (2018). Performance Optimization of Centrifugal Pump for Crude Oil Delivery. The Journal of Engineering Research [TJER]. 15. 88. 10.24200/tjer.vol15iss1pp88-101

[20] Bellary, Sayed Ahmed Imran &Samad, Abdus. (2017). An alternative approach to surrogate averaging for a centrifugal impeller shape optimisation. International Journal of Computer Aided Engineering and Technology. 9. 62. 10.1504/IJCAET.2017.080769.

IOP Conf. Series: Materials Science and Engineering **779** (2020) 012024 doi:10.1088/1757-899X/779/1/012024

The presence of the "locking effect" when the fluid flows through the cylindrical throttle channels

V Kuznetsov[1] and V Yarots[1,2]

[1]Bauman Moscow State Technical University

[2]E-mail: vyaroz@bmstu.ru

Abstract. A study was made for the continuous flow of the working fluid through a cylindrical throttle channel (so-called "locking effect"). The flow of the fluid into the atmosphere and the change in absolute pressure at the inlet to the throttle channel ranging between 0.221 MPa and0.24 MPa are studied. It was found that the pressure in the compressed section inside the throttle channel remains almost constant and equal to 4 kPa. A comparison is made of the experimental values of the flow rate determined for the same absolute pressure difference Δp in the separated and unseparated flow regimes. Knowing that the value of the flow coefficient μ in the zone of the existence of the "locking effect" can be determined from known data for the process of fluid outflow from holes with a sharp edge in a thin wall. Also, similar phenomena occur when fluid flows through small holes and slots in the guide, control and regulatory equipment of various hydraulic systems.

1. INTRODUCTION

Studies show that the hydrodynamic characteristics of fluid flows in hydraulic systems that contain throttle channel are very complex and have a significant impact on the performance of these systems [1–4].

The task that developers of hydraulic systems for various purposes often have to solve, is to provide control over the values of the working fluid pressure and its flow rate. One of the ways to control these parameters is the so-called throttle control method. It is customary to call a regulating device a throttle capable of establishing a certain relationship between the calculated differential pressure across the throttle and the flow rate [5–12]. In hydrodynamics, this relationship is expressed by the formula:

$$Q = \mu A \sqrt{2\Delta p / \rho} \tag{1}$$

Where Q — the volumetric flow rate through the throttle;

μ— flow rate coefficient;

A — area of throttle channel;

$\Delta p = (p_{in} - p_{out})$ — differential pressure in the throttle;

p_{in}— pressure at the inlet of the throttle;

p_{out}— pressure at the outlet of the throttle;

ρ— density of the working fluid.

It is known that when liquid flows through a throttle channel of a cylindrical shape, whereas the ratio of the length of the channel to its diameter in the range from two to four, two flow modes are possible — separated (Fig. 1a) and unseparated (Fig. 1b).

Content from this work may be used under the terms of the Creative Commons Attribution 3.0 licence. Any further distribution of this work must maintain attribution to the author(s) and the title of the work, journal citation and DOI.
Published under licence by IOP Publishing Ltd

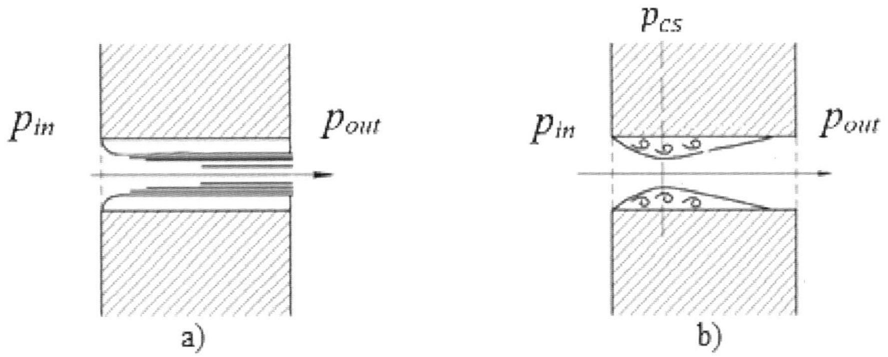

a) separated b) unseparated

Figure1. Expirationmodes

In the separated flow mode, the hydrodynamic parameters are close to the parameters of the flow from a hole in a thin wall. In the unseparated flow mode, the flow rate of the working fluid through the throttle at a given inlet pressure may depend on the pressure value at the outlet, or it may not depend (so-called "locking effect") [13–18]. The last mode represents a special particular interest to developers of hydraulic systems and control devices for their work. When the hydraulic actuator operates in the existing mode of the "locking effect", the speed of movement brought by it in the movement of the load ceases (up to certain limits) depends on the magnitude of this load and will remain constant.

One of the research goals is to develop recommendations on the use of formula (1) for applied calculations of throttle devices based on the use of a throttle channel of cylindrical shape with a ratio of channel length to its diameter in the range from two to four. From the analysis of formula (1), creates the following questions:

1. What is meant by area "*A*" of a cylindrical throttle channel?
2. What is meant by the pressure drop in separated and unseparated modes of expiration?
3. How to determine the value of the coefficient of flow μ during unseparated flow mode?

Some clarifications of well-known answers to the first question are given in the articles [12].

Below are the results of the research, which purpose was to find an answer to the second and third questions - what is meant by the pressure drop in formula (1) when it is used to calculate the parameters of the fluid flow through a cylindrical throttle channel.

2. METHODS

From the analysis of related literature, many authors of the studies assume fluid flows that are hydrodynamically similar in the area from the entrance to the sharp-edged hole to the most compressed section when flowing through the sharp-edged hole in the thin wall and in the area from the entrance to the most compressed flow section in unseparated flow through a cylindrical throttle channel.

Based on the foregoing, it can be assumed that the calculation of the parameters of the fluid flow through the cylindrical nozzles in the zone of the "locking" effect using formula (1) should be carried out with respect to the pressure drop Δp in the section between the nozzle inlet — $p_{in,abc}$.abs and the most compressed flow cross-section inside nozzle — $p_{cs,abc}$.

Experimental studies were carried out on a model of a cylindrical nozzle made of plexiglass. The drawing of the model is shown in Fig. 2.

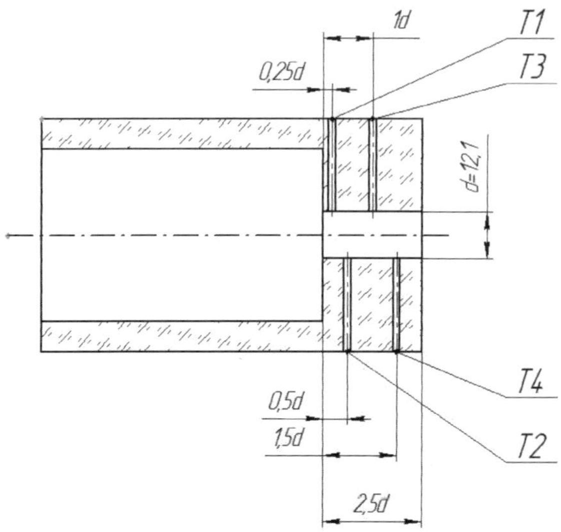

Figure 2. Experimental nozzle model

3. RESULTS

In fig. 3 shows an experimental graph of the absolute pressure variation p_{abc} in various sections of the flow inside the cylindrical nozzle as a function of the absolute pressure at the inlet of the nozzle $p_{in.abc}$. The experiment was conducted with the flow of liquid into the atmosphere. At the same time, atmospheric pressure p_{atm} equals 102124,652 *Pa.*

Overpressure was measured with an exemplary pressure gauge of accuracy class 0.15, and vacuum with an exemplary vacuum gauge of accuracy class *0.25*. A model barometer was used to measure atmospheric pressure, and a mercury laboratory thermometer was used to measure the temperature.

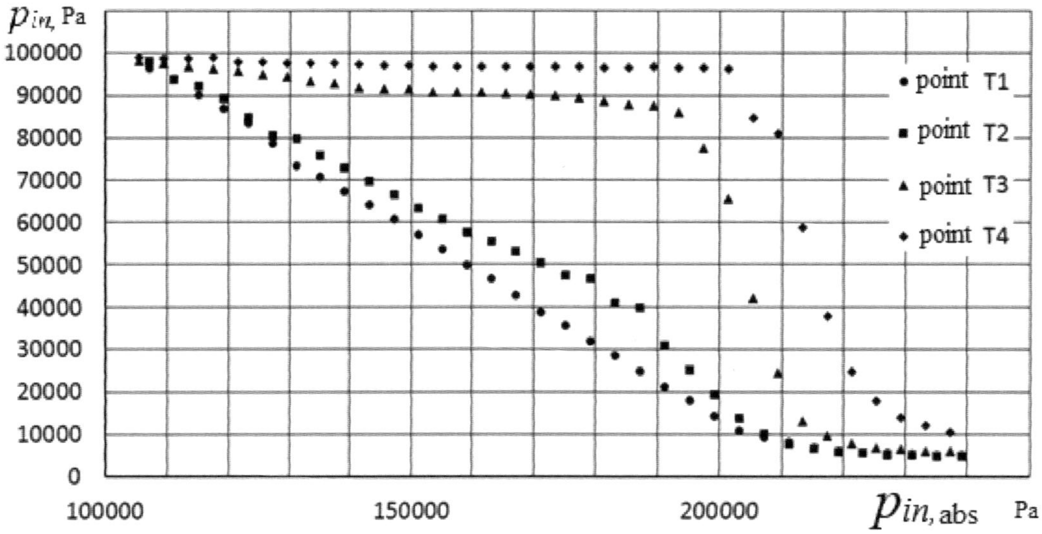

Figure 3. Graph of the absolute pressure values p_{abc} at points T1, T2, T3, T4 as a function of absolute pressure at the inlet of nozzles $p_{in\ abc}$

From the experimental data shown in Figure 3, it is noticed that when the absolute pressure at the inlet p_{in} changes in the range from 0.221 MPa to 0.24 MPa, the pressure p_{abc} at points *T1* and

*T2*remainedthe same and did not change much. Minimum Pressure Ratio at these points in the reduced range of its change to the pressure at the outlet of the nozzle (atmospheric pressure) was 0.46.

With a further increase in pressure at the inlet p_{in}, there was a sharp transition from an unseparated flow mode to a separated. The transition took place in the range of absolute values of this pressure from 0.245 to 0.25 MPa and with the ratio of the absolute pressure at the outlet to the absolute pressure at the inlet ranging from 0.417 to 0.408 MPa.

It was not possible to precisely establish the value of the pressure Pin, at which the transition from an unseparated to a separated mode occurred. The probable reason (as photography shows [5,6]) is the instability of the geometric parameters of the tail of the stream in the throttle channel of the nozzle and the total length of the vapor-gas cavity surrounding the stream inside the nozzle.

The above experimental data are in good agreement with the results of experimental studies presented in the book of B.N. Siova [13]. This result allows us to state that when the ratio of the absolute pressure at the outlet to the absolute pressure at the inlet varies from 0.46 to 0.408, a blocking effect occurs. In this zone, the pressure in the vapor-air cavity surrounding the fluid flow inside the throttle channel in the region of its most compressed section remains almost independent of the pressure at the outlet of it.

Based on the research results, it follows that it is expedient in formula (1) when calculating the flow parameters for Δp to take the difference in absolute pressures at the inlet to the throttle channel and in the vapor-air cavity surrounding the fluid flow inside the throttle channel in the region of its most compressed section ($\Delta p = p_{in.abs} - p_{1abs}$).It should be noted that according to the results of published experimental studies of various authors, the magnitude of the absolute pressures in the analyzed flow region varied in the range from 2 kPa to 4.5 kPa.

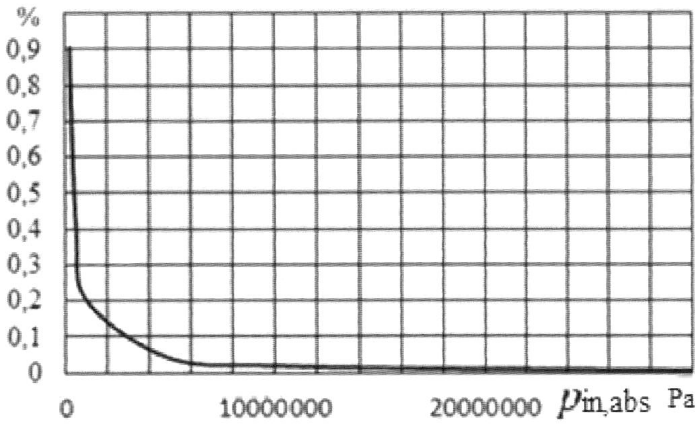

Figure 4. Evaluation of the degree of error in the percentage entered into the calculation $\sqrt{\Delta p}$ refusal of accounting$p_{cs.abs}$

Fig. 4 shows the results of assessing the degree of difference in the percent of root values from pressure drop $\Delta p = p_{in.abs} - p_{cs.abs}$, as a function of absolute pressure at the nozzle inlet$p_{in.abs}$. While taking in one case the value $p_{cs.abs}$ according to the data given in Fig. 3, and in another$p_{cs.abs}$ equal to zero. From the graph, it follows that in the range of pressure changes on a stream of the throttle channel starting from 0.5 MPa, the percentage of determination error Δ does not exceed 0.04%. Based on the foregoing, we can conclude that in applied calculations of hydraulic systems operating at high pressures at the inlet to the throttle, in formula (1), the pressure drop Δp with a fairly high degree of accuracy can be determined only by the magnitude of the absolute pressure at the inlet to the throttle channel.

Fig. 5 shows experimental plots of the flow rate of fluid Q flowing through the model of a cylindrical nozzle (Fig. 2) under separated and unseparated modes of flow as a function of $\sqrt{\Delta p}$. In

IOP Conf. Series: Materials Science and Engineering **779** (2020) 012024 doi:10.1088/1757-899X/779/1/012024

this case, Δp was determined in the separated mode of expiration as $(p_{in.\ abs.} - p_{atm})$, and in the unseparated mode as $(p_{in.abs} - p_{cs.abs})$.

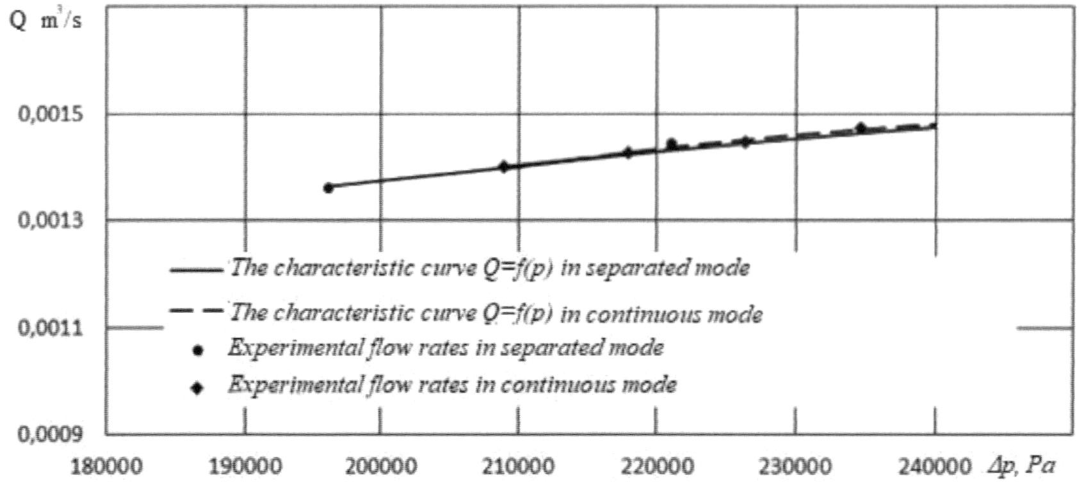

Figure 5. Flow characteristics of separated and unseparated modes of fluid flow

Analysis of the data shown in Fig. 5, shows that the difference in flow rates determined for the same pressure drop Δp in the range of its change from 0.22 to 0.24 MPa in the separated and unseparated flow regimes varies from 0.12 to 0.49%.

The obtained experimental results confirm the assumptions made in the literature [2, 7, 8, 9, 10] about the hydrodynamic similarity of the process of fluid outflow in the section from the inlet to the compressed section of the flow both when it flows through a hole with a sharp edge in a thin wall and when through cylindrical nozzles. analysis of the results of calculating the hydrodynamic parameters of the process of the flow of an ideal fluid through a flat model of a cylindrical nozzle, performed by the method of professor s. s. rudneva [6] showed that in the zone of the flow regimes corresponding to the existence of a "locking effect", the values of the flow coefficients practically become equal to their values when they flow out of round holes with a sharp edge in a thin wall. from the foregoing it follows that the value of the flow coefficient μ in formula (1) when it is used to calculate the parameters of fluid flow through cylindrical nozzles in the zone of existence of the "locking effect" can be determined from known data for the process of fluid flow from holes with a sharp edge in a thin wall [1].

4. CONCLUSION

1. Based on the above research results, it follows that it is expedient in formula (1) when calculating the flow parameters for Δp to take the difference in absolute pressures at the inlet to the throttle channel and in the vapor-air cavity surrounding the fluid flow inside the throttle channel in the region of its most compressed section $(\Delta p = p_{in.abs} - p_{1abs})$.

2. Evaluation of the degree of error in the percentage entered into the calculation efusal of accounting $p_{cs.abs}$ allows us to conclude that in applied calculations of hydraulic systems operating at high pressures at the inlet to the throttle in formula (1), the pressure drop Δp with a sufficiently high degree of accuracy can be determined only by the magnitude of the absolute pressure at the inlet to the throttle channel.

3. The value of the flow coefficient μ to the formula (1) when it is used to calculate the parameters of the fluid flow through the cylindrical nozzles in the zone of existence of the "locking effect" can be determined from known data for the process of fluid flow from holes with a sharp edge in a thin wall [1].

REFERENCES

[1] Altshul A.D. Hydraulic resistance // 2-nd revised edition M.: Nedra, 1982. 224 p.

[2] Hydraulics, hydraulic machines and hydraulic drives / Textbook. For engineering universities / T.M. Bashta, S.S. Rudnev, B.B. Nekrasov et al. 4-th stereotyped edition, M.: Publishing House "Alliance", 2010. 423 p.

[3] Charru F. Hydrodynamic Instabilities. Cambridge University Press, 2011. 391 p.

[4] Hibbeler R.C. Fluid Mechanics. Part 2. Pearson Pumtke Hall, 2015. 904 p.

[5] Kuznetsov V.S., Yarots V.V. An indirect method for measuring some geometric parameters of throttle channels of a round cylindrical shape // News of MSTU "MAMI". Vehicles and power plants. Scientific peer-reviewed journal. M.: MSTU "MAMI", 2015. №. 4 (26). T. 1.p. 42–49.

[6] Shablovsky A.S., Kuznetsov V.S., Yarots V.V. Comparison of the results of theoretical and experimental studies of the compression coefficient of the fluid flow in a flat slit-like channel // News of MSTU "MAMI". Series 3. Natural sciences. M.: MSTU "MAMI", 2013. №. 1 (15). T.3. p. 135–138.

[7] Blagoveshchensky I.G., Balykhin M.G., Blagoveshchenskaya M.M., Makarovskaya Z.V., Zhirov M.V., Shkapov P.M. Using global optimization methods in optimal control problems. University proceedings / Technology of the textile industry. 2018. №. 4 (376). p. 21–25.

[8] SalavatovT.Sh. Model representation of hydrodynamic systems such as compressible viscoplastic fluids in fluid and gas mechanics / In Collection: 2 nd International Conference on Innovations in Technical and Natural Sciences Procttdings of the Conference. 2018. p. 55–57.

[9] Bereslavsky E.N. On transformations of the travel time hodograph in some problems of fluid and gas mechanics / Engineering Physics Journal. 2015. V. 88. №. 5. p. 1182–1188.

[10] Kalyakin A.M., Chesnokova E.V., Sautkina T.N. Schemes, models and pseudo-paradoxes in the course "Fluid and Gas Mechanics" / Scientific and technical problems of improvement and development of gas and energy supply systems. 2018. №. 1. p. 224–229.

[11] Nikitin O.F. Some issues of static calculation when designing a highly efficient unregulated hydraulic drive / Engineering Bulletin. 2014. №. 6. p.1.

[12] Shakirzyanov A.A., Nikitin O.F., Bakiev T.A. The prospect of using a spring regulator of constant power supply as part of a hydraulic actuator in order to increase its energy efficiency / Electrical and information systems and systems. 2017. V.13. №. 2. p. 25–32.

[13] Siov B.N. Flow of fluid through nozzles. M.: Mechanical Engineering, 1968. 140 p.

[14] T Valiev and APetrov 2019 *IOP Conf. Ser.: Mater. Sci. Eng.*492 012038

[15] V Cheremushkin and V Lomakin 2019 *IOP Conf. Ser.: Mater. Sci. Eng.*492 012039

[16] V Cheremushkin and APolyakov 2019 *IOP Conf. Ser.: Mater. Sci. Eng.*589 012001

[17] I Borovik et al 2019 *IOP Conf. Ser.: Mater. Sci. Eng.*589 012002

[18] [A Protopopov et al 2019 *IOP Conf. Ser.: Mater. Sci. Eng.* **589** 012003

IOP Conf. Series: Materials Science and Engineering **779** (2020) 012025 doi:10.1088/1757-899X/779/1/012025

Avanepneumatic motor modification to reduce torque ripples

V Bozrov[1,3], V Ivlev[1] and S Semenov[2]

[1]Blagonravov Mechanical Engineering Research Institute of Russian Academy of Sciences (MERI of RAS), Moscow, Russia
[2]Bauman Moscow State Technical University

[3]E-mail:vmbzrv@bk.ru

Annotation. The problem of determining and possible ways to reduce the amplitude of torque ripples in a vane pneumatic motor without a significant reduction of its energy efficiency is considered. This problem is important for improving the ergonomic performance of manual mechanized pneumatic tools and improving the accuracy of pneumatic servosystems. The results are presented for the traditional layout scheme of a vane pneumatic motor, as well as for the proposed modified design. The decision was made using a detailed mathematical model of a pneumatic motor. The prospects of the modified pneumatic motor to reduce the torque ripples at higher energy efficiency indicators are shown, compared with the original model with an increased number of vanes.

Introduction

Vane pneumatic motors (PM) received the most widespread application among volumetric pneumatic motors: they occupy more than 75% of the PM market in power range from 0.1 to 18 kW. This is due to the simplicity and reliability of their design, relatively low cost and high developed capacity per unit of weight and volume. The most widespread segment of vane PMs is manual tools (screws, grinders, drills, mixers, etc.). Vane PMs are used in explosive industries, in the pharmaceutical and food industries, because this type of PM can be subjected to washing and sterilization. PMs in a special non-magnetic version are used as a part of servos for operation in strong magnetic fields [1].

One of the problems associated with the use of vane PM for some applications is the presence of torque ripples within one revolution of the shaft. The occurrence of these ripples is associated with a finite number of PM working chambers as well as with a sharp pressure change in chambers when they are connected to supply or exhaust lines. Torque ripples can cause vibrations that adversely affect both the resource of a PM itself and the device where it is installed. For manual tools, it is possible to reduce the quality of processed products as well as a harmful effect on the working personnel [2].

Note that similar problems with torque ripples are typical for different types of electric motors [3]. Here, to reduce the amplitude of these ripples, both constructive solutions are used, for example, changing the shape and number of poles, and more common solutions based on the formation of a certain law of current change in the windings of an electric motor with the help of a feeding device.

The catalogs of the main manufacturers of PMs (Atlas Copco, Ingersoll Rand, Deprag, etc.) usually contain their mechanical and flow characteristics, i.e. the dependence of the developed torque, power and compressed airflow on the steady angular velocity at a given supply pressure. The values of the torque ripples amplitude for each particular model are not given; that is this indicator is not regulated. The customer can choose only a model with a certain number of vanes. It is not possible to estimate possible torque ripples from catalogue data.

Content from this work may be used under the terms of the Creative Commons Attribution 3.0 licence. Any further distribution of this work must maintain attribution to the author(s) and the title of the work, journal citation and DOI.
Published under licence by IOP Publishing Ltd

Problem statement

For vane PMs (by analogy with electric motors), an approach associated only with design solutions is possible, since the speed of existing devices for controlling the pressure and flow of compressed pneumatic is not enough.

The simplest and most obvious way is to increase a number of vanes, i.e. increase a number of working chambers. But increasing a number of vanes, together with a decreasing torque rip pleamplitude, leads to increasing friction losses and decreasing energy efficiency. The original approach is presented in [4], where the stator has a non-circular profile, the geometry of which is chosen taking in account minimum torque ripple amplitude. But this requires a more complex and expensive stator manufacturing technology, and can also reduce the life of vanes due to the variable vane — stator contact area.

The purpose of this paper is to develop a method for solving the problem of estimating the level of vane PM torque ripples and the possibility of reducing them while maintaining sufficient energy efficiency. The energy efficiency index is extremely important here, because the cost of pneumatic energy is 5–6 times higher than electric energy. Here, as indicators of torque ripple, we will use its amplitude, and then-the specific flow of compressed air (flow per unit of developed power).

To solve this problem, it is necessary to produce several prototypes of PM with a different number of vanes and perform a sufficiently large amount of expensive experimental research. To reduce the costs and volume of these works, mathematical modeling methods have been developed that allow predicting the characteristics of vane PM depending on its geometric parameters and supply pressure [5]–[9]. At the same time, the models in these sources differ in the degree of detail of both structural details and in the description of thermodynamic processes in working chambers. On the other hand, mathematical models of a vane PM contain a number of parameters that are quite difficult to estimate experimentally or to obtain analytical expressions for their calculation. In our case, these are the parameters characterizing the friction in the various nodes of PMs, and the cross-sections leaks and overflows in the gaps of PMs. To determine them, the method of vector identification [10] was used which allows restoring the values of the required parameters on the basis of a limited number of experimental data and a mathematical model. Mathematical modeling is widely used for the analysis and synthesis of various machines based on volumetric and dynamic principle of action [11],[12].

Therefore, the solution of the problem was carried out in the following sequence:

–determination of mechanical and consumable characteristics of a specific PM model on a typical experimental stand;

– obtaining an adequate mathematical model of vane PM;

– carrying out the procedure of vector identification of effective cross-sections of channels of overflows and leaks values, and also coefficients of friction of vanes about a stator and rotors lotes;

–determination of torque ripples and energy efficiency of the original PM model with different number of vanes and parameters determined during the identification procedure, as well as its modified design.

It is reasonable to assume that the identified parameters do not change for PMs with different number of vanes.

The vane PM basic design and its modification

The РП42-55 PM model was used as an object of study, the structural diagram of which is shown in Figure1.

In the modification design the working volume of the PM is divided into two equal parts, and the rotor consists of two halves with slotes for vanes turned at an angle π/z, where z is the number of vanes. So, for $z = 4$ (see Figure 2) this angle is 45 °. The supply of compressed air to the chambers and exhaust are carried out through the channels in the stator as in the original design. The analysis of this scheme showed that all parts of the modified PM can be manufactured on the same equipment as for the original model, with the required quality of the working surfaces, followed by assembly into the finished product. The disadvantages of this scheme are the greater number of parts (correspondingly

IOP Conf. Series: Materials Science and Engineering **779** (2020) 012025 doi:10.1088/1757-899X/779/1/012025

slightly higher cost), as well as the appearance of additional channels for the flow of compressed air, which will negatively affect energy efficiency. The main geometric parameters of the initial PM model are given in [8].

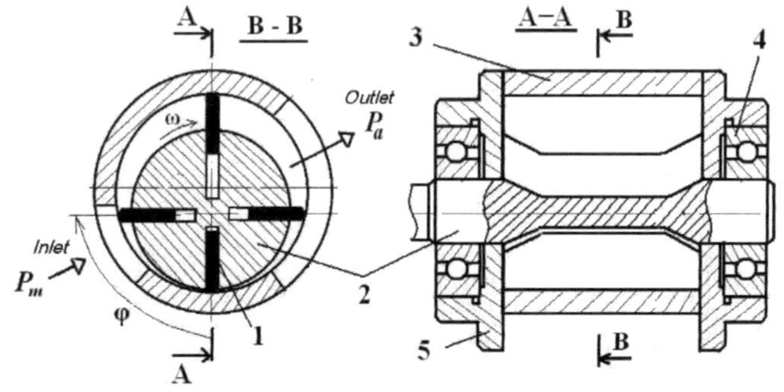

Figure1. Design scheme of PM model RP42-55.1 — plate, 2 — rotor, 3 — stator, 4 — bearings, 5 — covers

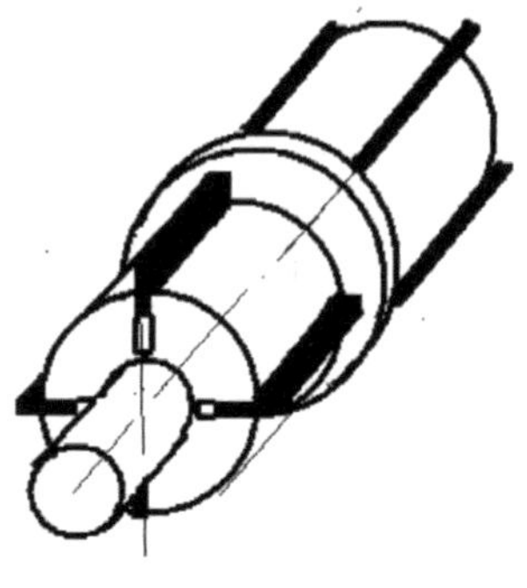

Figure2. The modified PMrotor

Test System and Experimental Results

The test rig constructive scheme of the mechanical and pneumatic part is shown in Figure3.The test rig consists of a rigid metal frame, on which the PM (model RP42-55), the torque sensor and the angular velocity TM-307, as well as the loading device, an electromagnetic brake ANV-12, are coaxially fixed. ShaftsofPM, sensor and brake are connected using compensating couplings MIC-5-2470.Sensor, brake and couplings are products of "Magtrol" (Germany).

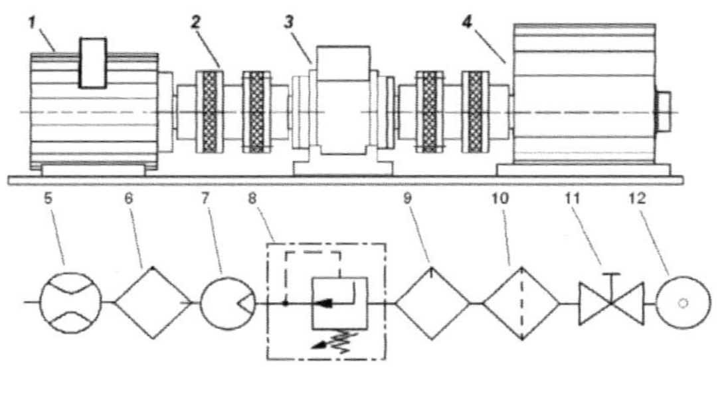

Figure 3. Scheme of test rig. 1 — air motor, 2— coupling, 3— torque sensor, 4— electromagnetic brake, 5— flow meter, oil separator, 7— air motor, 8— pressure regulator, 9— lubricator, 10— filter, 11— valve, 12— supply.

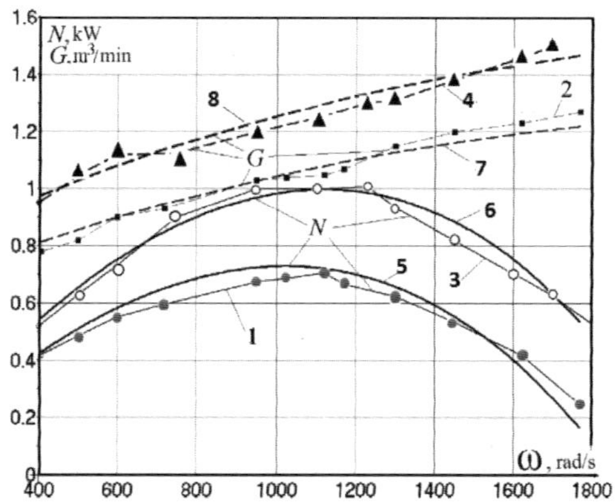

Figure 4. Experimental and calculated mechanical characteristics and air consumption performance.

Compressed air consumption was measured with a 'testo-6460" flow meter (Germany). The readings were taken in the angular velocity range from 400 to 1800 $s^{-1}$for supply pressures of 0.5 and 0.6 MPa. The experiments were carried out when adjusting the lubricator to supply 35–40 mm^3 of "BEZAN-pneumo" oil per 1 m^3 of compressed air (innormal conditions).Figure 4 presents experimental (marked with markers and connected by lines) mechanical and air consumption characteristics - dependences of developed power N and consumed compressed air consumption G on steady angular velocity.In Figure 4 curves 1 and 2 - respectively power, flow rate at a supply pressure of 0.5 MPa. Curves 3 and 4 are similar dependences at a supply pressure of 0.6 MPa.

Mathematical Modeling

Model consists of the following system of equations:

– two differential thermodynamic equations that describe the change in pressure and temperature of air in the working chamber are obtained based on the energy conservation law and equation of state for the compressed air in the chamber in differential form. Each working chamber of PM is considered as a cavity of variable volume with variable input and output flow zones;

– geometric ratios describing the law of change in the chambers volume and the position of the vanes as functions of the shaft angle;

– Saint-Venant ratios for compressed air consumption at inlet and outlet, as well as for the flow of compressed air through the gaps between the chambers and leaks into the atmosphere;

– logical relations describing the condition of connecting each working chamber to the pressure line and exhaust, as a function of the rotation angle and the angles of air distribution;

– equation of motion of the PM shaft;

– relations describing friction losses in bearings, as well as the forces acting on the vane, when it moves in the rotor slote.

Below we dwell in more detail on the equation of motion , because the proposed approach makes it possible to more accurately calculate friction losses in a vane PM and to describe torque ripples than in the models considered in [5]–[9].

The equation of motion of the PM rotor is compiled on the basis of the Lagrange equation:

$$\frac{d}{dt}\left(\frac{\partial T}{\partial \dot{\varphi}}\right)-\frac{\partial T}{\partial \varphi}=M \quad (1),$$

where φ is the angle of rotor rotation , measured from the vertical axis (see Figure 1), which also uniquely determines the position of the vanes.

The kinetic energy T is defined as the sum of the kinetic energies of rotor rotation and vanes in relative and figurative movements:

$$T=\frac{1}{2}\left[J_0\omega^2+\sum_{i=1}^{z}m\left(e\omega sin\varphi_i\right)^2+\sum_{i=1}^{z}m\left(R_1-ecos\varphi_i\right)^2\omega^2\right],$$

where J_o— the moment of inertia of the rotor, m the mass of the vane, e— the eccentricity, z— the number of vanes, φ_i— the angle determining the position of the i-th vane: $\varphi_i=\varphi_1+2\pi(i-1)/z$. The radius vector from the center of the rotor to the vane — stator contact pointis $R_{2i}=R_c-ecos\varphi_i$[9], where R_c— the stator radius. The quantity R_{1i} is the radius vector of the gravity center ofthe vane. Fora rectangular vane: $R_{1i}=R_{2i}-b/2$, b— the height of the vane. The equation $e\omega sin\varphi_i$— the speed of the i-th vane relative to the rotor slote.

The generalized torque M in formula (1) is calculated as $M=M_d-M_m-M_н$, where M_d is the driving torque, M_m— the friction torque, which consists of the friction torque values in the bearings M_r and the friction torque of the vanes on the stator and the rotor slots, $M_н$ is payload torque.

The driving torque created by the i-th vane from the pressure of compressed air in two adjacent chambers p_i and p_{i+1} (see Figure5) is $M_{di}=(p_i-p_{i+1})[(R_{2i})^2-r^2]l/2$, where $r$and l are the rotor radius and its length.

The friction torque in bearings is usually determined by empirical relationships given in the materials of manufacturers. Here we used the dependencies given in [13]:

$$M_p=f_1F_p\,{}^{d}\!/_{2}+0{,}97910^{-10}f_c\left(vn\right)^{2/3}D^3,$$

where $Fp\ (p_m,\omega)$— the radial load on the bearing as a function of supply pressure and angular velocity. The radial load on the rotor and, accordingly, on the bearings arises due to the fact that the pressure in the working chambers of the PM in the inlet and expansion zone (see Figure1) is higher than in the chambers communicating with the atmosphere. The resultant from the air pressure forces acting on the rotor and directed to its center along the symmetry axis of the i-thchamber will be equal to $F_i = 2p_i r l sin(\pi/z)$.

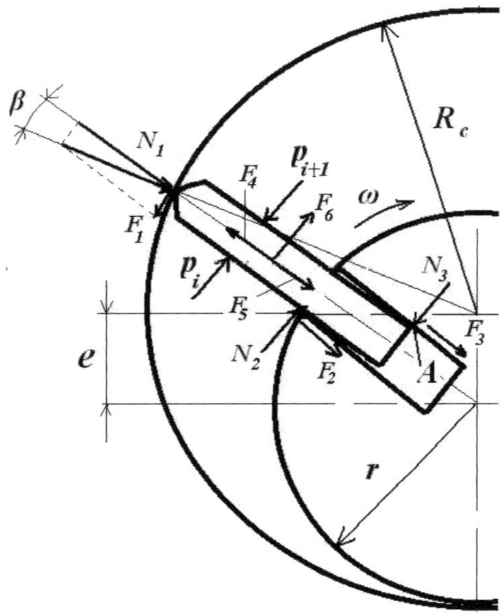

Figure5.A vane loading scheme

$$F_1 = k_1 N_1 \cos\beta + k_{11}\left(R_c - e\cos\varphi_i\right)\omega,$$

$$\beta = \arcsin\left(R_c^{-1} e \sin\varphi\right),$$

$F_2 = k_2 N_2$; $F_3 = k_2 N_3$—correspondingfriction forces, k_1 and k_2— the dry friction coefficients, k_{11}— the viscous friction coefficient;
$F_4 = m\omega^2(R_1 - e\cos\varphi_i)$ — centrifugal force;
$F_5 = me[(d\omega/dt)sin\varphi_i + \omega^2 cos\varphi_i]$ —inertial force from the radial movement of the vane;
$F_6 = 2me\omega^2\sin\varphi_i$— Coriolis force.

The radial load in the support of the rotor will be equal to half vector sum of forces F_i from all chambers, and its average value per revolution is(for the considered PM model, see [8] in more detail):

$$F_p = \frac{1}{2\pi}\int_0^{2\pi}\sum_{i=1}^{z} 2p_i(\varphi) rl sin\left(\pi/z\right) d\varphi$$

Where $d = 12mm$— the inner diameter of the bearing, $f_t = 0{,}002$ (a coefficient of friction for a ball radial bearing), f_c— a coefficient depending on the method of bearing lubrication, $v = 35\ mm^2/s$— the kinematic viscosity of the oil, n— the number of revolutions per minute, $D = 22mm$— the diameter passing through the centers of the rolling elements.

The given torque from the friction forces of the vane on the stator and the rotor slots is determined as follows. Figure 5 shows a diagram of forces acting on the vane. In determining the friction forces, we were guided by the following provisions. The value of the friction force in the general case is determined by the Stribek model. For vane PMs, more than 90% of friction losses are attributable to the vane — stator friction pair. Therefore, in the vane — rotor friction pair, the viscous component can be ignored, which should not lead to significant errors. In addition, in our case, the relative velocity in the vane — stator pair is significantly higher than those speeds where the Stribek effect is manifested.

Composing three equations of vane equilibrium, namely, the sum of the projections of the forces on the axis of the vane, on the orthogonal direction and the sum of the moments relative to point A, we obtain a system of equations for determining N_1, N_2 and N_3. The given moment from the friction forces acting on the vane:

$$M_t = F_1\left(R_c - e\cos\varphi_i\right) + \left(F_2 + F_3\right)e\sin\varphi_i.$$

IOP Conf. Series: Materials Science and Engineering **779** (2020) 012025 doi:10.1088/1757-899X/779/1/012025

Procedure of vector identification of PM parameters

The identification procedure is implemented in the software complex MOVI (Multicriteria Optimization and Vector Identification), which is based on the Sobol method of studying the parameter space. This method is widely used to solve problems of optimization of parameters of hydraulic and pneumatic machines [14]–[17].

As data for comparison, the calculated values of the power on the shaft $N_{1j}(\omega)$ and the consumption flow rate $G_{1j}(\omega)$ and the corresponding experimental values of these indicators (with index 2), at a supply pressure p_m= 0.5 MPa (j=1) and p_m = 0.6 MPa (j=2) are taken. The values of their maximum divergence are accepted as criteria of conformity:

$$\Delta_1 \leq max\left| N_{1j}(\omega i) — N_{2j}(\omega_i) \right|,$$

$$\Delta_2 \leq max\left| G_{1j}(\omega i) — G_{2j}(\omega_i) \right|,$$

where the index i means the value of the value at the point with the number, when sampling the studied interval of change in the angular velocity of the rotor ($i=1\text{-}12$). Friction coefficients k_1, k_2, k_{11} as well as effective cross — sectional areas of leakage channels from the chamber — f_u and flows between the chambers-$f_p$were taken as the required parameters of the mathematical model. At the first stage, when using the MOVI software package, the values Δ_1 and Δ_2 are set quite arbitrarily. If as a result of calculations in the test table there is no Pareto–optimal solution vector (k_1, k_2, k_{11}, f_u, f_p), then Δ_1 and $\Delta_2$should be increased. If there are several Pareto–optimal vectors, then it is necessary to reduce Δ_1 and $\Delta_2$so that there is one Pareto–optimal vector, which will give the desired values of the parameters. Several series of calculations were carried out with the number of tests from 256 to 2048 (for greater reliability) and the range of parameters k_1= 0.1-0.2; k_2= 0.1 — 0.2; k_{11}=0-0.12 Ns/m, f_u=0-37$^{10\text{-}6}$ m^2, f_p =0-37$^{10\text{-}6}$ m^2. As a result, the following values were obtained: k_1 =0.145, k_2 =0.132; k_{11}=0.026 Ns/m, f_u=4.$^{1210\text{-}6}$$m^2$; f_p =8.2$^{10\text{-}6}$$m^2$. In Figure4 the graphs of the developed power and consumed flow rate of 0.5 MPa (curves 5, 7) and 0.6 MPa (curves 6, 8) calculated according to the identified mathematical model are presented. The coincidence of the calculated and experimental curves can be considered quite satisfactory.

The main results and conclusions

After determining the required parameters for the PM with z = 4, using the mathematical model, we calculated the characteristics of the PM with a different number of vanes (from 4 to 9), as well as a modified PM, under the assumption that the identified parameters remain almost unchanged. The calculation results are presented in table 1 and in Figure6. The analysis shows that increasing the number of vanes leads to certain decreasing in the amplitude of torque ripples within one revolution of the shaft.

With the variation of the number of vanes, the optimal values of the air distribution angles can take slightly different values than for a motor with z = 4. Using the MOVI program, optimization calculations of the inlet end angle and the start exhaust angle for PM with vanes z = 4–9 were carried out. As quality criteria were selected: developed capacity $N \geq N_1$; specific consumption $q \leq q_1$; and amplitude of torque ripples $\delta \leq \delta_1$(criteria with index 1 relate to PM with air distribution angles). Figure 6 shows graphs illustrating the torque ripples in the nominal mode for the RP42-55 model with z = 4 (curve 1) and the modified PM, also with z = 4 (curve 2), as well as the PM with z = 5, respectively curves 3 and 4. We can state a significant decrease in the amplitude of torque ripples and a doubling of the frequency for the modified PM, with a slight increase in the specific consumption of compressed air.

The choiceof the number of vanes or use a modified design of a vane PM depends on the task, i.e. requirements for PM on the amplitude of the torque ripples and the specific flow rate of compressed air and for each specific case is decided individually. In the future, it is planned to carry out similar work with vane PM models having different ratios of geometric parameters and, accordingly, other nominal revolutions and powers, as well as with a special antifriction coating on the stator working surface and composite plate material providing greater bending strength with a smaller thickness and, accordingly, mass.

Table 1. Indicators in the nominal operation mode of the PM with optimized values of the air distribution angles

Number of vanes z	Corner speed ω, c^{-1}	Power N_{max},кWt	Specific consumption q,m^3/(min кWt)	M_{max}-M_{min},Nm	Intake end angle.grad.	Exhaust start angle,grad..
4	1000	0,72	1,42	0,128	70	221
5	950	0,7	1,51	0,08	90	214
7	850	0,614	1,77	0,023	97	206
8	800	0,54	1,96	0,016	89	223
9	750	0,435	2,27	0,011	78	243
			Modified pneumatic motor			
Z=4	1000	0,717	1,56	0,05	70	221
Z=5	950	0,694	1,63	0,028	90	214

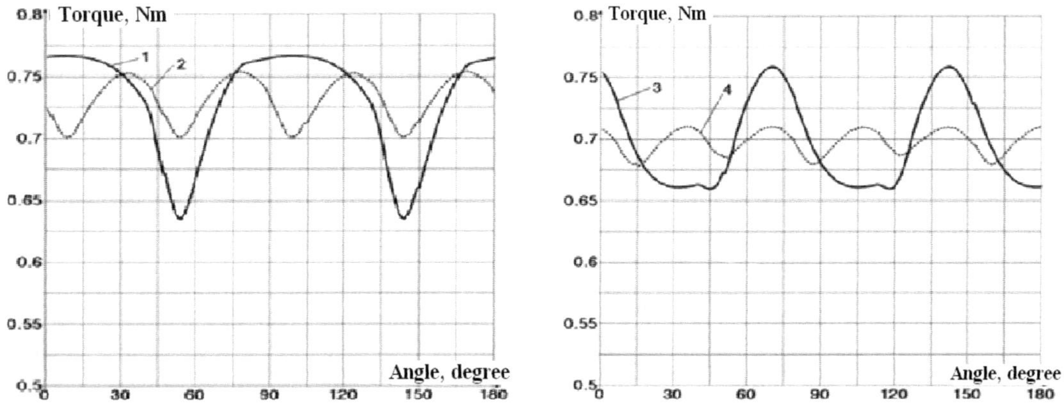

Figure6. The amplitude of fluctuations in torque for the original and modified PM

Select the number of plates or the use of a modified design of the plate depends on PM tasks, i.e. requirements for PM amplitude of fluctuations in torque and specific consumption of compressed air and each case is decided individually. In the future, it is planned to carry out similar work with models of plate PM, having other ratios of geometric parameters and, accordingly, other nominal revolutions and power, as well as with a special antifriction coating of the working surface of the stator and the composite material of the plates, providing greater bending strength at a lower thickness and, accordingly, weight.

The research was supported by Russian Foundation for Basic Research, project No. 18-29-10072 mk.

References

[1] Geurts P. Development of a vane motor servo system for presision positioning in an MR- and CT-environment.// MSc Report, 2017, University of Twente, Netherlands, 71p. https://essay.utwente.nl/73519/1/geurts_MA_EEMCS.pdf.

[2] Patvin J., Agnew M., Woert C. An ergonomic comparison of pneumatic and electrical pistol grip hand tools.// Int. Journal of Industrial Ergonomics. 2004, 34, p. 467–478.

[3] Hoang K., et al. Flux Control Range Broadening and Torque Ripple Minimization of a Double Excitation Synchronous Motor.// IEEE Transactions on Magnetics. 2017, v. 53, №1, pp. 1–10.

[4] Chen-Wei Cheng et al. Modeling and Design of Air Vane Motors for Minimal Torque Ripples.// J. of Mechanical Design. 2012, Vol. 134, p. 051003-1–051003-10.

[5] Carfagni M., Furferi R., Volpe Y. A novel tool for pneumatic spindle design.// Proc. of the IMProve 2011 Int. conf. on Innovative Methods in Product Design.2011, Venice, Italy, p.266–271.

[6] Luo X., Wang J., Spanin L. Development of a Mathematical Model for Vane-type Air Motors with Arbitrary N vanes.// Proc. of the world Congress on Engineering. 2008, V.1, London, p.362–367.

[7] Merkelboch S., Mas I., Murrenhoff H. Mathematical Modeling of pneumatic vane motor inMatlab/Simulink.// The 10-th IFPS Int. Symp. ofFluid Power. 2017,oct. 24–27. Fukuoka, Japan, p. 2B20; 1–10.

[8] Ivlev V.I., Bozrov V.M. Modifying a Plate-Type Pneumatic Motor to Operate on Compressed Air without the Spraying of Lubricant.// Journal of Machinery Manufacture and Reliability. 2012, Vol.41, №4, pp. 279–283. Allerton Press. ISSN 1052–6188

[9] Gerts E.V. Dinamica pnevmaticheskikh system mashin. Moscow, Mashinostroenie, 1985, 256 p.

[10] Sobol I.M., Statnikov R.B. Wybor optimalnykh parametrov v zadachakh so mnogimi kriteriami [Selection of optimal parameters in problems with many criteria]. Moscow, Drofa, 2006, — 176 p. (in Russian)

[11] N. Sosnovsky, D. Ganieva. Mathematical modeling and analysis of the electrohydraulic actuator with the throttle control.// IOP Conf. Series: Materials Science and Engineering 589 (2019) 012016.

[12] S. Semenov, D. Kulakov. Mathematical modeling of the mechanisms of volumetric hydraulic machints.// IOP Conf. Series: Materials Science and Engineering 492 (2019) 012042.

[13] Anuriev V.I. Spravochnik konstruktora-mashinostroitelja v 3-kh tomakh [Handbook of the designer-machine-builder: in 3 volumes. Volume 2].Moscow, Mashinostroenie, 2006,960p. (in Russian)

[14] E. Morozova, D. Bondareva and A. Shablovsky. Optimization of centrifugal pump guide vanes.// IOP Conf. Series: Materials Science and Engineering 589 (2019) 012012

[15] A. Protopopov, A. Mukhlaeva and E. Melnichuk. Optimization of axial volumetric drive aviation pump.// IOP Conf. Series: Materials Science and Engineering 589 (2019) 012004.

[16] V. Tkachyk, V Atamas and A. Shablovskiy. On the centrifugal pump optimization question.// IOP Conf. Series: Materials Science and Engineering 589 (2019) 012011.

[17] Ivlev V., et al. Parameterization of an Air Motor Based on Multiobjective Optimization and Decision Support.// Journal of Manufacture and Reliability. 2013, Vol. 42, №5, pp. 353–358.

Study of an inducer and axial vortex stage

A Ankudinov[1], D Klimenko[2], S Timushev[2,3] and A Regeda[1]

[1]Bauman Moscow State Technical University
[2]MAI (Moscow Aviation Institute)

[3]E-mail: irico.harmony@gmail.com

Abstract. Increasing the intake capacity and reducing vibration loads of high-speed pumps is an important task in the development of power systems in the aerospace industry, as well as in the application of such pumps in the energy and oil production sectors. With an improvement in cavitation qualities, pumps can operate at a higher shaft speed, and at a given shaft speed, with lower net positive suction head, that is, at reduced inlet pressure. With increasing shaft speed, the dimensions and weight of the pumps decrease. To increase the cavitation qualities of the pumps, an axial vortex stage (AVS), which is an object of comparison with the inducer, is considered as an upstream device. It is necessary to study the fact of pressure pulsations decrease when using the AVS. The results obtained with computational fluid dynamics to analyze pressure pulsations in the inducer and in the AVS are used in this article.

Introduction

Increasing the requirements for the characteristics of centrifugal pumps: cavitation qualities, reliability, the absence of low-frequency pulsations, the stability of operation in off-design modes, necessitate the search for means for their further improvement. The methods for improving the cavitation qualities of centrifugal pumps are known: the implementation of the first stage centrifugal impeller with an increased neck area at the inlet; the use of a booster pump with a lower shaft speed than the main pump; installation of an upstream jet pump; the use of an inducer. Most often, in order to improve cavitation qualities, an upstream axial screw impeller — an inducer, is installed at the entrance to the first stage centrifugal impeller. However, with good cavitation qualities, an inducer has significant disadvantages. When an inducer operates at flow rates less than 50% of the optimum, back flows are formed, amplifying with decreasing flow rate. The formation of back flows leads to increased vibration, unstable operation on the stall branch of the cavitation characteristic, low-frequency pressure pulsations. One of the promising ways to improve the cavitation qualities of the pump, reduce noise, vibration and low-frequency pulsations is to install the axial-vortex stage at the pump inlet [1]

An axial-vortex stage (AVS) has certain advantages over an inducer: a higher head coefficient, better cavitation qualities, and stability of work at low flow rates [2–4]. Experimental data show that the use of AVS reduces the level of pressure pulsations and vibrations. All these properties of the AVS determine the prospects of its use as an upstream stage of a centrifugal pump.

AVS is a pump consisting of an axial screw and a fixed helical cascade located on its periphery, the of the vanes of which is opposite to the agles of the blades of the screw. The fluid flow in the AVS [5] has a complex tree-dimensional character with a vortex flow at the periphery. Simplistically, the entire flow can be divided into two zones: the main axisymmetric flow located near an inducer hub, and the vortex flow in the peripheral part of the axial screw inducer and in the fixed helical cascade.

Content from this work may be used under the terms of the Creative Commons Attribution 3.0 licence. Any further distribution of this work must maintain attribution to the author(s) and the title of the work, journal citation and DOI.
Published under licence by IOP Publishing Ltd

The results of experimental studies of AVS

Tests of the AVS models [4–6] were carried out at various shaft speed with different geometry of the flow part. The tested AVS had a relative axial length $\overline{l} = l / D_s = 0{,}79 - 2{,}12$ and a relative diameter of the fixed helical cascade $\overline{D_p} = D_p / D_s = 1{,}11 - 1{,}33$; hub ratio $\overline{d_{sl}} = d_{sl} / D_s = 0{,}33 - 0{,}69$; the angles of installation of the inducer blades along the outer diameter of the inducer were $\beta_s = 8 - 15°$ and of fixed helical cascade were $\beta_g = 8 - 13°$.

The models were tested with a different shape of the inlet edge, a radial clearance between the inducer and the fixed helical cascade assembly, with a variable inducer hub and a variable pitch of its blades.

Fig. 1 shows the performance curves of several variants of AVS with different axial lengths.

For comparison with the performance curves of the AVS on the same graph shows the performance curve of the inducer with the same geometry, which does not have a fixed cascade on the periphery. The performance curves of the inducer without a fixed cascade, differing only in axial length were also studied. The inducer length was $l_s = 256$ mm; 400 mm; 200 mm; 150 mm. The performance curves of the inducer without a fixed cascade practically did not change when the axial length changed. It should be also noted that the head of the inducer without a fixed cascade at the periphery (variant 4) is significantly lower than the pressure of the AVS. The maximum flow rate of all variants, including a conventional inducer, has one value, which corresponds to the flow rate of an shocklessflow input to the inducer blades.

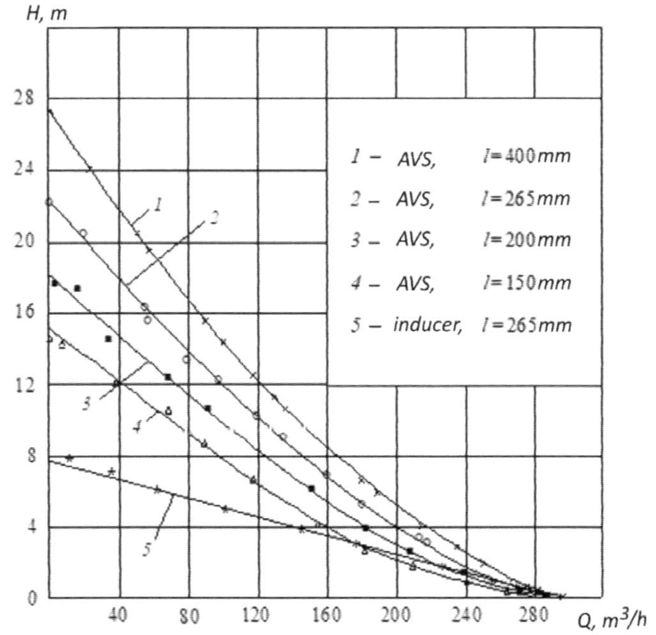

Fig. 1. The performance curves of the AVS with different axial inducer length, n=1500 rpm

Starting from the relative length, which corresponds to a length equal to the diameter of the inducer $\overline{l} = l / D_s = 1$, the pressure coefficient ψ of the AVS variants from the relative axial length when flow rate equal to zero increases almost linearly. This AVS property is similar to the characteristics of labyrinth pumps [7]. The use of AVS makes it possible to obtain the required head, increasing only the axial length. It allows having operating parameters in a larger range than for axial screws that do not have a fixed helical cascade on the periphery, at the same shaft speed.

The experiment confirmed the selected physical model of the AVS and the assumption that the head should increase with increasing mass of the fluid taking part in the secondary flow.

According to the accepted flow model, the theoretical head is proportional to the mass of the secondary flow taking part in the vortex interaction. The mass of the secondary stream depends both on the axial length and on the height of the blades of the movable and fixed helical cascade.

The pressure coefficient of the AVS for the variants $\overline{D_p} = D_p / D_s = 1{,}11; 1{,}22; 1{,}33$ of the fixed cascade changes linearly with the relative diameter at zero flow rate.

To study the development of cavitation in the AVS, visual studies were carried out to determine the nature and direction of the flow in the channels, various stages of the development of cavitation were noted. Fig. 2 is photograph of the fluid flow in the AVS channels under stroboscopic illumination with developed cavitation in the channels.

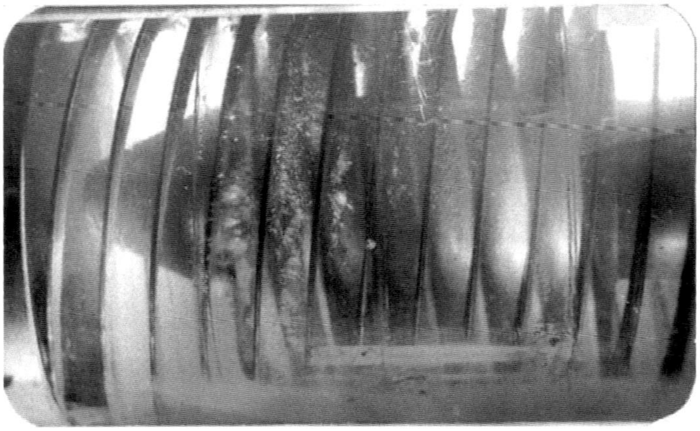

Fig.2. Fluid flow with developed cavitation in the channels.

The presented results of researches confirm the selected flow model, consisting of the main axisymmetric flow and the secondary vortex flow, on the periphery of the inducer and in the fixed channels.

As it was established in experimental studies, in the axial vortex stage in all operation modes of NPSH, there are no low-frequency pressure pulsations, with the development of cavitation, the levels of pulsation and vibration decrease. It is explained by the fact that the cavitation cavities are localized in a vortex core located in the central part of the helical channels. The collapse of the caverns does not occur on the walls of the casing and blades, as in an ordinary inducer, but in a fluid flow. With decreasing pressure, the volume of cavitation cavities increases. A vortex cavitation core extends over fixed channels over a greater distance, which leads to intense damping of pulsations in the liquid and reduces the vibration level.

Computational fluid dynamics methods

Nowadays, the use of computational fluid dynamics methods for analyzing and optimizing the pumps flow part [8–13] is being standard for engineers. Numerical simulation using computational fluid dynamics methods makes it possible to analyze the influence of the unsteady distribution of parameters in the inducer and AVS on the amplitude and spectrum of pressure pulsations. The model uses the continuity and Navier-Stokes (1,2)equations:

$$\frac{\partial \rho}{\partial t} + \nabla(\rho V) = 0 \tag{1}$$

$$\frac{\partial \rho \cdot V}{\partial t} + \nabla(\rho \cdot V \times V) = -\nabla P + \nabla((\mu + \mu_t) \cdot (\nabla V + (\nabla V)^{\tau})) \tag{2}$$

These equations are supplemented by the corresponding equations of the turbulence model [12]. The calculations were carried out using the turbulence model (kinetic energy of turbulence k – ε, where ε is dissipation rate), in which the turbulent viscosity μ_t is determined by following expression:

$$\mu_t = C_\mu \cdot \rho \cdot \frac{k^2}{\varepsilon} \tag{3}$$

$$C_\mu = 0.09 \tag{4}$$

Standard wall functions are used on the wall.

The numerical method [13,14] is realized on a rectangular mesh with local adaptation and mesh resolution in areas with complex geometry of the mesh computational domain. In the radial clearance zone, the "gap cells" method is used to replace the Navier-Stokes equation with the Poiseuille equation.

To combine numerical solutions in the rotor and stator of the pump, the sliding mesh method is used. The transfer of flow parameters from the rotating to the stationary calculation area is carried out by a special interface "sliding surface", which provides interpolation of the flow parameters taking into account the "virtual" angular displacement of the meshes.

Fig. 3.shows the computational mesh of the second adaptation level for the AVS. The border of the sliding surface between the fixed mesh and the inducer is highlighted in blue color.

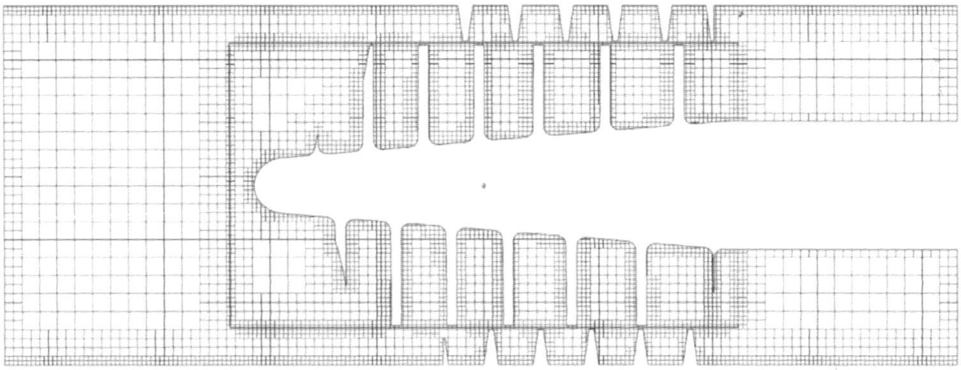

Fig.3. The computational mesh of the second adaptation level for the AVS

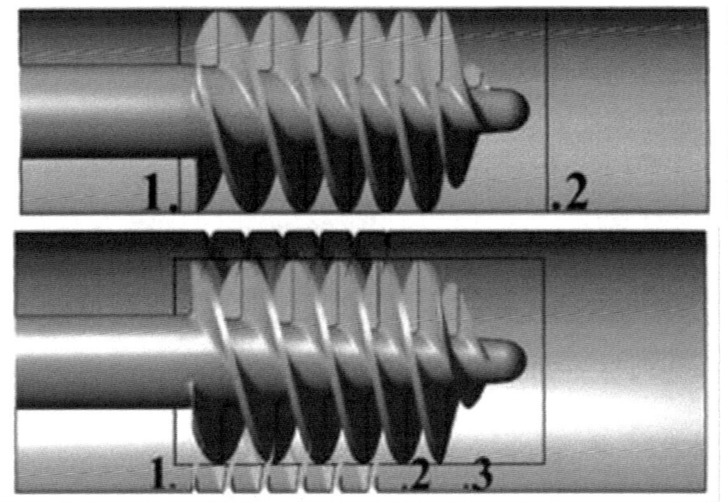

a) inducer

b) AVS

Fig.4. The selected points of pressure pulsations recording a) inducer b) AVS

Numerical modeling is carried out using the direct non-stationary method with the development of a solution from zero initial conditions to a quasi-stationary mode of operation, the output of which is controlled by the value of head. Typically, this mode is achieved within ten to fifteen full revolutions of the rotor. Data on pressure pulsations at selected points of the flow part are recorded after reaching the stationary mode.

The size of the computational mesh was around 500,000 cells, the time step of calculation was 4.4 e-5 s.

Results of numerical simulation of pressure pulsations in an inducer and AVS

The calculations were carried out at the optimum flow rate. For AVS, the results differ at different points. At the periphery, in the area of the fixed cascade of the AVS, there are zones of recirculation vortex flow.

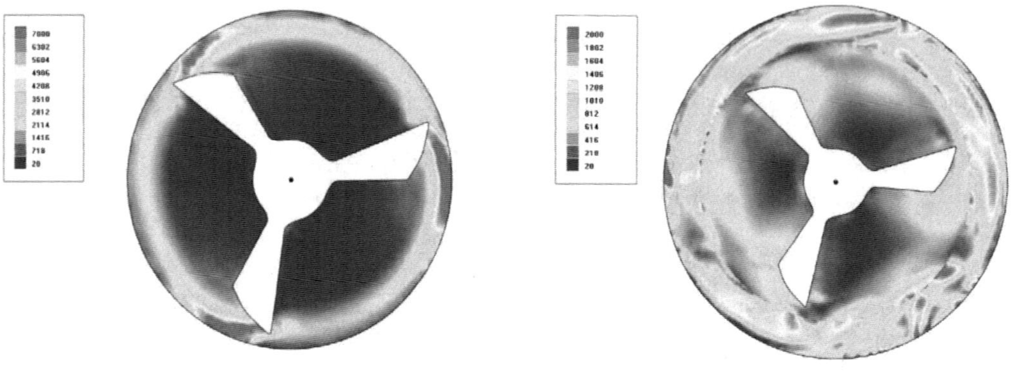

Fig. 5. Instant input vorticity field

Such velocity field inhomogeneity is also reflected in the inhomogeneity of the pressure field, especially at the periphery of the input part of the AVS. An inhomogeneous field of velocities and pressures generates random low-frequency pressure fluctuations at the input of the upstream AVS stage, which are well fixed at point 2 (Fig. 5). When moving away from the inlet of the stator channels, the low-frequency oscillations decrease, the discrete component of the blade passing frequency (BPF) of the inducer and its second harmonic prevail in the spectrum.

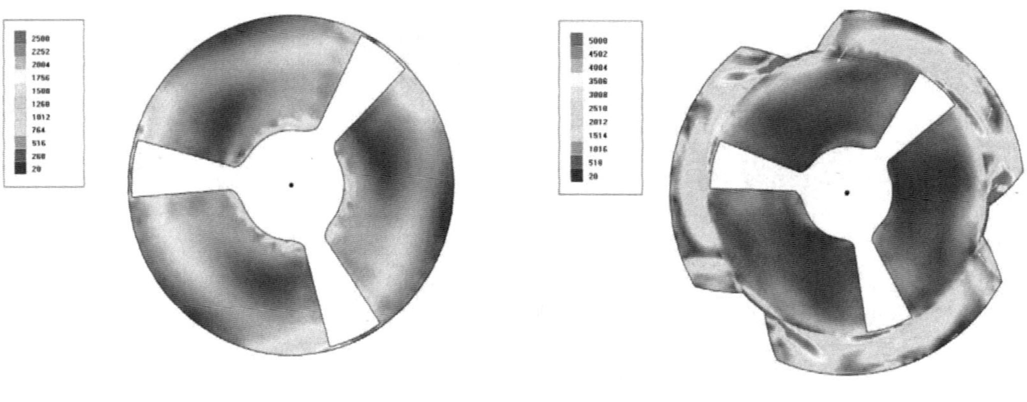

Fig. 6. Instant output vorticity field

IOP Conf. Series: Materials Science and Engineering **779** (2020) 012026 doi:10.1088/1757-899X/779/1/012026

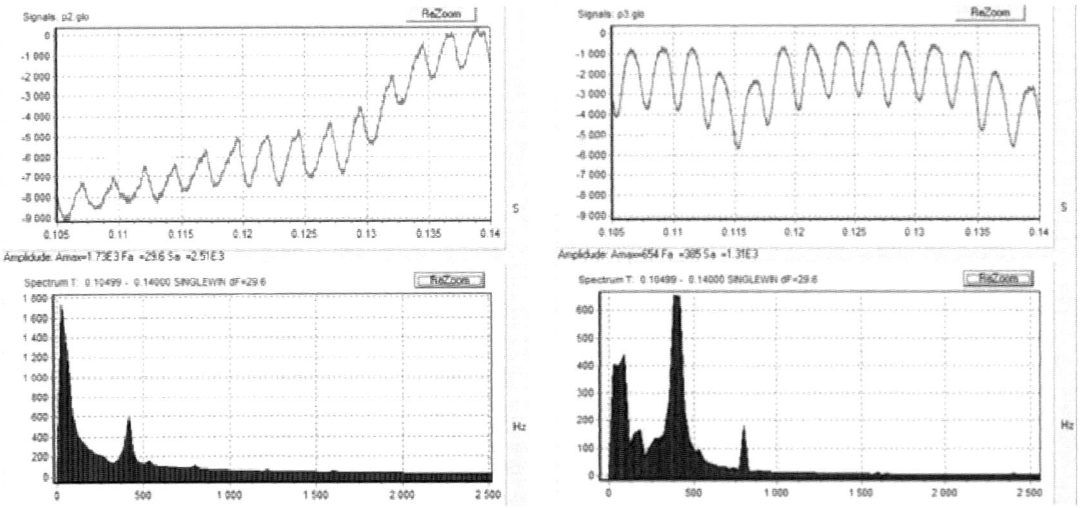

Fig. 7. Pressure pulsations at the inlet of the AVS at point 2 (on the left) and point 3

An analysis of the velocity and pressure fields in different sections of the flow part of the inducer and an AVS shows more equilibrium pressure distribution over the inducer blade grid in an AVS. This ensures a decrease in the BPF of the inducer at the outlet of the AVS compared with an single inducer almost in 2 times.

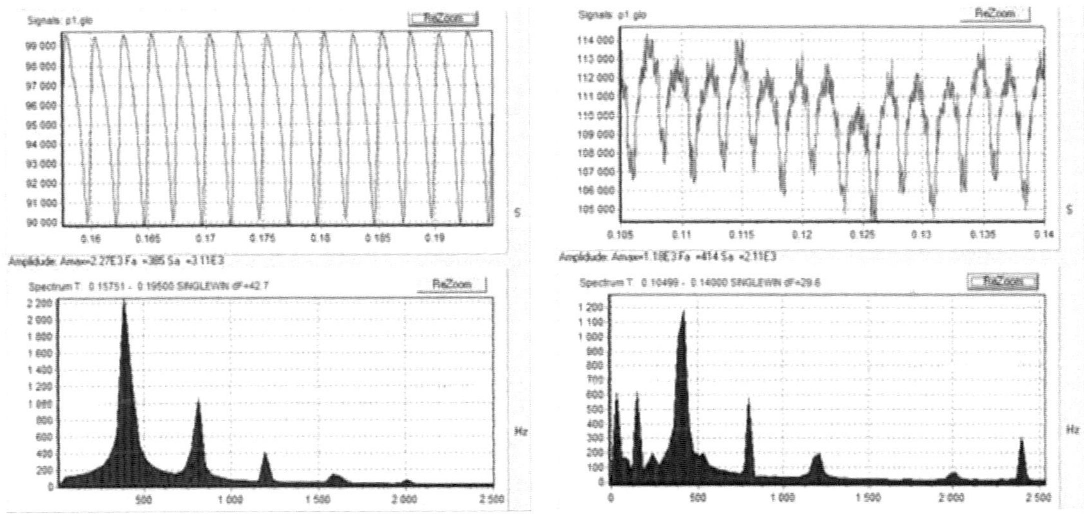

Fig.8. Pressure pulsations at the outlet of inducer (on the left) and AVS

Discussions

In the spectrum of pressure pulsations of the AVS at the inlet, low-frequency oscillations are recorded, which are provoked by vortex disturbances when the flow enters on the fixed cascade vanes. Oscillations at the shaft speed are also observed, which is explained by the circular asymmetry of the pressure field due to the interaction of the rotor with the fixed vane cascade of the AVS. Further studies of these phenomena are needed to optimize the position and number of vanes of the fixed cascade of the AVS.

IOP Conf. Series: Materials Science and Engineering **779** (2020) 012026 doi:10.1088/1757-899X/779/1/012026

List of reference

[1] Zotov B. N., Ankudinov A. A. Axial vortex pump. RF patent 2014509.
[2] Ankudinov A. A., Kirillov A. A. Improving the technical characteristics of the feed pump using an axial vortex stage. A look into the future — 2007. Materials NTK SPb: Publishing house of FSUE "Central Design Bureau MT" Rubin ", 2007.
[3] Ankudinov A. A., Zubov A. A., Misiagina IU. IU. Automated calculation and design of an axial vortex pump stage: a manual. — M.: Publishing house of BMSTU, 2017. — 44 p.
[4] Ankudinov A. A., Kuftov A. F. Energy characteristics of an axial vortex pump // News of Universities. — M.: Mechanical Engineering, No. 2, 1989.
[5] Ankudinov A. A. Calculation and design of the upstream axial vortex stage of a centrifugal pump: a Training manual. — M.: Publishing house of BMSTU, 2003. — 64 p.
[6] Ankudinov A. A., Turkova L.A. The use of the upstream axial vortex stage in the pumps of the Kaluga Turbine Plant. // Current status and development prospects of hydraulic engineering in the 21st century. // Proceedings of MNTK. —SPb: Publishing house of SPbSPU, 2003.
[7] Golubev A. I. Labyrinth screw pumps and seals for aggressive environments. — 2nd ed., Revised. and add. — M.: Mechanical Engineering, 1981. — 112 p.
[8] A Protopopov and V Vigovskij 2019 IOP Conf. Ser.: Mater. Sci. Eng. 492 012003
[9] N Egorkina and APetrov 2019 IOP Conf. Ser.: Mater. Sci. Eng. 492 012015
[10] V Lomakin and O Bibik 2019 IOP Conf. Ser.: Mater. Sci. Eng. 492 012037
[11] V Cheremushkin and V Lomakin 2019 IOP Conf. Ser.: Mater. Sci. Eng. 492 012039
[12] Wilcox, D. C., (1994) "Turbulence modeling for CFD" // DCW Industries, Inc. 460 p.
[13] Software package for gas and fluid flow simulation FlowVision. Version 2.5.0. Manual CAPVIDIA, 1999-2007 Leuven, Belgium
[14] FlowVision User's manual. TESIS LLC, 1999-2017, Moscow

IOP Conf. Series: Materials Science and Engineering **779** (2020) 012027 doi:10.1088/1757-899X/779/1/012027

Research of the relay-linear control of an electro-hydraulic servo drive on FESTO learning system

B Kulakov[1,2] and D Kulakov[1,2,3]

[1]Bauman Moscow State Technical University

[2]Peoples Friendship University of Russia (RUDN University), 6 Miklukho-Maklayastr, Moscow, 117198, Russian Federation

[3]E-mail: kulakov@bmstu.ru

Abstract. The task of a relay-linear control algorithm for an electro-hydraulic servo drive (EHSD) was considered. This algorithm provides reconfiguration of the state components feedback coefficients at the very initial stage of the control object movement, when the optimized control action does not depend on these coefficients. A FESTO hydraulic learning system combined with Matlab+Simulink Desktop Real Time was used for realization of the developed algorithm. Experimental results are included.

1. Introduction

One of the classical control theory task for a servo system is to follow out some reference signal that changes in time stepwise (some square wave for example). This task can be solved via several methods: as some relay law according to the L.S. Pontryagin maximum principle [1–8], or as some state feedback law [9]. There are various methods of relay-linear control [10–13]. A control law in such methods is a phase state function. The function describes some relay control law that changes during control object motion to some state feedback law nearby an origin of phase coordinates to exclude self-oscillations. N.A. Kuznetsov proposed to use specific EHSD relay-linear control law in his article [14]. The control law (that defines a hydraulic servo valve (HSV) control signal) changes from a relay law to state components feedback when a phase point of controlled hydraulic drive reachs specially built switching surface. Besides that, a relay law consists of a single constant-sign time-interval segment during an initial phase of EHSD step response. Figure 1 shows experimental results for such relay-linear control of the EHSD assembled on a hydraulic research system. (The research system based on the hydraulic learning system FESTO is described in [15]).

An experimental series have been produced to analyze the EHSD step response for a given step values. The EHSD reactions have been compared for the case of a state components feedback [9,15] and for the case of relay-linear control by the method [14]. It turned out that the feedback coefficients could be configured so that the relay-linear control can not significantly improve a quality of the obtained transients (a transient time minimization was considered). This can be explained by the fact that in this case the relay portion of the relay-linear control (Fig.1) coincides with the saturation zone of the control signal supplied to the HSV of the EHSD in the proportional feedback control case.

Figure 2 shows a block-diagram of a simple nonlinear model of the EHSD open loop (control object), realized on the hydraulic research system [15]. In the given bock-diagram the speed and position saturation of a HSV spool and a saturation of the control signal to the HSV are considered.

Content from this work may be used under the terms of the Creative Commons Attribution 3.0 licence. Any further distribution of this work must maintain attribution to the author(s) and the title of the work, journal citation and DOI.
Published under licence by IOP Publishing Ltd

IOP Conf. Series: Materials Science and Engineering **779** (2020) 012027 doi:10.1088/1757-899X/779/1/012027

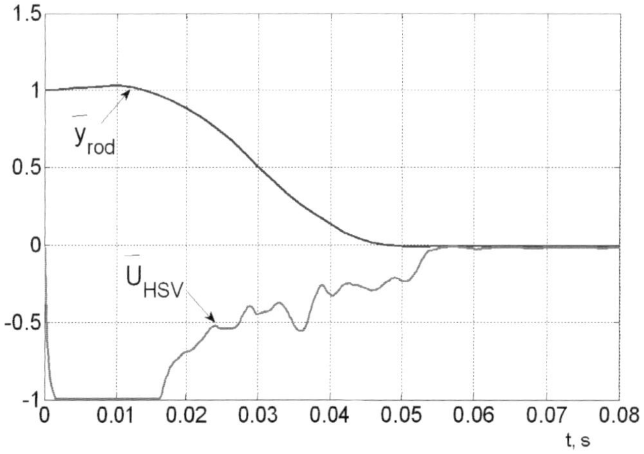

Fig. 1. The EHSD step response and hydraulic servo valve control signal (the signals are normalized, a required step input is M=0.2V) in a case of relay-linear control method described in [14].

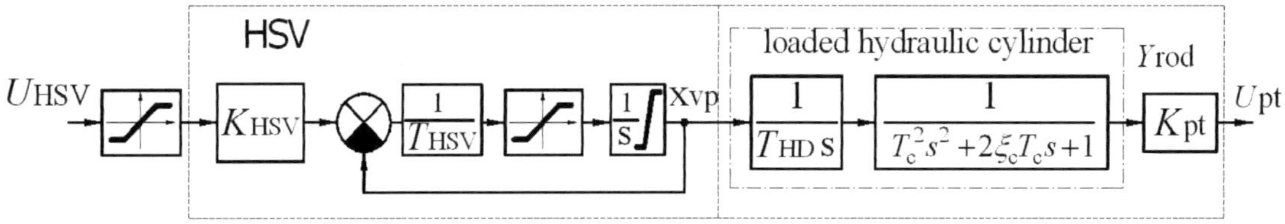

Fig. 2. The EHSD mathematical model block-diagram: K_{HSV} — the HSV gain; T_{HSV} — the HSV time constant; T_{HD} — the hydraulic drive time constant; T_c — the time constant of the loaded hydraulic cylinder; ξ_c — the hydraulic cylinder damping coefficient; K_{pt} — the potentiometric feedback sensor gain; Y_{rod} — the position of the hydraulic cylinder rod; U_{pt} — the rod position feedback signal.

A study on the mathematical model (Fig.2) for the state components feedback law and for a time-dependent relay control law was produced. The same ITAE criterion was used for obtaining the state components feedback coefficients and for defining the switching times of the control signal to the HSV respectively. The results of the numerically researches are shown in Figure 3.

In this paper, we consider a case in which the EHSD parameters determined by its inertial load are not known in advance and the drive with state components feedback follows some required position step input. The EHSD parameters are unchanged during the transition time. The idea of the synthesized controller is based on the two features. Firstly, as a rule, a HSV control signal enters a saturation zone in an initial phase of the EHSD working out the step in our case. In the initial phase an EHSD phase trajectory is determined by its parameters only, including the load's parameters. Secondly, the EHSD parameters determine optimized coefficients of the state components feedback. Therefore, we will first apply to the HSV the maximum possible control action (according to a relay law), with simultaneous analysis of the motion of the actuator in the phase space. As a result of this analysis, the feedback coefficients, optimal for the current EHSD parameters, are selected in real time and the control is switched from relay to linear with the selected coefficients. In common case, the drive may continue to move in the HSV control signal saturation mode after switching to linear feedback control and will come out of saturation mode naturally. That is, we do not determine the operating time of the controller in the saturation mode. We use this time to determine and set during it the optimal state feedback coefficients, depending on the current parameters of the control object.

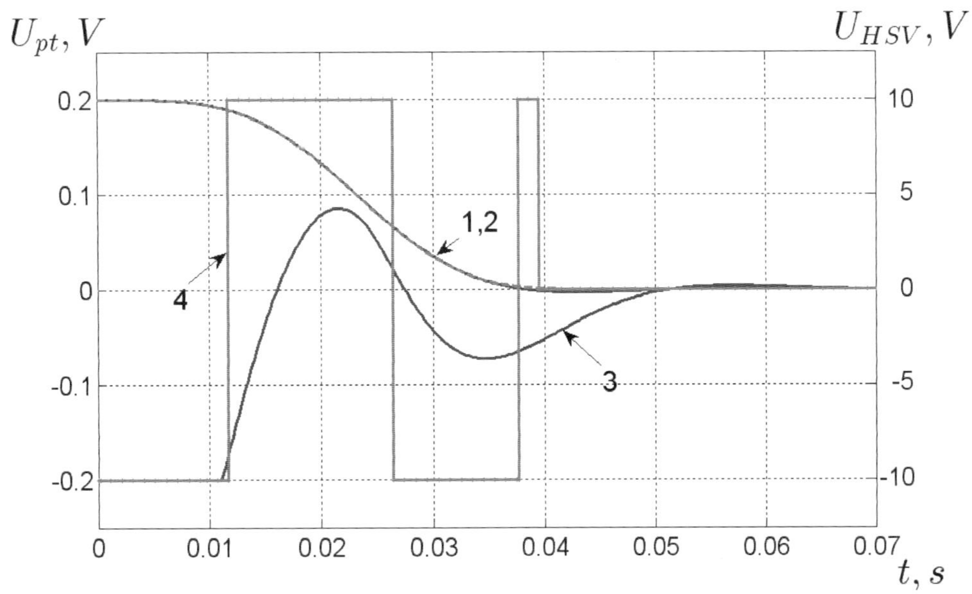

Fig. 3. The EHSD transients U_{pt} (the lines 1,2) and corresponding to them normalized control signals $\overline{U}_{\text{HSV}}$ (the lines 3,4), optimized according to ITAE criterion, for the case of the state components feedback control (the lines 1,3) and for the case of the time-dependent relay control (the lines 2,4).

Relay-linear control realization on FESTO hydraulic learning system

Research of the developed relay-linear control were carried out on the EHSD physical model [15] and on its mathematical model. A scheme of the research system — the EHSD physical model (Fig. 4) is shown in Figure 5.

Fig. 4. The hydraulic research system.

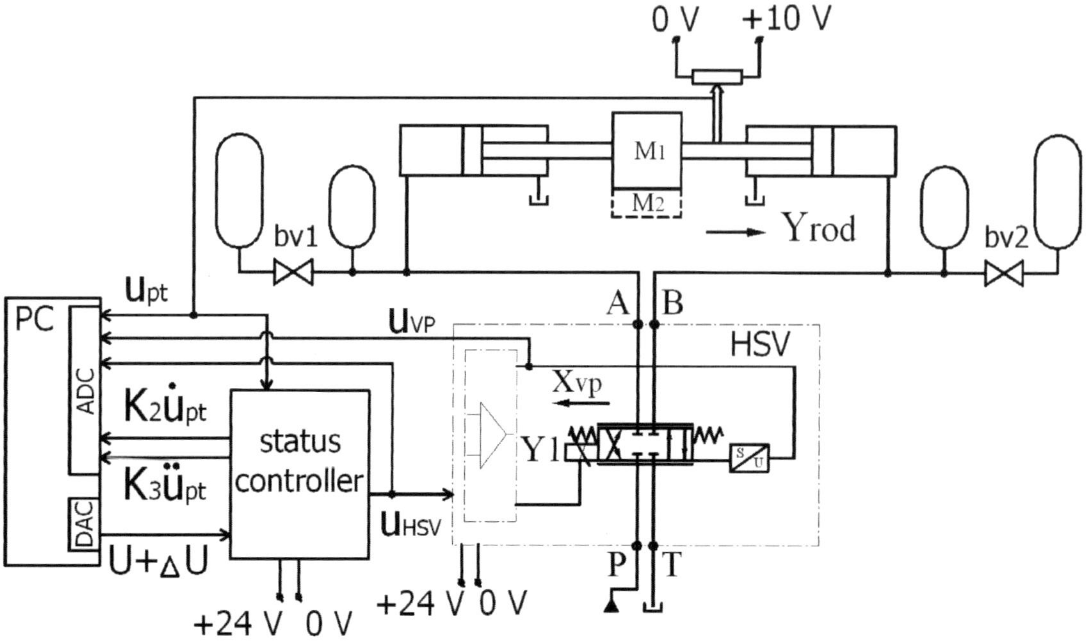

Fig. 5. The scheme of the research system for studying the EHSD with relay-linear control: bv1, bv2 — the ball valves; U_{HSV} — the HSV control signal (in effect a control piston setpoint position value); U_{VP} — the measured position of the HSV control piston; $M_1 = 27.5\,\text{kg}$, $M_2 = 10\,\text{kg}$, — the masses of the inertial load (main and additional, respectively). Status controller — the unit, shortly described below.

"Heavy" and "Light" EHSD physical models (that are named as H and L respectively) are realized on FESTO hydraulic research system. Two models require different optimized feedback coefficients while working out some step signal from a given range of amplitudes. The state components feedback controller of the investigated EHSD is realized by the unit (Part Number 162253), which is a part of FESTO hydraulic learning system. Its feedback coefficients are set manually using the unit potentiometers (Fig.6).

The studied relay-linear control algorithm implies a change in the state components feedback coefficients during the EHSD operation. To do this, an external computer produce dynamic adjustment of the control signal U, providing the required change in the state components feedback coefficients. An additive ΔU to the signal U, supplied to the input of the status controller, (Fig. 7) is calculated as a function of:

$$\Delta U = f_{\Delta U}\left(U, U_{pt}, \dot{U}_{pt}, \ddot{U}_{pt}, {}^{H}\bar{K}, \bar{K}\right).$$

Let us call a proposed control principle (Fig. 7) as a state components feedback with dynamic correction. State components feedback coefficients $\left[{}^{H}K_1 \quad {}^{H}K_2 \quad {}^{H}K_3\right]$ are tuned on the status controller as time-invariable set (fig.6, 7). Let's write the following form for the status controller:

$$U_{HSV} = f\left(U + \Delta U, U_{pt}\right) =$$
$$= \mathrm{K_{amp}} \cdot {}^{H}\mathrm{K}_1 \cdot \left[\mathrm{U} + \Delta\mathrm{U}\right] - \mathrm{K_{amp}} \cdot \left({}^{H}\mathrm{K}_1 + {}^{H}\mathrm{K}_2 \cdot s + {}^{H}\mathrm{K}_3 \cdot s^2\right) \cdot U_{pt}, \quad (1)$$

IOP Conf. Series: Materials Science and Engineering **779** (2020) 012027 doi:10.1088/1757-899X/779/1/012027

where ΔU — is an additive to the U signal supplied to the input of the status controller (fig.6,7). The aim of the ΔU adding is to obtain the required control law with state components feedback:

$$U_{HSV} \overset{\Delta}{=} \mathrm{K}_{amp} \cdot \mathrm{K}_1 \cdot U - \mathrm{K}_{amp} \cdot \left(\mathrm{K}_1 + \mathrm{K}_2 \cdot s + \mathrm{K}_3 \cdot s^2\right) \cdot U_{pt} \quad (2)$$

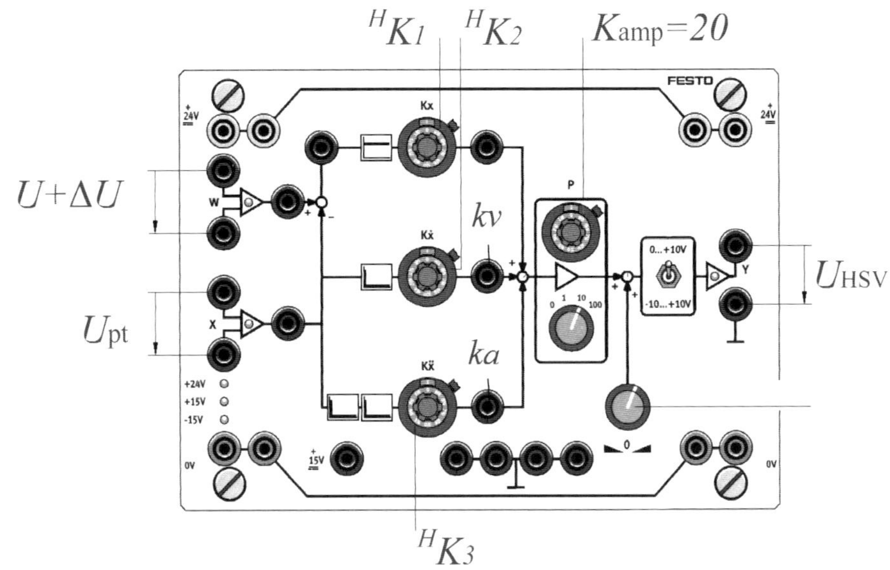

Fig. 6. Status controller front panel.

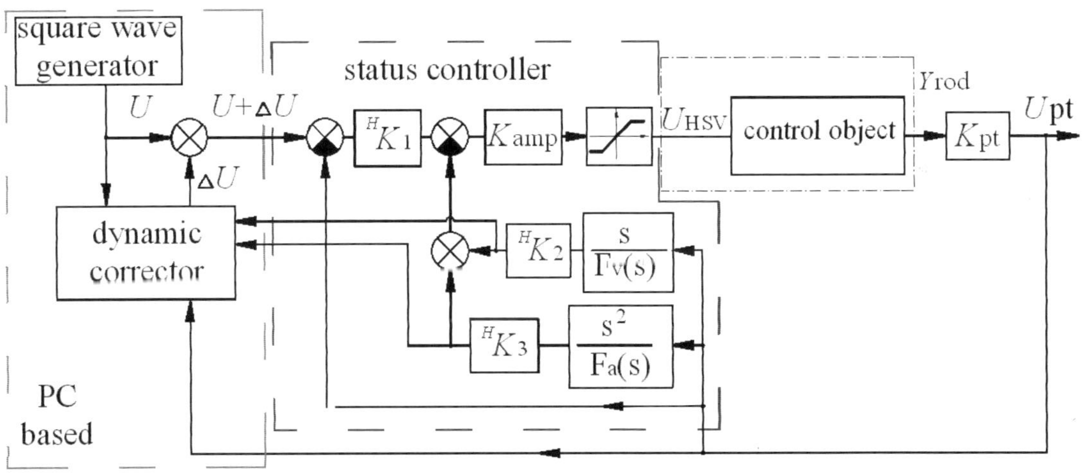

Fig. 7. A block scheme shows an adjustment of the state components feedback coefficients by the control signal dynamic correction: $\mathrm{F_v}(s), \mathrm{F_a}(s)$ — the low pass filters for derivative and double derivative computing.

The required law of ΔU is determined by the equation:

$$[\Delta U] = \frac{\left(\mathrm{K}_1 - {}^H\mathrm{K}_1\right) \cdot U + \left[\left({}^H\mathrm{K}_1 - \mathrm{K}_1\right) + \left({}^H\mathrm{K}_2 - \mathrm{K}_2\right) \cdot s + \left({}^H\mathrm{K}_3 - \mathrm{K}_3\right) \cdot s^2\right] \cdot U_{pt}}{{}^H\mathrm{K}_1} \quad (3)$$

Let's introduce notations: $kv = {}^{H}\mathrm{K}_2 \cdot \dot{U}_{pt}$, $ka = {}^{H}\mathrm{K}_3 \cdot \ddot{U}_{pt}$ — signals on the corresponding outputs of the status controller unit (Fig. 6), and we get the expression for the additive ΔU:

$$[\Delta U] = \begin{bmatrix} \dfrac{{}^{H}\mathrm{K}_1 - \mathrm{K}_1}{{}^{H}\mathrm{K}_1} & \dfrac{{}^{H}\mathrm{K}_2 - \mathrm{K}_2}{{}^{H}\mathrm{K}_1} & \dfrac{{}^{H}\mathrm{K}_3 - \mathrm{K}_3}{{}^{H}\mathrm{K}_1} \end{bmatrix} \cdot \begin{bmatrix} U_{pt} - U \\ \dfrac{kv}{{}^{H}K_2} \\ \dfrac{ka}{{}^{H}K_3} \end{bmatrix} \quad (4)$$

Figure 8 shows experimental transients of the hydraulic cylinder rod and the signals supplied to the input of the status controller unit when working out the step action U=0.7 V in two modes of the EHSD controller operation. The first mode: the coefficients of the state components feedback loops K =[1.64 0.00325 0.00016] are set by the potentiometers on the status controller unit. The second mode: ^{H}K =[1.58 0.0154 0.000353] are set the same way on the unit and the resulting feedback coefficients K are obtained by adding signal ΔU (4) to the input signal U. The input of the status controller is their sum.

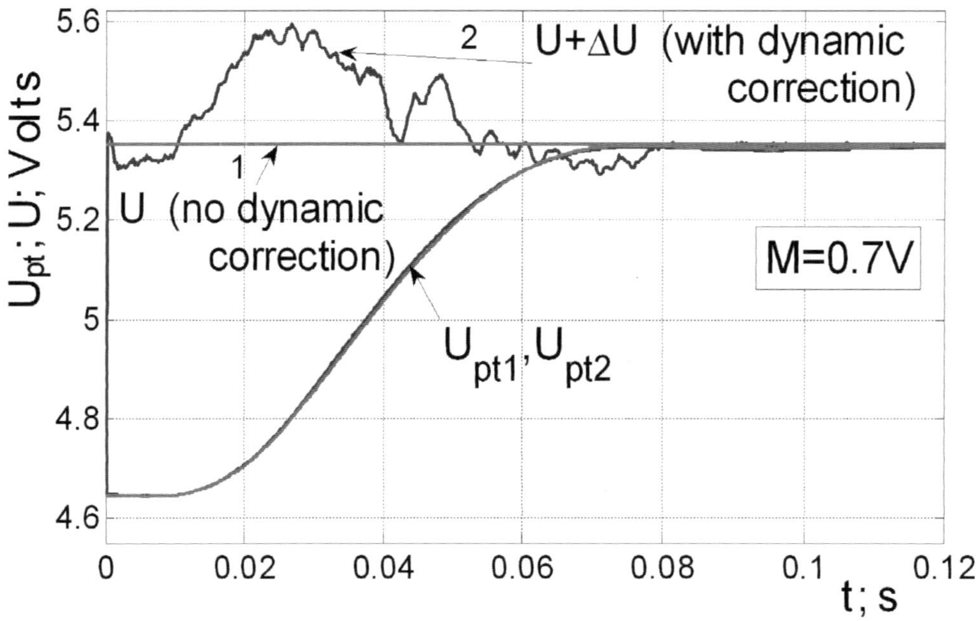

Fig. 8. The EHSD 0.7V step response and status controller input signal for the two cases: 1 — NO dynamic correction, 2 — WITH the dynamic correction (Figure shows experimental results for the EHSD physical L — model).

The resulting EHSD block-diagram with synthesized relay-linear controller is shown in Figure 9.

Figure 10 shows typical control signal U (Fig. 7) during some step change of desirable position of the hydraulic cylinder rod (a square wave with superimposed relay control). At time t_1 the control object phase-point reaches a switching hypersurface. The hypersurface defines switching from relay to linear control. The state components feedback coefficients are assigned as a result of location analysis of the point. Also, at time t_1 the feedback control is realized with selected coefficients by using the status controller with the dynamic correction (Fig.7).

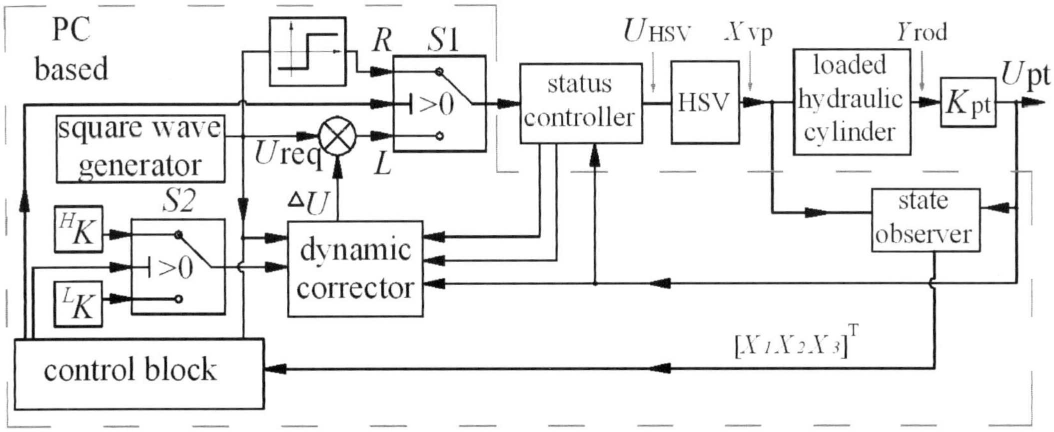

Fig. 9. The EHSD block-diagram with synthesized relay-linear controller.

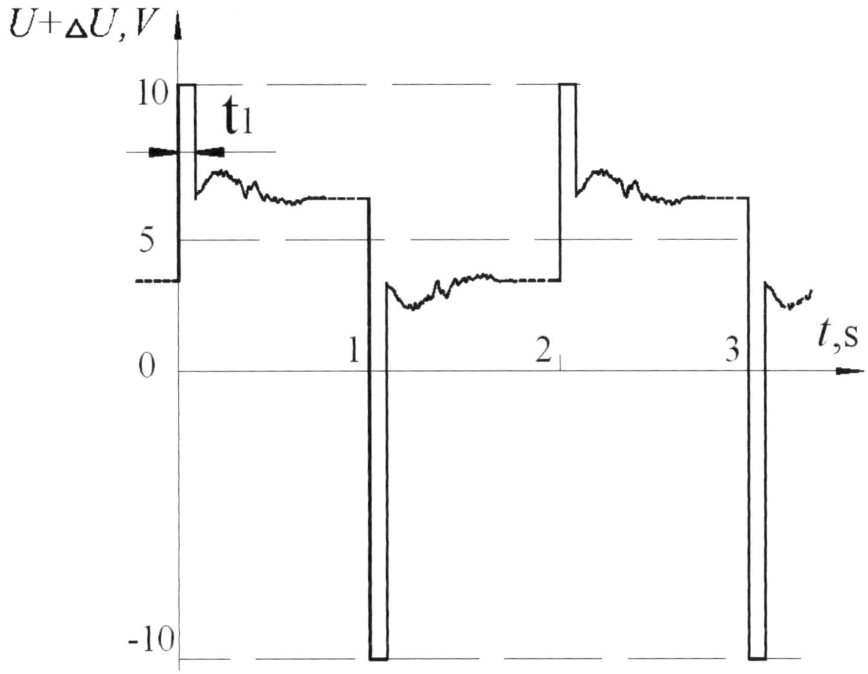

Fig. 10. Status controller input signal $U + \Delta U$ during relay-linear control.

Analysis of the relay-linear control work

We use the approach described in [15] to physically simulate the operation of loaded hydraulic drives on FESTO hydraulic learning system. Two physical models with different parameters were formed for the experimental study of the relay-linear control on the experimental complex described above. Next, we will denote them: L — is less loaded model and H — is more loaded. There are two differences between these physical models. Firstly, the different masses of the inertial load, moved by the hydraulic cylinders (M_1=27 kg and M_1+M_2=37 kg, respectively) are used. Secondly, the different additional volumes connected to the piston cavities of the hydraulic cylinders (0.00006m^3 and 0.000185m^3, respectively) are used (Fig.5).

We will conduct studies on two physical models (L and H) with two sets of state components feedback coefficients K_L and K_H optimized for their respective models. Optimized coefficients of the state components feedback for each physical model were determined in advance. The relay-linear control algorithm described in this article has a task to select the "correct" coefficients from a set of previously defined. The phase trajectory of the control object, which is used to select the "correct" set of the feedback coefficients, is estimated in real time using a state observer [9, 15], implemented in our experimental complex.

Figure 11 shows experimentally determined phase trajectory segments of the control object in physical models L and H, when working out the step action U=0.7 V.

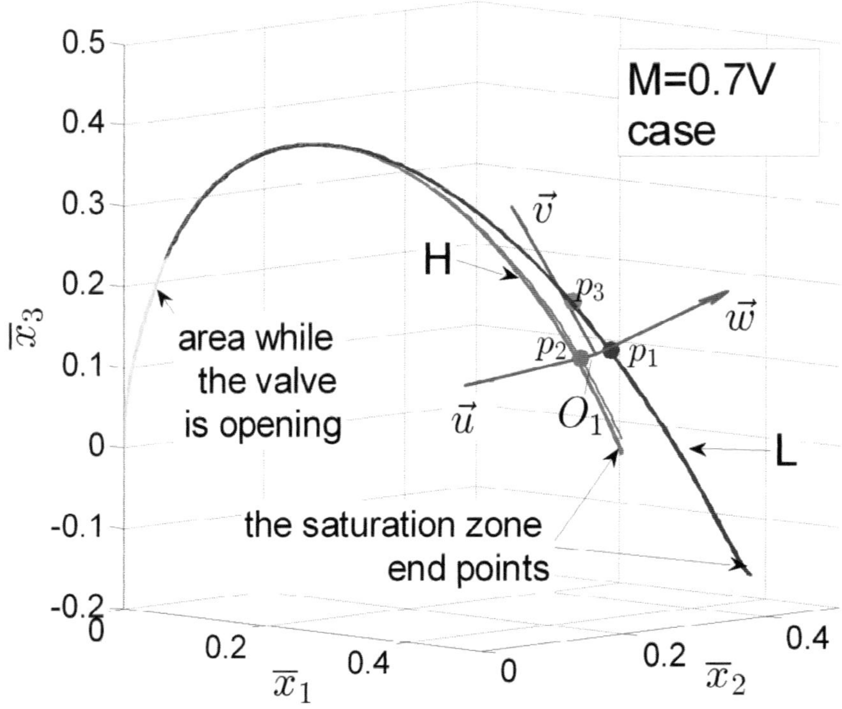

Fig. 11. The experimental phase trajectories while control signal U_{HSV} is in the saturation zone.

Figure 11 shows the parts of the phase trajectories, which are limited by the condition, that the signal U_{HSV} (Fig. 7, 10) is in the saturation zone only $(\overline{x}_1(t)=x_1(t)-x_1(0),\ \overline{x}_2(t)=x_2(t)/40,$ $\overline{x}_3(t)=x_3(t)/3000)$. Each trajectory shown in Figure 11 includes a motion phase trajectory section due to a relay control (before the intersection with a hyperplane O_1uw) and a motion phase trajectory section with a linear control with the state components feedback coefficients optimized for respective model (after that intersection). The HSV is in the saturation zone in the both sections.

The control block (Fig. 9) performs the following functions:

1 in the event of a step required input action, it switches the operation mode of the controller (s1) to relay control;

2 it tracks the phase trajectory of the control object;

3 when a phase point of controlled hydraulic drive reaches hyperplane O1uw (Fig.11):

3.1 it analyzes the position of the intersection point in the O_1UVW coordinate system;

3.2 it assigns the state components feedback coefficients (s2);

3.3 it switches back the operation mode of the controller (s1) to linear control with the selected coefficients.

One can judge the expediency and efficiency of switching to a linear control with "correctly" assigned state components feedback coefficients by the transients shown in Figure 12.

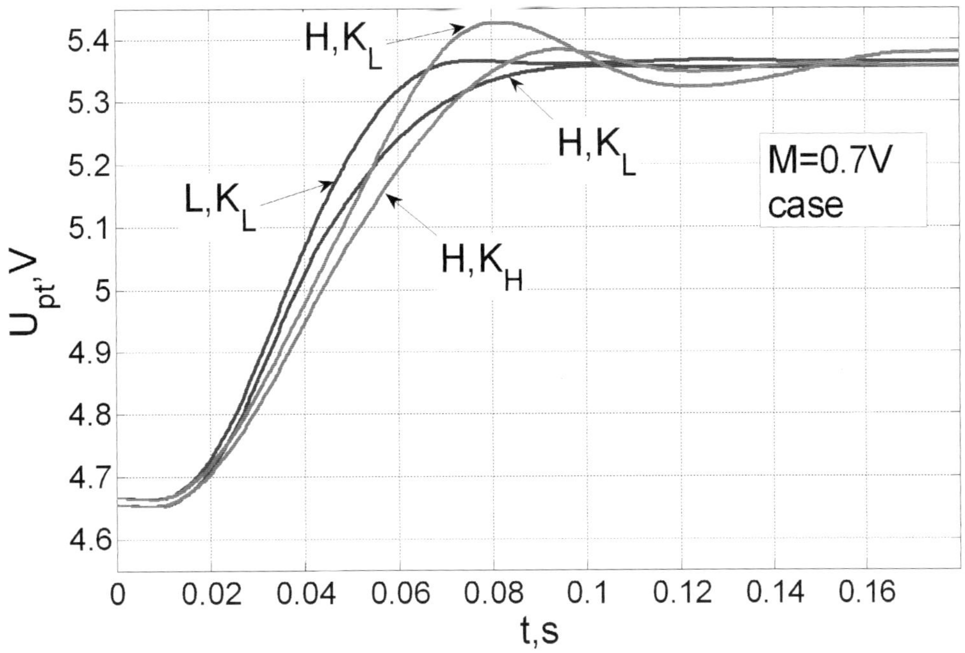

Fig. 12. The experimental step responses for the two EHSD physical models L, H and two status controller coefficients K_L and K_H.

The obtained transients for two physical models L and H are shown in different colors. For each model, optimized state components feedback coefficients K_L and K_H were previously determined. Each physical model was studied with both sets of feedback coefficients.

Solid/dashed lines shows the transients for the EHSD physical models, working with the "correct" feedback coefficients and the feedback coefficients configured for a different model respectively.

Conclusions

Two EHSD physical models with different inertial load parameters were realized on the research system based on the hydraulic learning system FESTO. These physical models require significantly different optimized state components feedback coefficients. The relay-linear algorithm was implemented and researched, which provides reconfiguration of the state components feedback coefficients at the very initial stage of the control object movement, when the optimized control action does not depend on these coefficients.

The experimental complex developed on the basis of FESTO hydraulic learning system with the use of Matlab+Simulink Desktop Real Time allows to carry out the study of linear and nonlinear EHSD control.

References

[1] PONTRYAGIN, L.S., BOLTANSKII, V.G., GAMKRELIDZE, R.V., AND MISHCHENKO, E. F., "The Mathematical Theory of Optimal Processes." Wiley (Interscience), New York, 1962.

[2] Pupkov K A, Egupov N D, et al. 2004Metody klassicheskoy isovremennoy teorii avtomaticheskogo upravleniya v 5 tomakh. Tom 4.Teoriya Optimisazii system avtomaticheskogo upravleniya [Methods of Classic and ModernControl Theory in 5 volumes

Vol. 4 Automatic Control Systems Optimization Theory](Moscow: Bauman MSTU Publ.) 744 p (in Russian)

[3] V. G. Boltyanskii Mathematical methods of optimal control (Balskrishnan-Neustadt series).Holt, Rinehart and Winston Publ.; 1st edition (1971). 272 p. ISBN-10: 0030722853

[4] Flugge-Lotz, I. Discontinuous automatic control (2015) Discontinuous Automatic Control, pp. 1–168 p. https://www.scopus.com/inward/record.uri?eid=2-s2.0-84954206760&partnerID=40&md5=98503b526955b4b6f685a993ea2a816

[5] Reshmin, S.A. The Threshold Absolute Value of a Relay Control Bringing a Satellite to a Gravitationally Stable Position in Optimal Time (2018) Doklady Physics, 63 (6), pp. 257–261.https://www.scopus.com/inward/record.uri?eid=2-s2.0-85050142824&doi=10.1134%2fS1028335818060101&partnerID=40&md5= DOI:10.1134/S1028335818060101

[6] Reshmin, S.A. Threshold Absolute Value of a Relay Control when Time-Optimally Bringinga Satellite to a Gravitationally Stable Position (2018) Journal of Computer and Systems Sciences International, 57 (5), pp. 713–722. https://www.scopus.com/inward/record.uri?eid=2-s2.0-85055038451&doi=10.1134%2fS106423071805012X&partnerID=40&md5= DOI:10.1134/S106423071805012X

[7] Reshmin, S.A. Estimate of the control threshold value in the problem on a time-optimal satellite attitude transition maneuver (2017) Mechanics of Solids, 52 (1), pp. 9–17. https://www.scopus.com/inward/record.uri?eid=2-s2.0-

[8] Almuzara, JosL.G.,Flügge-Lotz, I. Minimum time control of a nonlinear system (1968) Journal of Differential Equations, 4 (1), pp. 12–39. DOI: 10.1016/0022-0396(68)90046-6

[9] Goodwin G.C., Graebe S.F., Salgado M.E. Control system design. Upper Saddle River; L.: Prentice-Hall, 2001. 908 p.

[10] Moroz, A.I., Pupkov, K.A. ON ONE METHOD OF SYNTHESIS OF RELAY-LINEAR CONTROL LAWS. (1978) Engineering Cybernetics (English translation of TekhnicheskayaKibernetika), 16 (5), pp. 138-144.https://www.scopus.com/inward/record.uri?eid=2-s2.0-0018017880&partnerID=40&md5=a7e0b7c45cbf1feba1529c32e66e045e

[11] Pupkov, K.A., Moroz, A.I. On the Theory of Relay-Linear Control Laws. [Article@KTEORII RELEINO-LINEINYKH ZAKONOV UPRAVLENIYA.] (1976) Probl Control InfTheory, 5 (5–6), pp. 401–419 (engtransl, 1–18.

[12] B Kulakov and D Kulakov. Synthesis of the relay-linear control with the hysteresis region for the second order object example. 2019 *IOP Conf. Ser.: Mater. Sci. Eng.* **589** 012036

[13] Pupkov K A, KuiCh N 2006 Relay-linear control law in a system of a homing guidance.Problem imashinostroeniyaiavtomatizazii— Engineering and Automation Problems (4) pp 55–58 (in Russian)

[14] N. A. Kuznetsov, “Synthesis of a relay-linear algorithm for control by means of a throttled hydrodrive”, Automation and Remote Control, 1967, 1275–1280

[15] D Kulakov and B Kulakov Study of the dynamics of an electro-hydraulic servo drive physical model on FESTO learning system. 2019 *IOP Conf. Ser.: Mater. Sci. Eng.* **492** 012029

IOP Conf. Series: Materials Science and Engineering **779** (2020) 012028 doi:10.1088/1757-899X/779/1/012028

Justification of the concept of creating a perspective dump truck

A Kartashov[1,2], G Harutyunyan[1], A Kosolapov[1], E Shkarupelov[1]

[1]Bauman Moscow State Technical University

[2]E-mail: kartashov@bmstu.ru

Annotation. The article carried out a deep technical and economic analysis of existing samples of quarry vehicles. The modern technical level of the mining industry is determined by performing a patent search and review of literature on this topic. The main problems associated with the operation of mining vehicles in mining are considered. The basic requirements for the development of a perspective dump truck have been formed.

The development of the mining industries of the Russian Federation requires the implementation of new cost-effective and safe technologies for extraction and transportation of minerals [1]–[4]. This industry is important national economy and requires improvement. A method of open-cut mining is called if it involves working in open pits. Due to the rapid increase of the depth of open pits it become impossible to use railroad freight activity, scraper or conveying belt due to the complexity of organizing efficient and high-performance transportation of rocks. Wheel vehicle seems to be most suitable for working in open pits, as it is capable of ensuring the proper level of working productivity. One such vehicle is a dumper. A dumper is a self-unloading truck designed to transport running ground and other goods in off-road conditions.

To identify current trends in the development of dumper designs it is necessary to conduct a deep technical and economic analysis of existing samples of mining equipment. The high priority task of feasibility study is to determine the current technical level of the industry by performing a patent search and review of literature on this topic.

An analysis of the patent search gives reason to build a short-time prediction (for 5 years) for the development of career equipment. The main trend in world mining equipment is the course towards increasing energy efficiency, the main indicator of which is the specific fuel consumption.Problem of energy efficiency and control law of dump trucks are presented in[5]–[9]. One of the main ways to reduce specific fuel consumption is to reduce resistance to movement and transfer technology to a parallel hybrid or distributed hybrids.

We give an example of apatent RU 158 664 U1 «The control device for a set of traction electrical equipment for vehicles, mainly dump trucks». This patent describes a control device for electric traction drive (see Figure 1) for vehicles. The kit contains a housing and a liquid cooling system of power converters fixed therein, which are connected by buses.

We give an example of a patent US RE 36152 «Method and apparatus for controlling differentially driven wheel-slip for an articulated machine». This patent describes a Method and apparatus for controlling differentially driven wheel-slip for an articulated machine. The described device includes a brake mechanism, which has the ability to block each of the wheels individually, angular velocity

Content from this work may be used under the terms of the Creative Commons Attribution 3.0 licence. Any further distribution of this work must maintain attribution to the author(s) and the title of the work, journal citation and DOI.
Published under licence by IOP Publishing Ltd

sensors, a sensor for the mutual angle of rotation of the articulated parts of the frame, and a controller. The application of the traction control system (see Figure 2) on dumper can reduce wheel slip and at the same time reduce power consumption for driving.

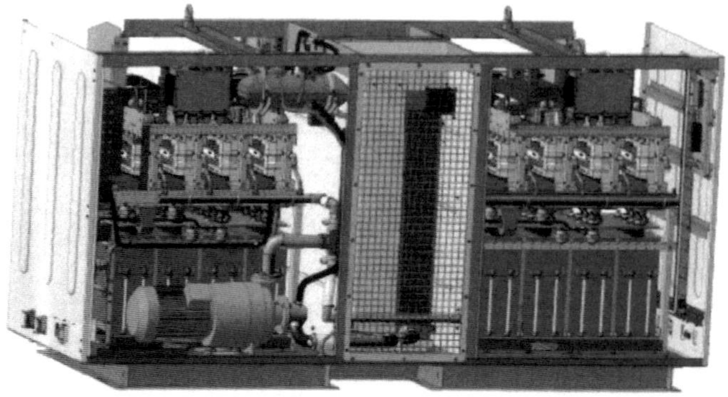

Figure1. Control device for an electric traction drive

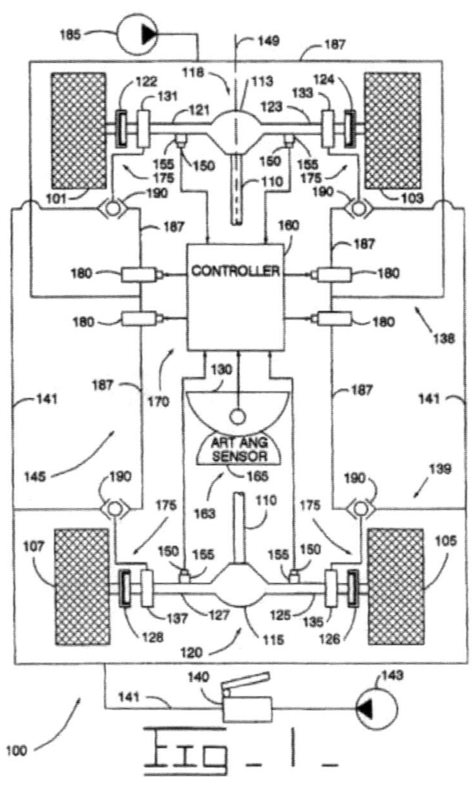

Figure2. Traction control system

After conducting a patent search, it is necessary to proceed to a review of literary sources, as well as sources of the Internet.

The efficiency of using hybrid transmission schemes is shown in [10]–[13]. Ways to reduce fuel consumption and harmful emissions into the atmosphere through the use of dump trucks with a hybrid power plant. Modeling was performed in MATLAB / Simulink of a dump truck transmission with a diesel-generator hybrid power plant. Similar model is presented in [14].

In the works [15]–[18] it was revealed that there is a tendency to deepen the career depth. Increasing depth leads to a deterioration in operating conditions. Gas contamination is increasing, especially in the lower part of the career. In the main works, the need to reduce harmful emissions from the operation of mining equipment is noted. In some works, it is proposed to use trolley cars that are powered by a contact network. The advantages of trolley cars over the classic diesel-electric circuit are examined in detail. Transfer to electric traction will reduce the level of harmful substances in the air, noise and vibration. The complete transition to the use of trolley cars is complicated by the lack of flexibility in moving the contact network. This approach requires the organization of fixed loading and unloading sites. To solve the problems of flexibility and mobility, it is proposed to use diesel trolley vehicles (see Figure 3).

Figure 3. BelAZ-7524-Э792

After analyzing the mining conditions for the use of trolley vehicles, the following operating conditions suitable for the use of trolley vehicles in open mines are distinguished: the remoteness of the trolley lines from the production site, the presence of a stationary loading and reloading point, the distance of transportation of the rock mass is 3–30 km, the gradient of roads is from 0 to 10%. However, with the development of technologies of electric batteries manufacturing, it became possible to create fully electric dump trucks [19]–[21].

The use of all-electric drivetrain in addition to reducing the cost of operating the transport allows manufacturers to significantly save on production due to the smaller number of transmission units and other elements. In addition, the electrified mining transport allows you to get the following advantages: 10–15% reduction in the mass of the dump truck, the exclusion of diesel exhaust and, therefore, the absence of gas contamination of the mine and the formation of fog, eliminating the costs of the purchase, storage and transportation of diesel fuel, reduction in the cost of the truck due to the exclusion of the diesel engine, reduction in tire costs due to the lower dead weight of the electric truck, ease of maintenance of the power unit, and immunity of the electric engine to frost, which especially distinguishes it in the northern regions from diesel (strong negative temperatures have a negative effect mainly on batteries).

One such example is the eDumper electric dump truck (see Figure 4). For several years, the Swiss government has been equipping heavy industry with electric-powered machinery. The largest in size is the eDumper electric dump truck weighing 45 tons, developed by Kuhn Schweitz AG, which has a length of 8.5 meters, a width and height of 4.2 m. Power is supplied by 600 kWh batteries which weighs 4.5 tons.

IOP Conf. Series: Materials Science and Engineering **779** (2020) 012028 doi:10.1088/1757-899X/779/1/012028

Figure4. Electric dump truckeDumper

Like all electric vehicles, eDumper recuperates the energy generated by braking. While braking occurs, the electric motor goes into reverse mode, producing electricity that charges the batteries. The eDumper is used in a mine near Biel (Switzerland). Ittransports 65 tons of rock to a cement factory down the mountain with a slope of 13%. When descending from the mountain, the dump truck has a large inertia, due to which the regenerative braking system chargesthe batteries. As a result, eDumper charges itself and does not require frequent network connections. Battery charge measurements were taken after ascending with a fully loaded body and after descending to the loading point in the quarry. At the beginning of the rise, the battery charge was 90%, after the rise it was 80%. After the descent due to the recovery of power, the batteries were charged up to 88%, which is almost equal to the initial charge of the batteries. A dump truck carries out about 20 descents and ascents daily.

The result of the implementation of electric transmission and engines is the saving of 50 tons of diesel fuel annually, which significantly increases the energy efficiency of work and makes mining dump trucks environmentally friendly.

Thus, a review and analysis of publications shows that development and research are currently focused on increasing the efficiency, reliability and safety of mining transport systems through the use of hybrid and electric transmissions.

Currently, autonomous vehicles are widely used in various fields [22]–[27]. The use of autonomous vehicles in industry can be dictated by dangerous working conditions, as well as the need to increase productivity by reducing vehicle downtime during rest and driver shifts, as well as working in the dark. The creation of autonomous mining dump trucks is one of the priority tasks in the development of mining equipment. This is primarily due to the safety when working in quarries, often accompanied by accidents associated with crashes, and the danger posed by harmful gases, especially radioactive radon. Also, the use of autonomous dump trucks provides increased energy efficiency associated with the use of hybrid and electric transmissions.

Creating fully autonomous cars is a complex task that has not yet been completely resolved at the current level of technology development. Mining trucks operate in limited areas on repeating routes. These features of their operation allows to create autonomous mining equipment using currently available tools.

At the moment, several autonomous mining equipment have been created by the leading companies BelAZ, Volvo, Komatsu, Caterpillar, Hitachi.

BelAZautonomous mining dump truck.

The Russian company VIST Robotics together with BelAZ created a prototype of an autonomous mining dump truck based on the BelAZ-7513R model, shown in Figure 5. The dump truck hasloading capacity of 130..136tones.can operate both in autonomous mode and in telecontrol mode, has an electromechanical transmission, and an 1194 kW engine.

Figure 5. Autonomous mining dump truck BelAZ-7513R

BelAZ-7513R works as follows: the dispatcher issues a shift task to the system, allocates the required number of pieces of equipment and the production route. Trucks leave the parking lot for the site and begin continuous operation - loading, road, unloading. The operator can reassign the machine to another route or send it to the parking lot: the robot will finish the current cycle and head to a new task.

In the event of difficult conditions, the operator takes over the remote control, who can simultaneously control several vehicles at once.

The system of autonomous traffic control of the BELAZ-7513R dump truck includes an on-board control unit located in front of the car next to the driver's cab. On the left and right sides of the car there are two receivers of satellite navigation systems.

For visual control, 11 video cameras are used, which are installed around the perimeter of the vehicle. The car uses Orlaco EMOS series security cameras with an Ethernet interface. Several Delphi ESR radars operating in the medium and long range are responsible for detecting people and other objects along the route. Three-dimensional and two-dimensional lidars are installed on the dump truck, which are responsible for scanning space in the horizontal and vertical planes, as well as for monitoring the safety zone around the car. Along the perimeter of the car, Banner QT50U ultrasonic sensors (sonars) are used to detect obstacles at short distances, the sensor detection range is 0.2 ... 8 m.

Komatsu unmanned dump truck.

Komatsu started testing autonomous vehicles at the Codelco copper mine in Chile in 2005 and three years later, in January 2008, in the same place, the first commercial use of AHS (Autonomous Haulage Systems), which increases safety, commercial and economic efficiency, began.In September 2017 on one of operating mining enterprises of Rio Tinto company were conducted successful tests of the AHS upgrade kit for electric powered dump truck Komatsu 830E with a diesel-electric transmission, 1835 kW engine and carrying capacity 231 tons.

The fleet management system AHS works without participation of truck driver. Information about their route and speed is sent wirelessly to the Central control room. The exact location is determined by GPS navigation.

The fleet management system controls all equipment through GPS and wireless network, including other equipment and vehicles that are man-operated, to prevent collisions. Integrated analysis system of environment helps trucks themselves to react on suddenly appearing objects and obstacles.

Japan manufacturer presented a prototype of the dump truck Komatsu IAHV (figure 6), which is part of the autonomous cargo complex management system, in 2016 at the Mine Expo in Las Vegas.

Figure 6. Komatsu IAHV dump truck

The main features of this model — are the ability to move forward and backward with a same speed, and steering of all four wheels. Thanks to these options the time for setting the dump truck for loading and unloading is reduced, as well as space in bottomhole and unloading zones is saved, since there is no need to turn the truck.Komatsu IAHV can move sideways. The center of mass is located in the middle part of the truck, what allowed to evenly distribute load on the wheels and to equalize grip coefficient to the road of all wheels. Today, the design is in a trial operation.

Volvo unmanned dump truck.

Volvo HX unmanned dump trucks are the part of the Volvo Electric Site project of development electrified mining company with a minimal harmful emission. The first Volvo HX1 prototype (figure 7) was used to refine the concept of unmanned electric dump truck. Volvo chose the concept that uses a large number of small dump trucks instead of several with a large carrying capacity and which allows to reduce losses in the event of failure of one of the dump trucks [28]–[30].

Figure7. VolvoHX1 dump truck

The second prototype Volvo HX2 uses more of Volvo's serial parts such as traction electric motors, batteries and power electronics. Three-dimensional lidars were added in the motion control system to detect people and obstacles. Dump truck has four-wheel drive and all-wheel steering, which allows to increase maneuverability and gives the possibility of sideway moving. The maximum speed is 40 km/h, both forward and backward. The dump truck is equipped with three electric motors. Two electric motors are used as traction, the third motor drives a hydraulic pump. High pressure hydraulic is used for steering and drives unloading cylinders.

Charging of the traction batteries of Volvo HX2 dump truck occurs due to pantograph, located on a special base below ground level. When truck stops, the pantograph rod rises above it and connects contacts to the dome, located under the bottom of the truck.

Conclusion.

Based on review of existing vehicle developments in the mining sphere, as well as patent researches, it is possible to reveal main tendencies of technology development in this sphere. These include hybridization or the complete electrification of power point and transmission of dump trucks, using of unmanned technologies and complete absence of cabs in the developed technique samples. Accordingly, two ways of development autonomous vehicles are formed: retrofitting of existing vehicles with a necessary aggregates or designing new vehicles taking into account the possibility of autonomous movement up to refusing manual control and excluding the cab from the vehicle construction, similar to the Volvo and Komatsu dump trucks.

Also diesel-trolley trucks are the one of the perspective directions of hybrid dump trucks. A contact network is used for movement along the main sections of the routes, and a diesel generator set is used to access loading and unloading areas.

The use of unmanned systems in dump trucks compared to road vehicles is simplified due to the following factors: the absence of many lanes, other road users, a variety of environmental objects, low speeds along standard routes. However, there are also negative factors, affecting on operating of unmanned dump trucks. These include lack of marking, deformability of the soil, uneven supporting surface at an angle to the horizon, lack of high-quality communication for GPS operation.

Cab-free design of dump trucks is the most perspective and preferred and can significantly increase the size of the loading platform, as well as to provide movement in both directions on the basis of a "push-pull" principle.

The work was done at Bauman Moscow State Technical University with financial support from the Russian Ministry of Education and Science under Agreement № 14.577.21.0287 (Unique identifier of the work:RFMEFI57718X0287).

List of used sources:

[1] Senkus, V.V., Ermakov, N.A., Senkus, V.V. Ventilation of underground excavations in open pit/underground coal mining (2018) Mining Informational and Analytical Bulletin, 2018 (5),pp. 89–94.

[2] Zenkov, I.V., Vokin, V.N., Kiryushina, E.V., Raevich, K.V. Remote monitoring data on opencast mining and disturbed land ecology in the bakal iron ore field (2018) Eurasian Mining, 2018 (2), pp. 29–33.

[3] Paithankar, A., Chatterjee, S. Open pit mine production schedule optimization using a hybrid of maximum-flow and genetic algorithms (2019) Applied Soft Computing Journal, 81.

[4] Khoyutanov, E.A., Gavrilov, V.L. Modeling coal deposits in hard-to-reach regions of Yakutia (2019) IOP Conference Series: Earth and Environmental Science, 262.

[5] Jin, C., Yi, T., Shen, Y., Khajepour, A., Meng, Q. Comparative study on the economy of hybrid mining trucks for open-pit mining (2019) IET Intelligent Transport Systems, 13 (1), pp. 201–208.

[6] Liu, G., Chai, S. Optimizing open-pit truck route based on minimization of time-varying transport energy consumption (2019) Mathematical Problems in Engineering, 2019.

[7] Bolshunova, O.M., Korzhev, A.A., Kamyshyan, A.M. Adaptive control system of dump truck traction electric drive (2018) IOP Conference Series: Materials Science and Engineering, 327.

[8] Jiang, L., Qiu, H., Huang, S., Ling, S., Ma, L., Huang, C. Differential Control Strategy Based on Hierarchical Control of Electric-Driven Articulated Dump Truck (2016) Proceedings — 2016 9th International Symposium on Computational Intelligence and Design, ISCID 2016, 1, pp. 255–260.

[9] Polack, P., Delprat, S., d'Andréa-Novel, B. Brake and velocity model-free control on an actual vehicle (2019) Control Engineering Practice, 92.

[10] Yanbiao Feng, Zuoming Dong, Jue Yang, Rui Cheng. Performance Modeling and Cost-benefit Analysis of Hybrid Electric Mining Trucks. / University of Science & Technology Beijing, Beijing, China. IEEE, 2016, 6 pp .: il. Access Mode: https://ieeexplore.ieee.org/document/7587102.

[11] Singh, K.V., Bansal, H.O., Singh, D. A comprehensive review on hybrid electric vehicles: architectures and components (2019) Journal of Modern Transportation, 27 (2), pp. 77–107.

[12] Zhang, R., Meng, K. Driving Force Coordinated Control of Separated Axle Hybrid Electric Dump Truck (2017) SAE Technical Papers, 2017-October.

[13] Yang, W., Yang, J., Zhang, N. A novel coaxial multi-mode hybrid transmission system for mining trucks (2019) Proceedings of the Institution of Mechanical Engineers, Part D: Journal of Automobile Engineering, 233 (10), pp. 2492–2501.

[14] Miroslaw, T., Szlagowski, J., Zawadzki, A., Zebrowski, Z. Simulation model of an off-road four-wheel-driven electric vehicle (2019) Proceedings of the Institution of Mechanical Engineers. Part I: JournalofSystemsandControl.

[15] Krivda VV. Justification of the rational parameters of development systems for deep iron ore quarries when using dump trucks of an improved design. The dissertation for the degree of candidate of technical sciences. / Ministry of Education and Science of Ukraine State Higher Educational Institution "National Mining University". Dnepropetrovsk, 2015.211 p. ill.

[16] Koptev, V.Y., Kopteva, A.V. Improving Pit Vehicle Ecology Safety (2018) Journal of Physics: Conference Series, 1015.

[17] V.L. Yakovlev. Actual problems of career transport and prospects for its development. / Institute of Mining, Ural Branch of the Russian Academy of Sciences. Ekaterinburg, 2017.5 p.: Ill.

[18] Safonov, B.A., Travnikov, A.N. Unmanned all-terrain cargo and passenger transportation system for operation conditions when automobile roads are unavailable (2019) Journal of Physics: Conference Series, 1177.

[19] Kotiev, G.O., Butarovich, D.O., Kositsyn, B.B. Energy efficient motion control of the electric bus on route (2018) IOP Conference Series: Materials Science and Engineering, 315.

[20] Smirnov, K.A. Simulation of rectilinear motion of a four-wheel car-like robot with an electromechanical drivetrain (2019) Mechanisms and Machine Science, 73, pp. 2671–2679.

[21] Khazin M. L., Shtykov S. O. Quarry electrified transport. / Ural State Mining University, Yekaterinburg, 2018. Bulletin of the State Technical University named after G.I. Nosova, 2018, v. 16, No. 1, 11–17 pp., Ill.

[22] Shorin, A.A., Karpukhin, K.E., Terenchenko, A.S., Kondrashov, V.N. Traction module of cabless unmanned cargo vehicles with electric drive (2018) International Journal of Mechanical Engineering and Technology, 9 (11), pp. 1903–1909.

[23] Dou, F., Liu, W., Huang, Y., Liu, L., Meng, Y. Modeling and path tracking for articulated steering vehicles (2017) Proceedings — 2017 Chinese Automation Congress, CAC 2017, 2017-January, pp. 5263–5268.

[24] Vdovin, D.S., Khrenov, I.O. Systems of the self-driving vehicle (2019) IOP Conference Series: Materials Science and Engineering, 534.

[25] Naumov, V.N., Mashkov, K.Y., Byakov, K.E. Autonomous tracked vehicles effectiveness estimation (2019) IOP Conference Series: Materials Science and Engineering, 534.

[26] Abosekeen, A., Karamat, T.B., Noureldin, A., Korenberg, M.J. Adaptive cruise control radar-based positioning in GNSS challenging environment (2019) IET Radar, Sonar and Navigation, 13 (10), pp. 1666–1677.
[27] Wolschke, C., Kuhn, T., Rombach, D., Liggesmeyer, P. Observation based creation of minimal test suites for autonomous vehicles (2017) Proceedings — 2017 IEEE 28th International Symposium on Software Reliability Engineering Workshops, ISSREW 2017, статья № 8109298, pp. 294–301.
[28] Shojaei, K., Yousefi, M.R. Tracking control of a convoy of autonomous robotic cars with a prescribed performance (2019) Transactions of the Institute of Measurement and Control, 41 (13), pp. 3725–3741.
[29] Volvo unveils next-generation autonomous hauler
URL: https://www.fluidpowerworld.com/volvo-unveils-next-generation-autonomous-hauler/
[30] Volvo, Skanska to test electric, autonomous site quarry for 10 weeks this fall
URL: https://www.equipmentworld.com/volvo-skanska-to-test-electric-autonomous-site-quarry-for-10-weeks-this-fall-video/

IOP Conf. Series: Materials Science and Engineering **779** (2020) 012029 doi:10.1088/1757-899X/779/1/012029

Technological heredity effect on fatigue strength of hydropower plant parts after combined processing

A Yakovleva[1,5], A Dubov[2], A Sobranin[2], E Karpovich[3] and A Marchenkov[4]

[1]Bauman Moscow State Technical University
[2]Energodiagnostika Co. Ltd
Yubileyny av., 8-1, Reutov, Moscow Region, 143965 Russian Federation,
[3]Orel State Agrarian University named after N.V. Parahin
General Rodin st., 69, Orel, 302019 Russian Federation
[4]National Research University Moscow Power Engineering Institute
Krasnokazarmennayast., 14, Moscow, 111250 Russian Federation

[5]E-mail: yakovleva525@mail.ru

The paper investigates the effect of combined processing on fatigue strength. We developed a plant for combined processing, processed 40X steel samples with various modes. Furthermore, we experimentally studied the fatigue strength of the samples by the metal magnetic memory method, and found the relationship between stress concentration zones and combined processing modes under the heredity factor effect. Finally, we gave recommendations on the use of combined processing for hardening parts of hydropower plants.
Keywords: combined processing, fatigue strength, power plants, technological heredity, metal magnetic memory method

Introduction

The main reasons for the failure of hydropower plants are corrosion, wear and fatigue fracture, such as dimensional wear or fractures of pump and hydraulic motor drive parts, i.e. bearings, gear teeth, shafts, couplings [1–10].

The last two reasons are no less dangerous, since they require a long repair or lead to a complete stop of the plant. Therefore, the tasks of reducing wear and increasing the fatigue strength of such parts are relevant [5, 7].

The operational properties of parts of hydropower plants are affected by: manufacturing accuracy, surface roughness, material grade, material structure and hardness, as well as methods and modes for producing parts. This is due to the fact that accuracy and roughness cannot characterize the physical state of the surface layer of the metal. Surfaces with the same accuracy and final roughness can have varying degrees of cold work, uneven nature and residual stresses, their crystalline structure can be distorted to different degrees, as well as surface integrity due to microcracks, burrs, loosening, etc. The chemical composition and structure of the surface layer of the metal may be different. If the surface is obtained as a result of a number of physical and mechanical operations with the release of a significant amount of heat in the technological zone, individual chemical components can burn out, resulting in a decrease in the operational properties of the metal [8–10].

Many methods have been developed to increase wear resistance and fatigue strength, all of them having their own advantages, disadvantages and applications [10]. However, it is possible to make a

Content from this work may be used under the terms of the Creative Commons Attribution 3.0 licence. Any further distribution of this work must maintain attribution to the author(s) and the title of the work, journal citation and DOI.
Published under licence by IOP Publishing Ltd

parameters increase reserve by the methods of combined processing (CP) [8, 11, 12].With combined processing methods, when technological factors sequentially one after another or jointly in parallel affect the work piece, technological heredity is clearly manifested. Moreover, heredity affects not only immediately after the finishing operations of the technological process, but it can also occur during operation as a result of the influence of certain surface quality parameters created in the surface layer of a part during its previous processing. From the standpoint of technological and operational heredity, it is possible to establish a connection between technological operations and transitions, including the cases when combined methods of machine parts are applied [11–14].

Description of the Experiment

The study introduces a combined processing method, which consists in sequential electromechanical processing and diamond smoothing [15, 16]. We developed a trial plant composed of a universal lathe, a power source, a quench, a current supply system, and a quenching area cooling system, in which combined processing modes were worked out [17, 18].

On the surface there was obtained a regular microrelief with oil pockets to hold the grease during friction (Fig. 1). During electromechanical processing (EMP), there appear overlapping zones of the quenching bands, which depend on the feed rate S, with a decrease in hardness being observed in these zones. After EMP, a diamond smoother passing through the surface finally smooths out the microroughness of the profile, and when it gets into the overlapping zone, it forms grooves.

In [15, 16], the effect of CP modes on the samples wear resistance was studied. Comparative wear tests of the samples processed by various methods were carried out. Findings of research showed that CP increases wear resistance compared to high-frequency hardeningby 55%, cementation— by 50%, nitriding— by 45%. The obtained results allow us to recommend combined processing for hardening parts of power plants operating in conditions of intense friction with lubricant.

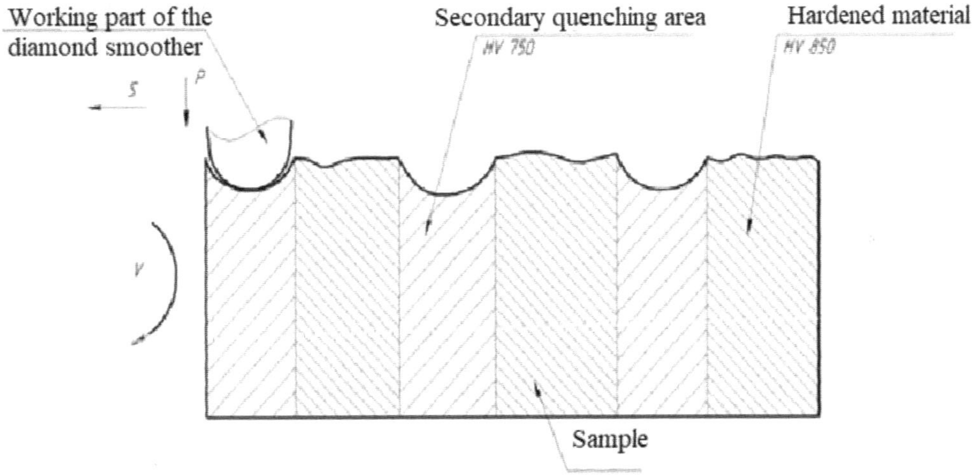

Fig. 1. Modeling the process of obtaining a regular microrelief: hardened material (HV 850), secondary quenching area (HV 750)

However, the overlap of quenching bands, which are the stress concentrators on the surface of parts, can significantly reduce fatigue strength and cause the destruction of the entire mechanism. Therefore, the task of diagnosing hidden stress concentrators and the center of crack propagation in power plants is relevant [10, 19–21].

The paper introduces the study of the combined processing effect on fatigue strength by the metal magnetic memory method (MMM) [22–24].

The MMM method is a non-destructive testing method based on the registration and analysis of the distribution of intrinsic scattering magnetic fields on the surface of the product in order to determine stress concentration zones (SCZ): defects, heterogeneity of the metal structure, and centers of crack nucleation.

For the study, we used samples made of steel grade 40X with a diameter of 12 ± 1 mm and a length of 120 ± 10 mm. To detect SCZ, we used an INK-3M-12 model device with a four-channel scanner which has two two-component sensors for measuring both normal and tangential components of the magnetic field, and a whole length counter with a wheel (Fig. 2).

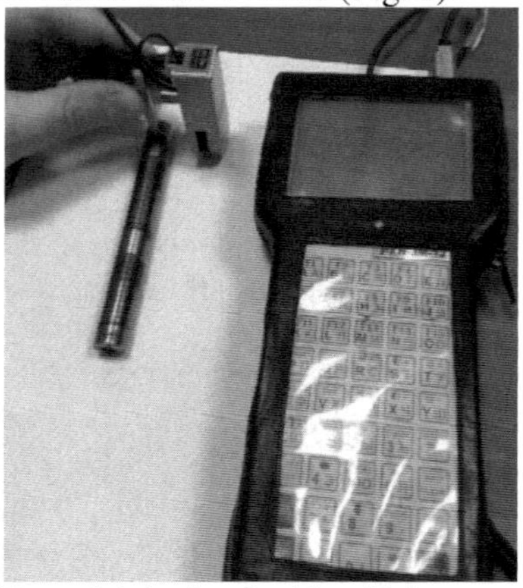

Fig. 2. The INK-3M-12 model device for identifying stress concentration zones

The study was carried out according to GOST R ISO 24497-1-2009, GOST R ISO 24497-3-2009, GOST R ISO 24497-3-2009.The manifestation of technological heredity can lead to both improvement and deterioration of the operational properties of parts [12].

The problems of mechanics of technological heredity of the quality of the machine parts surface layer as a result of the plastic flow are studied in [11, 14]. The life cycle of a metal from a workpiece to failure during operation is considered as a single process of exhaustion of metal. However, the issue related to the properties heredity by the surface layer after machining, thermal and finishing treatment is not well understood. The formation of stress concentrators on the surface is affected by technological heredity. Due to stress concentrators, cracks can form on the surface of parts, and lead to fatigue fractures of parts.

a b c

Fig. 3. Samples for stage 1 of the study for the presence of SCZ after CP

At stage 1, we examined only those samples which did not have hereditary SCZ before combined processing.

To do this, the SCZ control was carried out at all stages of sample preparation:

• before machining (turning) (Fig. 3 a);
• after machining (Fig. 3 b);
• after combined processing (Fig. 3 c).

At stage 2, we studied the samples having SCZ before combined processing. The results of the study were obtained on magnetograms.

Experimental Result

Findings of experiment 1 showed that after combined processing, SCZ on the surface of the parts were not found (Fig. 4, 5).

A comparison of the magnetograms before CP (Fig. 4 a, b) and after (Fig. 4 c, d) showed that the quenching bands overlap does not cause structural inhomogeneity and centers of crack nucleation on the surface.

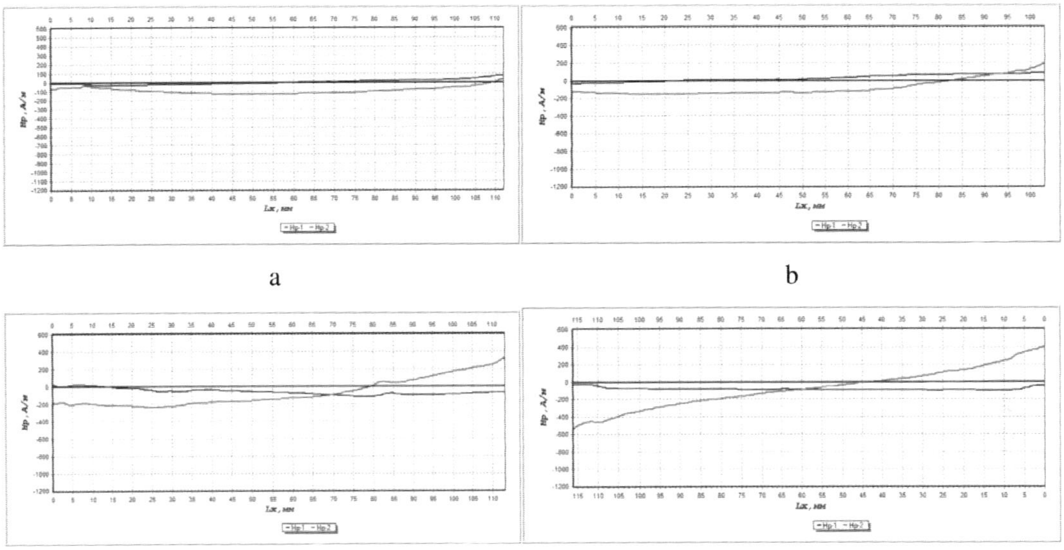

a b

c d

Fig. 4. Magnetograms of samples: a, b — before combined processing, c, d — after combined processing

For stage 2 of the study we chose samples with SCZ before CP (Fig. 5). The results of this stage showed that in some samples SCZ did not have abrupt jumps (Fig. 6 a), and in some samples SCS were absent (Fig. 6 b).

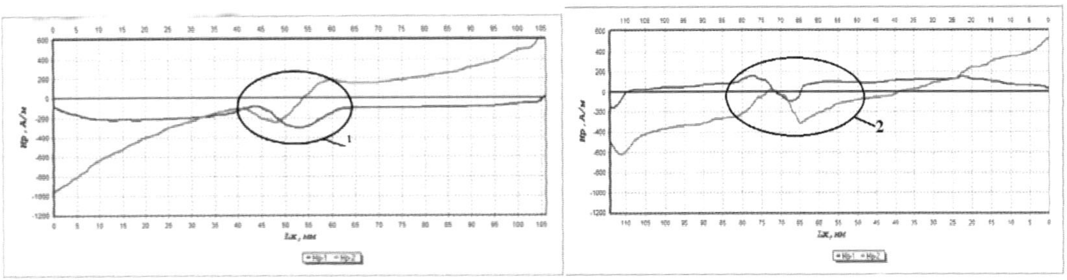

a b

Fig. 5. Magnetograms of samples with SCZ before CP: a — region 1, b — region 2

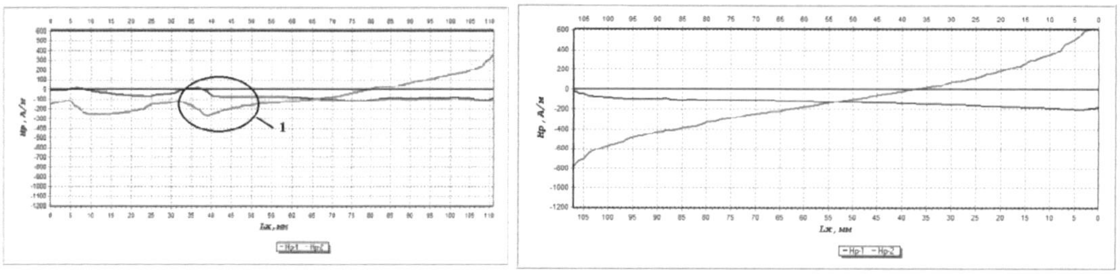

a b

Fig. 6. Magnetograms of samples with SCZ after CP: a — reduction in the risk of crack propagation (region 1), b — absence of SCZ

It is possible to control the process of technological and operational heredity so that properties positively affecting the quality of the part are preserved throughout the entire process, and properties affecting negatively are eliminated at the beginning.

Conclusions

By successive hardening, i.e. EMP, and diamond smoothing, it is possible to control the characteristics of surface microgeometry in a wide range, improve lubricity and increase the operational indicator — wear resistance. The effect of technological heredity on the fatigue strength of parts of hydropower plants after combined processing is established.

The quenching bands overlap after CP reduces fatigue strength, and various defects in the surfaces of parts, in particular microcracks, can "heal". After CP, stress concentrators are not formed in the surface layer of the samples, and CP is a technological barrier to SCZ.

Thus, the combined processing method can be recommended to increase the wear resistance and fatigue strength of parts of hydropower plants.

References

[1] Yashcheritsyn P.I., Kheyfets M.L., Chemisov B.P., et al. Tekhnologicheskie osnovy vysokoeffektivnykh metodov obrabotki detaley [Topological fundamentals of high-efficient methods of parts processing]. Novopolotsk, PGU Publ., 1996. (in Russ.).

[2] Vasil'yev A.S., Dal'skiy A.M., Kheyfets M.L., et al. Tekhnologicheskie osnovy upravleniya kachestvom mashin [Technological fundamentals of machines quality control]. Minsk, FTI NANB Publ., 2002. (in Russ.).

[3] Isaev N. The effect of design parameters of the closed type regenerative pump the energy characteristics. *IOP Conf. Ser.: Mater. Sci. Eng.*, 2019, vol. 492, no. 1, art. 012026. DOI: 10.1088/1757-899X/492/1/012026
URL: https://iopscience.iop.org/article/10.1088/1757-899X/492/1/012026

[4] Korsakova S., Protopopov A. Obtaining of dependence of sustained angular rotor speed of centrifugal pump with hydrostatic bearings. *IOP Conf. Ser.: Mater. Sci. Eng.*, 2019, vol. 492, no. 1, art. 012032. DOI: 10.1088/1757-899X/492/1/012032
URL: https://iopscience.iop.org/article/10.1088/1757-899X/492/1/012032

[5] Protopopov A., Jakovich C. Compromise resource-efficiency curve for a centrifugal pump. *IOP Conf. Ser.: Mater. Sci. Eng.*, 2019, vol. 492, no. 1, art. 012034.
DOI: 10.1088/1757-899X/492/1/012034
URL: https://iopscience.iop.org/article/10.1088/1757-899X/492/1/012034

[6] Shablovskiy A., Kutovoy E. Obtaining the head characteristic of a low flow centrifugal pump by numerical methods. *IOP Conf. Ser.: Mater. Sci. Eng.*, 2019, vol. 492, no. 1, art. 012035.
DOI: 10.1088/1757-899X/492/1/012035
URL: https://iopscience.iop.org/article/10.1088/1757-899X/492/1/012035

[7] Petrov A., Isaev N., Kuleshova M. Test bench flow straightener design investigation and optimization with computational fluid dynamics methods. *IOP Conf. Ser.: Mater. Sci. Eng.*, 2019, vol. 492, no. 1, art. 012036. DOI: 10.1088/1757-899X/492/1/012036 URL: https://iopscience.iop.org/article/10.1088/1757-899X/492/1/012036

[8] Lomakin V., Bibik O. Numerical prediction of the gas content effect on the cavitation characteristics of the pump using the simplified Rayleigh-Plesset equation. *IOP Conf. Ser.: Mater. Sci. Eng.*, 2019, vol. 492, no. 1, art. 012037. DOI: 10.1088/1757-899X/492/1/012037 URL: https://iopscience.iop.org/article/10.1088/1757-899X/492/1/012037

[9] Valiev T., Petrov A. Simulation of non-stationary loads on the rotor of a pumping unit with an assembly pipeline. *IOP Conf. Ser.: Mater. Sci. Eng.*, 2019, vol. 492, no. 1, art. 012038. DOI: 10.1088/1757-899X/492/1/012038 URL: https://iopscience.iop.org/article/10.1088/1757-899X/492/1/012038

[10] Cheremushkin V., Lomakin V. Development and research of hydraulic disk pump. *IOP Conf. Ser.: Mater. Sci. Eng.*, 2019, vol. 492, no. 1, art. 012039. DOI: 10.1088/1757-899X/492/1/012039 URL: https://iopscience.iop.org/article/10.1088/1757-899X/492/1/012039

[11] Petukhov A.N. Soprotivlenieustalostidetaley GTD [Fatigue strength of gas-turbine engine parts]. Moscow, Mashinostroenie Publ., 1993. (in Russ.).

[12] Mukhin V.S., Smyslov A.M., Borovskiy S.M. Modifitsirovaniepoverkhnostidetaley GTD pousloviyamekspluatatsii [Modification of gas-turbine engine parts surface according to operating conditions]. Moscow, Mashinostroenie Publ., 1995. (in Russ.).

[13] Petukhov A.N. specific requirements to production technology for turbine blades.*Nauchnyetrudy*, 2005, no. 9(81), pp. 107–111. (in Russ.).

[14] Petukhov A.N. Physical basis of technological heredity and methods for providing structural strength of main parts of gas-turbine engine.*Konversiya v mashinostroenii*, 2005, no. 1–2, pp. 44–58. (in Russ.).

[15] BlyumenshteynV.Yu.,Smelyanskiy V.M. Mekhanika tekhnologicheskogo nasledovaniya nastadiyakh obrabotki I ekspluatatsii detaley mashin [Technological heredity mechanics at machine parts processing and exploitation stages]. Moscow, Mashinostroenie Publ., 2007. (in Russ.).

[16] Shchedrin A.V., Kostryukov A.A. Application of tribotechnology on the basis of self-organization for the systematic improvement of the processes of cold plastic deformation. [Strengthening Technologies and Coatings], 2017, no. 11(155), pp. 495–499. (in Russ.).

[17] Garkunov D.N., Babel' V.G., Mel'nikov E.L., et al. New scientific discovery in tribology based on self-organization. *Remont.Vosstanovlenie.Modernizatsiya* [Repair, Reconditioning, Modernization], 2019, no. 6, pp. 18–25. (in Russ.).

[18] Fedorov S., Albagachiev A., Isaenkova M., Yakovleva A., Zaripov V., MinushkinR.The effect of combined processing schedules on the value and nature of residual stresses in the surface layer of cylindrical friction pairsВсборнике: IOP Conference Series: Materials Science and Engineering electronic edition. 2019. С. 012024.

[19] N.N. Zubkov, A.I. Ovchinnikov, S.G. Vasil'ev, Tool–workpiece interaction in deformational cutting. Russian Engineering Research , 36(3) (2016) 209–212.

[20] N.N. Zubkov, V. Poptsov, S. Vasiliev., A. Batako, Steel Case Hardening Using Deformational Cutting. Journal of Manufacturing Science and Engineering, 140(6) (2018) 061013.

[21] N. Zubkov, V. Poptsov, S. Vasiliev, Surface Hardening by Turning without Chip Formation. Jordan Journal of Mechanical and Industrial Engineering, 11(1) (2017) 13–19.

[22] ISO-24497-1:2007 Non-destructive testing - Metal magnetic memory - Part 1: Vocabulary.

[23] ISO-24497-2:2007 Non-des0tructive testing - Metal magnetic memory - Part 2: General requirements.

[24] ISO-24497-3:2007 Non-destructive testing - Metal magnetic memory - Part 3: Inspection of welded joints.

IOP Conf. Series: Materials Science and Engineering **779** (2020) 012030 doi:10.1088/1757-899X/779/1/012030

Variety of gas-dynamic turbulent configurations at internal flow in channels of structurally complex bodies

Yu Kochetkov[1], I Borovik[1], O Podymova[2] and A Protopopov[3, 4]

[1]Moscow Aviation Institute(MAI)
[2] Keldysh Research Centre (KeRC)
[3]Bauman Moscow State Technical University

[4]E-mail: proforg6@yandex.ru

Annotation. For the study of turbulent configurations in complex channels, it is proposed to use the experimental method of mass entrainment in combination with theoretical analysis and mathematical description of the results obtained. The paper presents the results of numerous experiments conducted on high-enthalpy gas flows in nozzles of complex shape. Images were obtained that visualize the picture of the torsion-wave dynamic movement of gas flows, the presence of stable structured turbulent flow configurations, and the abrupt transition from one form of flow stability to another.

Introduction

We are surrounded by a great many types of flows of liquids and gases. The study of these trends is given a lot of attention in the literature [1]–[11]. All these types are conventionally placed in two large groups: "laminar flows", when each particle of the liquid moves parallel to the main direction of the flow, "turbulent flows", when the flow is represented by a combination of a huge number of vortices. All gas dynamics, flight theory, and hydrodynamics are ultimately just an attempt to understand the mechanics of the formation and interaction of flows with various bodies, surfaces, and channels, as well as to find methods of mathematical description and calculation of these processes. The issues of heat transfer in the areas of separation and re-joining of compressed gas flows are directly related to the dynamics of motion and are considered by many researchers [12].

The analysis of various types of turbulence in complex channels confirms the validity of the following statements: turbulence is a system of stable forms of flow; turbulence is a combination of four types of simple flows (translational, vibrational, rotational and torsion).

Three-dimensional flow of liquid, gas and plasma inside chambers, nozzles, air intakes and other flow paths of complex shapes, as a rule, is turbulent. In General, in any, even in the simplest channel, which has a high degree of symmetry, the flow can be quite complex and depend largely on the mode that determines the combination of independent types of flow [13]. The presence of rotations, rotations, wall vibrations and the combined effect of these factors on the working body leads in addition to the complexity of the flow structure and the emergence of new stable forms, which include the classic Taylor-Getler pair vortices.

For flow paths we can conditionally distinguish three independent types of flows:

in axisymmetric channels;

in asymmetric channels;

in axisymmetric channels with asymmetric input conditions.

Content from this work may be used under the terms of the Creative Commons Attribution 3.0 licence. Any further distribution of this work must maintain attribution to the author(s) and the title of the work, journal citation and DOI.
Published under licence by IOP Publishing Ltd

These types of currents can be combined. For example, rotated about the axis of the combustion chamber of an elliptic supersonic nozzle with an oblique cut combines axisymmetrical current along the cylindrical chamber, the rotation in the supersonic part of the nozzle, the flow inside the nozzle with an elliptical orifice and flowing through a slanting cut with POS-now, the next sharp bend of the stream. Similar flows are those in turbines and LRE pumps, where the process is complicated by the additional rotating movement of the channel walls.

Complex forms of channels lead to the formation of various vortex structures of moving flows. Combinations of these three types of currents give rise to secondary, tertiary and multi-branched flows. In their mathematical description, we have to deal with increasingly complex ways of setting boundary conditions that adequately describe the flow.

Many flows in complex channels cannot be predicted analytically or computationally at present. An experimental approach is used to describe them. One of the effective experimental methods for observing traces of complex turbulent flow near the walls of various structures is the method of mass entrainment or the method of hot visualization. This method is used for research of high-enthalpy flows, such as the flow in the flow paths of solid- fuel rocket engine, the movement of descent vehicles in the atmosphere, as well as processes occurring in the designs of nuclear power propulsion systems.

Experimental methods, unfortunately, also do not always allow you to get the desired result. It is difficult, for example, to measure the angular velocity of the steam flow in the turbine nozzle. It is almost impossible to measure the pressure at the junction of the electric arcs of a three-phase plasma torch. It is a big problem to measure the velocity profile of a uranium plasma fuel element in a gas-phase nuclear reactor. In such cases, experimental and computational methods are used to model flows in complex channels.

Method of the study of the flow structure and installation

For the study of turbulent configurations in complex channels, it is proposed to use the experimental method of mass entrainment in combination with theoretical analysis and mathematical description of the results obtained.

The essence of the method is to organize the conditions under which a high-enthalpy gas flow leaves a trace on the investigated carried-away surface. At the same time, polymer materials that obey the laws of linear pyrolysis are used as the carried-away material, when the rate of destruction and entrainment of the material is almost directly proportional to the heat flow. Preliminary experiments are conducted for various materials with these properties: fluoroplast 4, polymethylmethocrillate, viniplast, as well as some fiberglass and graphite. The work done shows that fluoroplast-4 is the most suitable material for the created installation, the selected time and temperature conditions of the experiment.

An important aspect of research is the short duration of the process. The correct selection of the peak time provides a clear "instantaneous fixation" of the three-dimensional structure, eliminates the imposition of one gas-dynamic pattern on another, and also eliminates the influence of the formed traces in the material on the subsequent flow structure.

For experimental research, a model installation [14] was developed, which is a small-sized solid-fuel rocket engine (sfr) (Fig.1). It consists of a combustion chamber and a nozzle block. The chamber contains a solid fuel charge and a fuse. The dimensions of the charge and its shape are determined in advance to set the desired pressure and the burning time (exposure time to the sample). The nozzle is made collapsible and has a glass in which a fluoroplastic sample is laid, simulating a full-scale nozzle or channel. The liner of the critical section of the nozzle is made of a carbon composite material with a low entrainment rate compared to the fluoroplastic of the test sample. The described design allows you to explore a variety of configurations of channels and nozzles, repeatedly on one working part to produce launches of high-enthalpy flow. Left after exposure to hot gas prints make it possible to visualize the relief structures of the flow near the washed walls. Measurements of the geometry of the resulting patterns allow you to build profiles of entrainment along the generators.

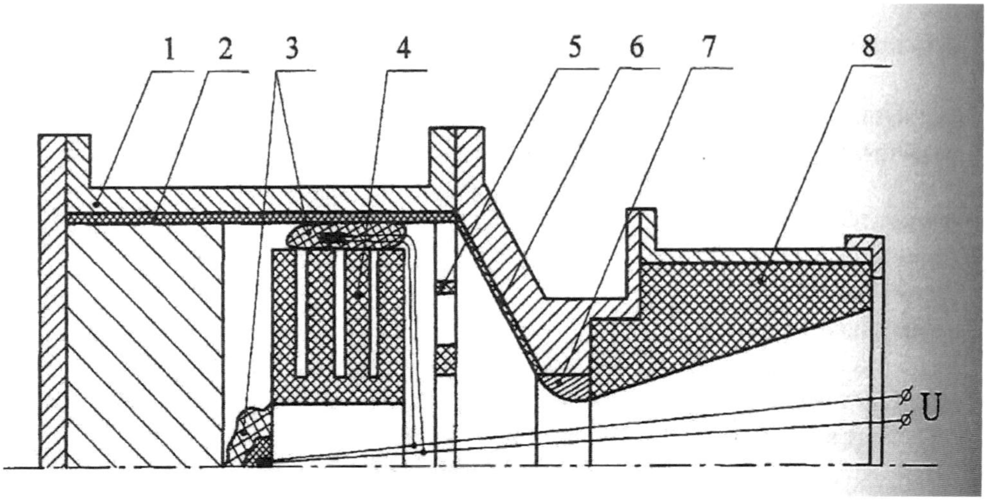

Figure 1. The scheme is a model SFR for the study of turbulent flow configurations. 1 — case; 2—heat-protective coating; 3—igniter; 4—charge; 5—diaphragm; 6—cone; 7—liner; 8—test sample of fluoroplast.

The results and generalization of numerous experiments conducted on the described installation on the study of the physics of turbulent flows in the channels of structurally complex bodies are given below. the gas flow Velocities vary from zero at the wall to supersonic in the core.

Results and discussion

1. Flow in an axisymmetric conical nozzle

Let's start by describing the flow structure in an axisymmetric conical nozzle. Research is carried out on the described model solid-fuel installation with a nozzle made of fluoroplast. Nozzles with different angles of opening of the supersonic part are tested. Experiments are carried out using different compositions of solid fuel, at different pressures in the combustion chamber and different percentages of condensed combustion products (mass fraction of condensate from 0 to 0.4). As a result of research, it was found that the peak in the supercritical region occurs in a stable gas-dynamic mode. At the point of transition from the non-combustible liner of the critical section to the inflaming fluoroplastic cone, a so-called "supercritical pit" occurs, in which a zone of subsonic flow is formed with the formation of Taylor-Gertler vortices. The vortex structure can be visualized by carrying away the mass of the fluoroplastic nozzle. In time, this structure (Fig.2) does not change, the number of vortices is preserved, and their size increases in proportion to the time of acceleration. It is established that the amplitude of the vortices $\Delta\delta$ obeys the empirical ratio $\Delta\delta/\delta max=0,2\ Go^{-0,2}$ and, as can be seen, depends on the Gertler number Go.

The amplitude of the vortices is also functionally related to the maximum depth of the "pit". The experimental dependence of the maximum entrainment value δ_{max} obtained for cones with different slopes of the generators is uniquely determined by the angle of entry into the supersonic part of the nozzle θ_{in}: $\delta_{max} = S_{max}\ (0,525 — 0,25\ tg\theta_{in})$, S_{max} -the coordinate of the maximum entrainment point along the forming nozzle.

The absolute value of the magnitude of the amplitude $\Delta\delta$ is approximately 10 % of the value δ_{max}. At sufficiently strong peaks downstream, cellular formations are visualized, evenly spaced around the circle in a staggered order. These are traces of torsion of the stream.

Photos of prints (Fig.2) shows that even in non-stationary processes, turbulent structures are not chaotic, but strictly ordered. The vortex structure of the flow is stable, once born, the prints of the vortex bundles are repeated in the next experiment on the same sample placed in the installation after measurement.

IOP Conf. Series: Materials Science and Engineering **779** (2020) 012030 doi:10.1088/1757-899X/779/1/012030

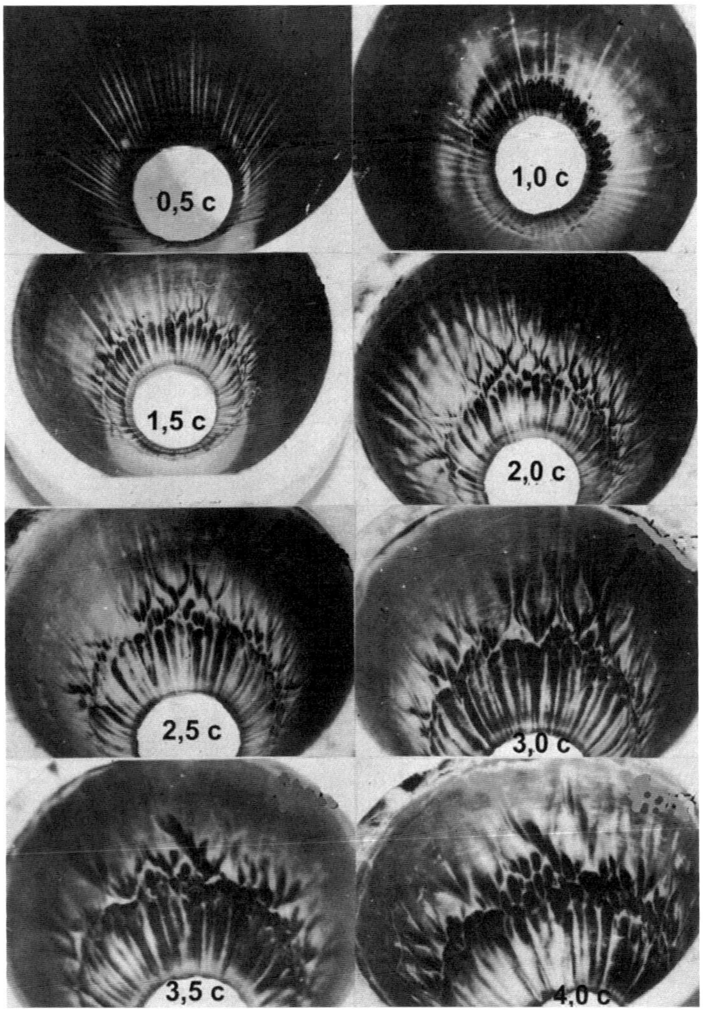

Figure 2. The imprint of the Taylor vortices of Getler in the area of the height of the conical nozzle

The resulting photo also demonstrates that after passing the "supercritical pit" the flow becomes laminar.

2. Turbulence in the prismatic nozzle

A prismatic or rectangular nozzle has either a rectangle or a square in its cross section. In this case, the angle of inclination of one pair of opposite planes is zero. If the height of the rectangle is much less than the width, the nozzle degenerates into a flat slit.

The peculiarity of the flow in a prismatic nozzle is that the expansion of the flow along adjacent planes is different and greater where the angle of their inclination to the axis is greater. This would lead to axial currents. But since there is no turbulence in the supersonic flow, a spatial system of jumps is formed in such a nozzle. Near the walls, at the height of the nozzle, subsonic zones are formed. Local pressures in these zones correspond to the degrees of expansion along each of the forming planes, where the corresponding axial flows are formed.

Experiments carried out on prismatic nozzles made of fluoroplast show that Taylor-Gertler vortices are formed on the faces of prisms. The values of the amplitudes of these vortices are inversely proportional to the values of the angles of inclination of the planes and obey the dependencies for the

IOP Conf. Series: Materials Science and Engineering **779** (2020) 012030 doi:10.1088/1757-899X/779/1/012030

conical nozzle. In accordance with these dependences, the vortex amplitudes of the studied nozzles differ from each other by about 20 %.

At the intersection of the planes, on the edges of the prism, a larger single vortex is formed, the size of 2.5-3 times larger than any of the vortices on the faces of the nozzle (Fig.3). It is obvious that the parity of this vortex is provided by vortex flows in the core of the flow at some distance from the wall.

Figure 3. The imprint of the vortex flow in the zone of the height of the prismatic nozzle

3. Turbulent flow in elliptical nozzles

A similar flow is obtained in experiments on an elliptical nozzle. Unlike a prismatic nozzle, there is no angular vortex on an elliptical nozzle. In this case, the size of the Taylor-Gertler vortices change inversely proportional to the angle of inclination of the generators. They are larger along the minor axis of the ellipse and smaller along the major axis (Fig.4). Analysis of flared nozzles allows us to establish the unevenness of cellular prints located downstream of the longitudinal vortices. Along the generatrix, which lies in the plane of the small axes, the cells of the imprint are larger, and the large ones are smaller. The cells are staggered with a half-period shift. These are prints of oppositely rotating paired bundles formed as a result of torsion of the stream. The analysis shows that the torsion region of the flow is located at the zero pressure gradient.

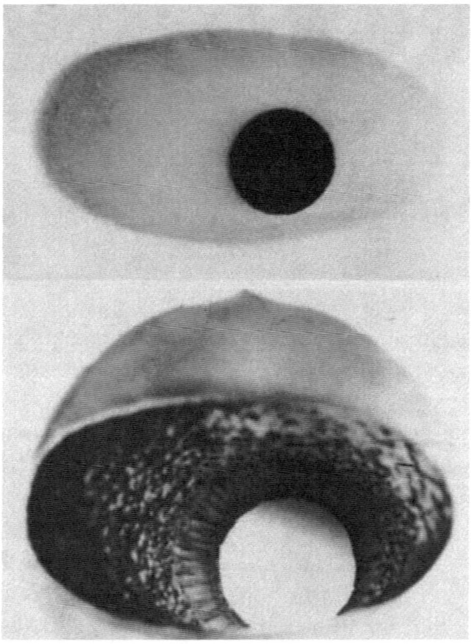

Figure 4. The imprint of the vortex flow in the zone of the height of the elliptical nozzle

IOP Conf. Series: Materials Science and Engineering **779** (2020) 012030 doi:10.1088/1757-899X/779/1/012030

4. Analysis of turbulent flows in a segment nozzle

Segment nozzles are used to remove the exhaust jets of auxiliary engines from the surfaces of the bodies and critical components of the rocket. Typically, these are conical nozzles with a flat section parallel to the axis of the camera. In the area where the nozzle is "cut off" the flow is deflected relative to the conical wall.

Experiments show (Fig.5) that areas of Taylor-Gertler turbulence additionally occur behind the fracture site on the plane. The intensity of the vortices is directly proportional to the difference between the angles of the forming cone and the slope of the cut plane. Vortices in this case arise after the flow of a complex turbulent secondary flow formed by a "supercritical pit" in the undeformed segment of the nozzle. The new vortex flow formed behind the fracture is tertiary, since it is obtained by the complication of the previous secondary one. But as you can see in the photo, this new current is as orderly as the secondary one. It is a vortex, and the size of the Taylor-Gertler vortices is comparable to the size of the secondary flow vortices.

In the area of the interface of the conical surface and the cut plane, single larger vortices are observed. Their nature is similar to the vortices in the edges of the prismatic nozzle.

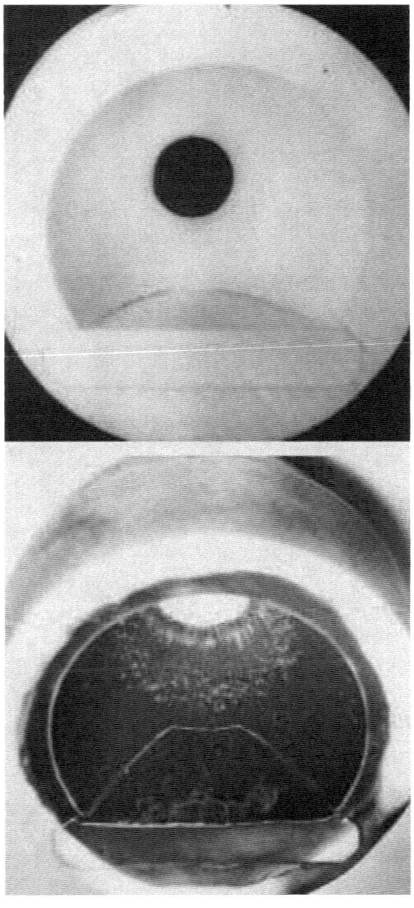

Figure 5. The imprint of the vortex flow in the zone of the peak of the segment nozzle

5. Current in the oblique nozzle

The oblique nozzle is used to sharply rotate the supersonic flow at an angle of up to 30-40° relative to the axis of the combustion chamber and create a constant control lateral force. Experiments carried out by the method of mass entrainment on conical nozzles with an oblique section show (Fig.6) that the

IOP Conf. Series: Materials Science and Engineering **779** (2020) 012030 doi:10.1088/1757-899X/779/1/012030

presence of an oblique cut does not have any effect on the turbulent structure of the previous supersonic flow. All forms of flow stability at the wall do not undergo any changes during the transition to the short, long and middle forms. In this case, the characteristic property of all supersonic flows is manifested, when there is no feedback up the flow, and the flow does not react to approaching obstacles and changes in the shape of the nozzle wall.

The presence of an oblique cut does not lead to positive pressure gradients if the longest part of the wall is operating in the design mode. If it operates in the mode of overexposure, then in this area there are local secondary spatial flows directed into the nozzle.

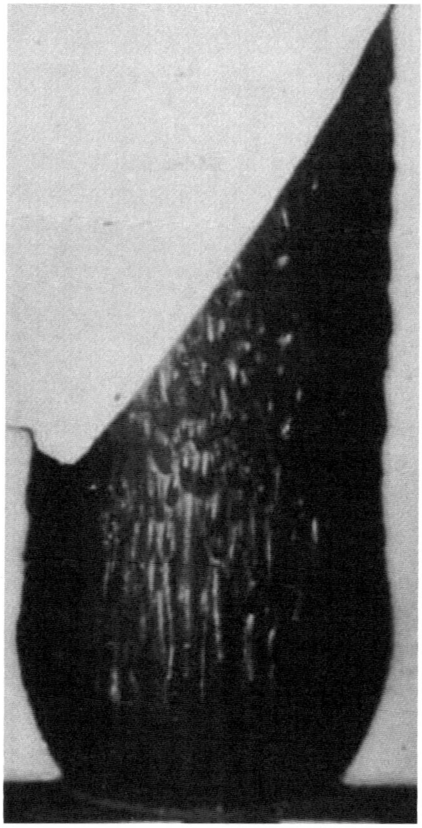

Figure 6. The imprint of the vortex flow in the zone of the height of the conical nozzle with an oblique section

6. The rotary control nozzle (RCN)

The turbulent flow in the rotary control nozzle depends significantly on the conditions at the entrance to the supersonic part, and depending on these conditions, as well as on the length and shape of the supersonic nozzle, the flow structure is formed on the windward and leeward part. The trivial rotation of the nozzle relative to the combustion chamber can not serve as a criterion for the appearance of the windward side in the area of the greatest geometric angle of attack. Along the wall of the nozzle there is a redistribution of pressure, which has an oscillatory character. Depending on the flow conditions, windward and leeward zones may be located successively on the same formation. In this case, the resulting lateral forces periodically change the sign. With a certain period, the heat flows change, leading to asymmetric entrainment. The gas-dynamic structure of a two-phase turbulent flow also changes and becomes asymmetric. For figure 7 shows a photo of asymmetric entrainment due to condensed phase deposition. On the windward side, there is a trace of asymmetric burnout of the nozzle wall.

IOP Conf. Series: Materials Science and Engineering **779** (2020) 012030 doi:10.1088/1757-899X/779/1/012030

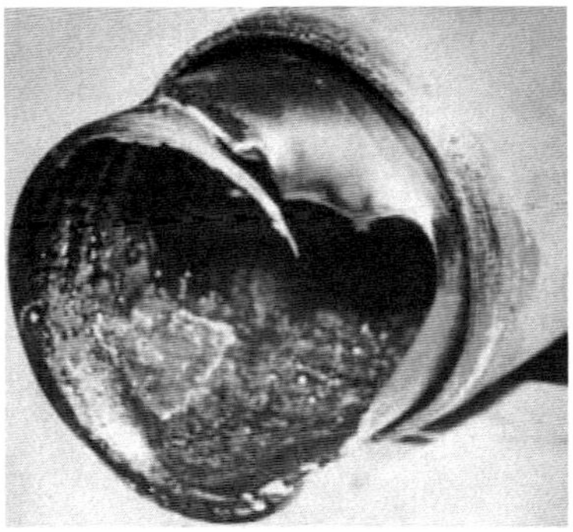

Figure 7. Burnout of the wall of the rotary control nozzle

7. Split control nozzle (SCN)

The split control nozzle differs from the rotary control nozzle in that the rotation of the flow is carried out in the supersonic part of it. Under the influence of steering machines, part of the nozzle is introduced into the stream, and part of it comes out, forming zones of windward and leeward currents, forming the resulting plot of pressure forces and lateral force.

For figure 8 traces of the influence of turbulent flow on the windward side of the SCN were recorded. The positive pressure gradient that occurs when the wall is rotated in the direction of the flow leads to the formation of Taylor-Gertler vortices. A similar picture is obtained on the segment nozzle. Tertiary currents are formed. But contrary to the expectation of a more complex flow pattern, when a highly turbulized flow at the wall runs into a wedge-like fracture and breaks in the area of impact against it, its structure on the contrary is simplified and becomes similar to the structure of the secondary flow. The sizes of Taylor-Gertler vortices still satisfy the dependences for the amplitude $\Delta\delta$, and their shape and number do not change over time.

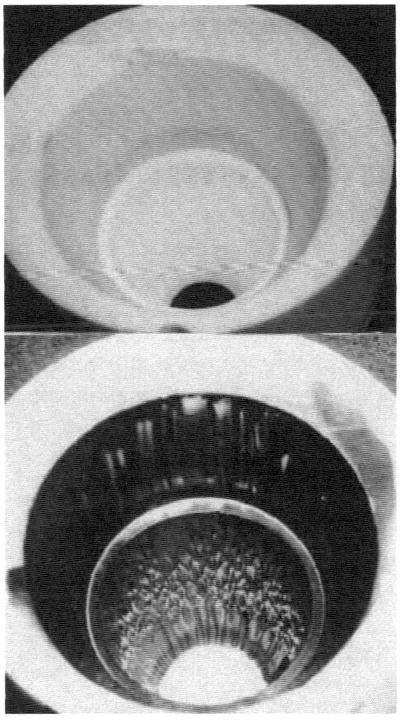

Figure 8. The imprint of the vortex flow in the zone of the height of the split control nozzle

8. Torsion structures in the high-altitude nozzle diffuser

When testing the high-altitude nozzle model LRE in the diffuser were recorded traces of three-dimensional flow on the wall, manifested in the form of prints of condensed combustion products of carbon-containing fuel (soot).

For figure 9 visible cellular formations on the surface of the diffuser, located in a staggered order with a shift of half a period. These formations occur as a result of contact of pairs of twisted bundles with the wall of the nozzle. This torsion flow is formed after the Taylor-Gertler longitudinal vortices flow into the gradient-free region. In this case, pairs of neighboring vortices form stable bundles (twists), lined on the surface of the wall.

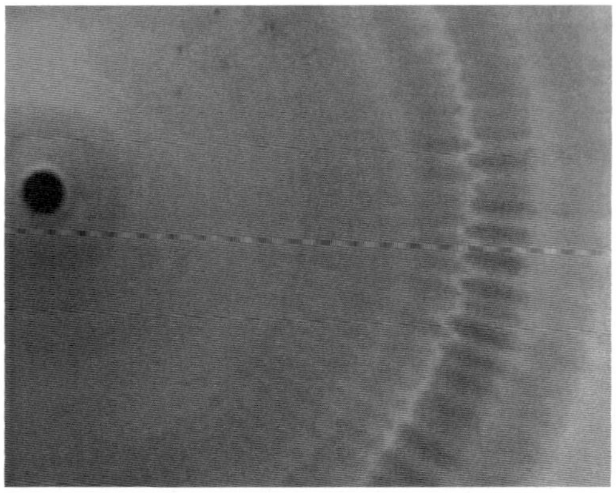

Figure 9. The imprint of the vortex flow on the surface of the LRE(liquid rocket engine) diffuser

Summary

The above examples of turbulent flow in channels of complex shape are the most significant. They clearly demonstrate the main provisions of the torsion-wave theory and confirm the fundamental statements.

1. Turbulence is a complex torsion-wave dynamics that reflects stable forms of highly structured flow of the working body in the channels and in the flow of complex aerodynamic forms. The transition from one form of stability to another occurs abruptly.
2. Any turbulent flow can be represented as a complex combination of four types of simple flows: translational, vibrational, rotational and torsion.

Maximum turbulence is realized at critical values of the flow, when its speed becomes equal to the speed of sound and almost equal to the speed of thermal (chaotic) motion of molecules. In criticism, the current jumps from turbulent to laminar. Channels of complex shape create conditions for the separation of the flow and the formation of vortex subsonic structures. Turbulence is impossible in a supersonic stream. Conditionally, it is triggered in the jumps of compaction, where there is a transition through the sound.

List of references

[1] Irfan, M., Alwadie, A., &Glowacz, A. (2019). Design of a novel electric diagnostic technique for fault analysis of centrifugal pumps. *Applied Sciences (Switzerland), 9*(23) doi:10.3390/app9235093
[2] Wang, L., Lu, J., Liao, W., Zhao, Y., & Wang, W. (2019). Numerical simulation of the tip leakage vortex characteristics in a semi-open centrifugal pump. *Applied Sciences (Switzerland), 9*(23) doi:10.3390/app9235244
[3] Guo, C., Wang, J., &Gao, M. (2019). A numerical study on the distribution and evolution characteristics of an acoustic field in the time domain of a centrifugal pump based on powell vortex sound theory. *Applied Sciences (Switzerland), 9*(23) doi:10.3390/app9235018

IOP Conf. Series: Materials Science and Engineering **779** (2020) 012030 doi:10.1088/1757-899X/779/1/012030

[4] Rapur, J. S., & Tiwari, R. (2019). Experimental fault diagnosis for known and unseen operating conditions of centrifugal pumps using MSVM and WPT based analyses.*Measurement: Journal of the International Measurement Confederation,147* doi:10.1016/j.measurement.2019.07.037
[5] P Chaburko and Z Kossova 2019 *IOP Conf. Ser.: Mater. Sci. Eng.***492** 012011
[6] V Lomakin et al 2019 *IOP Conf. Ser.: Mater. Sci. Eng.***492** 012012
[7] A Gouskov et al 2019 *IOP Conf. Ser.: Mater. Sci. Eng.***492** 012013
[8] N Egorkina and APetrov 2019 *IOP Conf. Ser.: Mater. Sci. Eng.***492** 012015
[9] K Dobrokhodov and APetrov 2019 *IOP Conf. Ser.: Mater. Sci. Eng.***492** 012016
[10] N Isaev 2019 *IOP Conf. Ser.: Mater. Sci. Eng.***492** 012026
[11] M Saprykina and V Lomakin 2019 *IOP Conf. Ser.: Mater. Sci. Eng.***589** 012017
[12] Kochetkov, Yu.M., Gubertov, A.M., Mironov, V.V. Similarity of heat and mass transfer processes in supersonic flow past steps, cavities and joints (1996) Heat Transfer Research, 27 (1), pp. 179–183.
[13] Kochetkov, Yu.M.,Kravchik, T. N., Podymova, O. A. Five Theorems of Turbulence and Their Practical Application (2019) Russian Engineering Research, 39 (10), pp. 855–861.
[14] Hubertov a.m., Mironov V. V., Kochetkov Yu. M. and others. gas-Dynamic and thermophysical processes in solid fuel rocket engines / / M. Mashinostroenie, 2004. P. 511.

IOP Conf. Series: Materials Science and Engineering **779** (2020) 012031 doi:10.1088/1757-899X/779/1/012031

Famous trumpet effects

Yu Kochetkov[1], T Kravchik[1] and A Protopopov[2,3]

[1]Moscow Aviation Institute (MAI)
[2]Bauman Moscow State Technical University

[3]E-mail: proforg6@yandex.ru

Annotation. An analytical review of the effects obtained in the course of studies of flows in pipes is presented. There are descriptions of the famous experiments of Reynolds, who put on the agenda the problem of turbulence. The description and analysis of the discovery of Taylor vortices, which gave the received stimulus to the development of spatial hydro-gas dynamics, is given. The original invention of the Shebeko gas-dynamic tube, which allowed solving the problem of high-altitude testing of the Burya rocket, is presented. The original experiments with the thermal acoustic effect of Riike and Bossch — Riess are described, which stimulate new views on the stability of processes in the future. The invention of Gershman, which gave a practical impetus to the use of resonators in technology, is described. The description of the Rank — Hilsh pipe is given, the effect of which is widely used in the chemical and oil industry. The effect of a heat pipe used in industry for space heating is described.

Introduction

The most popular configuration in Hydrogas dynamics is the pipe. And this is no accident, it is the pipe that is the simplest channel and determines the internal flow in critical conditions. Why critical, because it is a geometric shape, when passing the working body through which the flow does not experience a pressure gradient at the same time, grad P = const. Therefore, often in hydro-and gas-dynamic problems use this situation, and then correct the solution using coefficients. It was on the pipes that many magnificent effects were obtained, which were the standard of comparison of processes. In themselves, seemingly simple at first glance, the processes in the pipes are very complex and fraught with many discoveries and surprises. Always newly discovered effects on pipes led to a great surge of scientific activity and were a source of inspiration for naturalists. Often these discoveries resolved the most complex, sometimes contradictory ideas; led them to a logical understanding and gave a clear explanation of the processes taking place. For example, the experiments of O. Reynolds on the tube resolved the dispute between Dalembert in Alliance with Euler and Robert Fulton, the inventor of the steamship [1]. Turbulence was discovered. Taylor's experiments inspired a whole galaxy of scientists dealing with vortex currents to new research. Riike's experiments on the humming trumpet had an absolutely stunning effect. Contemporaries appreciated the discovery of thermal separation in the Ranke pipe. The invention of a gas-dynamic device for testing in high-altitude nozzles (Shebeko pipe) by Ivan Fedorovich Shebeko was of great practical importance for computational engineering.

The people who discovered these effects were the first. They were followed by followers. They improved, implemented designs, expanded the range of parameters. But the beginning was always behind the effects, the first, hitherto unknown results.

Content from this work may be used under the terms of the Creative Commons Attribution 3.0 licence. Any further distribution of this work must maintain attribution to the author(s) and the title of the work, journal citation and DOI.
Published under licence by IOP Publishing Ltd

Below we will consider some, of course not all, pipe effects that have played an outstanding role in the study of hydro-gas dynamics [2]–[6], acoustics and heat transfer.

1. The Experiments Of Reynolds

Reynolds carried out his magnificent experiments on interlocked tubes, with which he observed the flow pattern in them. He showed that a thin string of dye introduced into the water, flowing in a glass tube, quickly stretched into a long, sharply defined, not mixing with water strip, moving parallel to the walls of the tube. The water seems to move in semi-centric layers, like metal tubes embedded in one another: the inner one is faster, the one adjacent to it is a little slower, the next one is even slower. Layered-laminar - Reynolds calls this flow. With increasing speed, immediately, a sharp jump changes the movement of the stream. You can see how fast and erratic swirls mix paint with water throughout the volume of the tube-laminar flow lost its stability, turned into a vortex flow, for which later Lord Kelvin came up with a great name-turbulent flow.

Reynolds showed that turbulence in a liquid develops the faster it is heavier, the lower its viscosity, the larger the tube diameter, and the greater the velocity. He realized that the stability of the laminar flow depends on the ratio of the forces of viscosity and inertia. Then he got a dimensionless quantity — the famous Reynolds number, which just controls the movement of viscous liquids in pipes. The resulting criterion is valid for any liquids, and they were shown later — gases. This is water, air, mercury, honey, oil, hydrogen, etc. Thus, when a certain value is reached (Re ~2300), the flow loses stability and passes from the laminar state to the turbulent one (Fig. 1) [7].

2. Taylor's Whirlwinds

Apparently, it all started with a desire to repeat Kuett's experiments on coaxial pipes with a flow between two parallel walls, one of which could make movements relative to the other. This design allowed to increase the path of movement conditionally to infinity. For the first time, such viscous fluid flows were experimentally investigated by a remarkable English oceanographer, a future member of the Royal society of London, Geoffrey Ingram Taylor. He devoted most of his creative work to the development of the theory of turbulence and developed his own theory of the stability of the flow of a viscous liquid. But the main merit of this scientist is the discovery of an essentially incredible effect, his famous Taylor vortices. This effect is the basis of a solid, solid Foundation of knowledge of all gas dynamics and especially the theory of turbulence. One can only imagine the violent reaction that Taylor's results produced among researchers. Many scientists, judging by their numerous publications of those years, literally clung to this problem. New experiments and theories began to appear. For figure 2 an image of Taylor vortices between two transparent coaxial cylinders is presented. Here clearly visible toroidal vortices that fill the entire space between the cylinders. The rotation of neighboring vortices occurs in the opposite direction and is a natural torus with the formation of wall lines and spreading at the contact boundaries.

Under the impression of this discovery, another magnificent discovery was made. Essentially similar vortices on curved concave walls were predicted and then experimentally repeatedly obtained. These Taylor-Gertler vortices were named in honor of the famous canadian scientist Henry Gertler [8].

The practical significance of these vortices is clearly illustrated by the example of the destruction of the inner walls of the nozzle blocks of the SFR(solid-fuel rocket engine). [9,10].

3. Shebeko Pipe

When creating the Burya rocket, the entire complex of research works was assigned to the research Institute-1 [11]. The main problem was experimental confirmation of design solutions. In accordance with the resolution of the Council of Ministers of the USSR of 1954 on the creation of the Burya rocket, responsible managers and specialists were appointed. Academician M. V. Keldysh was appointed scientific Director of the development. The government resolution also approved a group of specialists, heads of the main scientific departments. In this group was Ivan Fedorovich Shebeko. He was entrusted with the development of experimental methods for the study of pvrd on aerodynamic

IOP Conf. Series: Materials Science and Engineering **779** (2020) 012031 doi:10.1088/1757-899X/779/1/012031

stands. Naturally, he not only directed, but also directly participated in various tests. One of the most difficult problems was the problem of simulating high-altitude conditions when working with pvrd. At the beginning of the work on this topic, there were no such stands in the country, the TsAGI stands were used, but they allowed for high-altitude testing for a very limited time. The problem was solved by engineer Shebeko. He invented a simple tube that could be used for long-term testing. The pipe provided the necessary injection, which was realized with an accurate analytical calculation. Systemically selected the necessary passageways for external incoming air and gaps. The pipe was practically not profiled. It was assembled from available materials. In the future, this pipe allowed to solve the problem of high-altitude testing of elements of the Burya rocket, and its design was taken as the basis for the creation of new GDP(gas-dynamic pipes), which are equipped with experimental laboratories at the present time.

4. Riike and Bossch–Riess experiments.

Many theoretical foundations of the problem of instability in the LRE [12] are based on two fundamental experiments: Riike, with a sounding pipe when the lower zone is heated, and Bossch-Riess — when the upper zone is cooled. [13].

The first experiment set by Riike was an example of the formation of sound as a result of heating. He demonstrated a striking phenomenon-the sound of a vertically installed pipe after bringing heat energy to it (Fig. 3). The energy came from a heated to red metal flat grid with a frequent small cell in the area with a cross section located at a distance of a quarter of the length of the pipe from the lower part. The pipe was positioned vertically for the purpose of organizing convective flows inside it. The through-thrust produced in the tube produced a sound of considerable force. Reike found that the location of the heated grid in any other place, in the lower or upper section did not lead to sound.

The phenomenon discovered by Riike has inspired many scientists to study this process. After some time, the scientists Bossam and RISSA was revealed another phenomenon is similar in its significance to the Rijke phenomenon. They put a similar experiment on the pipe, but the grid was placed at three-fourths of the height of the pipe. At the same time, the grid was not heated, but rather cooled. Streams of warm air came from the bottom of the pipe and crossed the cooled grid. The trumpet also sounded. It should be noted that in both experiments, the heat was either supplied to the grid or withdrawn from it. After experiments, it became obvious that heat is the main source of energy that feeds the oscillatory system.

Analyzing the experiments of Rieke and Bossch — Riess, Rayleigh first proposed the possibility of excitation of acoustic oscillations due to the energy of heat supply. Assuming in advance the frequency of receipt (removal) of heat to the fluctuating mass of air in the pipe, Rayleigh pays attention to the phase of oscillations at which they occur. Thus, if the heat is transmitted to the air at the moment of greatest compression and is not taken away from it at the moment of greatest rarefaction (maximum and minimum pressure), the oscillations are amplified. On the contrary, if the heat is supplied at the moment of the greatest rarefaction of the air, the vibrations are weakened. The results of the analysis of the experiment carried out by Rayleigh, allowed him to formulate a criterion for the occurrence of oscillations: if the phase shift between the oscillatory component of the heat supply and the oscillatory component of the pressure is an absolute value of less than $\pi/2$, then acoustic oscillations are excited in the system.

At present, the Rayleigh criterion is one of the few designed to predict the beginning of oscillations, but, unfortunately, it is suitable for the case of oscillations in the pipe. In LRE, this criterion must be used with significant stipulations.

5. Hartmann's Whistle.

Julius Hartmann discovered the phenomenon of resonance in the whistle during experimental studies of the axial distribution of Pitot pressure in a supersonic jet. He observed strong vibrations in the Pitot tube when he placed it in certain areas containing the shock zone of the free jet. Based on this, the researchers named some areas downstream of the jet where the resonance occurred as "instability

IOP Conf. Series: Materials Science and Engineering **779** (2020) 012031 doi:10.1088/1757-899X/779/1/012031

areas". Hartmann also experimented with a large Helmholtz resonator instead of a Pitot tube. This configuration, known as the "Hartmann pulsator", resonates at very low frequencies (on the order of 1-100 Hz), which makes it possible to examine the vibrations using x-ray systems (Fig. 4). The main part of the generator is the nozzle, from which a supersonic gas jet flows, in which there are waves of compaction and rarefaction. If the resonator is placed coaxially with the nozzle at a certain distance in the form of a tube with a sealed end, then when the jet is decelerated before the resonator there is a disconnected jump of the seal. As a result of the interaction of the main jet and the jet flowing out of the resonator at a certain distance between the nozzle and the resonator, the section of the jet behind the jump becomes a source of sound and ultrasonic waves. The frequency of the emitted sound depends on the distance between the nozzle and the resonator, as well as on the size of the resonator. The most favorable radiation conditions occur when the diameter of the nozzle outlet and the length of the resonator are equal to each other, and the diameter of the cavity of the resonator is 1.3 — 1.5 times the diameter of the nozzle. The power of the acoustic radiation of the Hartmann generator reaches several tens of watts, and the efficiency is 3-5%. When using compressed air, frequencies from 1 — 2 to 60 kHz are obtained. Using hydrogen instead of air, you can get frequencies up to 180 kHz.

6. The effect of the Rank-hilsh vortex tube.

Invented by the French engineer Georges Joseph Rank in 1931, the device for heat separation, the so-called Rank pipe, and later, in 1941, careful research and optimization of the design on the same principle by the German Professor Rudolf Hilsch (therefore the Rank–Hilsch pipe), is by far the most elegant and cheap way to get cold. The flow in the Ranka–hilsha pipe is still the subject of research by many scientists and engineers. Simple in execution and uncomplicated in configuration, the design of the pipe conceals many excellent opportunities for its application in many industries.

Its main property is the thermal stratification of the flow with its division into very cold and quite hot flows. The main reason for this stratification is the centrifugal flow of the main flow through the tangential channels into the device of the impeller type, followed by a vortex movement in opposite directions along the left and right border of the pipe into the atmosphere. The initial spin creates conditions for the appearance of a stable vortex in the feed zone, which is further divided into a peripheral and Central. In this case, the peripheral part of the vortex according to the geometry is cut off by the coaxial body, and the Central part is collected in a smaller pipe and output. Due to the centrifugal forces in the center of the vortex, the pressure decreases (in proportion, first of all, to the value of the centrifugal force) and depends on its value at the entrance to the channel. In this case, the mass of the Central part of the vortex begins to leave the center, creating a vacuum in this area. There is an adiabatic expansion of the gas, and the temperature in the center of the vortex drops sharply. At the same time, the temperature at the periphery increases in accordance with the increase in pressure: $T= P/PR$ It is clear that a secondary convective effect occurs simultaneously in the vortex. The cooled particles have a higher density than the hot ones, and according to the laws of hydrodynamics, the less dense ones must float, that is, move completely in the opposite direction relative to the centrifugal forces. But this effect at high speeds is very weak in comparison with the main centrifugal and may not be taken into account in practical calculations.

Depending on the design of the pipe, the capacity of cold and heat production will significantly change, since it (the design) determines the ratio of mass expenditure in one and the other direction. The thermal stratification will also be significantly affected by the shape of the channels. Therefore, the main task of engineers and researchers of the processes occurring in the Rank–hilsh pipes is to optimize the design in terms of obtaining both maximum stratification and maximum performance of either cold or heat.

As the flows were studied, various authors developed many methods for calculating (mainly engineering) processes occurring in the Rank–hilsh pipes; many experimental results were obtained on various working bodies and various structures; the possibility of obtaining the effect of heat stratification not only on gaseous but also on liquid working bodies was shown. But despite all efforts, at the moment there is no generally recognized physical and mathematical model of the phenomenon of energy separation.

In [14], experimental and computational studies were carried out to study the properties of the Rank-hilsh pipe (Fig. 5). As a result, several effects were found that characterize the features of this device.

First of all, this is the effect of thermal stratification. It was obtained in a wide range of parameters, primarily temperatures. The effect of injection. This effect was recorded from the side of the cold flow and consists in the fact that when the area of the output section of the hot vortex increases, the flow first stops, and then begins to move in the opposite direction. The vortex tube starts working as an injector, sucking air out of the atmosphere from the cold side of the tube. It was found that the injection started at the size of the area on the hot air vortex relative to the entrance area at the level of 10%.

The next effect that was recorded by calculation is the effect of thermal inversion. According to the results of the calculations, it was possible to observe the movement of one vortex inside another. Moreover, the internal vortex reflected from the wall rotated in the same direction as the external one, but the speed in the axial direction changed to the opposite. As a result of the collision of the flow with the wall, an inversion occurred — the sign near the axial speed changed to the opposite. It was curious that the axial velocity of the flow on the axis was zero, which is the condition for the stable existence of two screw flows, when one is inside the other.

Then a discrete tangle effect was discovered. When the right exit was closed, spherical vortices (up to three) were concentrated in the end region, which rotated in opposite directions. The multiplicity of vortices was determined by the length of the dead-end zone, and the diameter of the ball — the diameter of the pipe.

The vortex effects obtained experimentally and by calculation made it possible to formulate the principles of calculating the design of vortex tubes.

7. The effect of the heat pipe.

Another effect that is observed in the pipe and is widely used in the space industry is the thermal effect. The tube acts here as an element of the cooling system, the principle of which is based on the fact that in closed tubes of heat-conducting metal (for example, copper) there is a light-boiling liquid. Heat transfer occurs due to the fact that the liquid evaporates at the hot end of the tube, absorbing the heat of evaporation, and condenses on the cold, where it moves back to the hot end. An example of the construction of such a tube is shown in (Fig. 6).

Heat pipes are of two types: smooth-walled and with a porous coating on the inside [15]. In smooth-walled tubes, the condensed liquid returns to the evaporation zone under the action of gravity alone — in other words, such a tube will only work in a position where the condensation zone is higher than the evaporation zone, and the liquid is able to flow into the evaporation zone. Heat pipes with filler (wicks, ceramics, etc.) can work in almost any position, since the liquid returns to the evaporation zone through its pores under the action of capillary forces, and gravity plays a minor role in this process.

Materials and refrigerants for heat pipes are selected depending on the application conditions: from liquid helium for ultra-low temperatures and even indium for high-temperature applications. However, most modern tubes use ammonia, water, methanol, and ethanol as the working fluid.

Summary

Scientific and practical significance of the effects that occur in the pipes.

Obviously, the significance of trumpet effects is priceless. This is a benchmark for comparison with complex design configurations. This is an opportunity to generalize experimental research, which can often be understood only through localized areas in the form of pipes. Many hydraulic and gas-dynamic regularities are based on measurements in pipes. All the great scientists — hydraulics worked with pipes: Prandtl, Nikuradse, Blasius. A lot of useful laws were obtained for practice, which form the basis for the design of almost all industrial machines, devices, and devices.

And more. The pipe is the simplest mathematical approximation in Hydrogas dynamics. The problem of motion in this case is reduced to an axisymmetric one. The heavy — weight Navier-Stokes equations are greatly simplified. This makes it easier to set boundary conditions. Yes, it (the pipe) does not solve all problems, and often, using the pipe approximation, you can emasculate the essence of the problem. But pipes are irreplaceable at verification of the most difficult software complexes intended for calculation of modern problems of science and technology.

List of references

[1] G. V. Smirnov. Born of a whirlwind. M. Edition. One thousand nine hundred eighty two
[2] P Chaburko and Z Kossova 2019 IOP Conf. Ser.: Mater. Sci. Eng. 492 012011
[3] V Lomakin et al 2019 IOP Conf. Ser.: Mater. Sci. Eng. 492 012012
[4] A Gouskov et al 2019 IOP Conf. Ser.: Mater. Sci. Eng. 492 012013
[5] N Egorkina and a Petrov 2019 IOP Conf. Ser.: Mater. Sci. Eng. 492 012015
[6] K Dobrokhodov and a Petrov 2019 IOP Conf. Ser.: Mater. Sci. Eng. 492 012016
[7] M. Van Dyke. Album of liquid and gas flows. M. Publishing World. One thousand nine hundred eighty six
[8] G. Gertler. Three-dimensional instability of a plane flow with critical accuracy in the presence of vortex-like perturbations. Boundary layer problems and heat transfer issues. M. State energy publishing house. One thousand nine hundred sixty
[9] 4. G. F. Glotov. The aerodynamics of the aircraft in the photos. Zhukovsky. TSAGI. 2003.
[10] A.V. Mezentsev, V. I. Smyslov, V. I. Honichev. Periodic structure of mass transfer on the ablating surface in the supersonic part of the nozzles / / heat and mass Transfer. Volume 3. Minsk. One thousand nine hundred eighty four
[11] V. N. Akimov, Yu. g. Demyashko, P. S. Kurskov and others. 70 years at the forefront of rocket and space technology. M. Mechanical Engineering. Two thousand three
[12] Yu. M. Kochetkov. Turbulence. The occurrence of instability in a rocket engine // Engine # 2. 2012.
[13] Rijke, Bosscha, Riess, Pogg. Ann., Vol. CVII. 1859.
[14] Relay. The theory of sound. Volume 2. M. State publishing house of technical and theoretical literature. One thousand nine hundred fifty five
[15] Yu. m. Kochetkov, I. N. Borovik, O. A. Podymova and others. Vortex effects in vortex tubes of Rank-hilsh / / Vestnik MAI. Volume 23.No. 4. Two thousand sixteen

IOP Conf. Series: Materials Science and Engineering **779** (2020) 012032 doi:10.1088/1757-899X/779/1/012032

Electro-hydromechanical processes in mineral oil flow

V Pilgunov[1] **and K Efremova**[1,2]

[1]Bauman Moscow State Technical University

[2]E-mail: efremova.k.d@gmail.com

Abstract: The results of experimental studies of electrical processes occurring in the mineral oils in narrow channels and slots of hydraulic devices are considered. The theoretical substantiation of the detected phenomena is given.

Introduction

In hydraulic devices of automatic type various types of throttles are widely used as shut-off and control elements [1]. In diaphragm throttles the flow of the pressure fluid has a pronounced turbulent character and high speed causes cavitation and the associated release of bubbles of undissolved air and steam in narrow section. Linear and angular deformations of a liquid particle can lead to its polarization. In the flow of a moving at high speed fluid can occur the concentration of energy at which there are local flows with kinetic energy significantly exceeding the average flow energy values [2–4]. When studying the flow of mineral oil in transparent models of a diaphragm throttle at its inlet edge, localized stable light emission was detected in the blue part of the color spectrum which was called hydrodynamic luminescence [5–8]. The brightness of light emission increased during the speed increasing and stopped to change achieving of certain value of the Reynolds number. As there is a limited amount of information about light emission in the cavitation zone at low pressure on the sharp part of the sharp inlet edge of hole in scientific and technical literature [9–15], the additional studies in order to clarify the physical and physic-chemical effects and nature of light emission in a flow of dielectric liquid in narrow slots were conducted.

An experimental study of electro-hydromechanical processes in a flow of mineral oil

Photos of the studied transparent models of throttles are presented in Fig. 1. Models were manufactured of polished organic glass with pronounced dielectric properties. The diameter and width of the diaphragm control throttle hole were 1 mm and 0,5 mm. The pressure fluid was mineral oil brand "SHELL - 20".

The parameters characterizing the fluid motion in the flow cross sections in the diaphragm and capillary throttles are presented in Fig. 2.

For quantitative estimation of the electrical processes in the throttle sections a string manufactured of nickel-chromium wire with diameter of 0,05 mm in three-layer enamel insulation was stretched along its axis (Fig. 1a).

A local stripping of enamel insulation was performed on the string as a result of which a mark with width of 0,1 mm was obtained which provided detection of the electric potential in the selected flow section. A string with mark was extended along the axis of the channel by tension device (Fig. 3) and detected electrical signals in the flow cross sections.

Content from this work may be used under the terms of the Creative Commons Attribution 3.0 licence. Any further distribution of this work must maintain attribution to the author(s) and the title of the work, journal citation and DOI.
Published under licence by IOP Publishing Ltd

IOP Conf. Series: Materials Science and Engineering **779** (2020) 012032 doi:10.1088/1757-899X/779/1/012032

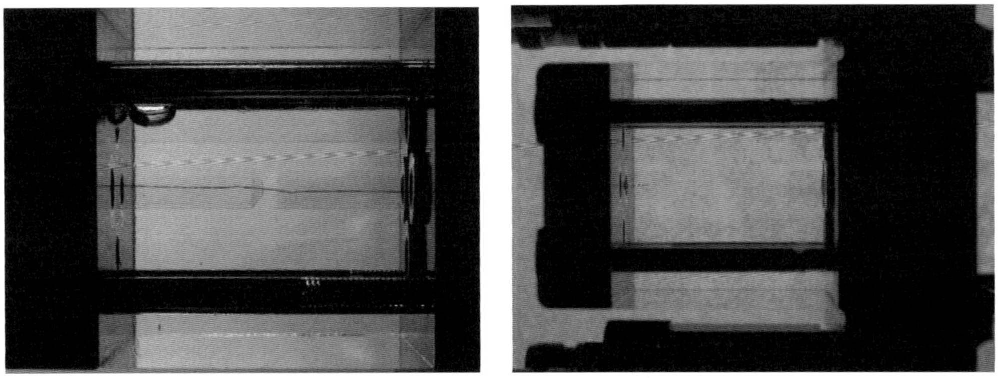

Figure 1. The tested models of throttles
a) Diaphragm throttle b)Capillary throttle

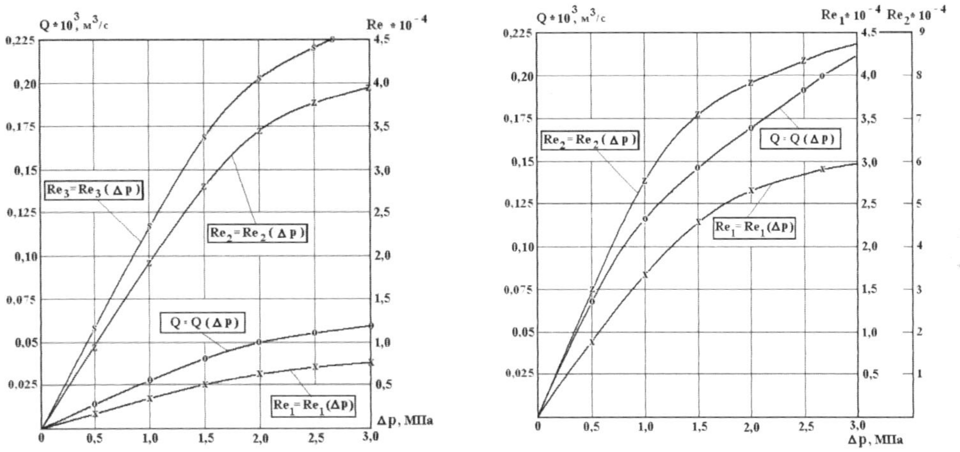

a) Diaphragm throttle b) Capillary throttle

Figure 2. Flow parameters in throttle modules

Figure 3. String tension device and mark position recorder

High-frequency electrical signals of complex shape in the flow cross sections were detected (Fig. 4).

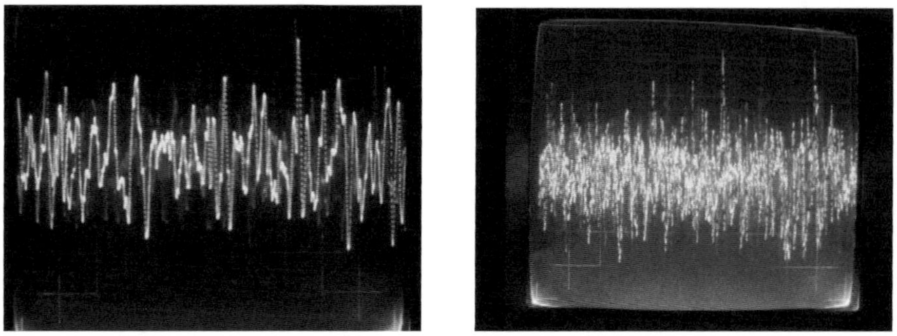

a) Strong string tension b) Released string tension

Figure 4. The shape of the electrical signal at the input edge of the diaphragm control throttle

The numerical results of electrical signals in the flow cross sections are presented in the summary diagram on Fig. 5. Positions of marks 1, 2, 3, 4 and 5 correspond to sections of the throttle module "inlet", "light emission", "cavitation core", "cavitation flare" and "outlet".

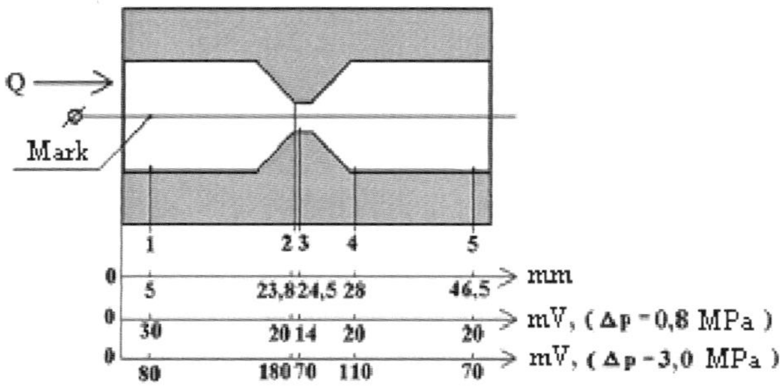

Figure 5. The span of the electrical signals in the flow cross sections

The integrated spectral characteristic of electrical signals was studied using a spectrum analyzer (Fig. 6).

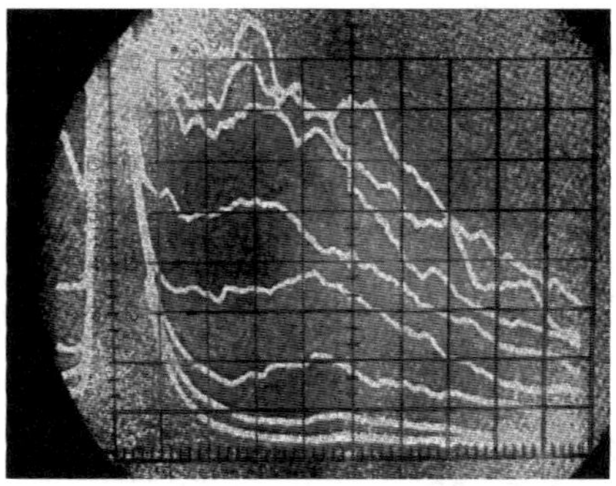

Figure 6. The spectrograms of electrical signal on the inlet edge of the throttle

Spectrograms were recorded at various inlet pressure valuesp_1 = 1.0; 1.5; 2.0; 2.5; 3.0; 4.0 MPa: the lower lines in the spectrograms correspond to the minimum pressure values, the upper lines – to the maximum. The frequency spectrum of the signals is in the range D_f = 2 ... 10 kHz. In cold mineral oil in the cross section along the input edge of the throttle the integrated voltage value at a frequency of

IOP Conf. Series: Materials Science and Engineering **779** (2020) 012032 doi:10.1088/1757-899X/779/1/012032

2 kHz reached a maximum value of U_{2kHz} = 200 mV. Heating the oil to t^0C = + 50^0C caused a decrease of the voltage integrated value at this frequency on the average in 1.5 times.

Significant span of electrical signals detected in the inlet and outlet of the diaphragm throttle necessitated the study of a cavitation-free flow of mineral oil in the capillary throttle (inlet diameter d_p = 6 mm, diameter of the capillary d_c = 2 mm (Fig. 1b). The velocity of the fluid in the capillary wasV_2 = 70 m / s. The shape of the electric signal in the middle of the length of the capillary is shown in Fig. 7.

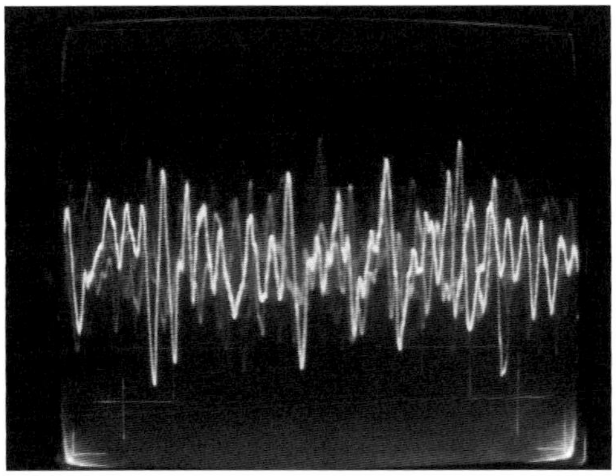

Figure 7. The shape of the electric signal in the middle of the length of the capillary throttle

Corrected span values of electrical signal are shown on summary diagram in Fig.8.

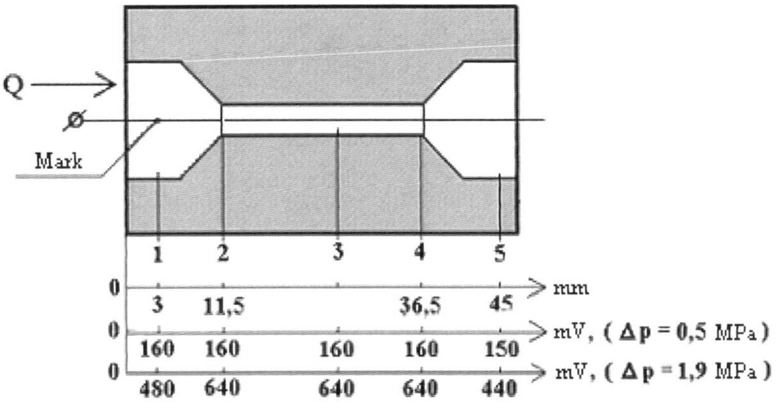

Figure 8. Span values of electrical signal of capillary throttle cross sections

The frequency spectrum of the signals is in the range D_f = 2 ... 10 kHz. In the inlet, the value of the integral voltage is slightly higher than in the outlet which caused by the loss of voltage along the length of the channel. In order to find out the reasons for the presence of electrical processes in the mineral oil flow and the influence of the material of the channel walls on them a model of a combined throttle was manufactured (Fig. 9) in which the inlet and diaphragm throttle were manufactured of brass (CuZn) and grounded.

The corrected signal spans presented in the summary diagram (Fig. 10). From the diagram it follows that the electrical signal reaches its maximum value of 270 mV at the entrance to the capillary (section 3).

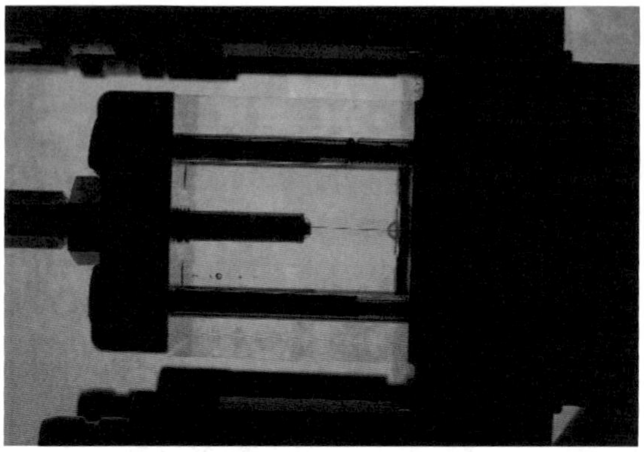

Figure 9. Combined throttle

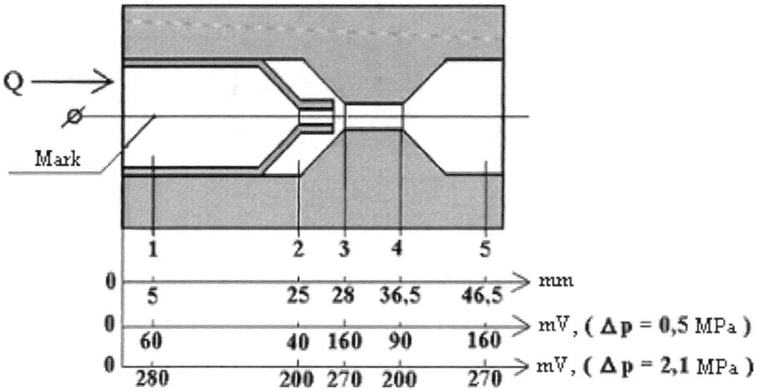

Figure 10. Signal span in the combined throttle cross sections

The occurrence of significant charges in the inlet channel indicates that the detected electrical processes are not associated with friction during the relative movement of two dielectrics.

For a more detailed study of electro-hydromechanical processes in the flow of a dielectric fluid a module was studied in which a metal plain washer was installed with a round hole with diameter of 0.9 mm with sharp inlet edge (Fig. 11).

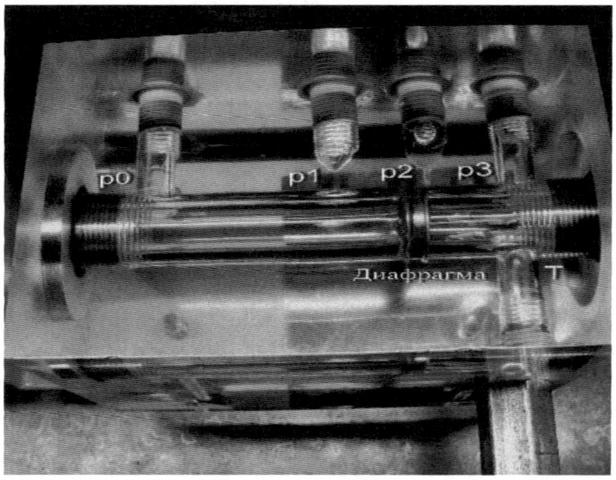

Figure 11. Module with a metal plain washer

IOP Conf. Series: Materials Science and Engineering **779** (2020) 012032 doi:10.1088/1757-899X/779/1/012032

The researches were carried out at the experimental test bench (Fig. 12).

Figure 12. The experimental test bench for the study of a module with a metal plain washer

The electro-hydromechanical processes in the module were quite intense (Fig. 13).

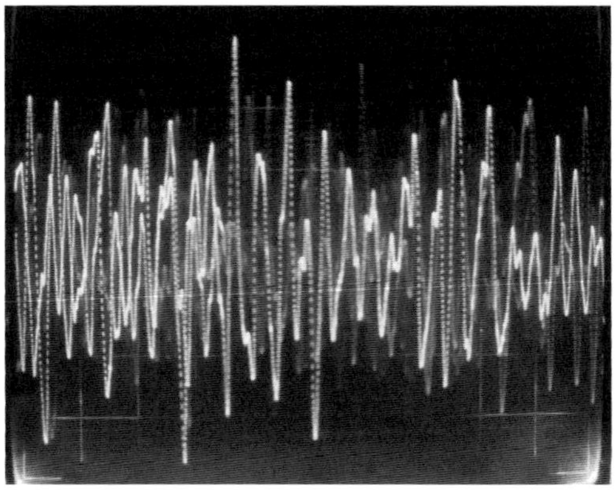

Figure 13. Oscillogram of electrical signals in a module with a metal plain washer

The corrected span values of electrical signal are presented in the summary diagram (Fig. 14).

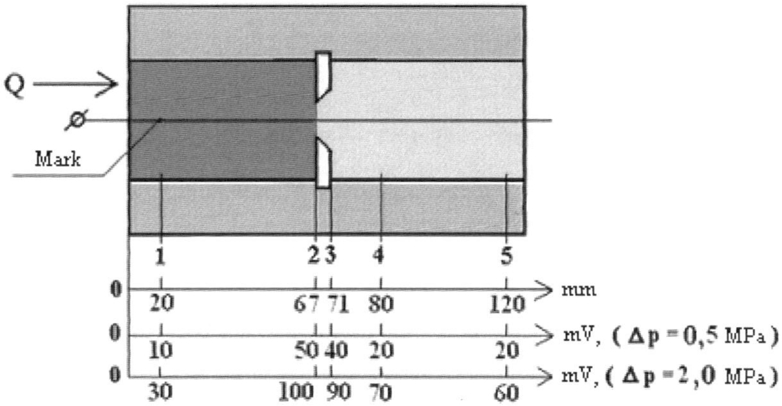

Figure 14. Span values of electrical signals in module with a metal plain washer

Discussion of the experimental results

An analysis of oscillograms and spectrograms allows us to establish the occurrence of electrical processes in the flow of mineral oil which are observed both in the inlet channels and in local compressions of the flow (diaphragm and capillary throttles) while their occurrence does not depend on the material of the walls restricting the flow. The frequency range of electrical signals is in the range D_f = 2 ... 10 kHz while low frequencies are more intense (Fig. 6). It can be assumed that in a metallized inlet channel a thin laminar sub layer of a dielectric fluid plays the role of insulation. The spectral characteristics (Fig. 7) indicate that the Reynolds number has a significant effect on the intensity of electrical processes. In accordance with the data given in [16–21], a double electric layer forms at the interface between two media ("solid wall - flow of dielectric fluid")which is a common property of dielectric fluids [22–24]. The fluid motion in a narrow channel can be considered in the OXYZ 3-coordinate system in which the origin of coordinates is associated with an arbitrary point on a solid interface between two media, the X axis is directed normal to the interface towards the flow, the Y axis is parallel to the tangent to the interface and the "Z" axis is parallel to the main flow vector (Fig. 15). The solid wall has an electric potential equal to the thermodynamic potential φ_T. The dense part of the double electric layer with thickness of Xc equal to 5 ... 10 molecular diameters ("sliding boundary") borders the interface between the two media.

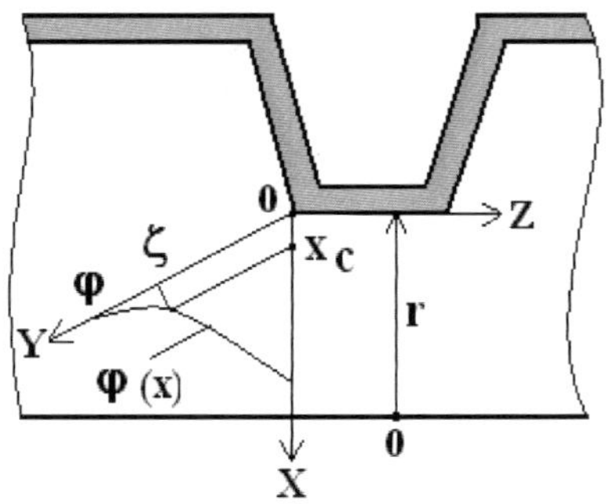

Figure 15. Binding OXYZ 3-coordinate system

In the remote from the interface surface of the diffuse part of the double electric layer ions can freely move and be carried away by the fluid flow. At values of X <X_c single electric charges can only move together with the surface layer. After flushing the diffuse layer by the liquid flow these charges form an uncompensated electric charge on the interface between the two media.

According to the Poisson equation the distribution of the density of the electric charge in a dielectric fluid is determined by the following expression:

$$\rho(x) = -\varepsilon_a \cdot \varepsilon_o \left(\frac{\partial^2 \varphi}{\partial x^2} \right) = -\varepsilon \cdot \varepsilon_o^2 \left(\partial^2 \varphi / \partial x^2 \right), \tag{1}$$

where ε_a — absolute dielectric constant, $\varepsilon = \varepsilon_a / \varepsilon_o$ — relative dielectric constant, $\varepsilon_0 = 8.85 \times 10^{-12}$ $F \cdot m^{-1}$ (Farads per meter)–vacuum permittivity(epsilon zero).

In accordance with Newton's law of viscous friction the velocity distribution of elementary streams over the flow cross section is determined by the equality:

$$V(X) = F(X - X_c) / 2\pi\mu rL, \tag{2}$$

where F — Newton's force of friction, μ — dynamic fluid viscosity, r — inner radius of the channel, L — channel length. In the laminar sub layer at the interface between two media the transverse velocity gradient dV /dX and the force F are almost constant. From equality (2) follows the expression for the force F:

$$F = 2\pi\mu r\, LV(X)/(X - X_c). \tag{3}$$

The amount of electricity carried by the diffuse layer per unit time is determined by the double integral over the surface:

$$I = \iint_{(S)} V(X)\rho(x)dX\, dY, \tag{4}$$

where (S) — is the area of the "real" section of the stream. Taking into account equalities (1) and (2) equation (4) is equated to following expression:

$$I = \left(\varepsilon \cdot \varepsilon_0^2 F / 2\pi\mu r L\right) \underset{0\ldots2\pi r}{H} dY \int_{Xc}^{X} X\left(\frac{d^2\varphi}{dX^2}\right)dX. \tag{5}$$

If we take into account that the first integral within the integration limit 0 ... 2πr takes the value equal to 2πr and the second integral within the integration limit from X = Xc to X = r is equal to the "zeta potential" ς, then equality (5) is significantly simplified and takes the form

$$I = \varepsilon \cdot \varepsilon_0^2\, F\zeta / \mu L. \tag{6}$$

The equation of the dynamics of the flow flushing the diffuse part of the double electric layer, taking into account equality (3) and the smallness of the slip boundary X_c<< r is determined by the expression:

$$F = 2\pi\mu r\, LV(X/(X\ \ Xc) = \pi r^2 \Delta p, \tag{7}$$

whereΔp — is the pressure drop across the areas of the "real" cross sections of the flow flushing the diffuse layer.

After substitution the value of Newton's force in equality (6) get the following equation:

$$I = \pi\varepsilon\varepsilon_o^2\, \zeta\Delta p\, r^2 / \mu L. \tag{8}$$

The flow rate of the uncompensated charge Q = Q (t) through the "real" flow section is dQ/dt = I – i, where i = $Q/\varepsilon\varepsilon_0^2\, \lambda$ is the conduction current determined by the specific electric resistance of the fluid λ (for mineral oil λ = 1×10^7 kOhm× m). Thus, the flow rate of the uncompensated charge will be determined by the equality:

$$dQ / dt = \pi\varepsilon\varepsilon_o^2\zeta\Delta p\, r^2 / \mu L - Q / \varepsilon\varepsilon_o^2\lambda. \tag{9}$$

Temporal integration of charge consumption (9) determines its accumulated value:

$$Q(t) = \pi\varepsilon^2\varepsilon_o^4\zeta\lambda\Delta\, p\frac{r^2}{\mu L}\left[1 - \exp\left(-t / \varepsilon\varepsilon_o^2\lambda\right)\right]. \tag{10}$$

For t →∞ and dQ /dt→ 0 the accumulated charge Q becomes equal

$$Q_\infty = \pi\varepsilon^2\varepsilon_o^4\, \zeta\lambda p\, r^2 / \mu L. \tag{11}$$

This charge determines the electric field near the solid wall E = $Q/4\pi\varepsilon\varepsilon_0^2 r$ L and the stationary charge density:

$$\sigma = 2E\varepsilon\varepsilon_o^2 = Q_\infty / 2\pi r\, L. \tag{12}$$

Will give a theoretical estimate of the values of the accumulated uncompensated charge Q_∞(11) and the stationary charge density (12) for the investigated transparent throttle modules.

Diaphragm throttle

Source data:

$r = 0{,}5\times 10^{-3}$ m; $L = 0{,}5 \times 10^{-3}$ m; $\varepsilon = 2{,}2$; $\varepsilon_0 = 8{,}85 \times 10^{-12}$ F·m^{-1}(Farads per meter); ζ= 1V;

$\lambda = 1\times10^7$kOhm×m; μ= 170 kg/ms; Δp = 3 MPa.

Calculation results: $Q_\infty = 1{,}4\times10^{-10}$ C(Coulomb); σ= $0{,}9 \times10^{-4}$C/m^2.

Capillary throttle

Source data:

$r = 1\times 10^{-3}$ m; $L = 25\times 10^{-3}$ m; $\varepsilon = 2{,}2$; $\varepsilon_0 = 8{,}85 \times 10^{-12}$ F·m^{-1}(Farads per meter); ζ= 1V;

$\lambda = 1\times10^7$kOhm×m; μ= 170 kg/ms; Δp = 3 MPa.

Calculation results: $Q_\infty = 0{,}11\times10^{-10}$ C (Coulomb); σ= $0{,}9 \times10^{-6}$C/m^2.

As it follows from the calculation results, the accumulated uncompensated charge in the diaphragm control throttle is 12,5 times higher and the stationary density is in 100 times higher than in the capillary one.

The expression in square brackets of equality (10) determines the attenuation decrement of the process of increasing the flow rate of an uncompensated charge and is 0,37 s^{-1}. For a time $t \approx 2,7$ s, due to the draining of the resulting charge and in accordance with equality (10), the damping ratio decreases 2,7 times. During this time, in the cavitation flow of mineral oil an explosion of a cavitation bubble filled with a vapor-air mixture can occur. The explosion voltage of the bubble is much less than the critical tension of mineral oil which can cause light emission at the input edge of the diaphragm throttle.

The stationary charge density in the capillary throttle is 100 times less than in the diaphragm throttle, therefore, light emission is not observed. The composition of the vapor bubble is closely related to the base of mineral oil and a set of corrective additives. Due to this, with identical parameters of the cavitation flow light emission at the inlet edge of the diaphragm throttle may not be observed [25 - 27].

A comparative analysis of the flow parameter graphs (Fig. 2a and 2b) and the range of electrical signals (Fig. 5 ... 9) shows that the electrical signals in the module leads become observable at numbers Re > 0.25×10^4.

List of reference

[1] Wave Processes Regulators Optimisation in Hydraulic Systems D N Popov, N G Sosnovsky and M V, Siukhin 2018 *IOPConf. Ser.: Mater. Sci. Eng.* **468** 012014.htpps://iopscience.iop.org/issue/1757-899X/468/012014.

[2] Margulis M.A. Sonochemistry and cavitation. Gordon & Breach, London, 1995. — 543 p.

[3] Bagrov V.V., Desiatov A.V. and others. Water: effects and technologies. Publishing House ONIC "Engineer", LLC "Oniko-M", M. — 2010, 488 p.

[4] Gertsenshtein S. IA.Monakhov A. A. Electrification and fluorescence of a fluid in a coaxial channel with dielectric walls. Fluid Dynamics A Journal of Russian Academy of Sciences, No. 3, M. — 2009, p. 114–119.

[5] Margulis M.A., Pilgunov V.N. About the mechanism of occurrence of luminescence and electrification during the flow of liquids in a narrow channel. Journal of Physical Chemistry, Volume 83, No. 10, M. — 2009, p. 1975–1979.

[6] Koldomasov A.I. Plasma formation in cavitating dielectric fluid. Journal of Theoretical Physics. Volume 61, M. — 1991, p. 256–263.

[7] Vysotskii V. I. and others. The generation of intense directional emission during the rapid movement of a jet of liquid through dielectric channels. Fluid Dynamics A Journal of Russian Academy of Sciences. Surface. X-ray, synchrotron and neutron studies. No. 3, — 2007, p. 55–60.

[8] IA. I.Frenkel. About electrical phenomena associated with cavitation due to ultrasonic vibrations in a liquid. Journal of Physical Chemistry, Volume 14, M. — 1940 p. 305.

[9] M.A. Margulis. The study of electrical phenomena associated with cavitation. Journal of Physical Chemistry, Volume 55, M. — 1981, p. 154–158.

[10] Ring-shaped fluorescence of mineral oil on the front edge of the choke coil V Pilgunov and K Efremova 2019 *IOPConf. Ser.: Mater. Sci. Eng.* **492** 012028. htpps://doi.org/10.1088/1757-899X/492/1/012028.

[11] Measurements of acoustic flow parameters in the orifice on non — linear regimes. A Bykov, A Komkin and V Moscalenko 2019 *IOPConf. Ser.: Mater. Sci. Eng.* **589** 012015. htpps:// iopscience.iop.org/issue/1757-899X/589/1/012015.

[12] M.A. Margulis. Electrical phenomena during the explosion of cavitation bubbles. Journal of Physical Chemistry, Volume 71, M. — 1997, p. 1885–1889.

[13] Independent component analysis of hydrodynamic bubbles propertiesSGavrilov, O Ivanova and T Decaen 2019 *IOPConf. Ser.: Mater. Sci. Eng.* **492** 012014. htpps://doi.org/10.1088/1757-899X/492/1/012014.

[14] Regulation of bubbles descending speed in water. B Ksenofontov, M Ivanov, S Shauhan and I Schaikhiev 2019 *IOPConf. Ser.: Mater. Sci. Eng.* **492** 012023. htpps://doi.org/10.1088/1757-899X/492/1/012023.

[15] Analysis of hydrodynamic properties of a gas phase in water by hydroacoustic method. S Gavrilov and M Ivanov 2019 *IOPConf. Ser.: Mater. Sci. Eng.* **492**012030. htpps://doi.org/10.1088/1757-899X/49 2/1/012030.

[16] M.A. Margulis, I.M. Margulis. Electrical phenomena on the surface of pulsating cavitation bubbles. Journal of Physical Chemistry, Volume 71, M. — 1997, p. 1890–1895.

[17] C.V. Seghal, R.E. Verral. A review of the Electrical Hypothesis of Sonoluminescens, Ultrasonics,V.20, 1982, p. 37–39.

[18] T. Lepoint, D. De Pauv, F. Lepoint-Mullie, M. Goldman. Sonoluminescens: an Alternative Electrodynamic Hypothesis, *J. Acoust. Soc. Amer.,* V. 101, 1997, p. 2012–2030.

[19] Sonoluminescence. Milia A Margulis 2000 *Phys.-Usp.* **43** 259. https://doi.org/10.1070/PU2000v043n03ABEH000455

[20] Sonoluminescence and acoustic cavitation Pak-Kon Choi 2017 *Jpn. J. Appl. Phys.***56** 07JA01. https://doi.org/10.7567/JJAP.56.07JA01.

[21] M.A. Margulis, I.M. Margulis. Theory of Local Electrification of Cavitation Bubbles: New Approaches. Sonochemistry, V. 6, 1999, p. 15–20.

[22] Research of centrifugal gas-liquid separator G Budaev, D Danilov, A Kuznechov, V Lomakin and V Cheremushkin 2019 *IOPConf. Ser.: Mater. Sci. Eng.***589** 012035. htpps: // iopscience.iop.org/issue/1757-899X/589/1/012035.

[23] Electric Characteristic and cavitation bubble dynamics using under water pulsed discharge. Minglei SHAN. Bingyan CHEN. Cheng YAO. Qingbang HAN. Changping ZHU and Yu YA/ 2019 *Plasma Sci.Technol.* **21** 074002. https://doi.org/10.1088/2058-6272/abOb62.

[24] Electrical pulses produced by cavitation bubbles G.Gimenez 1979 *J.Phys.D: Appl. Phys.* **12** L25. https://doi.org/10.1088/0022-3727/12/4/001.

[25] E. Kh. Isakaev, A. S. Tyuftyaev, M. Kh. Gadzhiev, N. A.Demirov and P. L. Akimov. Influence of ultrasound on the electrical breakdown of transformer oil. 2018 *J. Phys.: Conf. Ser.***946**012175. https://doi.org/10.1088/1742-6596/946/1/012175.

[26] Measurement of near-wall stratified bubbly flows using electrical impedance. J Cho. M Perllin and S I Ceccio 2005 *Meas. Sci. Technol* **16** 1021. https://doi.org/10.1088/0957-0233/16/4/015.

[27] A liquid dielectric in an electric field. Mikhail N Shneider and Mikhail Pekker. Pages 4-1 to 4-30. Chapter 4 from Liquid Dielectrics in an Inhomogeneous Pulsed Electric Field (Second Edition).

IOP Conf. Series: Materials Science and Engineering **779** (2020) 012033 doi:10.1088/1757-899X/779/1/012033

Calculation of the hydro-pneumatic suspension damper

S Britsin[1], M Ryabinin[1,2] and E Sarach[1]

[1]Bauman Moscow State Technical University

[2]E-mail: ryabinin@bmstu.ru

Annotation. The paper considers general approaches to the design and calculation of the characteristics of hydropneumatic dampers that can be used in the suspensions of mobile cars. It is paid particular attention to the determination of the main hydraulic characteristics of the flow part of the damper, including a direct-acting belt valve, based on numerical simulation.

Introduction

The desire to increase the efficiency of transport equipment requires increasing its average speed.

As it is well known, a machine speed is limited and is determined by the movement conditions and the technical characteristics of the machine. Restrictions depending on driving conditions include restrictions on traction, on the uncontrolled movement danger and on the ride smoothness [1]. An increase in the specific power of modern machines and the correct choice of traction engines made it possible to remove a number of restrictions on the traction force [2]. The subject of work are the methods to reduce the likelihood of uncontrolled movement [3]. In this regard, the problem of improving the suspension systems of transport vehicles aimed at increasing the speed of movement on bumps is relevant.

An integral element of the suspension system of the transport vehicle is the damper. It serves to damp the vibrations of the sprung body when driving on rough roads.

The modern approaches to transport vehicles suspension theory require the selection of damping characteristics according to the criteria of smoothness using a simulation mathematical model of the movement of the vehicle along rough roads. The methods for determining the damping characteristics and mathematical models of the movement of transport vehicles for the study of ride smoothness are widely presented in the literature [4]–[14], therefore, in this work, the power characteristic of the damper is taken as the initial data. Particular attention in this work is paid to the design of throttle damper system valves, since their design feature allows to implement a given power characteristic.

The results obtained in this paper can be extended to the damping elements of both uncontrolled and controlled suspension systems, since the analysis of literature on controlled suspension systems [15]–[31] shows that it is advisable to use a passive suspension system with controlled damping for transport vehicles.

Initial data

The initial data for the engineering calculation of the damper will be considered a given power characteristic, suspension design and geometrical dimensions of the main surfaces, obtained on the basis of similar successful design solutions. Figure 1 shows the required power characteristic of the hydropneumatic spring damper, which must be ensured by calculating its main elements: throttles and valves. In this paper, we restrict ourselves to considering only the forward stroke of this diagram

Content from this work may be used under the terms of the Creative Commons Attribution 3.0 licence. Any further distribution of this work must maintain attribution to the author(s) and the title of the work, journal citation and DOI.
Published under licence by IOP Publishing Ltd

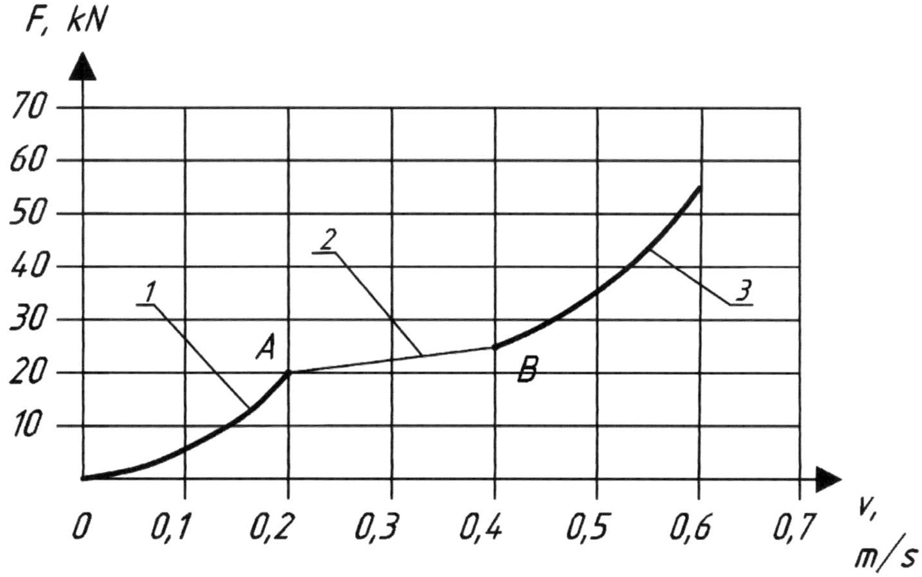

Figure 1. Power characteristic of the damper (straight stroke)

The point A corresponds to the beginning of the bypass valve opening, and the point B to the valve output at the maximum opening and the maximum value of the valve gap area f (Fig. 2). The line 1 is provided with a constant throttling channel, which is available in the design, the line 3 is determined by the total conductivity of the fully open valve and constant throttling channel, and the form of curve 2 connecting the points Φ and B depends mainly on the increase degree in the flow rate through the valve from its opening. This dependence is largely determined by the spring deformation ratio when the valve h rises to its preliminary value and the emerging hydrodynamic force ratio to the spring force, i.e. on the pressure drop across the valve and the valve design features. The design of a direct acting hydraulic starting valve consists of a locking element with a diameter d_2 resting on a saddle with a diameter d_1, having a sealing belt with a significant radial length "a", springs, and also a body with a supply channel with a pressure p_1 and discharge channel with a pressure p_2. The force of the fluid pressure acting on the locking element of the valve is balanced only by the force of the spring. Due to the presence of an extended sealing belt, such a valve design is able to provide a shock-free fit on the saddle with a rapid increase in pressure p_2 in the internal space. The softness of the landing is achieved due to the intensive deceleration of the locking element when it approaches the surface of the saddle. The reason for this is the increase in pressure from the side of the fluid displaced from the radially extended sealing gap.

The soft fit and, consequently, the low wear of the sealing surfaces led to the widespread use and high reliability of such valves for systems where the number of their operating cycles is hundreds or thousands per minute. Flat starting valves allow the sealing element to move in the plane of contact with the saddle and therefore do not require an exact guide element. This explains the small size and simplicity of the design of such valves, they are used in hydropneumatic dampers. It is necessary to know the valve conductivity at its various openings for the suspension dynamic calculations of the machine, since the valve is an integral part of the flow channels of the damper.

There are methods for conducting dynamic calculations of the damper, taking into account the change in the conductivity of the valve from the Reynolds number [32]. It isn't always possible to use these data and it is necessary to observe the similarity between the flow part of the damper being developed and a sample with known characteristics. It is also a non-trivial task to take into account analytically the mutual influence of various hydraulic resistances forming the flow part of the damper.

IOP Conf. Series: Materials Science and Engineering **779** (2020) 012033 doi:10.1088/1757-899X/779/1/012033

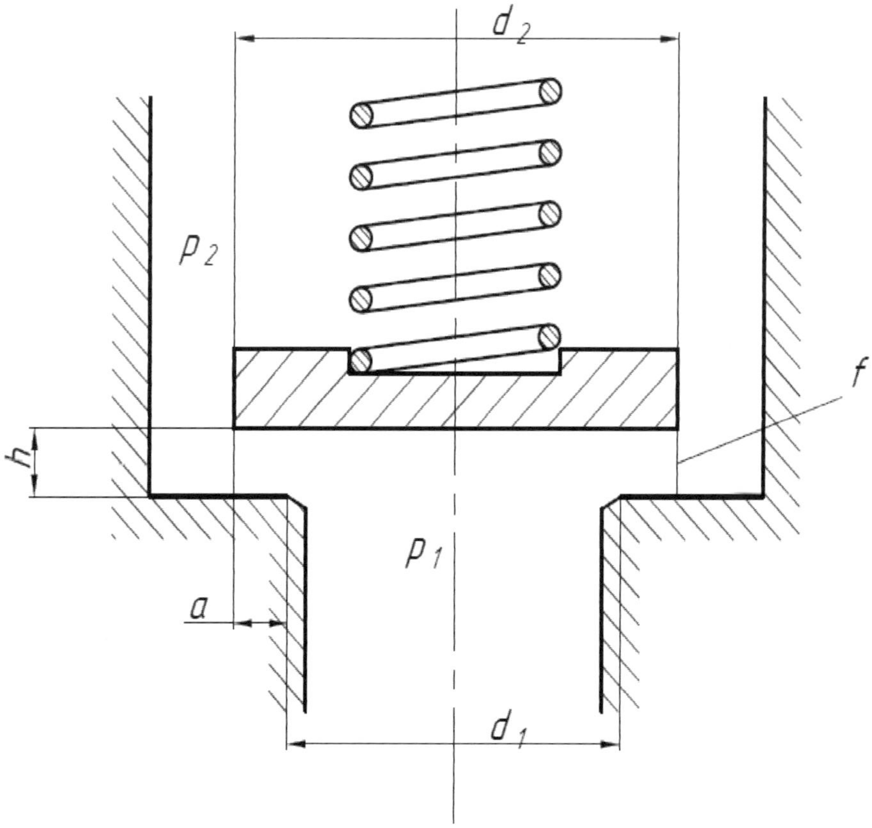

Figure 2. Design diagram of a direct-acting starting valve

Methods

It is proposed to determine the resistance of the damper flow part in all the prescribed regimes using numerical simulation and to use the obtained values for further dynamic calculations of the suspension characteristics of the machine.

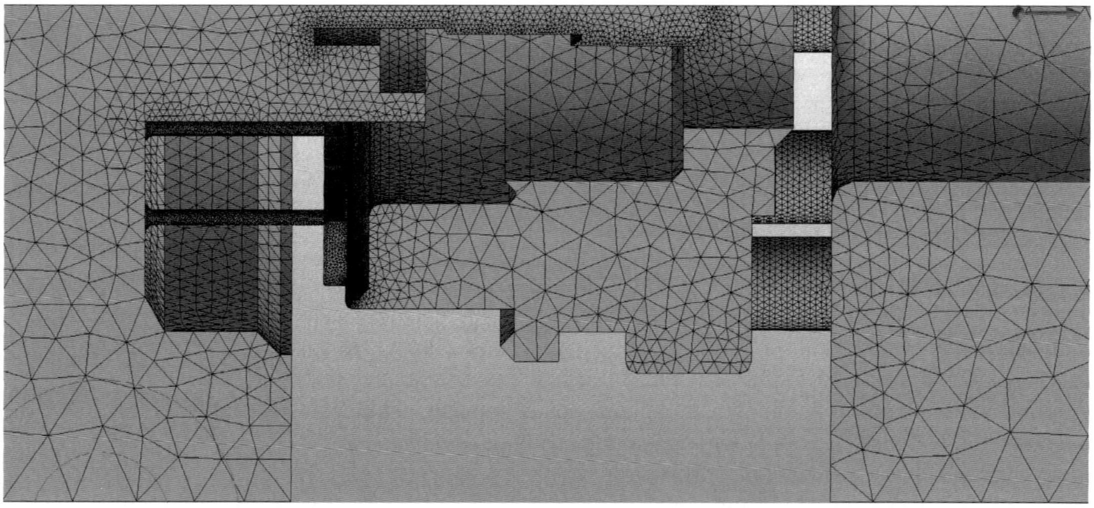

Figure 3. Mesh of the flow part of the damper with an opening height of 0.1 mm

IOP Conf. Series: Materials Science and Engineering **779** (2020) 012033 doi:10.1088/1757-899X/779/1/012033

For a numerical analysis of the flow part of the damper, it is used the CFD simulation software package ANSYS Fluent, which allows to simulate complex hydrodynamic processes in various devices. In order to obtain a series of results corresponding to different lift heights and flow rates for the valve, a parametric model was created with variable geometry and boundary conditions (the creation of so-called design points). Modeling the flow for different valve heights in narrow channels and slots requires a special approach for a computational grid with a number of cells from 4.2 to 5 million cells (Figure 3).

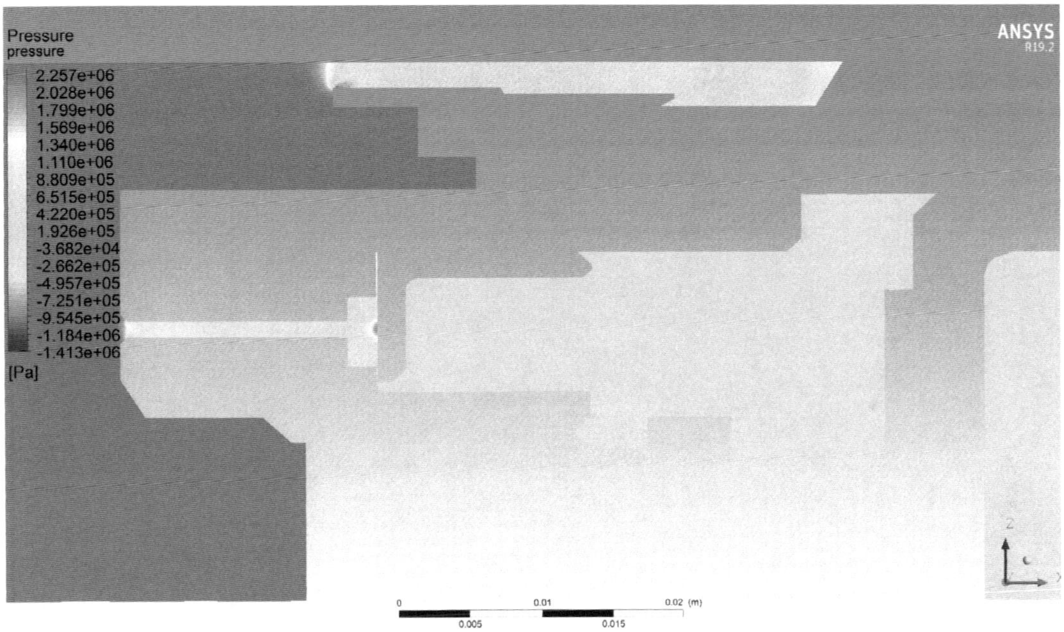

Figure 4. The field of distribution of static pressure at $h = 0,1мм$ and $v = 0,2м/s$

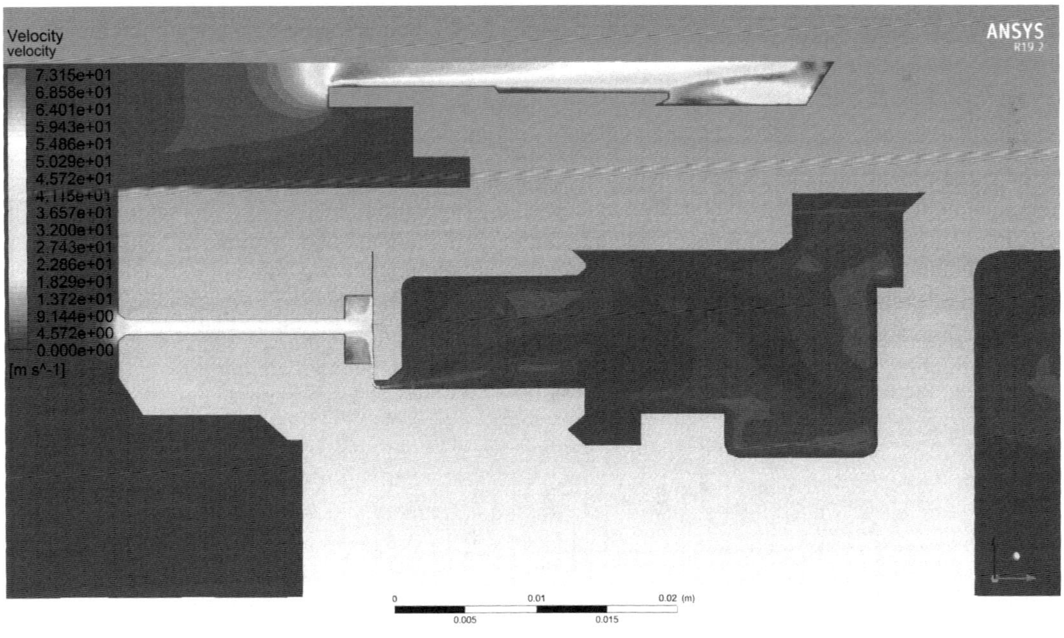

Figure 5. The field of velocity distribution at $h = 0,1мм$ and $v = 0,2м/s$

In order to describe the flow in the flow part, it is used the SST model of turbulence based on the turbulence kinetic energy transfer equation and the specific dissipation equation. This model describes well the near-wall layers of the flow and is able to describe flows at large pressure gradients

As boundary conditions was selected an inlet velocity of $v = 0{,}1 м / s$ to $v = 0{,}5 м / s$ м/с and damper outlet pressure $p = 0 Pa$

The simulation results are illustrated in Figures 4 and 5.

Conclusions

The comparison between the obtained values for flow coefficient of the flow part from the Reynolds number with the data given in the literature [32] shows their qualitative coincidence. The quantitative difference (Figure 6) is explained by the hydraulic resistance of the channels leading to the valve and the mutual influence of these resistances.

Also, the similarity of phenomena may be disrupted due to differences in the geometry of the flow part.

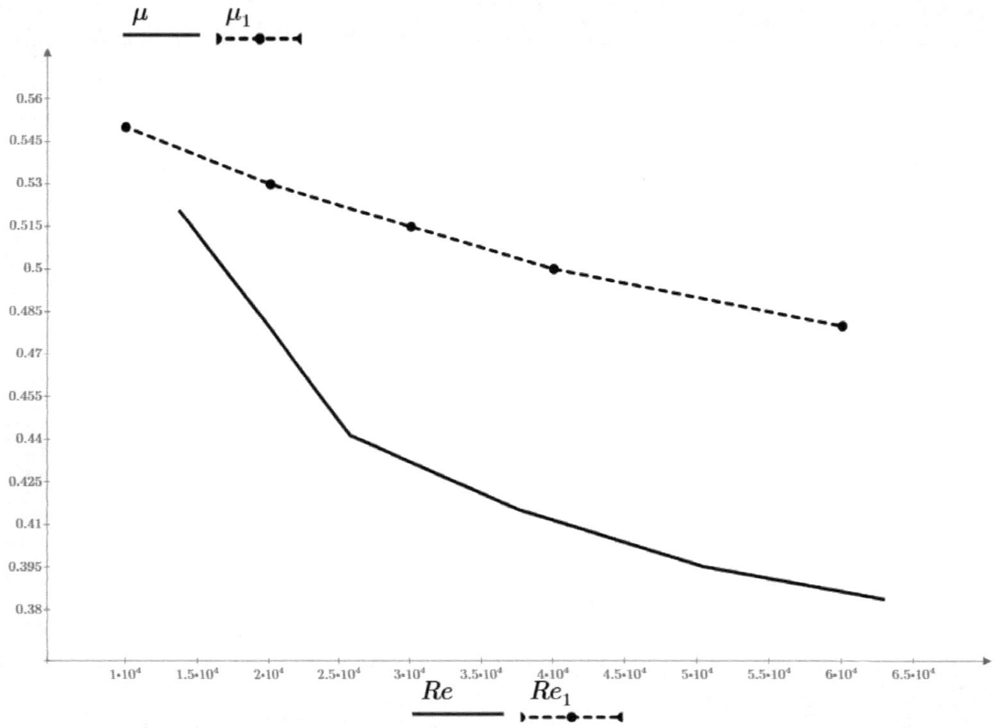

$\mu_1(\mathrm{Re}_1)$ — Experimental data from the reference [32]

$\mu(\mathrm{Re})$ — Data obtained by numerical modeling

Figure 6. Comparison of flow coefficient values

References

[1] Dmitriev A. A., Chobitok V. A., telminov A.V. Theory and calculation of nonlinear systems for springing tracked vehicles. - Moscow: Mashinostroenie, 1976. — 207 p.

[2] Kotiev G O, Miroshnichenko A V, Stadukhin A A and Kositsyn B B 2019 Determination of mechanical characteristics of high-speed tracked vehicles traction motor with individual drive wheels (Journal of Physics: Conference Series) vol 1177(1)) p 012058

[3] Ergin A A, Kolomejtseva M B and Kotiev G O 2004 Antiblocking control system of the brake drive of automobile wheel (PriboryiSistemyUpravleniya) vol 9 pp 11–13

[4] Dyadchenko M. G., Kotiev G. O., sarach E. B. Construction and calculation of suspensions of high-speed tracked vehicles: Studies.stipend. - Moscow: MGTU publishing House. N. E. Bauman, 2007. - 40C.
[5] Dyadchenko M. G., Kotiev G. O., Naumov V. N. Bases of calculation of systems of podressorivaniye of tracked cars on the computer: the Textbook. — M.: Publishing house of MGTU them. N. E. Bauman, 2002. - 52C.
[6] DhirAandSankarS 1994 Analiticalmodelfordynamicsimulationofoff-roadTrackedVehicles (Vehiclesystemdynamics) vol24 pp 26–49
[7] Dhir A and Sankar S 1997 Analitical wheel models for dynamic simulation of off-road tracked vehicles (Vehicle system dynamics) vol 27 pp 37–63
[8] Duym S, Stiens R and Reybrouck K 1997 Evaluation of shock absorber models (Vehicle system dynamics) vol 27 pp 109–127
[9] Lee K 1997 Numerical modeling for the hydraulic performance prediction of automotive monotube dampers (Vehicle system dynamics) vol 28 pp 25–39
[10] Wong J Y 1997 Dynamics of Tracked Vehicles (Vehicle System Dynamics) vol 28 pp 197–219
[11] Sarach E, Kotiev G and Beketov S 2018 *Methods for road microprofile statistical data transformation* (MATEC Web of Conferences) vol 224 p 04009
[12] Klubnichkin V E, Klubnichkin E E, Kotiev G O, Beketov S A and Makarov V S 2018 *Interaction between elements of the track ground contacting area with the soil at curvilinear motion of the timber harvesting machine* (IOP Conference Series: Materials Science and Engineering) vol 386(1) p 012016
[13] Weiss R A 1981 *Terrain Microroughness the Dynamic Response of Vehicles* (West Point (New York) ARO Report 82–1) 6 p
[14] Shinozuka M 1971 *Simulation of a multivariate and multimensional random processes* (Journal of the Acoustical Society of America) vol 49 № 1 pp 357–368
[15] Venhovens P J Th 1991 *Semi-Active Suspension for Automotive Application*. (Delft: Univ. of technology) 28 p.
[16] Venhovens P J Th 1994 *The Development and Implementation of Adaptive Semi-Active Suspension Control* (Vehicle System Dynamics) vol 23 pp 211–235
[17] Karnopp D and Sang-Gyun So 1998 *Energy Flow in Active Attitude Control Suspensions: A Bond Graph Analysis* (Vehicle System Dynamics) vol 29 pp 69–81
[18] Redfield R C 1991 *Performance of Low-bandwidth, Semi-Active Damping Concepts for Suspension Control* (Vehicle System Dynamics) vol 20 pp 245–267
[19] Foag W A 1989 *Practical Control Concept for Passenger Car Active Suspension with Preview* (Proc. Instn. Mech. Engrs (London (UK))) p 203
[20] Prem H A 1987 *Laser-based Highway-speed Road Profile Measuring System* (Proc. 10th IA VSD) p 37
[21] Zhileykin M M, Kotiev G O and Nagatsev M V 2018 *Comparative analysis of the operation efficiency of the continuous and relay control systems of a multi-axle wheeled vehicle suspension* (IOP Conference Series: Materials Science and Engineering) vol 315(1), 012030
[22] Zhileykin M M, Kotiev G O and Nagatsev M V 2018 *Synthesis of the adaptive continuous system for the multi-axle wheeled vehicle body oscillation damping* (IOP Conference Series: Materials Science and Engineering) vol 315(1), 012031
[23] Crosby M J and Karnopp D C 1973 *The Active Damper - a New Concept for Shock and Vibration Control* (43rd Shock and Vibration Bulletin) vol H, June, pp 46–73.
[24] Dhir A and Sankar S 1994 *Ride Dynamics of High-Speed Tracked Vehicles: Simulation with Field Validation* (Vehicle System Dynamics) vol 23 pp 379–409.
[25] Gordaninejad F and Breese D 2000 *Heating of Magnetorheological Fluid Dampers* (Journal of Intelligent Materials, Systems and Structures, accepted, in press) vol 4 pp 45–56.
[26] Han-Shue Tan and Bradshaw T 1998 *Validation of the Harshness Problem of Automotive Active Suspensions* (Vehicle System Dynamics) vol 30 pp 89–115.

IOP Conf. Series: Materials Science and Engineering **779** (2020) 012033 doi:10.1088/1757-899X/779/1/012033

[27] Hrovat D, Margolis D and Hubbard M 1988 *An Approach Toward the Optimal Semi-Active Suspension* (Journal Dynamics System. Measurement and Control) vol 110 № 3 pp 288–296.
[28] Wang X and Gordaninejad F 2000 *Flow Analysis of Field-Controllable, Electro- and Magnetno-Rheological Fluids Using Herchel-Bulkley Model* (Journal of Intelligent Materials, Systems and Structures, accepted, in press) vol 2 pp 87–121.
[29] Yokoyama M, Hedrick J and Toyama S 2001 *A Model Following Sliding Mode Controller for Semi-Active Suspension Systems with MR Dempers* (Proceedings of the American Control Conference, Arlington, VA June 25-27) P. 53–58.
[30] Yu F and Crolla D 1998 *An Optimal Self-Tuning Controller for an Active Suspension* (Vehicle System Dynamics) vol 29 pp 51–65.
[31] Yu F and Crolla D 1998 *State Observer Design for an Adaptive Vehicle Suspension* (Vehicle System Dynamics) vol 30 pp 457–471.
[32] Kirillovsky Yu. L. Hydraulic valves. Study guide for the course and diploma projects. M: MSTU publishing House. N. E. Bauman, 1977. -64 p.: Il.

IOP Conf. Series: Materials Science and Engineering **779** (2020) 012034 doi:10.1088/1757-899X/779/1/012034

Mathematical modeling of a hydraulic drive for leveling the support table of a robotic drilling platform

E Kolotvin

Bauman Moscow State Technical University

E-mail: egor.kolotvin.95@gmail.com

Abstract. The principle of moving a hydraulic drilling platform is described. To study the dynamics of transient response processes of hydraulic drive for leveling, a mathematical model was created. The structural diagrams of mathematical models of the actuator and hydraulic control valve are given. The time dependences of the spool positions of the hydraulic control valve and the hydraulic cylinder rod are presented. The dependences of pressure changes in the rod and piston cavities of the hydraulic cylinder are presented.

1 Introduction

Nowadays, there are works around the world to create various extreme robotics tools for various industries. Traditionally, these are mobile robots with a single manipulator mounted on a moving platform with a wheeled or caterpillar propulsion. Their wide distribution is due to the simplicity of not only a mechanical design, but also a control system that does not require the use of complex algorithms. However, the functionality of such robots is much less than that of robots with several manipulators, as well as robots with a wheel-walking propulsion [1–4].

Fig. 1.3D model of a robotic drilling platform.

Content from this work may be used under the terms of the Creative Commons Attribution 3.0 licence. Any further distribution of this work must maintain attribution to the author(s) and the title of the work, journal citation and DOI.
Published under licence by IOP Publishing Ltd

IOP Conf. Series: Materials Science and Engineering **779** (2020) 012034 doi:10.1088/1757-899X/779/1/012034

To carry out geological exploration works on the seabed, it is proposed to use a robotic drilling platform with a stepper mover (utility model patent No. 166446 "Walking drilling rig", utility model priority: 07/04/2016) (Fig. 1). The unit includes a farm, supports with platforms, hydraulic cylinders of supports, hydraulic cylinders of the supports of the supporting table, working body, control system. The load-bearing frame is made of two parallel pipes with long carriages and is equipped with eyelets pivotally connected to the support platforms, while the ends of the pipes are rigidly connected by bridging beams with blocks placed on them. The working body is made in the form of a trolley with rollers connected with the long carriages of the farm, with a drilling rig rigidly fixed on it and two draw-works equipped with flexible traction elements covering blocks of bridging beams, one end of which is fixed on the winch drum of draw-works and the other on the trolley of working body [5–6].

2 Methods

Description and principle of operation:

There are 3 groups of hydraulic actuators:

1st group. Basic hydraulic cylinders of a drilling platform;

2nd group. Hydraulic cylinders for leveling the support table;

3rd group. The hydraulic motor rotates the load-bearing frame relative to the support table. Of interest is a group of cylinders for maintaining the the support table horizontally (Fig. 2).

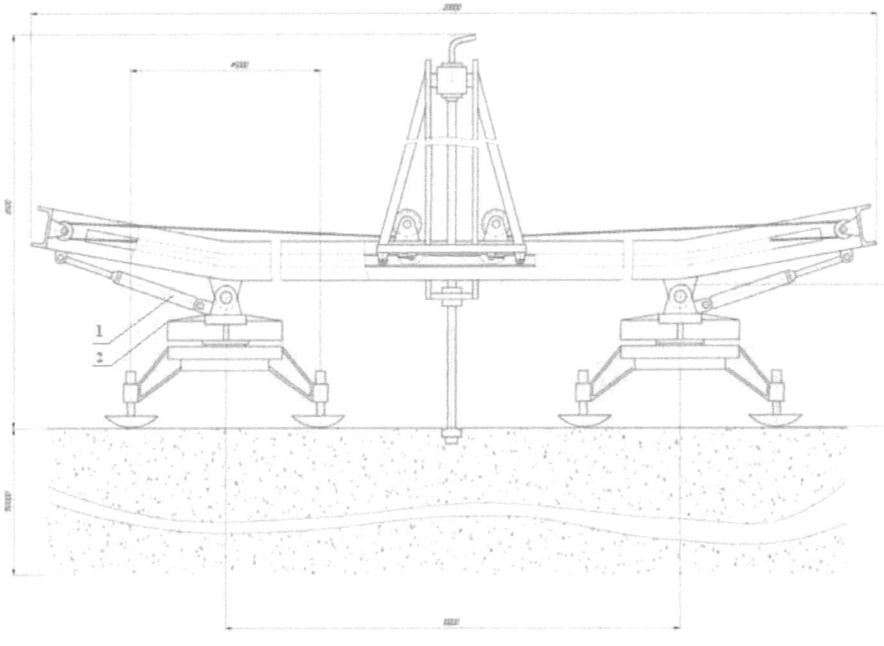

Fig. 2. Schematic construction of a robotic drilling platform.

Where: 1 — hydraulic cylinder for leveling the support table; 2 — supporting table.

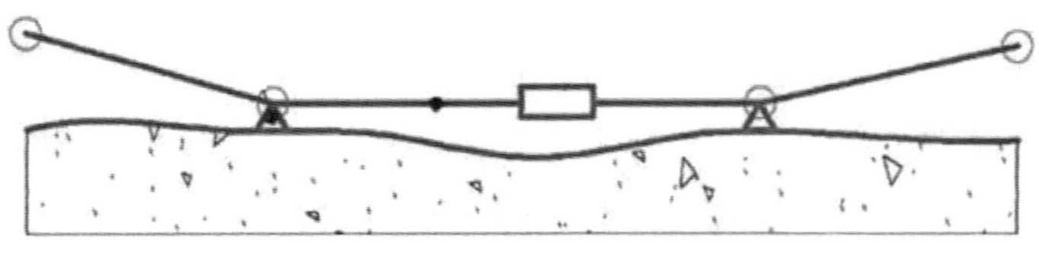

Fig. 3. Drilling platform in neutral position.

IOP Conf. Series: Materials Science and Engineering **779** (2020) 012034 doi:10.1088/1757-899X/779/1/012034

When the platform is in the neutral position (Fig. 3), the working body (drilling assembly) is capable to drill the geological exploration wells. Then the draw-works is turned and the platform with the counterweight moves to the cantilever section along the long carriages, for example, from left to right. The frame will rotate around the horizontal axis of the right support table by an angle of 12 degrees and raise the left support up (Fig. 4). Then, the frame will be rotated using a hydraulic drive (Fig. 5) around the vertical axis by an angle of 20 degrees. [7–10].

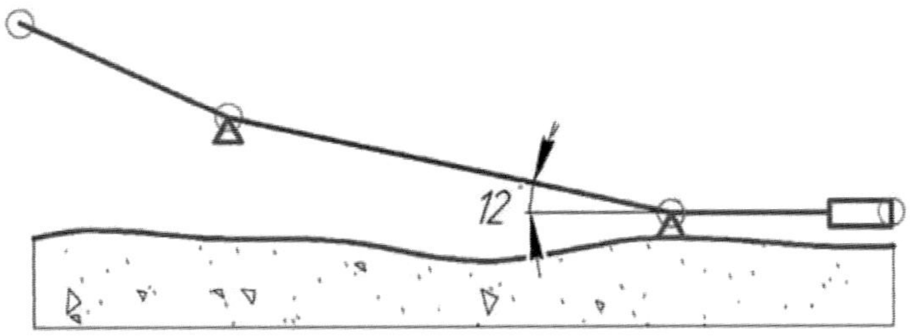

Fig. 4. Change in the center of mass of the system.

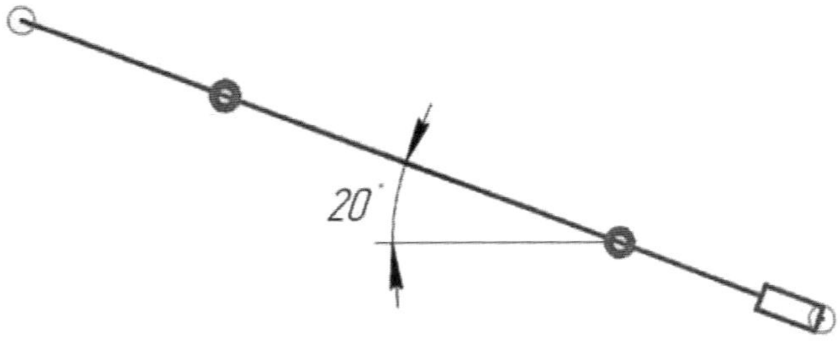

Fig.5. Rotation around the vertical axis

The process is repeated. The drilling assembly with a counterweight returns to its middle position and the left support table is brought down. Extended to the middle position, the rods of the support hydraulic cylinders at various times touch the bottom surface. The downward movement of the support table occurs until each of the hydraulic cylinders has touched the bottom surface and the pressure changes in the piston cavities of the hydraulic cylinders become equal to each other.

During downward movement of the left support table and equalization of pressure changes in the piston cavities of the hydraulic cylinders, the forces act on the table from the side of the cylinders for leveling (Fig. 6), which do not allow it to deviate from the horizon. To study the dynamics of transient response processes in the hydraulic drive for leveling(Fig. 7), its mathematical model was created.

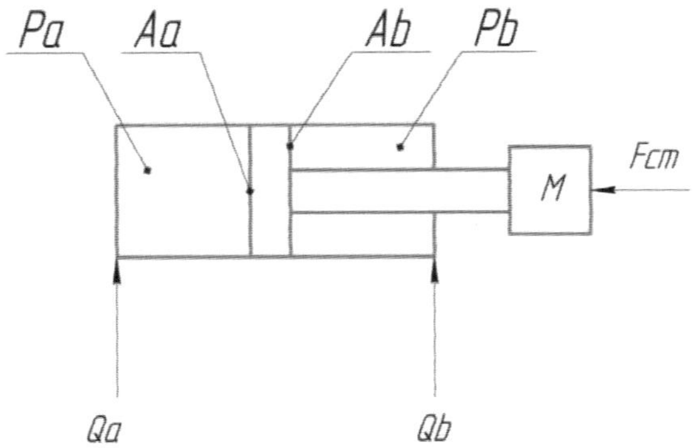

Fig. 6. The design scheme

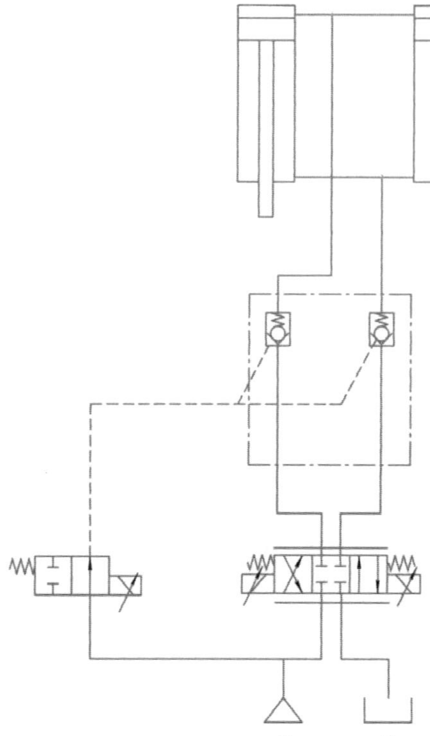

Fig. 7. The scheme of hydraulic drive for leveling the support table

The equation of motion is written:

$$M = p_A A_A l - p_B A_B l - bxl - F_{cm} l_{cm}$$

The pressure in the cavities is written taking into account the fluid compressibility:

$$p_A = \frac{E}{V_A}\left(Q_A - A_A x\right)$$

$$p_B = \frac{E}{V_B}\left(Q_B - A_B x\right)$$

A block diagram is drawn down (Fig. 8).

At the input, the flow rates QA and QB have to be given, the force F must also be indicated. The forces developed by the hydraulic cylinder, the arms of the forces, the areas of the piston and rod cavities of the hydraulic cylinder must be written.

3 Mathematical model of the control valve

A hydraulic control valve 4/3 model with a closed center is created and a variant of its design scheme is shown in the Fig. 9.

A cylindrical spool with three fillets is inserted into the barrel with the bores. Supply pressure is supplied to the central cavity, and the extreme ones are connected to the drain. The cavity between the fillets of the spool can be connected to the corresponding cavities of the hydraulic cylinder. The edges of the spool form throttling slots with the barrel bores, the areas of which depend on the displacement of the spool. [11]

IOP Conf. Series: Materials Science and Engineering **779** (2020) 012034 doi:10.1088/1757-899X/779/1/012034

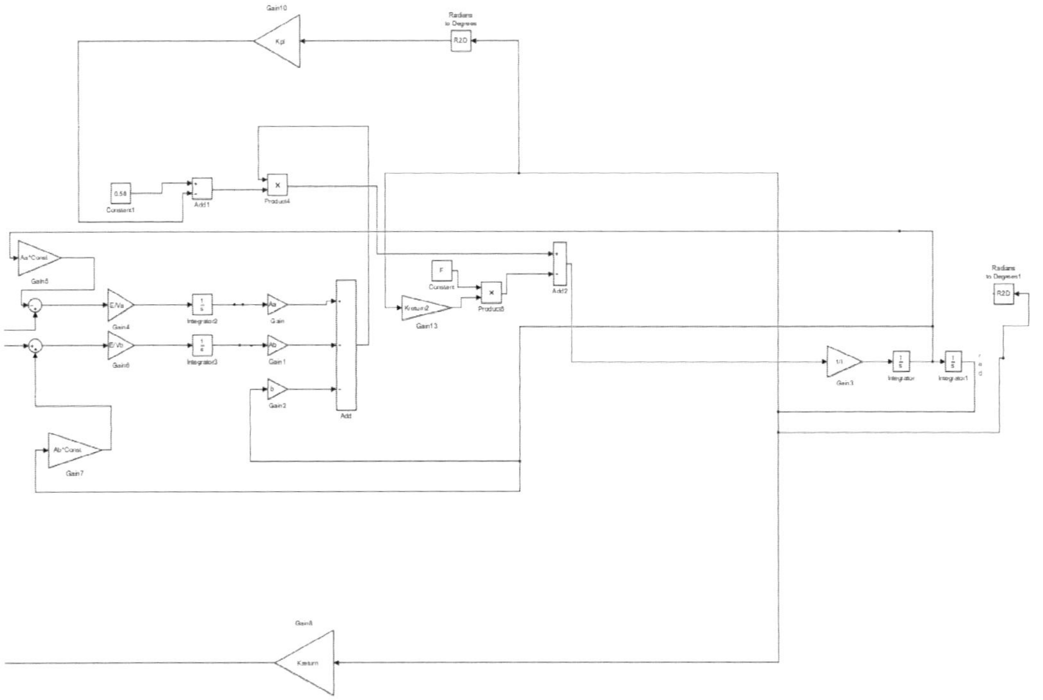

Fig.8. Structural scheme of mathematical model of actuator

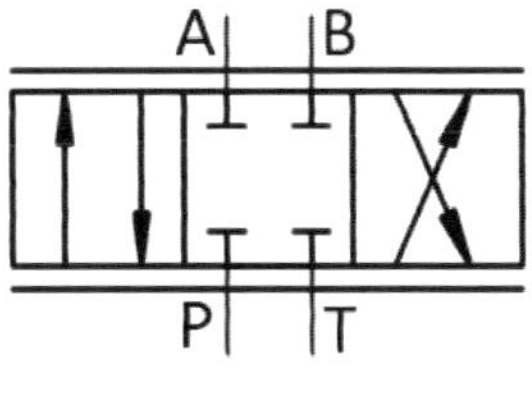

Fig.9. The scheme of hydraulic control valve 4/3 model with a closed center

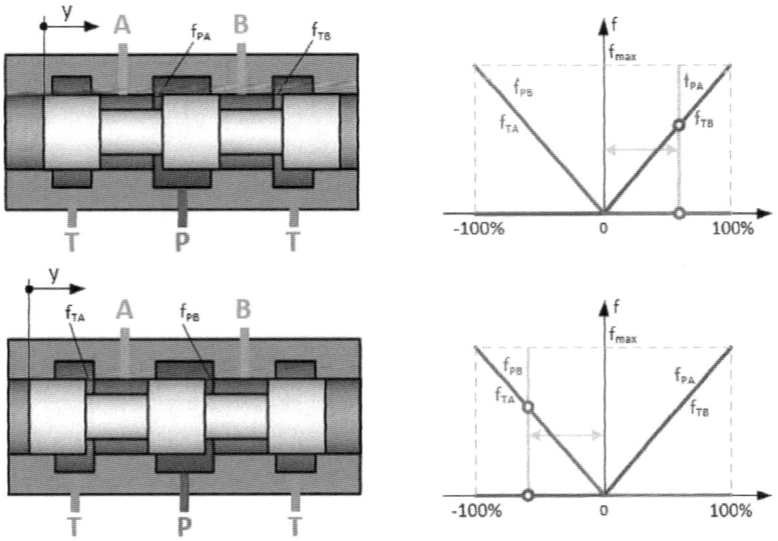

Fig. 10. The change in the opening area of the slot of hydraulic control valve

TA — connects the cavity A with drain;
PA — connects the cavity P with line A;
PB — connects the cavity P with line B;
TB — connects the cavity with a drain.
The dependence of the change in the opening area of the slot is shown in Fig. 10.

As in the case of the hydraulic cylinder model, all flow rates are movingt in one direction. The sign of flow rate will be determined based on the sign of the differential pressure. The formulas for the costs are:

$$Q_{PA} = \mu_S f_{PA}\sqrt{\frac{2}{\rho}\left|p_n - p_A\right|}\cdot sign(p_n - p_A);$$

$$Q_{PB} = \mu_S f_{PB}\sqrt{\frac{2}{\rho}\left|p_n - p_B\right|}\cdot sign(p_n - p_B);$$

$$Q_{TB} = \mu_S f_{TB}\sqrt{\frac{2}{\rho}\left|p_T - p_B\right|}\cdot sign(p_T - p_B);$$

$$Q_{TA} = \mu_S f_{TA}\sqrt{\frac{2}{\rho}\left|p_T - p_A\right|}\cdot sign(p_T - p_A).$$

To find all the necessary flow rates coming in or out of the control valve, it is necessary to algebraically add the corresponding components [8]:

$$Q_A = Q_{PA} + Q_{TA};$$
$$Q_B = Q_{PB} + Q_{TB};$$
$$Q_P = Q_{PA} + Q_{PB};$$
$$Q_T = Q_{TB} + Q_{TA}.$$

The structural scheme of the mathematical model of the control valve is shown in Fig. 11:

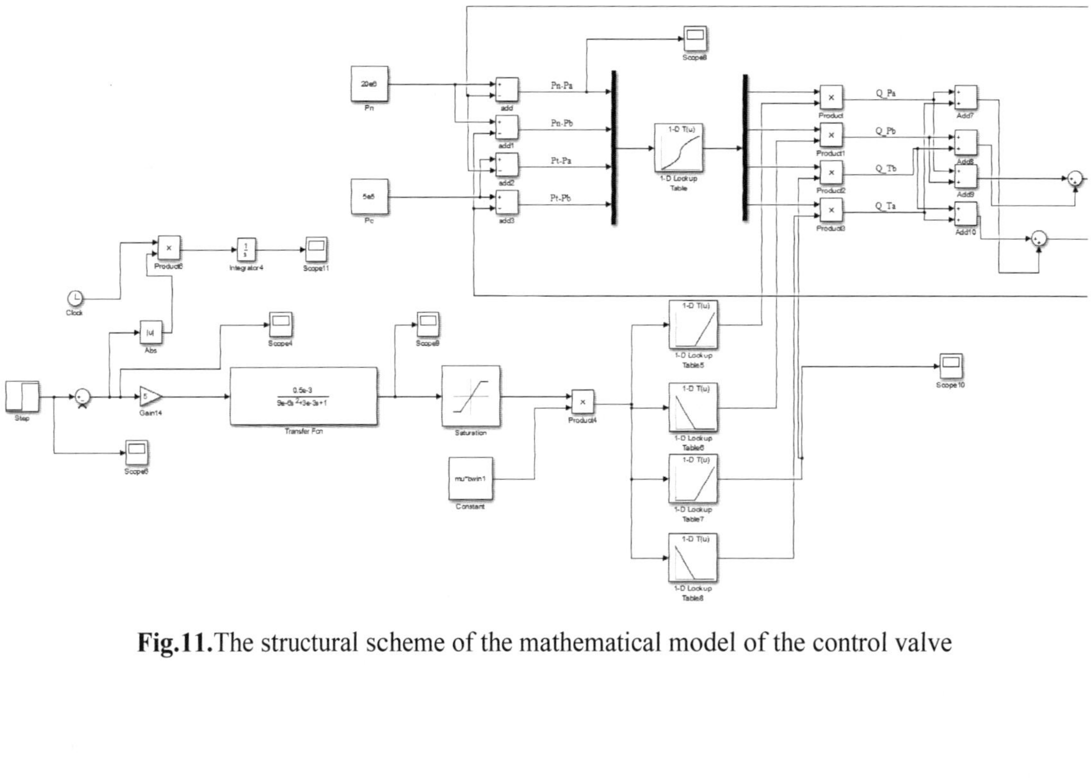

Fig.11. The structural scheme of the mathematical model of the control valve

4 Results

When the maximum electric signal is applied to the inlet of the hydraulic control valve, the spool will displace by 1 mm (Fig. 12). Figure 13 shows the dependence of the position of the rod of the hydraulic cylinder on time. The pressures in the rod and piston cavities of the hydraulic cylinder are shown in Fig. 14.

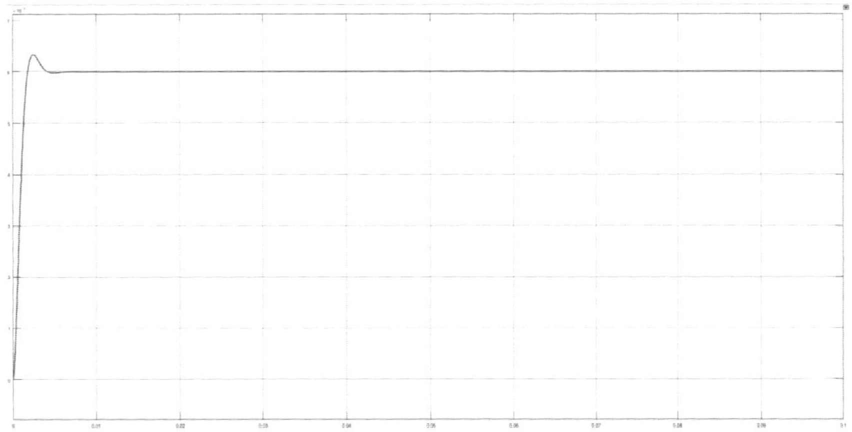

Fig.12. The displacement of the control valve spool

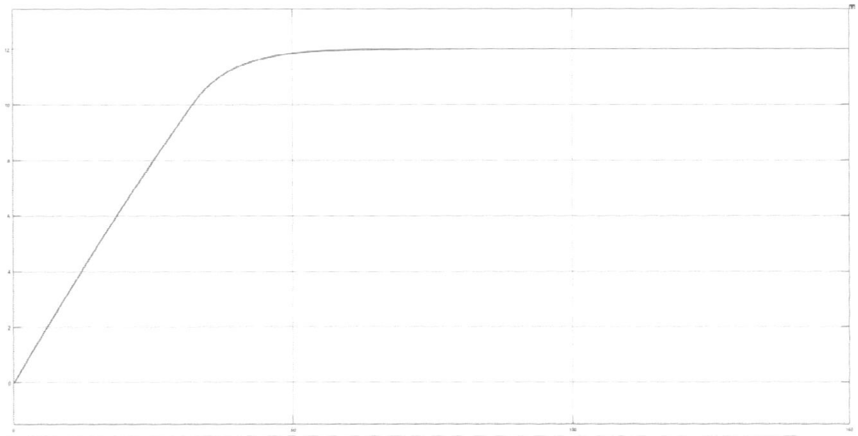

Fig.13. The extension of the rod in the hydraulic cylinder

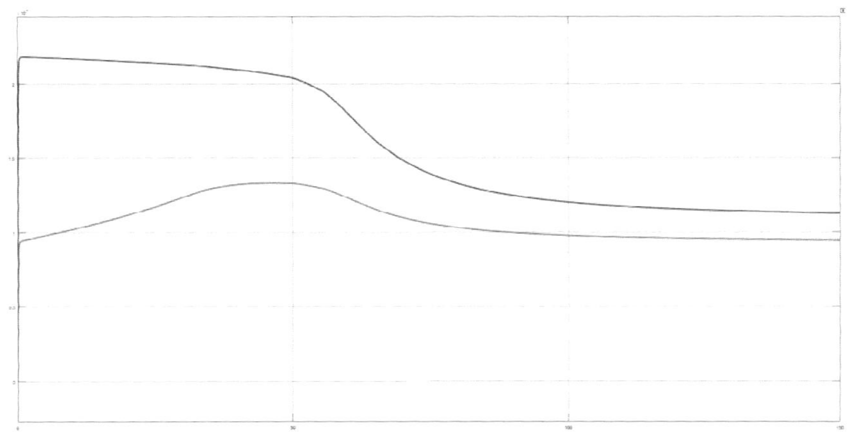

Fig.14. Pressure in the cavities of the hydraulic cylinder

List of reference

[1] Karginov L.A., Kovalchuk A.K., Kulakov D.B., Semenov S.E., IArots V.V. Design of drive systems for walking robots with a tree kinematic structure, 2013.

[2] Semenov S. E. 2015 Mathematical modeling of the electro-hydraulic actuation systems of the machines with tree-like kinematic structure Proceedings of 2015 International Conference on Fluid Power and Mechatronics, FPM 2015 article no 7337184 pp 583-92;

[3] Kulakov D B, Semenov S E, Kulakov B B, Shcherbachev P V and Tarasov O I 2015 Hydraulic Bipedal Robots Locomotion Mathematical Modeling Procedia Engineering 106 62-7;

[4] E Kolotvin 2019 IOP Conf. Ser.: Mater. Sci. Eng. 589 012019;

[5] RF patent № 2648365.13.06.06.2017. Timofeev I.P., Sokolova G.V., Bolshunov A.V., Vasilev N.I., Ignatev S.A., Kolotvin E.V., Avdeev A.M. Walking installation for transportation and laying of oil and gas pipes on the seabed.

[6] RF patent № 166446, 07.04.2016. I.P. Timofeev, G.V. Sokolova, G.A. Kolton, I.A. Korolev, E.V. Kolotvin. Walking drilling installation

[7] Semenov S E, Kovalchuk A K and Kulakov D B 2010 Mathematical modeling of mechanisms with a tree-arborescence kinematic structure Drive Technology;

[8] Timofeev I.P. 1987 Walking machines for the development of seabed resources.

[9] S Semenov and D Kulakov 2019 IOP Conf. Ser.: Mater. Sci. Eng. 492 012042

[10] Nikolenko I V and Zheglova V M 2018 Design and calculation of axial piston hydraulic machines: monograph (Simferopol: ARIAL Publ.) 300 p;

[11] Andreev M. A. 2017 Mathematical modeling of hydraulic drive.

IOP Conf. Series: Materials Science and Engineering **779** (2020) 012035 doi:10.1088/1757-899X/779/1/012035

Calculation of a three-stage electro-hydraulic amplifier

Li Jianhui[1] and N Sosnowski[1,2]
[1]Bauman Moscow State Technical University

[2]E-mail: sosn@bmstu.ru

Annotation. On this paper is analyzed the research state of three-stage electro-hydraulic amplifiers, the modeling features of such amplifiers are described. The parameters of the electro-hydraulic amplifier included in the servo hydraulic drive for the control system of the hovercraft are calculated. A mathematical model is compiled, the static and dynamic characteristics of the amplifier are obtained. Modeling is carried out in the MATLAB / Simulink software package. The amplifier was optimized for speed and dynamic transient error by varying the gain of the proportional controller in the direct circuit of the device.

Introduction

The electro-hydraulic amplifier is an integral part of the servo hydraulic drive, it connects the executive and information subsystems, contains various mechanical, electrical, hydraulic, control devices and converts the input electrical signal into a hydraulic signal to control the load of any object. Electro-hydraulic amplifiers are widely used in various hydraulic systems [1]–[5]. For example, an electro-hydraulic amplifier plays an important role in controlling an hovercraft. Hovercraft can be used in various difficult terrestrial conditions [6], and with the development of such vessels, requirements both to the control systems themselves and directly to the characteristics of hydraulic drives used in the systems increase.

The three-stage electro-hydraulic amplifier has the advantages of high flow rate and speed. It has a small size, compact structure, high sensitivity, small input signal, large output power and high reliability [7]–[9]. Currently, many engineers are studying three-stage electro-hydraulic amplifiers. So, through modeling, it was found that the rigidity of the feedback rod and the area of the spool of the output stage have a significant impact on the dynamic characteristics of a three-stage electro-hydraulic amplifier. Reasonable rigidity of the feedback rod and correctly selected spool diameter can not only improve the performance of the amplifier, but also increase the stability of its operation. Under the same conditions, the rectangular feedback rod improves the dynamic characteristics of the amplifier to a greater extent [10]–[12]. Van Dongway developed a simulation model of a three-stage electro-hydraulic amplifier using the AMESIM software package and obtained graphs reflecting its characteristics, which confirmed the accuracy of the developed model. Are particularly interesting the studies of typical failure phenomena during obliteration of the spool throttle distributor of a three-stage electro-hydraulic amplifier, as well as in the presence of a large amount of air in the working fluid, which leads to a deterioration in the performance of the amplifier [13]. At the same time, the researchers analyzed the optimal design of the amplifier [14]–[16]. When analyzing the operation of the spool of a three-stage electro-hydraulic amplifier, factors that affect the spool performance, such as hydrodynamic force and friction force, are also taken into account.

Content from this work may be used under the terms of the Creative Commons Attribution 3.0 licence. Any further distribution of this work must maintain attribution to the author(s) and the title of the work, journal citation and DOI.
Published under licence by IOP Publishing Ltd

IOP Conf. Series: Materials Science and Engineering **779** (2020) 012035 doi:10.1088/1757-899X/779/1/012035

The three-stage electro-hydraulic amplifier consists of an electromechanical control signal converter, a hydraulic signal pre-amplification stage (nozzle-damper type device), a spool throttle valve of the first spool (second stage), a spool throttle valve of the second spool (third stage) and a spool position sensor. The combination of an electromechanical converter, a hydraulic signal pre-amplification stage and a first spool valve is essentially a two-stage electro-hydraulic amplifier. A schematic diagram of a three-stage electro-hydraulic amplifier is shown in Fig. 1, where the number 1 shows the electric signal amplifier; 2 — electromechanical converter; 3 — the first spool; 4 — second spool; 5 — spool position sensor (feedback sensor); 6 — two-stage electro-hydraulic amplifier.

A two-stage electro-hydraulic amplifier converts an input electrical signal into an output fluid flow to control a second spool. The second spool for positioning uses electric feedback on the position of the spool, in addition, the feedback coefficient is easy to adjust. The role of feedback is to determine the position of the second spool using a sensor, the output signal from which, after transmission and amplification, is converted into a feedback signal proportional to the position of the second spool and is directed to an electric signal amplifier, where it is subtracted from the input (control) electric signal.

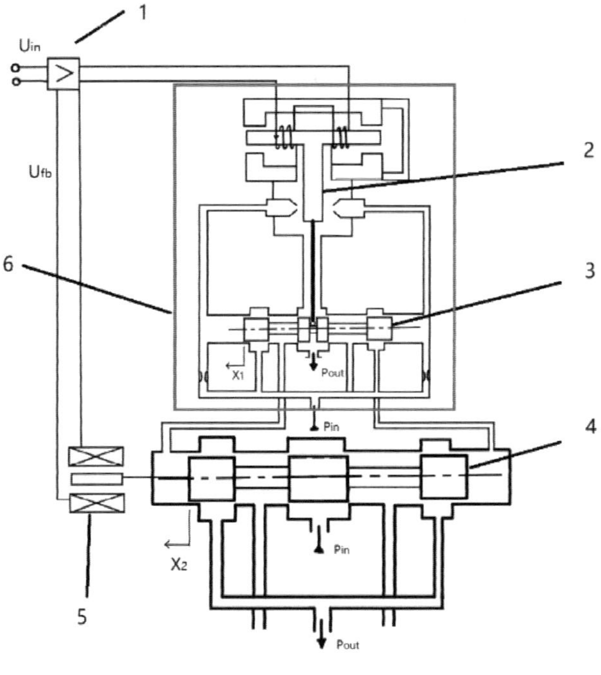

Fig. 1. Schematic diagram of a three-stage electro-hydraulic amplifier

The mathematical model of an electro-hydraulic amplifier

Two-stage electro-hydraulic amplifier is widely used at the present, its research results of are very developed, so in this paper it does not paid to it much attention. Here is discussed mainly about the calculation of the second spool and the characteristics of the entire three-stage electro-hydraulic amplifier [17]–[20].

Maximum flow at maximum displacement of the second spool and zero load

$$Q_z^{max} = \pi \cdot d_z \cdot k_p \cdot \mu_z \cdot x_z^{max} \cdot \sqrt{\frac{P_{in} - P_{out}}{\rho}}$$

Where x_z^{max} — maximum spool offset relative to neutral; d_z — spool diameter; k_p — spool perimeter utilization factor.

IOP Conf. Series: Materials Science and Engineering **779** (2020) 012035 doi:10.1088/1757-899X/779/1/012035

In the case of study, the flow rate through the spool valve at idle follow-up hydraulic actuator for the control system of the hovercraft is $Q_z^{max} = 5,8 \cdot 10^{-3} \dfrac{m^3}{s}$.

The spool perimeter utilization factor is taken equal to $k_p = 0,6$.

It's taken $x_z^{max} = 0,1 d_z$,

$$d_z = \sqrt{\frac{Q_z^{max}}{\pi \cdot k_p \cdot \mu_z \cdot 0,1 \cdot \sqrt{\frac{P_{in} - P_{out}}{\rho}}}} = \sqrt{\frac{5,8 \times 10^{-3}}{\pi \times 0,6 \times 0,65 \times 0,1 \times \sqrt{\frac{(32 - 0,3) \times 10^6}{850}}}} = 0,0157\,m,$$

Where ρ=850 $\dfrac{\text{kg}}{\text{m}^3}$— fluid density

And $d_z = 0,016\,m$,

Then the perimeter utilization factor of the second spool is

$$k_p = \frac{Q_z^{max}}{\pi \cdot d_z^2 \cdot \mu_z \cdot 0,1 \cdot \sqrt{\frac{P_{in} - P_{out}}{\rho}}} = \frac{5,8 \times 10^{-3}}{\pi \times 0,016^2 \times 0,65 \times 0,1 \times \sqrt{\frac{(32 - 0,3) \times 10^6}{850}}} = 0,5773$$

The maximum possible displacement of the second spool from the neutral position is assumed to be equal to

$$x_z^{max} = 0,1 \cdot d_z = 0,1 \times 0,016 = 0,0016\,m$$

The total width of the window casings

$$b_{ok\,\Sigma} = k_p \cdot \pi \cdot d_z = 0,5773 \times 3,14 \times 0,016 = 0,0290\,m.$$

Width of each window of the sleeve (there are 4 windows in the sleeve)

$$b_{okna} = \frac{b_{ok\,\Sigma}}{4} = \frac{0,0290}{4} = 0,00725\,m$$

Then $b_{okna} = 7,3\,mm$.

And $b_{ok\,\Sigma} = 4 \cdot b_{okna} = 4 \times 7,3 = 29,2\,mm$.

$$k_p = \frac{b_{ok\,\Sigma}}{\pi \cdot d_z} = \frac{29,2}{\pi \cdot 0,016} = 0,5809$$

Conductivity of the spool valve

$$k_z' = \pi \cdot d_z \cdot k_p \cdot \mu_z \cdot \sqrt{\frac{2}{\rho}} = \pi \times 0,016 \times 0,5809 \times 0,65 \times \sqrt{\frac{2}{850}} = 9,2067 \cdot 10^{-4}$$

We specify the diameter of the spool

$$d_z = \sqrt{\frac{Q_z^{max}}{\pi \cdot k_p \cdot \mu_z \cdot 0,1 \cdot \sqrt{\frac{P_{in} - P_{out}}{\rho}}}} = \sqrt{\frac{5,8 \times 10^{-3}}{\pi \times 0,5809 \times 0,65 \times 0,0016 \times \sqrt{\frac{(32 - 0,3) \cdot 10^6}{850}}}} = 0,016\,m.$$

The spool diameter is finally taken equal to $d_z = 0,016\,m$.

IOP Conf. Series: Materials Science and Engineering **779** (2020) 012035 doi:10.1088/1757-899X/779/1/012035

Since the second spool works by receiving an output hydraulic signal from a two-stage electro-hydraulic amplifier, it can be considered as a load on a two-stage electro-hydraulic amplifier. For the second spool, the input signal represents the flow rate of the first spool, and the output signal represents the offset of the second spool.

(1) Flow equation for controlling the second spool:

$$Q_y = F_z \frac{dx_z}{dt} + \frac{V_y}{2B_r} \frac{dP_y}{dt},$$

Where F_z – second spool area (m^2), x_z – the displacement of the second spool (m),

B_r – volumetric elasticity modulus of the liquid (Pa),

V_y – volume for controlling the second spool (m^3),

p_y – differential pressure for controlling the second spool (Pa).

(2) Motion equation of the second spool

$$m_z \frac{dx_z^2}{dt^2} + k_{tp} \frac{dx_z}{dt} + 2C_{gd} x_z = F_z P_y$$

m_z – weight of the second spool (kg), k_{tp} – coefficient of friction,

C_{gd} – the rigidity of the hydrodynamic spring.

The block diagram of the amplifier in MATLAB / Simulink in accordance with the equations described above is shown in Fig. 2.

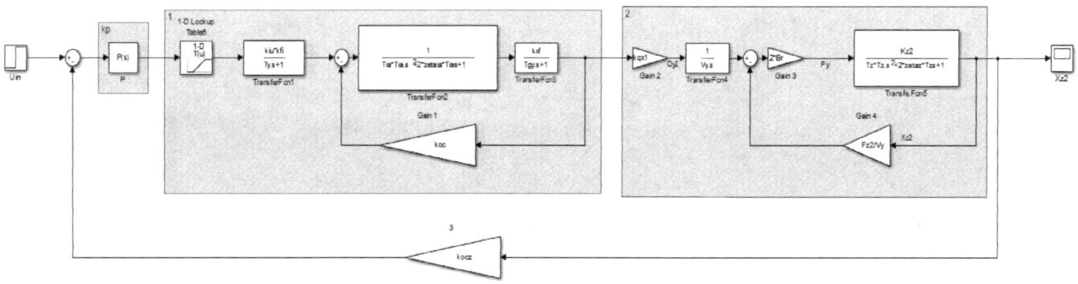

Fig. 2. Structural diagram of a three-stage electro-hydraulic amplifier: 1 — two-stage electro-hydraulic amplifier, 2 — second spool, 3 — feedback sensor.

The numerical values of the coefficients are accepted as follows: k_{iu}=*4e-3 ;*k_{fi}=*0.7750;Ty=4.8e-3; Tia=2.839e-4; zetaia=0.2676; Tgy=0.4502;* k_{xf}=*35.1613;koc=14.4351;* k_{qx1}=*0.1666; Vy=1.6085e-6; Br=1.4e9; Tz=3.0996e-4; zetaz=0.3; Kz2=1.865e-9;* F_{z2}=*8.0425e-4;kocz=1875.*

Simulation results

The simulation results are presented by the static and dynamic characteristics of the amplifier. Fig. 3 shows the static characteristic of the second spool (input electric current in the control winding of the electromechanical converter — moving the spool)

To calculate the flow-differential characteristics of the second spool valve in a three-stage electro-hydraulic amplifier, the following equation is used:

$$Q_z(p_h, x_{z2}) = k_z^{'} * x_{z2} * \sqrt{\frac{p_{in} - p_{out} - p_h * sign(x_{z2})}{2}}.$$

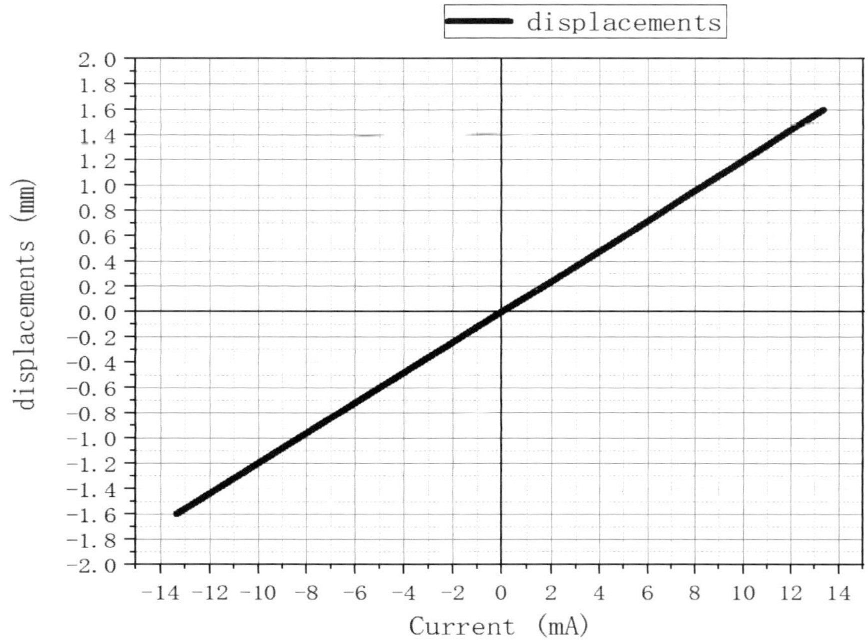

Fig. 3. Static characteristic of the second spool

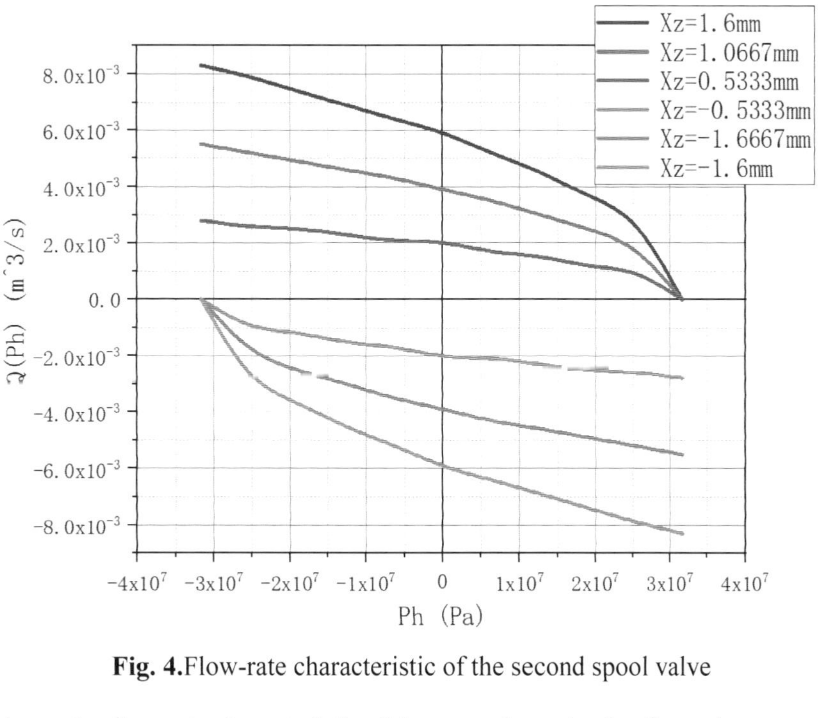

Fig. 4. Flow-rate characteristic of the second spool valve

Figure 4 shows the flow-rate characteristic of the second spool valve for various movements of the spool.

According to the above model, the transients are calculated according to the position of the second spool with an input signal of 3 V (Fig. 5), different results were obtained by setting different values of the proportional controller Kp ($Kp_1=0,9$; $Kp_2=0,75$; $Kp_3=0,5$).

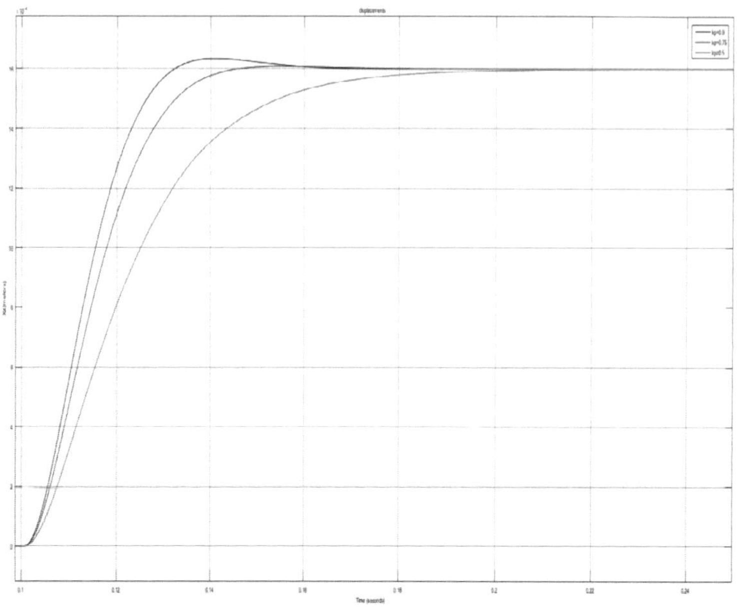

Fig.5. Transient process of the second spool

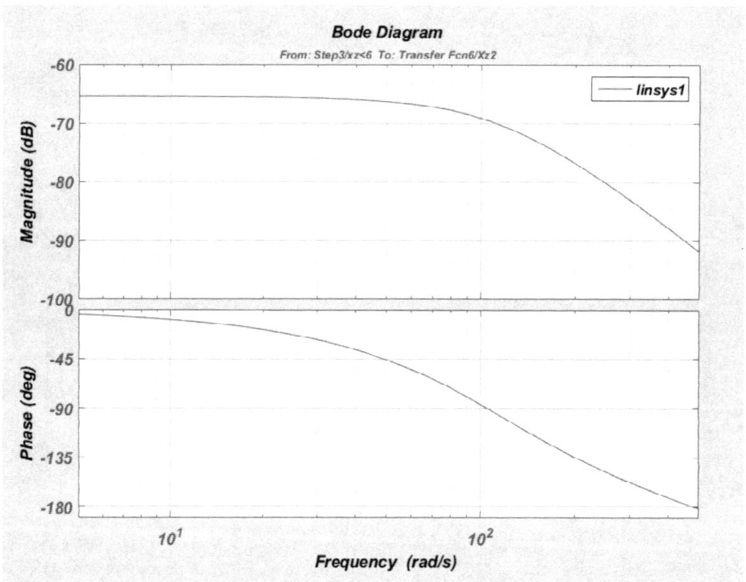

Fig. 6. Bode Chart

Figure 6 shows the calculated Bode diagram of the electro-hydraulic amplifier at Kp = 0,75. The frequency characteristics of the amplifier turned out to be quite satisfactory.

When varying the gain of the proportional controller in the direct circuit of the device, the amplifier was optimized for the quality of the transition process. At Kp = 0.75, a fairly fast transition process was obtained, which has a small overshoot (less than 2%).

Conclusions

In this work, the parameters of a three-stage electro-hydraulic amplifier included in the servo hydraulic drive for the control system of a hovercraft are calculated and a mathematical model of such an amplifier is created. Using the MATLAB / Simulink software package, modeling was performed, and the static and dynamic characteristics of the amplifier were obtained. The amplifier was optimized for speed and dynamic transient error by varying the proportional controller in the direct circuit of the device to obtain better system dynamics. It can be seen from the results that the mathematical model of the electro-hydraulic amplifier meets the requirements of the system and can be used for further studies of servo hydraulic drives.

List of references

[1] Zhang Liping. Principle, Use and Maintenance of Hydraulic Valves [M]. Chemical Industry Press. 2005. pp. 29–31, ISSN:1672-0121.
[2] Xinbei, L., Vladimir, L., Songjing, L. Performance and Flow Field Analysis of Flapper Deflection Servo Valve. (2018) 2018 Global Fluid Power Society PhD Symposium, GFPS 2018.DOI: 10.1109/GFPS.2018.8472394
[3] Guo, Q. Analysis of hydraulic mechanical control system for wind power generation. (2019) Revue de l'Energie, (643), pp. 1–7.
[4] B Kulakov and D Kulakov 2019 IOP Conf. Ser.: Mater. Sci. Eng. 492 012029
[5] Zhao, L., Zhang, H., Yin, J., Chen, H. Design and application of versatile automatic bin with valve splint slide way (2018) MATEC Web of Conferences, 228, № 03010. DOI: 10.1051/matecconf/201822803010
[6] Vavilov, I. S. Expediency of operation of air-cushion transport in the conditions of the polar region and the Far North. Omsk scientific Bulletin. 2007. no. 3 (60). Pp. 109-114.
[7] Sosnovsky N. G., Popov D. N., Siukhin M. V. Sensitivity of dynamic characteristics of an electrohydraulic servo (2018) international conference on industrial design, application and production, ICIEAM 2018, article no. 8728669.DOI: 10.1109 / ICIEAM.2018.8728669
[8] Song Xiaobo. Dynamic performance simulation of three-stage electro-hydraulic servo valve. Chinese Society of Metals. 2014: 5.
[9] Guo, H., Lin, P., Pan, X., Wang, G., Zhang, H., Zhang, Q. Development of an Automatic Grinding System for Servo Valve Spool Throttling Edge (2019) 16th International Conference on Ubiquitous Robots, UR 2019, № 8768715, pp. 718–722. DOI: 10.1109/URAI.2019.8768715
[10] Zhang Lei, Chen Kuisheng, Wu Yi, Zhan Congchang. Modeling and dynamic characteristics simulation of jet tube three-stage electro-hydraulic servo valve. Hydraulic and Pneumatic, 2018. №. 06. pp. 66–72.
[11] Wang Dongwei. Flow field characteristics and fault simulation analysis of three-stage servo valve. Harbin Institute of Technology. 2013.
[12] D V Ivanov, I L Sandler, E A Burtseva, V N Vlasova,.Identifation of slide valve dynamics with errors in variables (2019) IOP Conf. Ser.: Mater. Sci. Eng. 560 012021
[13] Jin, B.-X., Zhai, Y., Gu, H.-T., Zhang, T. Optimized method for shift valve used in integrated transmission (2016) BinggongXuebao/ActaArmamentarii, 37 (4), pp. 591–597. DOI: 10.3969/j.issn.1000-1093.2016.04.003
[14] I.L. Krivts. Optimization of Performance Characteristics of Electropneumatic (Two-Stage) Servo Valve[J]. Journal of Dynamic Systems. Measurement and Control. 2004. C. 416–420.
[15] Wang, Z., Liu, Z., Liu, F., Yu, X., Feng, Q. Research on operating characteristics of single slide valve capacity control mechanism of the single screw refrigeration compressor (2014) Proceedings of the Institution of Mechanical Engineers, Part A: Journal of Power and Energy, 228 (8), pp. 965–977.DOI: 10.1177/0957650914547605
[16] Lu, Q., Ruan, J., Li, S., Lu, P. Effects of Cutting Groove on Radial Clamping Force of a Slide Valve's Spool (2018) Hsi-An Chiao Tung Ta Hsueh/Journal of Xi'an Jiaotong University, 52 (6), pp. 76–83. DOI: 10.7652/xjtuxb201806012
[17] Liu Zengguang, Yue Daling, AnLinchao, Bai Guixiang. Modeling and Simulation of Force Feedback Two-stage Electro-hydraulic Servo Valve Based on MATLAB [J]. Hydraulic and

Pneumatic, 2015. №. 05. pp. 83–85.
[18] N Sosnovsky and D Ganieva 2019 IOP Conf. Ser.: Mater. Sci. Eng. 589 012016
[19] S Semenov and D Kulakov 2019 IOP Conf. Ser.: Mater. Sci. Eng. 492 (1) 012042.
[20] I Kolodin and M Ryabinin 2019 IOP Conf. Ser.: Mater. Sci. Eng. 589 012018

IOP Conf. Series: Materials Science and Engineering **779** (2020) 012036 doi:10.1088/1757-899X/779/1/012036

Conceptual features of improving the flow-through parts of gas separators of submersible electric pumps systems f for the production of formation fluid in order to improve the separating properties, energy efficiency and reliability

A Trulev[1], **S Timushev**[2] **and V Lomakin**[3,4]

[1]Rimera Group of companies, Moscow, Russian Federation
[2]Moscow Aviation Institute, Moscow, Russian Federation
[3]Bauman Moscow State Technical University

[4]E-mail: Lomakin@bmstu.ru

Abstract. This article analyzes the design of the power part of gas separators of various manufacturers, reveals the causes of backflows, which can lead to hydro-abrasive wear of the elements of the flow-through part, reduction of the separation properties as a result of the gas-liquid mixture flow dispersion, and decrease of pressure gradient in the separation chamber. A mathematical model was selected, hydrodynamic calculations of design options and upgrading were carried out, and animproved flow part of the separator was developed. Bench tests were conducted to confirm the improved parameter of the product.

Objective

The objective is to analyze the design of the power part of gas separators produced by various manufacturers; to reveal the causes of backflows, which can lead to the reduction of the separation properties as a result of the gas-liquid mixture flow dispersion, and decrease of pressure gradient in the separation chamber; to determine the ways to eliminate relevant hazards; to confirm the effectiveness of the conceptually new design of the gas separator.

The studies were carried out using numerical simulation of the operation of gas separators on water and on gas-liquid mixture. An experimental verification of the numerical calculations was carried out.

Introduction

At the moment, there is a steady tendency to the increase of depleted wells that currently account for about 30-40% of the total well stock in the Russian Federation. In order to increase the oil recovery rate (ORR), the bottom hole pressure is reduced, respectively, the content of free gas and solids at the pump intake increases.

Works are underway to intensify the flow from the bottom hole zone and increase the oil recovery; hence, some part of the wells is switched to the operating mode with low bottom hole pressures — about of less than 3MPa.

The growth rate of the equipment MTBF (mean time between failures) began to decline, and no it is clear that most of the wells the ESPs with the current wear resistance group can no longer provide a significant increase in the MTBF, since they do not correspond to the changed operating conditions [1].

Content from this work may be used under the terms of the Creative Commons Attribution 3.0 licence. Any further distribution of this work must maintain attribution to the author(s) and the title of the work, journal citation and DOI.
Published under licence by IOP Publishing Ltd

The serial oilfield submersible pumping equipment is often incapable to operate in such abnormal conditions.

An acute problem is the introduction of an effective technology for their operation with MTBF of at least 600 days [2].

In this regard, the development, study and pilot introduction of innovative technologies for oil production in abnormal conditions with the use of submersible pump systems becomes urgent.

In oil-producing wells with a high gas content, an increase in the efficiency of the electric submersible pumps is achieved by completing them with centrifugal gas separators, in which major portion of free gas separated from the pumped fluid is collected and drawn off to the annular space before the fluid reaches the pump intake.

The efficiency of a centrifugal gas separator significantly depends on the dispersion rate of the gas-liquid structure in the pumped fluid, on the size of gas bubbles, water cut, the presence of surface active agents, the pump intake and discharge pressure [3, 4]. The average diameter of the gas phase bubbles ranges from 40 to 300 μm, for crude oil with high water content — over 130 μm [5]–[11].

In order to increase the efficiency of the oil production process in depleted wells, Rimera Group of Companies has developed and offers a number of new conceptual equipment that is mainly commercialized at present. A conceptual design of pump, pump stages, protectors and gas separators was developed for marginal wells[12]–[20].

The gas separator is considered to be an unreliable element of a pump system due to hydro-abrasive cutting of the body and other elements of the flow-through part. Backflows are formed at the inlet of the axial wheel, auger, provided the pump discharge is more than half the estimated optimal feed [21]. The backflows are a trap for solids — their concentration is rapidly increasing, and the rotating abrasive ring can cut through the body and other elements of the gas separator. [22]–[25].

Analysis of structural schemes

Fig. 1 shows the design option of a complete gas separator. The following elements are installed in sequence at the intake: auger 4, axial wheel 3, diffuser 2, axial wheel 1.

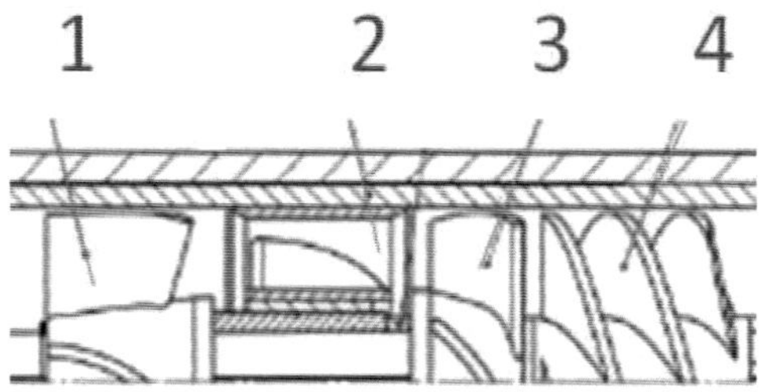

Fig. 1. The design option of a complete power part of gas separator.
Axial wheel 1, diffuser 2, axial wheel 3, auger 4.

An alternating stroke auger can be installed instead of auger 4 and an axial wheel 3.

Analysis of the vortex gas separators of various manufacturers showed that the following structural schemes of the flow part of the gas separator at the intake to the separation chamber are found:

1. Only a fixed stroke auger.
2. An auger and an axial wheel or an alternating stroke auger.
3. A fixed or alternating stroke auger, a diffuser, an axial wheel.
4. An auger, a diffuser of diagonal type, an axial wheel placed inside the cone sleeve, see Fig. 2, [26], [27].

Separation drums with radial fans are installed in the separation chamber of the centrifugal gas separator at the outlet of axial wheel 1. They increase pressure in the peripheral part of the separation chamber, but do not fundamentally change the flow scheme in the vortex separators with axial wheels.

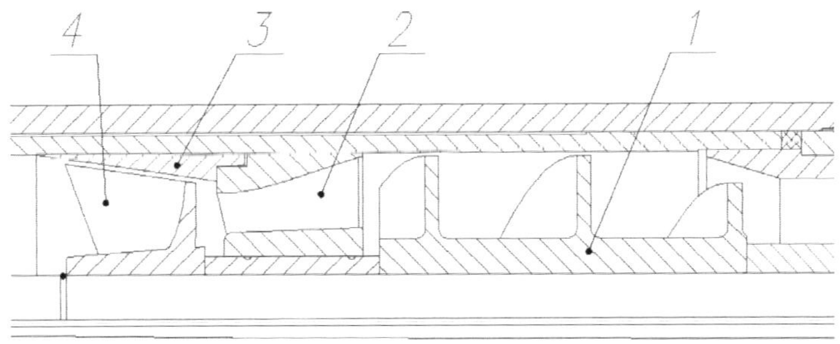

Fig. 2.A scheme of the power part of the conceptual design of the gas separator. Auger 1, diagonal diffuser 2, Cone sleeve 3, Axial wheel 4.

Mathematical model

In this article, we use the model of a multiphase incompressible fluid flow (ρ = const). Numerical simulation is based on solving discrete analogs of the basic hydrodynamic equations.

The calculation is carried out on the basis of a mathematical model of a divided multiphase flow. That is, for each phase, the equations of mass and momentum transfer are solved separately, but the pressure field is the same for all phases.

The mathematical model consists of a set of differential and algebraic equations:

1. The volume of the i-th phase in each calculated cell is calculated as:

$$Vi=\int \alpha i V dV$$

whereαi— is the concentration of the i-th phase in the cell.

The sum of the concentrations of all phases in the cell is one

$$\sum_{i6}^{n} \alpha_i = 1$$

2. The equation of conservation of mass (continuity equation):

$$\frac{\partial}{\partial t}\int_V \alpha_i \rho_i dV + \int_A \alpha_i \rho_i \vec{V_i} d\vec{a} = 0$$

whereρi—i-th phase density

$\vec{Vi}$—i-th phase velocity (in the case of turbulent flow modeling by a RANS-type model averaged over time)

3. The equation for the change in the amount of motion:

$$\frac{\partial}{\partial t}\int_V \alpha_i \rho_i dV + \int_A \alpha_i \rho_i \left(\vec{V_i}\vec{V_i}\right) d\vec{a} = -\int_V \alpha_i \nabla p dV + +\int_V \alpha_i \rho_i \vec{g} dV + \int_A \left[\alpha_i \left(T_i + T_i^t\right)\right] d\vec{a} + \int_V \vec{M_t} dV$$

where ($\vec{Vi}\vec{Vi}$)— tensor product of the velocity vectors of the i-th phase.

p— pressure

g— mass intensity vector (in this case, the gravity force is 9.81 m / s2 and the inertial force due to the rotation of the computational domain)

Ti— molecular viscosity stress tensor

Tit— turbulent stress tensor

$\vec{M}$—vector of total intensity of interfacial interaction forces per unit volume, for a vector $\vec{Mi}$fair equality: $\sum \vec{M_t} = 0$

Vector $M\vec{i}$ characterizes all the forces that separate phases interact with each other.

$$M_i = \sum_{i \neq j} \left(\overrightarrow{F_{ij}^D} + \overrightarrow{F_{ij}^{VM}} + \overrightarrow{F_{ij}^L} + \overrightarrow{F_{ij}^{TD}} + \overrightarrow{F_{ij}^{WL}} \right)$$

Where
$FijD\vec{}$— resistance force, $FijVM\vec{}$ power of the virtual mass
$FijL\vec{}$— lifting force
$FijTD\vec{}$— turbulent dispersion force
$FijWL\vec{}$— turbulent dispersion force

Power of resistance
In the case of modeling the flow of continuous and dispersed media, the resistance force acting on the dispersed medium I from the side of the continuous medium j is equal to:

$$\overrightarrow{F_{ij}^D} = \Lambda_D \overrightarrow{V_r},$$

Where
AD— linearized drag coefficient
$Vr\vec{} = Vj\vec{} - Vi\vec{}$— relative velocity of one medium relative to another

$$A_D = C_D \frac{1}{2} \rho_C \left|\overrightarrow{V_r}\right| \left(\frac{a_{cd}}{4} \right),$$

Where
CD— drag coefficient
ρC— continuous phase density
acd— interaction area of the phases (in this case, the member$acd4$ is the projected area of the spherical particle on the plane).
The drag coefficient is based on the relation:

$$C_D = f_D C_{D\infty},$$

Where
$CD\infty$ — coefficient of resistance of a single spherical particle moving in an infinite flow
fD— coefficient taking into account the concentration of particles

$$C_{D\infty} = \frac{24}{Re_d} \left(1 + 0.15 Re_d^{0.687} \right) npu Re_d < 1000,$$

and

$$C_{D\infty} = 0.44 npu\, Re_d > 1000$$

$$Re_d = \frac{\rho_C |V_r| l}{\mu_C}$$

l— characteristic interaction length or bubble size
μC— dynamic viscosity coefficient of continuous medium
Coefficient $fD$is the ratio:

$$f_D = \alpha_C^{nD}$$

αC— continuous phase concentration
nD=–8.3 for spherical particle

Virtual mass

The inertia of the surrounding fluid affects the acceleration of the particle immersed in the fluid. This effect is modeled by adding mass to the dispersed particle.

The force from the virtual mass acting on phase I, moving rapidly with respect to phase j is found as:

$$\overrightarrow{F_{ij}^{VM}} = C_{VM}\rho_C\alpha_C\left(\overrightarrow{a_j} + \overrightarrow{a_i}\right)$$

$aj,i^{\to}$— acceleration of jth and ith phase

CVM=0,5 — virtual mass coefficient for a spherical particle

Lift force

In the case where the flow surrounding the dispersed particle is non-uniform or the particle is swirling, it experiences a force perpendicular to the relative velocity.

$$\overrightarrow{F_{ij}^{L}} = C_{Leff}\rho_C\alpha_d\left(\overrightarrow{V_r} \times \left(\nabla \times \overrightarrow{V_c}\right)\right)$$

$Vr^{\to}$—relative phase velocity

$Vc^{\to}$— continuous phase speed

αd— dispersed phase concentration

CL=0,25 — lift force coefficient

Turbulent dispersion force

Additional change in phase concentrations caused by flow turbulence is modeled as the force of turbulent dispersion

$$\overrightarrow{F_{ij}^{TD}} = A_D\overrightarrow{V_{TD}}$$

$FijTD^{\to}=ADVTD^{\to}$

A_D—dragforce coefficient

$VTD^{\to}$—relative slip speed

$$VTD^{\to}=DTD(\nabla\alpha d\alpha d-\nabla\alpha c\alpha c)$$

$D_{TD}=C0vct\sigma\alpha I$– turbulent diffusion tensor

C_0=1

vct—turbulent viscosity coefficient kinematic $\sigma\alpha$– turbulent Prandtl number

I— unit tensor

$$\sigma_\alpha = \sigma_0\sqrt{1 + C_\beta\xi^2}\,\frac{1+\eta}{B+\eta}$$

$\sigma\alpha=\sigma0\sqrt{}1+C\beta\xi21+\eta b+\eta$

$\sigma0$=1 — unmodified turbulent Prandtl number

$C\beta$=1,8 — correction factor

ξ— particle sliding velocity related to the speed of turbulent fluctuations

η— particle interaction time related to relaxation time

b— the ratio of the accelerations of the continuous / dispersed phase

$$b = \frac{1 + C_{VM}}{\dfrac{\rho_d}{\rho_c} + C_{VM}}$$

$$\eta = \frac{\tau_I}{\tau_R}$$

IOP Conf. Series: Materials Science and Engineering **779** (2020) 012036 doi:10.1088/1757-899X/779/1/012036

τI— characteristic time of interaction of a particle and a turbulent vortex

τR— particle relaxation time

$$\tau_I = \frac{\tau_T}{\sigma_0} \frac{1}{\sqrt{1 + C_\beta \xi^2}}$$

$$\tau_T = \frac{2}{3} C_\mu \frac{k_c}{\epsilon_C}$$

ϵc-continuous phase turbulence energy dissipation rate

$$\tau_R = \tau_D \left(1 + \frac{\rho_c}{\rho_d} C_{VM} \right)$$

$\tau_D = \frac{\rho_d d^2}{18\mu_c}$—characteristic time scale for a dispersed particle

$$\xi = \frac{\left|\vec{V_r}\right|}{\sqrt{\frac{2}{3} k_c}}$$

kc— kinetic turbulence energy of the continuous phase

Wall strength

Being located near the solid wall, the gas bubble undergoes an asymmetrical action from the liquid. The force per unit volume that a gas bubble experiences is equal to:

$$\overrightarrow{F_{ij}^{VL}} = -C_{WL} \gamma_W \alpha_D \rho_c \frac{\left|\vec{V_r}\right|^2}{d_p} \vec{n}$$

Where

yw— distance from the wall

αD— dispersed phase concentration

ρC— dispersed phase density

dp— bubble diameter

CWL— coefficient is a function of the distance from the wall and decreasing with increasing distance

$\vec{n}$— unit normal to the wall at the point closest to the bubble

$Vr,\vec{\tau}$— tangent to the wall component of the relative velocity

Coefficient $CWL$is like:

$$C_{WL} = max\left(C_{w1} + \left(\frac{C_{w2}}{\gamma_w} \right) d_p, 0 \right)$$

Coefficients are $CW1$=–0,01, $CW2$=0,05.

Thus the force caused by the influence of the wall disappears at a distance from the wall equal to the five diameters of the bubble.

Mathmodelling results

To carry out hydrodynamic calculations, the geometry of the flow section of the vortex gas separators of various manufacturers was made in accordance with the structural schemes of the Analysis.

IOP Conf. Series: Materials Science and Engineering **779** (2020) 012036 doi:10.1088/1757-899X/779/1/012036

Fig.3 shown the places of hydro-abrasive wear of the gas separator body (a), the backflows in the flow part are calculated based on the numerical simulation.

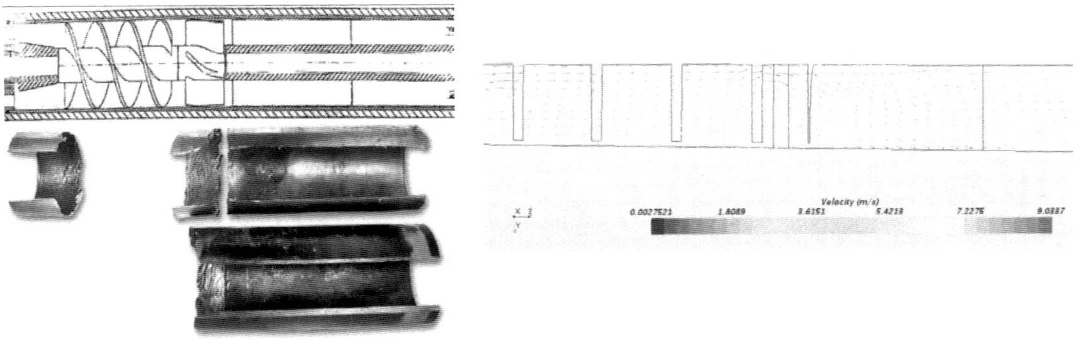

a b

Fig.3. The places of hydro-abrasive wear of the gas separator body (a), the backflows in the flow part, obtained by numerical modeling (b).

The auger and the axial wheel have two are as of the rotor with different pressures:

1. The feed auger has about 20 degree inclination angle of the blades at the outlet and, accordingly, a relatively low head.

2. The axial wheel has about 90 degree inclination angle of the blades at the outlet and, respectively, a higher head and pressure in the peripheral part.

There are backflows along the peripheral part from the axle wheel back to the auger. Backflows are a trap for solids, the concentration of which increases rapidly, and a rotating abrasive ring of a toroidal vortex can cut through the body due to the hydro-abrasive wear.

In the case of using an auger of alternating stroke, backflows occur at the modes which are twice or more times lower than the optimal mode, since at large angles of attack the pressure at the inlet decreases.

In the case of using a screw and an axial wheel, backflows can occur in all operating modes.

Fig. 4 shows the velocity plots in the designs of gas separators, the power part of which includes: a fixed stroke an auger, a diffuser, and an axial wheel. In Fig. 4a, the gas separator has an axial diffuser, and in Fig. 4b –a diagonal one with a conical sleeve, inside of which an axial wheel is installed.

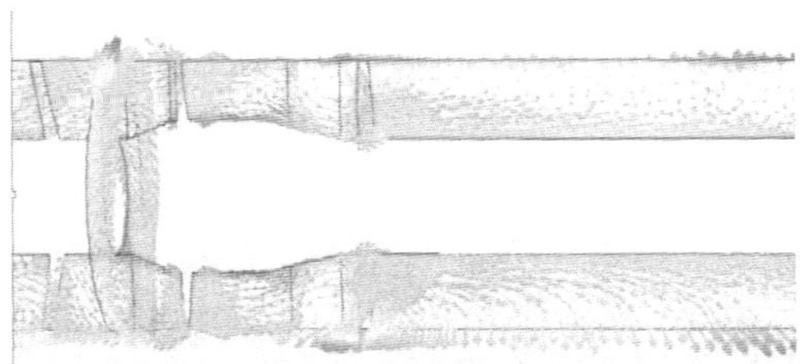

a

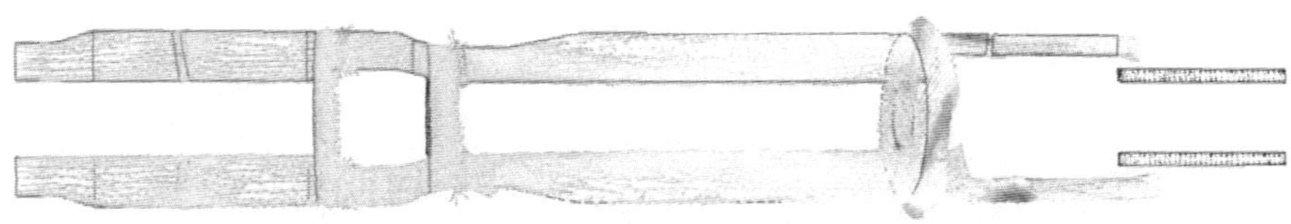

b

Fig.4. Velocity Plots in the design without cone sleeve (a) and with cone sleeve (b).

IOP Conf. Series: Materials Science and Engineering **779** (2020) 012036 doi:10.1088/1757-899X/779/1/012036

The cone sleeve separates the low-pressure auger and the high-pressure separation wheel, thus back flows and wear of the sleeve and auger are eliminated.

An analysis of the numerical calculation on water shows that in the design without a cone sleeve, backflows are clearly visiblealong the periphery from the separation chamber to the inlet of the diffuser. The backflows can lead to hydro-abrasive wear and cutting of the elements of the flow part, the diffuser grid and the housing, to a decrease in the separation properties due to dispersion of the gas-liquid mixture flow and a decrease in the pressure gradient in the separation chamber.

Installing a cone sleeve eliminates, or at least significantly reduces the backflows.

A separation process takes place in the separation chamber a rotating wheel is formed with a partially separated liquid on the periphery. Previously, this wheel pressed on the blades of a stage supplying a gas-liquid mixture, which led to backflows and dispersion. Now the wheel is pressing on the fixed axial bearing in the form of a cone sleeve.

Fig. 5 shows pressure plots in the above designs of gas separators. An analysis of the numerical calculation on water shows that in a structure without a cone sleeve has a lower pressure gradient compared to the structure with the sleeve due to the backflows. The higher is the radial pressure gradient, the more efficient is the separation of free gas from the gas-liquid mixture pumped into the annulus.

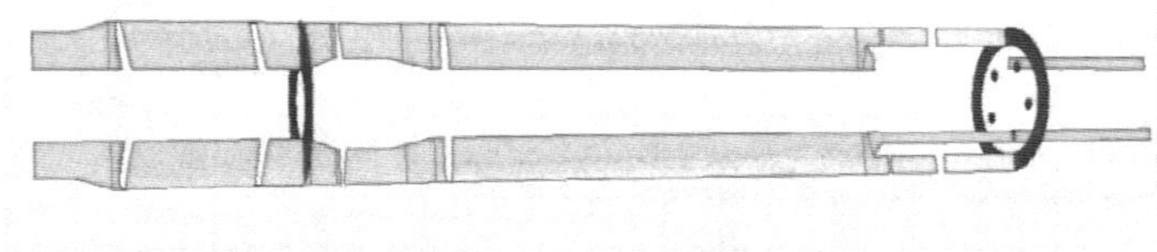

a

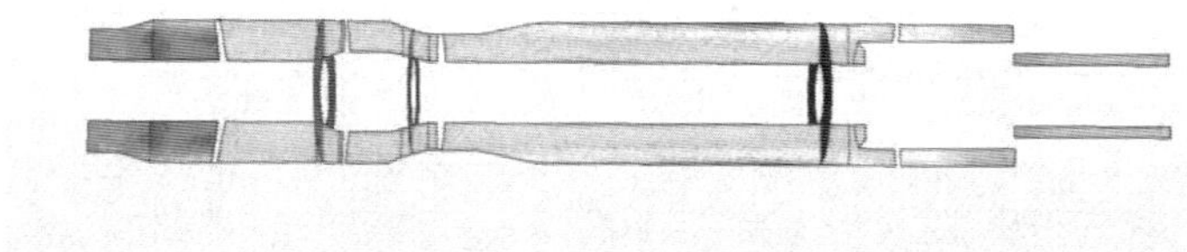

b

Fig.5. Pressure Plots in the design without cone sleeve (a) and with cone sleeve (b).

Fig. 6 shows the results of the calculation of a gas separator with a cone sleeve are presented for gas-liquid mixture.

An analysis of the numerical calculation for the gas-liquid mixture shows that in the separation chamber the process of separation of liquid and free gas occurs. A rotating wheel with a partially separated liquid is formed on the periphery. Previously, this wheel pressed on the blades of a stage supplying a gas-liquid mixture, which led to the backflows, which, in turn, dispersed the gas-liquid mixture, and reduced the pressure in the separation chamber. Now the rotating wheel presses on the fixed axial bearing in the form of a cone sleeve.

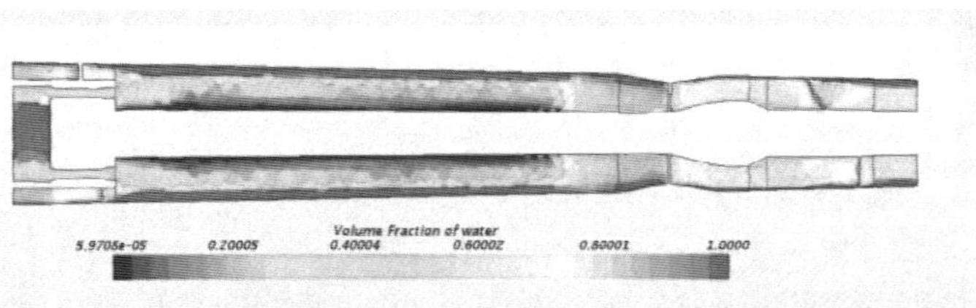

Fig.6. Volume fraction of water Plots in the design without cone sleeve

The new conceptual design of submersible gas separators (see Fig. 2)includes the structural separation by a cone sleeve of two areas: the area of delivery of the gas-liquid mixture to the separation chamber and the area of the separation chamber. The axial wheel (auger) delivering the gas-liquid mixture works with maximum feed and minimum pressure, the flow part of the separation area - with maximum head, providing the maximum pressure gradient.

It is necessary to maintain a certain ratio of pressure and feed in the auger. This is due to the fact that gas bubbles can pass through the flow part of the auger without significant separation and the formation of gas locks, which can lead to the delivery failure only at a certain ratio of the surface friction forces associated with the flow velocity and mass inertia forces associated with the pressure gradient.

Vortex and centrifugal gas separators with replaceable elements of the flow part, replaceable augers have been developed for certain supply ranges together with the specialists of the RGU NG Center for Research and Development of RIMERA Group of Companies. At low flow rate, it is proposed to install the augers obtained by reducing the diameter of the serial screws, by cutting, while maintaining the optimal input angles and the range of components.

One of the obvious advantages of gas separators with replaceable augers at the inlet for low rate systems include less power - this saves energy.

For the calculation, the StarCCM software package was used. As the surface mesh, the previously constructed geometry of the flow part was imported.

The following parameters are used as a physical model:

- Multiphase interaction, scale of interaction length 0. 1mm.
- Water H2O and gas Air are taken as phases.
- Initial liquid / gas distribution 1/0
- The inlet is set to stagnation at the inlet with a total pressure of 0 Pa and a liquid / gas ratio of 4/1
- At the exit boundary of the fluid, a negative speed is set at 4.5 m / s.
- Atmospheric pressure is set at the gas outlet boundary.

The following parameters were taken as the studied parameters: head, separation coefficient (water fraction at the separator outlet) and power depending on the flow rate.

Bench test results

Fig. 7 presents the results of bench tests for gas-liquid mixtures in the laboratory of the RSU NG Department of Technology for the development of oil fields.

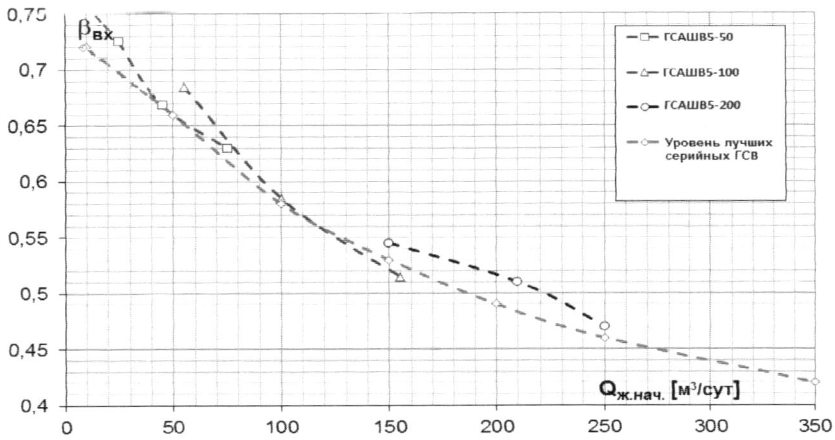

Fig. 7. The results of bench tests for gas-liquid mixtures in the laboratory of the RSU NG Department of Technology for the development of oil fields.

Characteristics of a gas separator with replaceable augers for nominal flow rates of 50, 100, 200 cubic meters per day.exceed the best serial products [28]–[31].

It is obvious that at low feeds a gas separator with a replaceable screw designed for low flow rate will have less power and higher reliability compared to a conventional gas separator, the power part of which is designed for maximum flow rate.

Findings

The novelty lies in the fact that the cone sleeve separates two areas: the area of delivery of the gas-liquid mixture to the separation chamber and the area of the separation chamber. The axial wheel (auger) delivering the gas-liquid mixture works with maximum feed and minimum pressure, the flow part of the separation area- with maximum pressure, providing the maximum pressure gradient.

The results of numerical modeling show that:

1. A new conceptual design of gas separators has been developed.
2. Back flows are eliminated or at least significantly reduced in thepower elements of the flow part in the new design; such backflows reduce the separation properties and reliability, decrease hydraulic efficiency, and increase power.
3. The results of bench tests for gas-liquid mixtures in the laboratory of the RSU NG Department of Technology for the development of oil fields show, that characteristics of the new gas separator with replaceable augers exceed the best serial products.
4. The conceptual ideas inherent in the design can serve as the basis for the development of the gas separators of other standard sizes.

List of references:

[1] Vakhitova R.I., Saracheva D.A., Urazakov D.R., Dumler E.B., Povysheniееffektivnostirabotypogruzhnykhelektrotsentrobezhnykhustanovokpridobychenefti s vysokimgazosoderzhaniem (Improving the efficiency of submersible electric centrifugal plants inoil production with high gas content), Al'met'evsk: Publ. of ASPU, 2019, 104 p.

[2] Yakimov S.B., Shportko A.A., ShalaginYu.Yu., Ways of improving gas separators reliability used to protect electric centrifugal pumps (ESP) in the deposits of PJSC "NK "Rosneft" (In Russ.), Oborudovanieitekhnologiidlyaneftegazovogokompleksa,2017, no. 1, pp. 33–39.

[3] Yakimov S.B. O vozmozhnostyakhoptimizatsiiklassov iznosoustoychivostielektrotsentrobezhnykhnasosovnamestorozhdeniyakh PAO "Orenburgneft"': nauch.-tekhn. vestnik. –PAO "NK "Rosneft"', 2015. – № 3. – S. 85–92.

[4] Drozdov A.N., Tekhnologiyaitekhnikadobychineftipogruzhnyminasosami v oslozhnennykhusloviyakh (The technology and technique of oil production by submergible pumps in the complicated conditions), Moscow: MAKS press Publ., 2008, 312 p.

[5] Den'gaev A.V., Povysheniееffektivnostiekspluatatsiiskvazhinpogruzhnymitsentrobezhnyminasosamipriotkachkegazozhidkostnykhsmesey (Improving the efficiency of well operation by submersible centrifugal pumps during pumping of gas-liquid mixtures): thesis od candidate of technical science, 2005.

[6] Lomakin V O, Chaburko P S, Kuleshova M S 2017 Multi-criteria Optimization of the Flow of a Centrifugal Pump on Energy and Vibroacoustic Characteristics *Procedia Engineering* **176** pp 476–482

[7] Lomakin V O, Kuleshova M S, Bozh'eva S M 2016 Numerical Modeling of Liquid Flow in a Pump Station*Power Technology and Engineering* **5**pp 324–327

[8] Lomakin V O 2015 Proceedings of 2015 International Conference on Fluid Power and Mechatronics

[9] Lomakin V O, Kuleshova M S, Kraeva E A 2015 Fluid Flow in the Throttle Channel in the Presence of Cavitation *Procedia Engineering* **106** pp 27–35

[10] Shargatov, V.A., Gorkunov, S.V., Il'ichev, A.T. Dynamics of front-like water evaporation phase transition interfaces (2019) Communications in Nonlinear Science and Numerical Simulation, 67, pp. 223–236.
[11] Arefyev, K.Y., Prokhorov, A.N., Saveliev, A.S. Study of the breakup of liquid droplets in the vortex wake behind pylon at high airspeeds (2018) Thermophysics and Aeromechanics, 25 (1), pp. 55–66.
[12] Trulev A V 2015 New equipment of CJSC Rimera for wells with complicated operating conditions, Proceedings of the conference on mechanized production Oil and Gas Vertical (NeftegazovayaVertikal)19
[13] Trulev A V 2015 New equipment of OJSC Alnas for wells with complicated operatingconditions Engineering Practice (Inzhenernayapraktika) (12) pp 71–73
[14] Trulev A V, Sabirov A A, Sibirev S V, Verbitsky V S and Timushev S F 2017 Submersible ESP units with wide channels in the flow part for the production of formation fluid from marginal wells with a high content of mechanical impurities Engineering Practice (Inzhenernayapraktika) (1–2) pp 60–63
[15] Trulev A V 2017 Submersible ESP unit of new generation. Innovative solutions for combatting complicated conditions in case of the marginal field, Oil and Gas Journal Russia (4) pp 30–34
[16] Trulev A V 2017 New submersible centrifugal pumps with gray cast iron stages and between bearing design Engineering Practice (Inzhenernayapraktika) (5) pp 74–76
[17] Trulev A V and Grachev V V 2018 Bench tests of a submersible gas and mechanical impurities separator in model mixtures Gas Industry (Gazovayapromyshlennost') (8) pp 20–24
[18] A.V.Trulev, V.F.Loginov, S.I.Gorbunov, S.F.Timushev, A.A.Sabirov, V.O.Lomakin, evelopment and pilot operation of ESP systems of a conceptually new design for the operation of depleted wells with a high content of free gas and mechanical impurities. Composite volume the laureates of the International Competition of Scientific, Technical and Innovative Developments aimed at the development of the fuel and energy and extractive industries in 2019.
[19] A Trulev, V Verbitsky, S Timushev, P Chaburko Electrical submersible centrifugal pump units of the new generation for the operation of marginal and inactive wells with a high content of free gas and mechanical impurities. Hydraulics, IOP Conf. Series: Materials Science and Engineering 492 (2019) 012041 IOP Publishing doi:10.1088/1757-899X/492/1/012041.
[20] A.V. Trulev, S.F. Timushev. A New Generation of Submersible Electrical Motor Protectors with a Dynamic Labyrinth for Operation in Aggressive Environment with High Concentration of Free Gas and Solids doi: 10.18698/0536-1044-2019-7-59-65.
[21] Borovskij B.I., Petrov V.I. Vysokooborotnyelopatochnyenasosy. M., «Mashinostroenie», 1975.
[22] Gouskov, A.M., Lomakin, V.O., Banin, E.P., Kuleshova, M.S. Minimization of Hemolysis and Improvement of the Hydrodynamic Efficiency of a Circulatory Support Pump by Optimizing the Pump Flowpath (2017) Biomedical Engineering, 51 (4), pp. 229–233. DOI: 10.1007/s10527-017-9720-9
[23] Arefyev, K.Y., Voronetsky, A.V., Suchkov, S.A., Ilchenko, M.A. Computational and experimental study of the two-phase mixing in gas-dynamic ignition system (2017) Thermophysics and Aeromechanics, 24 (2), pp. 225–237. DOI: 10.1134/S086986431702007X
[24] Lomakin, V.O., Kuleshovav, M.S., Bozh'eva, S.M. Numerical Modeling of Liquid Flow in a Pump Station (2016) Power Technology and Engineering, 49 (5), pp. 324–327. DOI: 10.1007/s10749-016-0623-9
[25] Arefyev, K.Y., Voronetsky, A.V. Modelling of the process of fragmentation and vaporization of non-reacting liquid droplets in high-enthalpy gas flows (2015) Thermophysics and Aeromechanics, 22 (5), pp. 585–596.
[26] A.V.Trulev, G.M. Makrushin, Patent No. 2616331 of the Russian Federation "A method for the efficient operation of submersible blade pumps when pumping off formation fluid with a

high content of gas and abrasive particles and a gas separator of ESP system for its implementation". Application date: 31.12.2015.

[27] A.V.Trulev Patent No. 2503808 of the Russian Federation "Gas separator for a downhole pump". Application date: 08.07.2011. DOI: 10.1134/S0869864315050078

[28] Belov, P.A., Kobets, L.P., Borodulin, A.S. Impregnation kinetics of fibers with liquids: Simulation within the generalization of Navier-Stokes equations (2014) Inorganic Materials: Applied Research, 5 (4), pp. 403–406.

[29] Loginov V.F. Kompressionnyenasosydlyaneftedobychi v oslozhnennyhusloviyah. // Oilandgasjournal. 2016, № 6. C. 54–56.

[30] D. C. Wilcox (1994). "Turbulence modeling for CFD" // DCW Industries, Inc., 460 pages.

[31] ArefEv, K.Yu., Voronetskii, A.V., Il'Chenko, M.A. Dynamic characteristics of a resonant gas-dynamic system for ignition of a fuel mixture (2013) Combustion, Explosion and Shock Waves, 49 (6), pp. 657–661.

IOP Conf. Series: Materials Science and Engineering **779** (2020) 012037 doi:10.1088/1757-899X/779/1/012037

The calculation of multiphase flows in flowing parts of centrifugal pump

M Saprykina[1,2] and V Lomakin[1]

[1]Bauman Moscow State Technical University

[2]E-mail: marias108@mail.ru

Abstract. The effect of gas content on the changing of head of a centrifugal pump is researched. The method of numerical hydrodynamic modeling is used for solving this problem. The using mathematical models are described and compared. The results of the simulation of fluid flow, which are depended on the air content, are presented. The data on the changing of head and the termination of its work are obtained.

Introduction

The variety of pumped fluids by phase composition can be classified into single-phase (contain only the liquid phase), two-phase (contain liquid and solid or liquid and gaseous phases) and three-phase (contain liquid, solid and gaseous phases). Therefore, the problem of pumping hydraulic mixtures is important for many industries. [1]–[5]

The development of a number of industries which are related to the processing and movement of gas-saturated suspensions (oil producing, oil refining, food, microbiological, etc.) requires the creation of special pumping equipment capable of efficiently pumping products with a high level of gas content in the pumped mixture. At the same time, the main operational requirements for the specified pumping equipment are: a) the maximum gas content of the pumped medium, b) profitability, c) overall dimensions. [6]–[7] Based on this, it follows that the main criterion that determines the level of production costs for the consumer is the maximum gas content of the pumped medium, pushing the indicator of pump efficiency to the background.

The operation of pumping equipment is complicated by the presence of free gas. It is received with the submersible pump as a part of the gas-liquid mixture as a result of the process of oil degassing. Free gas can have a negative effect on the operation of the pump, which will lead to increasing in the consumed electric power in comparison with the data of the rating characteristics, to decreasing in the mean time between failures of both individual elements and the entire pumping unit as a whole. It will also lead to decreasing in the technological efficiency of oil production from wells. [8]–[10]

In the case of pumping mixtures of liquid and gas, these pumps have several disadvantages, which include a low efficiency and disruption of the parameters with excessive gas content in the pumped medium. In this regard, special attention is required to analyze the concept of the quality of work of dynamic pumps in hydraulic mixtures in general and in GHS separately. Whenliquids with a high gas factor are pumped, the total technical and economic indicators of the quality of the pumping equipment are most acceptable not only in relation to the usual efficiency when the pump is running on clean liquid. Today, other indicators require greater attention, including the “stable operation” of a dynamic pump. The concept of "stable operation" refers to the ability of a pump to operate on a GHS

Content from this work may be used under the terms of the Creative Commons Attribution 3.0 licence. Any further distribution of this work must maintain attribution to the author(s) and the title of the work, journal citation and DOI.
Published under licence by IOP Publishing Ltd

without losing its performance at high levels of gas content in the pumped mixture. Loss of performance occurs due to disruption of the pump parameters (there is a complete cessation of its supply). [11]–[12]

The problem of stalling parameters is relevant for all vane pumps. An analysis of the technological regimes of the well operating stock shows upthat now there are a number of unresolved tasks for the effective control of the pump in high gas conditions. Thus, when analyzing the existing literature on this topic, it was revealed that the process of oil production from wells under free gas conditions is relevant and requires additional research. [13]

Mathematical model of multiphase fluid flow

In this paper, we use the model of a multiphase incompressible fluid flow (ρ = const). Numerical simulation is based on solving discrete analogs of the basic hydrodynamic equations.The two approaches of calculation multiphase fluid flow were used in this research.

The first calculation is carried out on the basis of a mathematical model of a divided multiphase flow. That is, the equations of mass and momentum are solved separately for each phase. Yet, the pressure field is the same for all phases. The developed pump is used in the system of formation water purification. It supplies the working fluid to the flotation machine. It pumps a two-phase medium: water-air. Numerical hydrodynamic modeling was performed in the StarCCM + software package. [14]–[15]

The mathematical model consists of differential and algebraic equations:

1. The volume of the i-th phase in computional cell is calculated as:

$$V_i = \int_V \alpha_i dV,$$

where α_i — the concentration of the i-th phase in the cell.

The sum of the concentrations of all phases in the cell is one.

$$\sum_{i6}^{n} \alpha_i = 1.$$

2. The equation for conservation of mass (continuity equation):

$$\frac{\partial}{\partial t}\int_V \alpha_i \rho_i dV + \int_A \alpha_i \rho_i \vec{V_i} d\vec{a} = 0,$$

where ρ_i — the i-th phase density; $\vec{V_i}$ — the i-th phase velocity (in the case of turbulent flow modeling by a RANS-type).[16]–[18]

3. The equation of the changing in the amount of motion:

$$\frac{\partial}{\partial t}\int_V \alpha_i \rho_i \vec{V_i} dV + \int_A \alpha_i \rho_i \left(\vec{V_i}\vec{V_i}\right) d\vec{a} = -\int_V \alpha_i \nabla p dV +$$

$$+\int_V \alpha_i \rho_i \vec{g} dV + \int_A \left[\alpha_i \left(T_i + T_i^t\right)\right] d\vec{a} + \int_V \overline{M_i} dV,$$

where $\left(\vec{V_i}\vec{V_i}\right)$ — the tensor product of the velocity vectors of the i-th phase; p — the pressure; $\vec{g}$ —the mass intensity vector (in this case, the gravity force is 9.81 m / s2 and the inertial force due to the rotation of the computational area); T_i — the molecular viscosity stress tensor; T_i^t — the turbulent stress tensor; $\vec{M}$ —the vector of the total intensity of interfacial interaction forces per unit volume, for the vector $\overrightarrow{M_i}$ the equality is:

$$\sum_i \overrightarrow{M_i} = 0.$$

The vector $\overrightarrow{M_i}$ characterizes all the forces that separated phases interact with each other.[19]–[22]

$$\overrightarrow{M_i} = \sum_{i \neq j}\left(\overrightarrow{F_{ij}^{D}} + \overrightarrow{F_{ij}^{VM}} + \overrightarrow{F_{ij}^{L}} + \overrightarrow{F_{ij}^{TD}} + \overrightarrow{F_{ij}^{WL}}\right),$$

where $\overrightarrow{F_{ij}^{D}}$ — the resisting force; $\overrightarrow{F_{ij}^{VM}}$ — the power of the virtual mass; $\overrightarrow{F_{ij}^{L}}$ — the ascensional power; $\overrightarrow{F_{ij}^{TD}}$ — the turbulent dispersive force; $\overrightarrow{F_{ij}^{WL}}$ — the force, which is caused by wall effects.

In fig. 1 the computational grid for hydrodynamic modeling is shown.

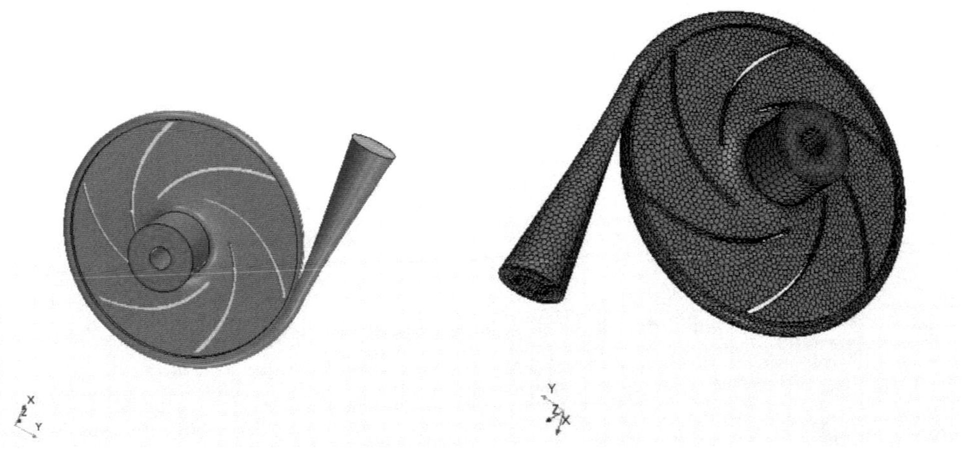

ab

Fig. 1. The computational grid
a — 3d model; b — section.

The VOF Multiphase model is used to solve problems involving immiscible fluid mixtures, free surfaces, and phase contact time. In such cases, there is no need for extra modeling of inter-phase interaction, and the model assumption that all phases share velocity, pressure, and temperature fields becomes a discretization error.

The Volume of Fluid (VOF) multiphase model implementation in Simcenter STAR-CCM+ belongs to the family of interface-capturing methods that predict the distribution and the movement of the interface of immiscible phases. This modeling approach assumes that the mesh resolution is sufficient to resolve the position and the shape of the interface between the phases.

Thus, this model combines air and liquid in one phase with the same speed. After the creating an air bubble and the beginning of the simulation process, it was revealed that this model has a more stable result and the solution converges. It was decided to calculate with this method.

IOP Conf. Series: Materials Science and Engineering **779** (2020) 012037 doi:10.1088/1757-899X/779/1/012037

Calculation results

In fig. 2 the visualized data on the volume of air in the calculated pump are shown. The approach of calculation was **Multiphase Segregated Flow**.

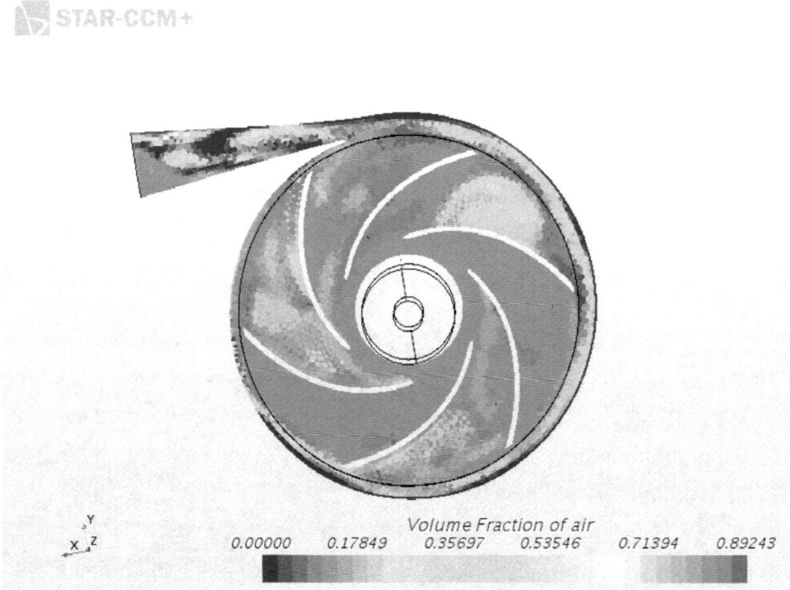

Fig. 2. The results of simulation

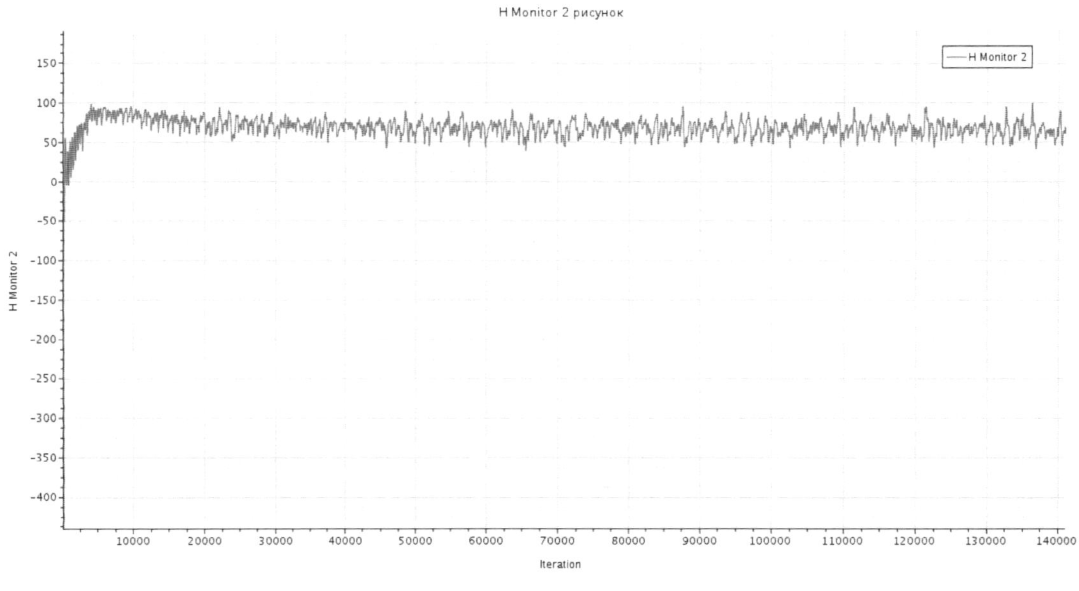

Fig. 3. Theheadgraph

The graph shows that the head practically does not decrease at a percentage of air around 13%. The solution is unstable. Air bubbles do not accumulate but they dissolve throughout the fluid. In this case the pump does not notice the air and works correctly. According to this fact the present method of calculation does not bring rights results.

It was decided to calculate with Volume of Fluid. The figure 4 shows the simulation results at a percentage of air 13%.

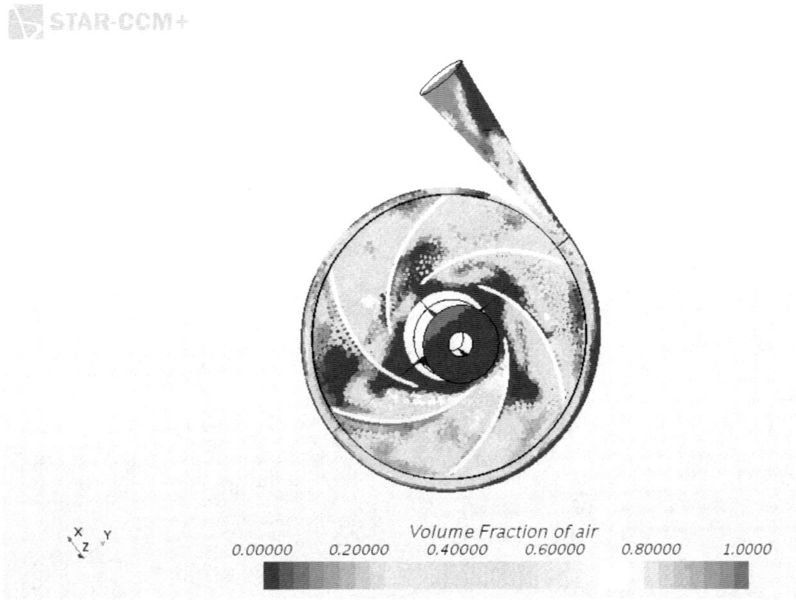

Fig. 4. VOF'ssimulation.

Infig. 5 the head's dropping can be seen.

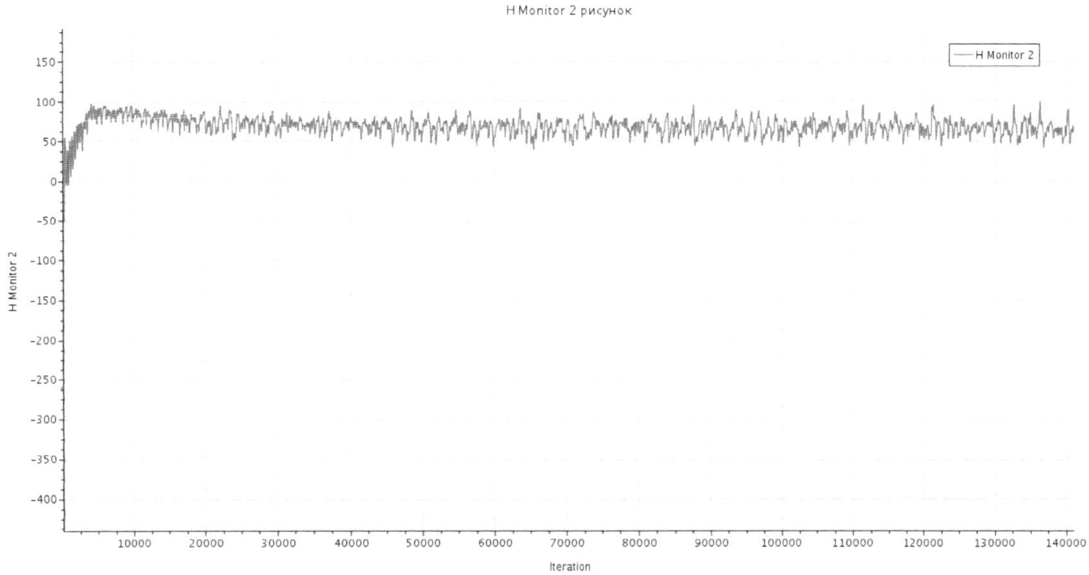

Fig. 5. Theheadgraph

In fig. 6-8 many different simulations-visualizations of percentage of air are shown. Besides, the data of head is shown too.

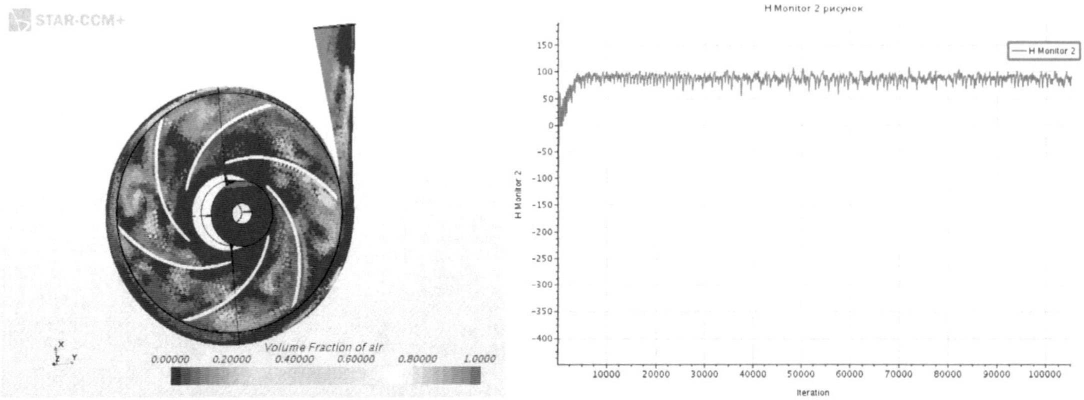

Fig. 6. The volume of air 5%.

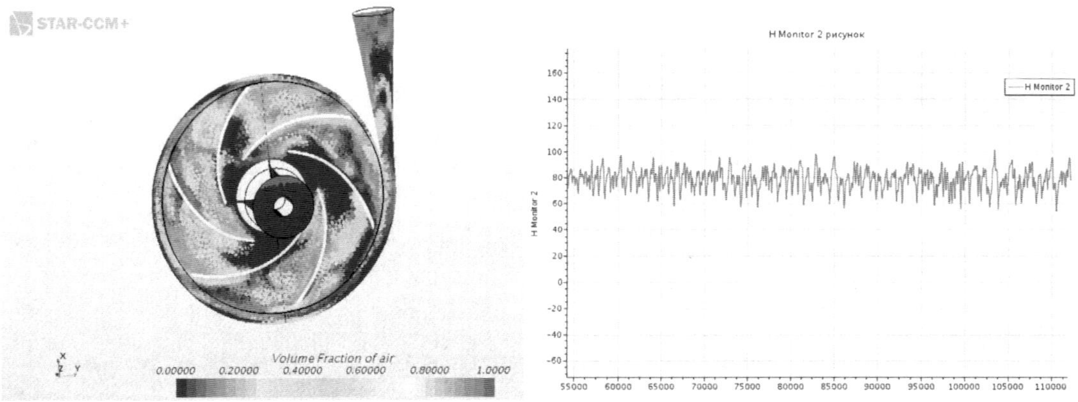

Fig. 7. The volume of air10%.

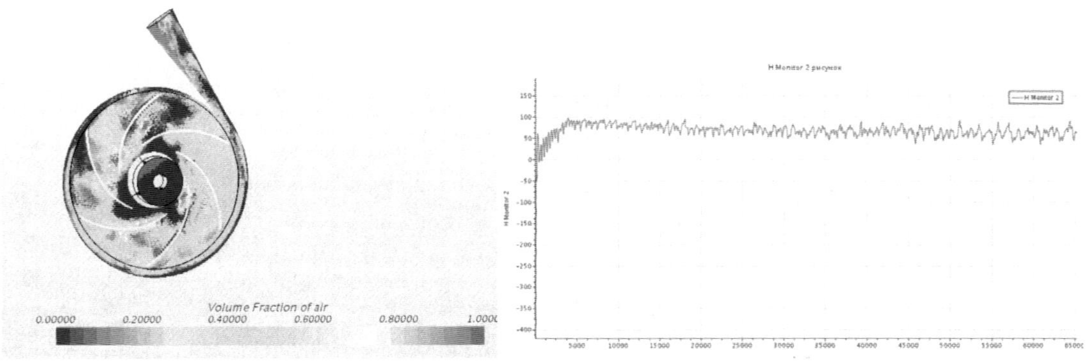

Fig. 8. The volume of air15%.

Fig. 9 shows a sharp head's drop- the breakdown.

IOP Conf. Series: Materials Science and Engineering **779** (2020) 012037 doi:10.1088/1757-899X/779/1/012037

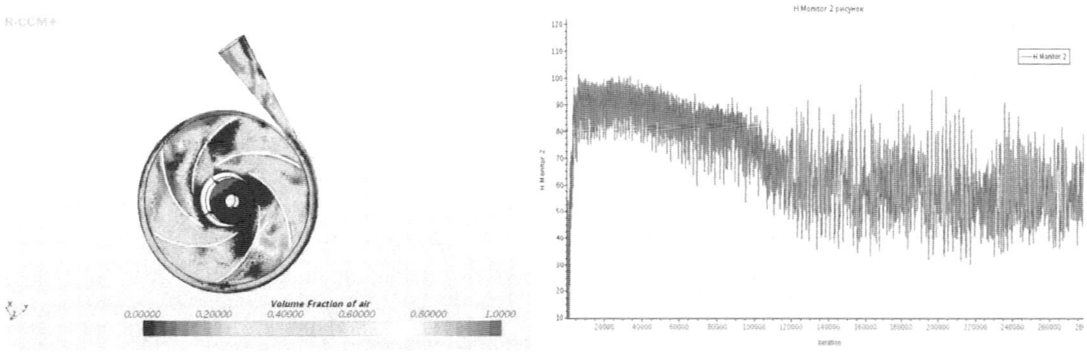

Fig. 9. The volume of air 17%.

When study was finished, the graph was obtained reflecting the dependence of the head on the air content in the pumped liquid and also an estimated breakdown point.

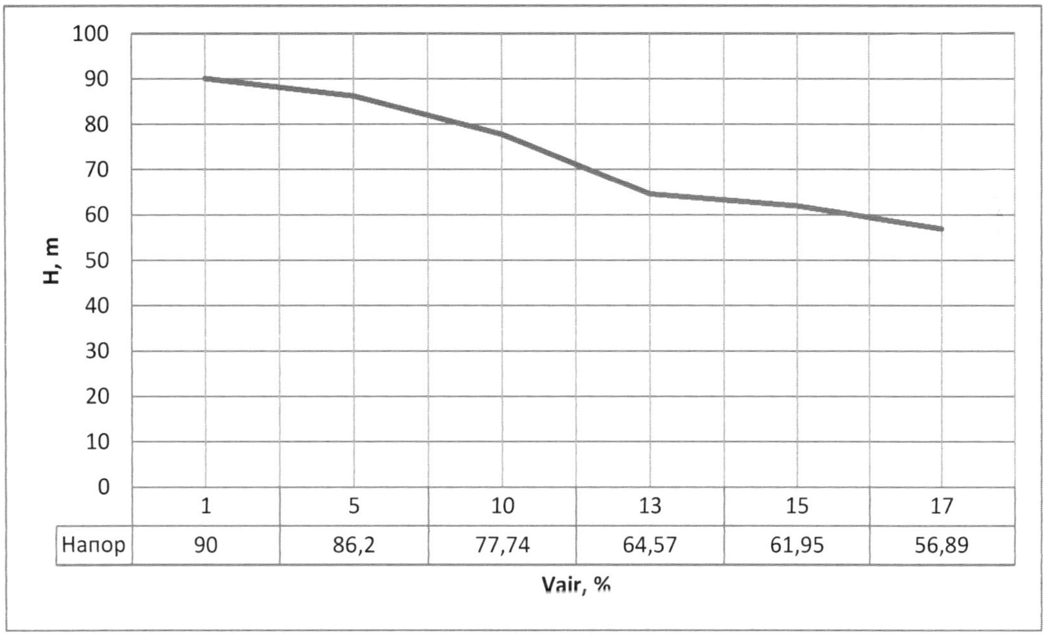

Fig. 10. The graph shows dependence between the volume of air and the head.

Conclusion

The analysis showed the effectiveness of the mathematical model VOF.The calculation showed that this method correctly reflects the behavior of the pump when the gas content is changing. Besides, it can also be widely used to find the breakdown point.

The proposed method of calculation and setting the boundary conditions allows to achieve a significant dropping of the head with low gas content at the inlet. The head dropped by a third with a gas content of 15 %.

References

[1] Antonenko S.S., Kolisnichenko E.V. The possibility of using centrifugal pumps for pumping gas-liquid mixtures. Vestnik SymGY, series «Technical science», №4, 2009г.(In Russian)

[2] Wilson B. L. Gas Separation: A New Generation, A New Twist. — 2005 ESP WORKSHOP, The Woodlands, Texas, April 25–29, 2005.

[3] Labah N., Development of methodological and technological solution for the selection of upstream modules of an electric centrifugal pumps in complicated well operating conditions, 2016, Moscow.

[4] Lomakin, V.O., Kuleshova, M.S., Bozh'eva, S.M. Numerical Modeling of Liquid Flow in a Pump Station (2016) Power Technology and Engineering, 49 (5), pp. 324–327. DOI: 10.1007/s10749-016-0623-9

[5] N IsaevThe effect of design parameters of the closed type regenerative pump the energy characteristics2019 IOP Conf. Ser.: Mater. Sci. Eng. 492 012026

[6] Lomakin V.O. Investigation of two-phase flow in axial-centrifugal impeller by hydrodynamic modeling methods Proceedings of 2015 International Conference on Fluid Power and Mechatronics, FPM 2015 no 7337302 pp 1204–6 DOI: 10.1109/FPM.2015.7337302

[7] Chaburko P, KossovaZWet part hybrid optimization method of hermetic pump 2019 IOP Conf. Ser.: Mater. Sci. Eng. 492 012011

[8] Lomakin V O, Petrov A I and Kuleshova M S 2014 Investigation of a two-phase flow in an axial centrifugal wheel by methods of hydrodynamic modeling Science and education: electronic scientific edition of Bauman MSTU (9) pp 45–64 (in Russian)

[9] Brennen C.E. Fundamentals of Multiphase Flows. Cambridge: Cambridge University Press, 2005. 410 p.

[10] Lozovetskii, V.V., Pelevin, F.V. Resistance of spherical fills in the case of flow of single- and two-phase media (2009) Journal of Engineering Physics and Thermophysics, 82 (2), pp. 277–282. DOI: 10.1007/s10891-009-0194-9

[11] Arefyev, K.Y., Voronetsky, A.V., Suchkov, S.A., Ilchenko, M.A. Computational and experimental study of the two-phase mixing in gas-dynamic ignition system (2017) Thermophysics and Aeromechanics, 24 (2), pp. 225–237. DOI: 10.1134/S086986431702007X

[12] Perevezentsev, V.V. Experimental study of the hydrodynamics of gas-solids flows in transient (accelerating) motion of the solid particles (1988) Fluid mechanics. Soviet research, 17 (3), pp. 1–6.

[13] J. Fan , J. Eves, H.M. Thompson, V.V. Toropov, N. Kapur, D. Copley and A. Mincher "Computational fluid dynamic analysis and design optimization of jet pumps" Computers & Fluids 46 (2011) pp.212–217.

[14] Gouskov A M, Lomakin V O, Banin E P and Kuleshova M S 2017 Minimization of Hemolysis and Improvement of the Hydrodynamic Efficiency of a Circulatory Support Pump by Optimizing the Pump Flowpath Biomedical Engineering 51 pp 229–233 DOI: 10.1007/s10527-017-9720-9

[15] Timushev S F,Knyazev V A,Panaiotti S S,Soldatov V A and Rohatgi U S 2009 Numerical procedure for assessment of centrifugal pump cavitation erosion ASME International Mechanical Engineering Congress and Exposition, Proceedings 14 pp 73–78 DOI: 10.1115/IMECE2008-67177

[16] Kochevsky A N, Kozlov S N, Aye K M, Schelyaev A Y and Konshin V N 2005 Simulation of flow inside an axial-flow pomp with adjustable guide vanes Proceedings of the American Society of Mechanical Engineers Fluids Engineering Division Summer Conference 1B pp 1397–1404

[17] Yamaguchi T., Tanaka K. Torque converter transient characteristics prediction using computational fluid dynamics. Proc. of the 26th IAHR Symposium on Hydraulic Machinery and Systems. 2012, 19-23 August, Beijing, China. Beijing, 2012. DOI: 10.1088/1755- 1315/15/4/042020

[18] Petrov A, Isaev N, Kuleshova M Test bench flow straightener design investigation and optimization with computational fluid dynamics methods IOP Conf. Series: Materials Science and Engineering 492 (2019) 012036 IOP Publishing DOI:10.1088/1757-899X/492/1/012036
[19] Zharkovsky A A and PospelovA Yu 2014 The use of 3D methods for the flow calculation, characteristics prediction and optimization of the shape of flow parts of hydraulic turbines Hydraulic engineering (11) pp 104–109
[20] Patankar S 1984 Numerical methods for solving problems of heat transfer and fluid dynamics (Moscow: Energoatomizdat Publ.) 152p
[21] Lomakin V O and Petrov A I 2012 Verification of the calculation results in the package of hydrodynamic modeling STAR-CCM+ of the flow part of the centrifugal pump AX 50-32-200 HigherEducational Institutions Bulletin. Sociology. Economics. Politics p 6
[22] ProtopopovA, Vigovskij V Development of methods for high-speed centrifugal pump analysis2019 IOP Conf. Ser.: Mater. Sci. Eng. 492 012003

IOP Conf. Series: Materials Science and Engineering **779** (2020) 012038 doi:10.1088/1757-899X/779/1/012038

Analytical study of the dynamics of a pumping unit, consisting of a fixed pump and a flow rate regulator

A Lepeshkin[1,2,3] and E Sarach[2]

[1]Moscow Polytechnic University, Moscow, Russia
[2]Bauman Moscow State Technical University

[3]E-mail:a.v.lepeshkin@mospolytech.ru

Abstract. A mathematical model of a pumping unit, consisting of a fixed pump and an automatic flow rate regulator, intended for use as part of a mathematical model of a volume hydraulic drive that requires a constant supply pressure for its operation is described in article. As the carried out studies showed (their results are given in this article), the proposed model describes the physical processes during the operation of the pump unit quite well. In addition, based on these studies, recommendations are given on the need to include a hydropneumatic accumulator in the structure of the unit, which will significantly improve its dynamic characteristics.

Introduction

One of the main characteristics of transport machines is the average speed, which is limited and determined by the traffic conditions and technical characteristics of the machine. Limitations, depending on traffic conditions include limitations on thrust force, on the danger of uncontrolled movement and on smoothness of movement [1]. An increase in the specific power of modern machines and the correct choice of traction motors made it possible to remove a number of limitations on the thrust force [2]. Methods to reduce the probability of occurrence of uncontrolled movement are the subject of [3]. In this context, the problem of improving the cushioning systems of transport machines for an increasing the motion speed on rough road is being important.

Recently, hydropneumatic cushioning systems are widely used in transport, which allow obtaining progressive suspension characteristics and changing the position of the coachbuilder in space. Determining the characteristics of the elements included in the hydropneumatic cushioning system is impossible without the use of simulation mathematical simulation.

Mathematical models of the movement of transport machines for studying smoothness of movement are widely presented in the literature [4–14], they are used to obtain the characteristics of elastic and damping elements of the suspension. In this paper, a mathematical model of the pumping unit of the coachbuilder position change system is presented. The quality of work and the system operation speed influence on the possibility of its use when the machine is moving, which is typical for controlled active cushioning systems [15–31].

The paper proposes a mathematical model of a pumping unit, consisting of a fixed pump and an automatic flow rate regulator, intended for use as part of a mathematical model of a volume hydraulic drive that requires a constant supply pressure for its operation.

Content from this work may be used under the terms of the Creative Commons Attribution 3.0 licence. Any further distribution of this work must maintain attribution to the author(s) and the title of the work, journal citation and DOI.
Published under licence by IOP Publishing Ltd

Methods and instruments of research

Based on the analysis of the current pumping units used in practice [32–35], which provide a constant supply pressure at the inlet to the hydraulic system, it is proposed to use a unit including a fixed axial-piston pump and an automatic flow rate regulator when developing this mathematical model. The corresponding design model is shown in Fig.1.

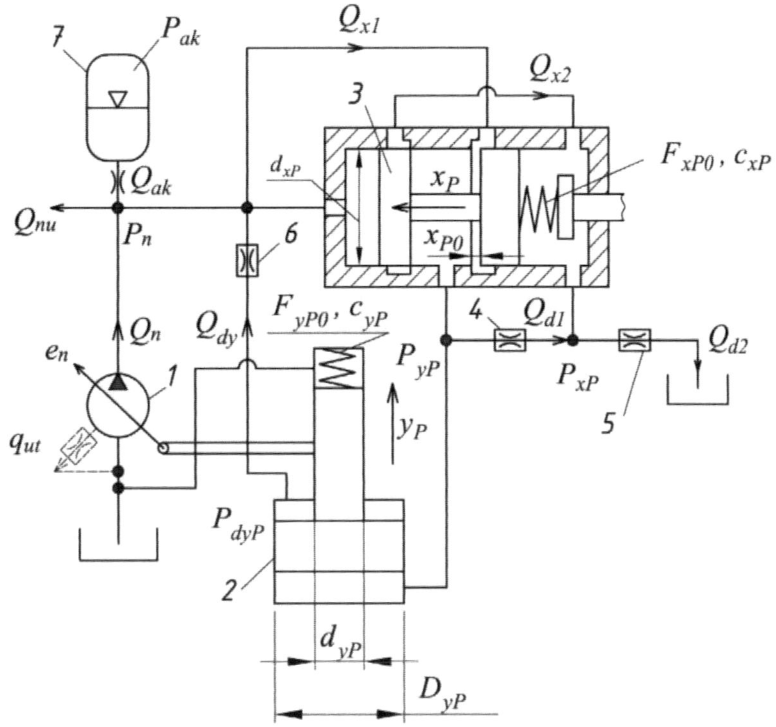

Fig. 1. Design model of the pumping unit: 1 — fixed axial-piston pump; 2 — differential piston of the flow rate regulator; 3 — spool throttle control valve; 4, 5, and 6 — tuned throttles; 7 — hydropneumatic accumulator

The mathematical description of the adopted version of the pumping unit includes:

1) equation characterizing the change in manometer pressure p_n at the pump outlet:

$$\dot{p}_n = \frac{B_g}{W_n} \cdot \left(Q_n - Q_{nu} - Q_{ak} - Q_{x1} + Q_{dy} + \dot{x}_p \cdot \frac{\pi d_{xp}^2}{4} \right)$$

2) equation characterizing the change in manometer gas pressure p_{ak} in the hydropneumatic accumulator:

$$\dot{p}_{ak} = \frac{\left(p_{zap} + p_{at} \right) \cdot n \cdot W_{g\max}^n \cdot Q_{ak}}{\left(W_{g\max} - W_{ak} \right)^{n+1}}$$

3) equation characterizing the piston movement y_p in regulator:

$$m_{y_p} \cdot \ddot{y}_p = p_{y_p} \cdot \frac{\pi D_{y_p}^2}{4} - p_{dy_p} \cdot \frac{\pi}{4}\left(D_{y_p}^2 - d_{y_p}^2 \right) - \left(F_{y_p 0} + c_{y_p} \cdot y_p + k_{y_p} \cdot \dot{y}_p \right)$$

4) equation characterizing the change in manometer pressure p_{y_p} in the piston cavity of the regulator:

$$\dot{p}_{y_p} = \frac{B_g}{W_{y_p}} \cdot \left(Q_{x1} - Q_{x2} - Q_{d1} - \dot{y}_p \cdot \frac{\pi D_{y_p}^2}{4} \right)$$

5) equation characterizing the change in manometer pressure p_{dy_p} in the rod cavity of the regulator:

$$\dot{p}_{dy_p} = \frac{B_g}{W_{dy_p}} \cdot \left(\dot{y}_p \cdot \frac{\pi}{4} \left(D_{y_p}^2 - d_{y_p}^2 \right) - Q_{dy} \right)$$

6) equation characterizing the movement of the spool x_p of the regulator:

$$m_{x_p} \cdot \ddot{x}_p = \left(p_{x_p} - p_n \right) \cdot \frac{\pi d_{x_p}^2}{4} + F_{x_p 0} - \left(c_{x_p} \cdot x_p + k_{x_p} \cdot \dot{x}_p \right)$$

7) equation characterizing the change in manometer pressure p_{x_p} in the end cavity of the regulator spool:

$$\dot{p}_{x_p} = \frac{B_g}{W_{x_p}} \cdot \left(Q_{x2} + Q_{d1} - Q_{d2} - \dot{x}_p \cdot \frac{\pi d_{x_p}^2}{4} \right)$$

In these equations, in addition to the notations used in Fig. 1, the following notations are agreed:
p_{zap} —manometer pressure of gas charging of the hydropneumatic accumulator;
p_{at} —atmosphere pressure, $p_{at} = 0{,}1$ MPa;
$W_{g\,max}$ —constructive volume of hydropneumatic accumulator;
n— gas polytropic coefficient, n=1,4;
W_{ak} —volume of fluid in the hydropneumatic accumulator at present sampling time:

$$W_{ak} = W_{ak0} + \int_0^t Q_{ak} \cdot dt$$

W_{ak0} — initial volume of fluid in the hydropneumatic accumulator;
B_g —fluid modulus of inelastic buckling, B_g=1000 MPa;
W_n, W_{y_p}, W_{dy_p}, W_{x_p} — fluid volumes reduced to the corresponding node points;
m_{y_p} — reduced mass of the regulator piston;
k_{y_p} —coefficient of viscous friction of the regulator piston;
m_{x_p} —regulator spool mass;
k_{x_p} —coefficient of the spool viscous friction of the regulator.

Taking into account the goal and prospects for further use, the mathematical model developed in the MATLAB Simulink software package based on the mathematical description of the pumping unit above is formed as a block «Pumpy» (Fig. 2). The input signals for this block are the pump shaft speed $n_n = \text{nn}$ and the flow rate consumed by the connected hydraulic system from the pumping unit $Q_{nu} = \text{Qnu}$, and output signals — are the pressure $p_n = \text{pn}$ created by the pump, and the resulting moment of resistance to rotation of the pump shaft $M_n = \text{Mn}$.

IOP Conf. Series: Materials Science and Engineering **779** (2020) 012038 doi:10.1088/1757-899X/779/1/012038

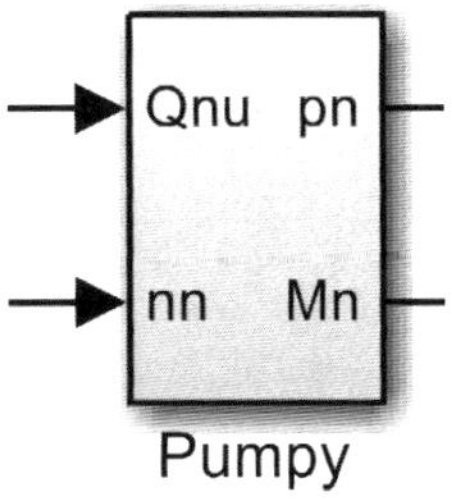

Fig.2. Block «Pumpy»

The composition of the block «Pumpy» is shown in Fig. 3.

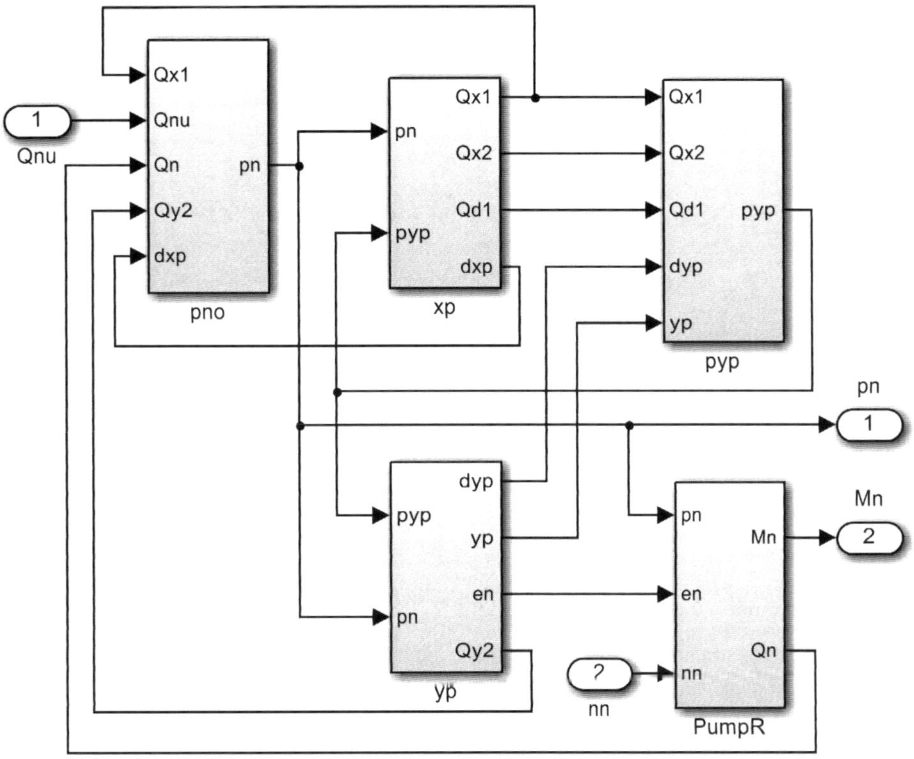

Fig.3. The composition of the block «Pumpy»: pno— block specified pressure p_n; xp— block specified movement x_p; pyp— block specified pressure p_{y_p}; yp— block specified movement y_p and regulation parameter of pump e_n; PumpR— block specified the flow rate of variable pump Q_n and the shaft torque M_n

The initial data (values of design parameters), that are needed for the analysis of the operation of the pumping unit were taken from design considerations, as well as on the basis of the conditions for obtaining the required static characteristics. In addition, the resulting dynamics of the transient response processes accompanying the operation of the pump unit were taken into account. The type of static characteristics obtained as a result of the calculation and taken as the basis is shown in Fig. 4 (abscissa axis — l/s; left ordinate axis — MPa; right ordinate axis is dimensionless).

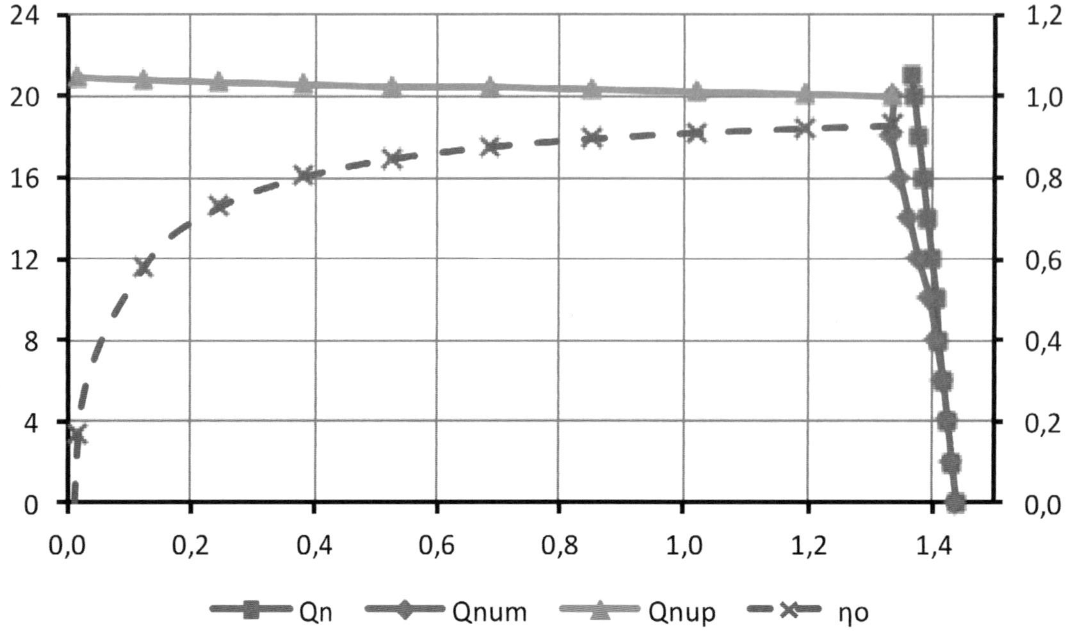

Fig.4. Static characteristics of pump unit: Qn—pq pump curve; Qnum - pq pump curve with $p_n < 20$ MPa; Qnup—pq pump curve with $20 < p_n < 21$ MPa; Efficiency —the change in volumetric efficiency of the pumping unit with $20 < p_n < 21$ MPa (right axis values)

It should be also noted that further studies of the developed mathematical model of the pumping unit under consideration suggest that the initial steady state of its moving elements corresponds to those shown in Fig. 1. This operation mode of the pumping unit occurs when the pump shaft rotational speed is constant and there is no fluid consumption of the hydraulic system connected to it. Therefore, before carrying out the planned studies, it is necessary to determine the appropriate initial conditions.

For this purpose, the simulation of the process of turning on the engine driving of the pump was completed.

The calculation results are shown in Fig. 5 (along the abscissa axis — time in seconds).

In the simulation, it was assumed that the pump shaft rotational speed increases linearly from zero to the adopted $n_n = 1920$ rpm in 1 second (Fig. 5a), and the design volume of the hydropneumatic accumulator $W_{g\,\max} = 2$ l.

The graphs show that approximately 6 seconds after turning on the pump drive, all transient responses associated with this are completed. The resulting steady-state values of the variable values were used as initial conditions for further mathematical simulation the operating mode of the pump unit under study. These initial conditions are the following variable values: $p_n = p_{ak} = p_{dy_p} =$ $= 20{,}902$ MPa, $p_{y_p} = 20{,}1$ MPa, $p_{x_p} = 0{,}0034$ MPa, $W_{ak} = 1$ l, $x_p = 0{,}0084$ mm, $y_p = 28{,}025$ mm, with $e_n = 0{,}066$.

IOP Conf. Series: Materials Science and Engineering **779** (2020) 012038 doi:10.1088/1757-899X/779/1/012038

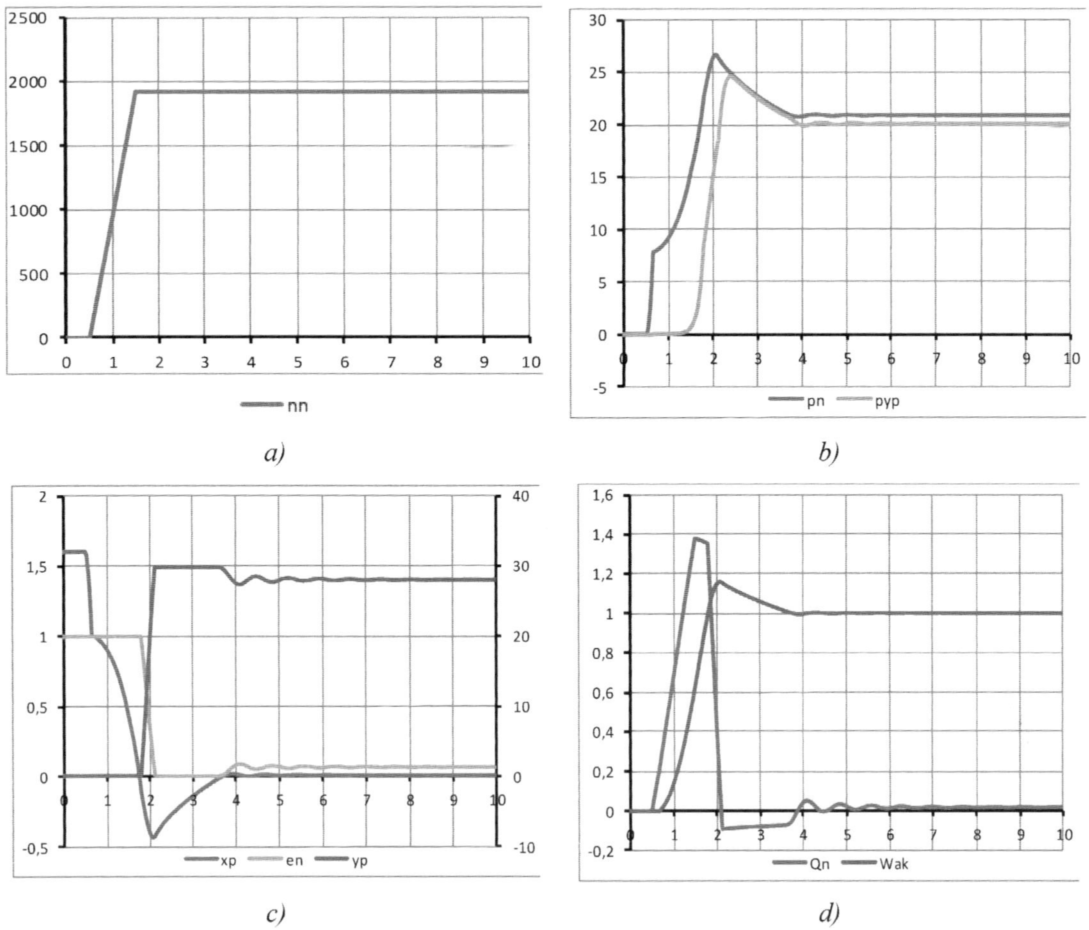

Fig. 5. The results of simulation of the turning on the engine drivin of the pump: *a*— the initial frequency change law of the pump shaft speed n_n (rpm); *b*— pressure changes p_n and p_{y_p} (MPa); *c*— movement of the spool x_p (mm) and piston y_p (mm on the right axis) of the regulator and the corresponding change in the regulation parameter e_n; *d*— changes in the pump flow Q_n (l / s) and the volume W_{ak} (l) of fluid in the hydropneumatic accumulator

Research Results

In the study of the dynamic processes accompanying the operation of the pump unit, the behavior of the change in the amount of fluid flow consumed from it, shown in Fig. 6 (the abscissa axis shows time in seconds; the ordinate axis l / s), is taken as the initial external disturbance. Fig. 7 and Fig. 8 show the results of mathematical simulation of the pumping unit in question. For comparison, the figures show the results for a pump installation without a hydropneumatic accumulator (Fig. 7a and Fig. 8a) and with its presence (Fig. 7b and Fig. 8b).

IOP Conf. Series: Materials Science and Engineering **779** (2020) 012038 doi:10.1088/1757-899X/779/1/012038

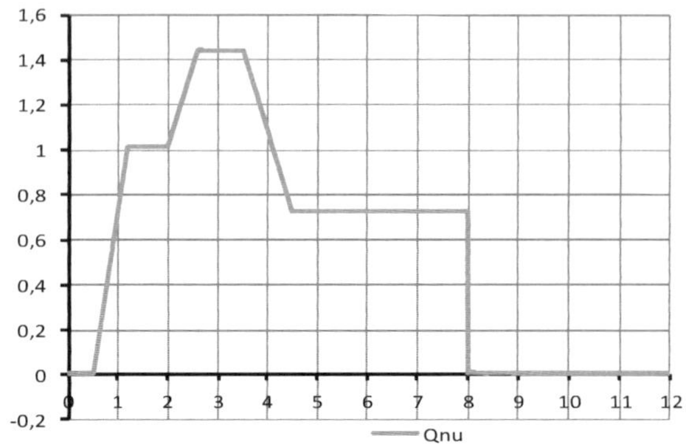

Fig. 6. The behavior of the change in flow rate Q_{nu} adopted in the simulation

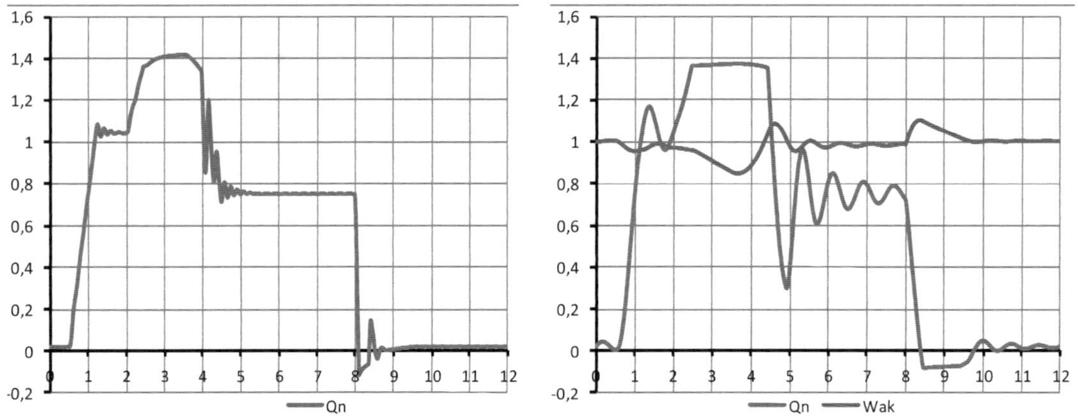

a) without accumulator *b)* with accumulator

Fig. 7. The behavior of the change in the flow rate Q_n (l / s) of the pump and the volume of fluid W_{ak} (l) in the accumulator

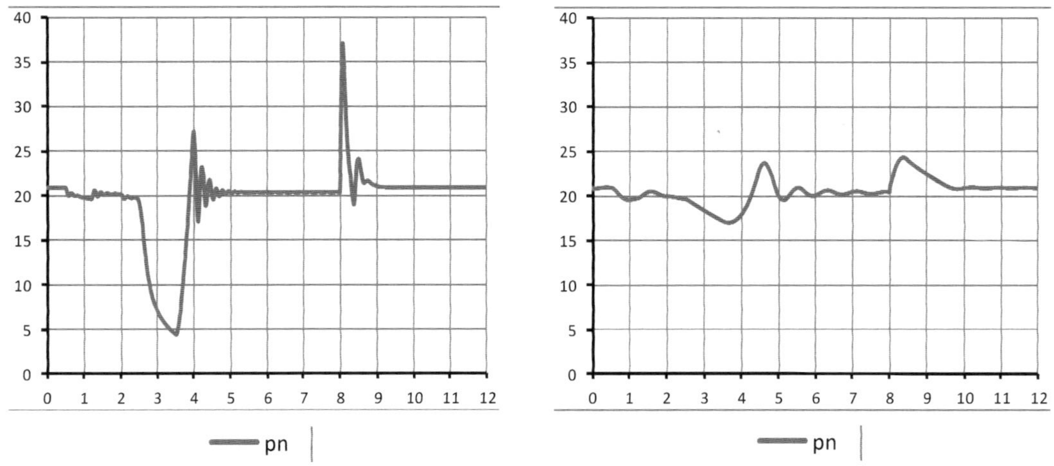

a) without accumulator *b)* with accumulator

Fig.8. The behavior of the pressure p_n (MPa) change at the pump outlet

Conclusion

An analysis of the results of mathematical simulation of the pumping unit operation in question shows that:

1) the developed mathematical model adequately reflects the physical processes that accompany the operation of the pumping unit in question under simulated modes;

2) this mathematical model can be recommended for simulation the operating modes of the pumping unit in question with a connected to it hydraulic system, which requires a significantly changing flow rate of pressure fluid for its operation;

3) the use of a hydropneumatic accumulator at the outlet of the pump unit in question significantly improves the quality of its operation in the conditions of implementation of transient response dynamic processes. That is, the use of a hydropneumatic accumulator at the output of the pump installation should be recognized as necessary.

List of reference

[1] Dmitriev A. A., CHobitok V. A., Telminov A. V. Theory and calculation of nonlinear suspension systems for tracked vehicles. — M.: Mechanical Engineering, 1976. — 207 p.
[2] Kotiev G O, Miroshnichenko A V, Stadukhin A A and Kositsyn B B 2019 Determination of mechanical characteristics of high-speed tracked vehicles traction motor with individual drive wheels (Journal of Physics: Conference Series) vol 1177(1)) p 012058
[3] Ergin A A, Kolomejtseva M B and Kotiev G O 2004 Antiblocking control system of the brake drive of automobile wheel (PriboryiSistemyUpravleniya) vol 9 pp 11–13
[4] Diadchenko M. G., Kotiev G. O., Sarach E. B. Design and calculation of suspensions of high-speed tracked vehicles: Textbook. — M.: Publishing House of BMSTU, 2007.— 40 p.
[5] Diadchenko M. G., Kotiev G. O., Naumov V. N. Fundamentals of calculating the suspension systems of tracked vehicles on a computer: a Training manual. — M.: Publishing house of BMSTU, 2002, 52 p.
[6] Dhir A and Sankar S 1994 Analitical model for dynamic simulation of off-road Tracked Vehicles (Vehicle system dynamics) vol 24 pp 26–49
[7] Dhir A and Sankar S 1997 Analitical wheel models for dynamic simulation of off-road tracked vehicles (Vehicle system dynamics) vol 27 pp 37–63
[8] Duym S, Stiens R and Reybrouck K 1997 Evaluation of shock absorber models (Vehicle system dynamics) vol 27 pp 109–127
[9] Lee K 1997 Numerical modeling for the hydraulic performance prediction of automotive monotube dampers (Vehicle system dynamics) vol 28 pp 25–39
[10] Wong J Y 1997 Dynamics of Tracked Vehicles (Vehicle System Dynamics) vol 28 pp 197–219
[11] Sarach E, Kotiev G and Beketov S 2018 *Methods for road microprofile statistical data transformation* (MATEC Web of Conferences) vol 224 p 04009
[12] Klubnichkin V E, Klubnichkin E E, Kotiev G O, Beketov S A and Makarov V S 2018 *Interaction between elements of the track ground contacting area with the soil at curvilinear motion of the timber harvesting machine* (IOP Conference Series: Materials Science and Engineering) vol 386(1) p 012016
[13] Weiss R A 1981 *Terrain Microroughness the Dynamic Response of Vehicles* (West Point (New York) ARO Report 82–1) 6 p
[14] Shinozuka M 1971 *Simulation of a multivariate and multimensional random processes* (Journal of the Acoustical Society of America) vol 49 № 1 pp 357–368
[15] Venhovens P J Th 1991 *Semi-Active Suspension for Automotive Application.* (Delft: Univ. of technology) 28 p.
[16] Venhovens P J Th 1994 *The Development and Implementation of Adaptive Semi-Active Suspension Control* (Vehicle System Dynamics) vol 23 pp 211–235
[17] Karnopp D and Sang-Gyun So 1998 *Energy Flow in Active Attitude Control Suspensions: A Bond Graph Analysis* (Vehicle System Dynamics) vol 29 pp 69–81

[18] Redfield R C 1991 *Performance of Low-bandwidth, Semi-Active Damping Concepts for Suspension Control* (Vehicle System Dynamics) vol 20 pp 245–267
[19] Foag W A 1989 *Practical Control Concept for Passenger Car Active Suspension with Preview* (Proc. Instn. Mech. Engrs (London (UK))) p 203
[20] Prem H A 1987 *Laser-based Highway-speed Road Profile Measuring System* (Proc. 10th IA VSD) p 37
[21] Zhileykin M M, Kotiev G O and Nagatsev M V 2018 *Comparative analysis of the operation efficiency of the continuous and relay control systems of a multi-axle wheeled vehicle suspension* (IOP Conference Series: Materials Science and Engineering) vol 315(1), 012030
[22] Zhileykin M M, Kotiev G O and Nagatsev M V 2018 *Synthesis of the adaptive continuous system for the multi-axle wheeled vehicle body oscillation damping* (IOP Conference Series: Materials Science and Engineering) vol 315(1), 012031
[23] Crosby M J and Karnopp D C 1973 *The Active Damper - a New Concept for Shock and Vibration Control* (43rd Shock and Vibration Bulletin) vol H, June, pp 46–73.
[24] Dhir A and Sankar S 1994 *Ride Dynamics of High-Speed Tracked Vehicles: Simulation with Field Validation* (Vehicle System Dynamics) vol 23 pp 379–409.
[25] Gordaninejad F and Breese D 2000 *Heating of Magnetorheological Fluid Dampers* (Journal of Intelligent Materials, Systems and Structures, accepted, in press) vol 4 pp 45–56.
[26] Han-Shue Tan and Bradshaw T 1998 *Validation of the Harshness Problem of Automotive Active Suspensions* (Vehicle System Dynamics) vol 30 pp 89–115.
[27] Hrovat D, Margolis D and Hubbard M 1988 *An Approach Toward the Optimal Semi-Active Suspension* (Journal Dynamics System. Measurement and Control) vol 110 № 3 pp 288–296.
[28] Wang X and Gordaninejad F 2000 *Flow Analysis of Field-Controllable, Electro- and Magnetno-Rheological Fluids Using Herchel-Bulkley Model* (Journal of Intelligent Materials, Systems and Structures, accepted, in press) vol 2 pp 87–121.
[29] Yokoyama M, Hedrick J and Toyama S 2001 *A Model Following Sliding Mode Controller for Semi-Active Suspension Systems with MR Dempers* (Proceedings of the American Control Conference, Arlington, VA June 25-27) P. 53–58.
[30] Yu F and Crolla D 1998 *An Optimal Self-Tuning Controller for an Active Suspension* (Vehicle System Dynamics) vol 29 pp 51–65.
[31] Yu F and Crolla D 1998 *State Observer Design for an Adaptive Vehicle Suspension* (Vehicle System Dynamics) vol 30 pp 457–471.
[32] A.V.Lepeshkin, A.A.Mihajlin, A.A. SHejpak,*Gidravlika i gidropnevmoprivod. Gidravlicheskie mashiny i gidropnevmoprivod* [Hydraulics and hydropneumatic drive. Hydraulic machines and hydropneumatic drive] (Moscow, Infra-M Publ., 2017. 446 p.)www.dx.doi.org/10.12737/21024
[33] https://dc-us.resource.bosch.com/media/us/products_13/product_groups_1/ mobile_hydraulics_4/pdfs_6/re-a92703.pdf
[34] https://www.psm-hydraulics.ru/gallery/product/251/29_3.pdf
[35] A.V. Lepeshkin, A.S. Medvedev, A.A. Mihajlin.Prospects for the use of pumping units driven by an autonomous internal combustion engine in hydraulic systems with a constant supply pressure.*Izvestiya MGTU MAMI*. 2019. No 2(40), pp. 41–47 (in Russ.).DOI: 10.31992/2074-0530-2019-40-2-41-47.

IOP Conf. Series: Materials Science and Engineering **779** (2020) 012039 doi:10.1088/1757-899X/779/1/012039

Method for leakage measurement in the recirculation path of a hermetic pump

P Chaburko[1,2] **and A Kuznetsov**[1]

[1]Bauman Moscow State Technical University

[2]E-mail: chaburko@bmstu.ru

Abstract. The article is devoted to the measurement of leakage in the recirculation path of a hermetic pump. The field of application of hermetic pumps is described. The mathematical model used in computer simulation and the mesh conditions are given. The results of the numerical modeling and the theoretical calculation are presented. The graph of the dependence of the leakage in the recirculation path on the pump head is shown. The scalar scenes of velocities and pressures in the c path are given.
Key words: hermetic pump, leak measurement.

Introduction

Nowadays, we need machines capable of pumping large volumes of chemically active liquids in large quantities. There are many different types of chemical pumps. Its main feature is to transfer fluids containing aggressive or explosive substances. This property is especially important for such industries as oil and gas, pharmaceutical, food and cosmetic industries. Examples of studies of these units are presented in[1–5].

Fig.1. Chemical pump.

Content from this work may be used under the terms of the Creative Commons Attribution 3.0 licence. Any further distribution of this work must maintain attribution to the author(s) and the title of the work, journal citation and DOI.
Published under licence by IOP Publishing Ltd

IOP Conf. Series: Materials Science and Engineering **779** (2020) 012039 doi:10.1088/1757-899X/779/1/012039

The most popular among pumping equipment is the centrifugal type of devices, and hermetic centrifugal pumps make up a significant part in the segment of chemical units (Figure 1), as a result, there are a large number of works on the development of this type of machine [6,7], which can provide complete isolation of the pumped fluid, protecting the environment from possible leaks of harmful mixtures, vapors of which can form explosive compositions with air.

A distinctive feature of a chemical centrifugal pump is impermeability, which is ensured in most cases by a magnetic coupling that requires the use of sliding bearings. Permanent fluid supply for lubrication and cooling purposes is required to maintain their performance. These conditions must be provided by recirculation paths (Fig. 2), which must be pre-calculated in order to procure the working of the pump. Otherwise, overheating in the coupling can cause the magnets to fail, thereby stopping the transmission of torque [8,9] and the bearings to lose their bearing capacity [10,11,12,13].

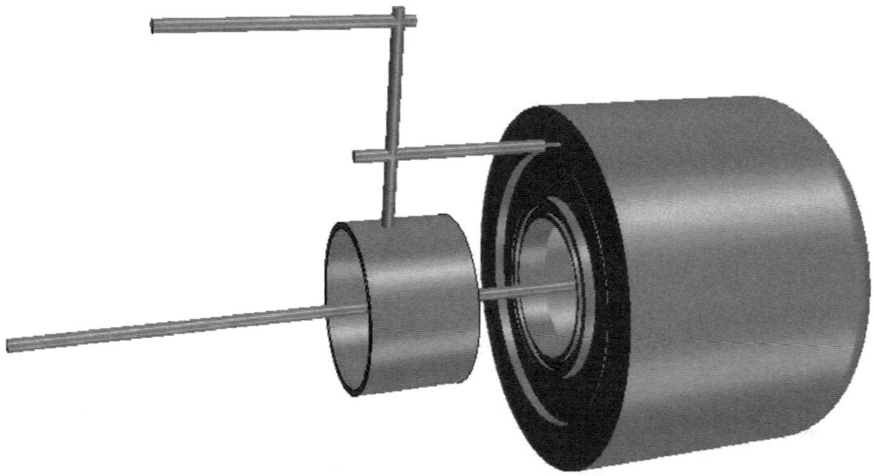

Fig. 2. Recirculation paths model.

The purpose of this work is with the help of previous studies [14] to calculate the leakage in the supply channels by means of numerical methods and theoretical calculation, also to compare the obtained results. In addition, it is necessary to reveal the dependence of the flow rate in the recirculation paths of a particular pump on its head.

1. Mathematical model and methods

Hydrodynamic simulation methods allow to determine the resistance coefficients of the channel elements without conducting tests or using inaccurate analytical dependencies. In this paper, the model of incompressible fluid [15,16] is used, and all calculations are based on discrete analogues of basic equations of hydrodynamics [17].

Mass conservation equation (continuity equation):

$$\frac{\partial \overline{u_x}}{\partial x} + \frac{\partial \overline{u_y}}{\partial y} + \frac{\partial \overline{u_z}}{\partial z} = 0,$$

where $\overline{u_i}$ — is the time averaged projections of fluid velocities on the corresponding axes;

Equation of the change in the amount of motion

$$\rho\left[\frac{\partial \overline{u_i}}{\partial t}+\overline{u_j}\frac{\partial \overline{u_i}}{\partial x_j}\right]=-\frac{\partial \overline{p}}{\partial x_i}+\frac{\partial}{\partial x_i}\left[T_{ij}^{(v)}-\rho u_i u_j\right],$$

where $\overline{u_i}\,\overline{p}$ — average speed and pressure;

$\tilde{T}_{ij}^{(v)}=2\mu\tilde{s}_{ij}$ —viscous stress tensor for incompressible liquids;

$\tilde{s}_{ij}=\frac{1}{2}\left[\frac{\partial \overline{u_i}}{\partial x_j}+\frac{\partial \overline{u_j}}{\partial x_i}\right]$ — strain rate tensor;

$\rho u_i u_j$ —Reynolds tension.

The system is closed using k-ω SST turbulence model [18,19,20]. In the wall area k-ω model is used, and in the flow kernel — k-ε model.

This model includes two equations of turbulence parameters transfer:

–the transfer equation of the turbulence kinetic energy:

$$\frac{\partial k}{\partial t}+\overline{u_j}\frac{\partial k}{\partial x_j}=P_k-\beta^* k\omega+\frac{\partial}{\partial x_i}\left[\left(\nu+\sigma_k\nu_T\right)\frac{\partial k}{\partial x_j}\right],$$

where $k=\frac{\left(u_x'^2\right)+\left(u_y'^2\right)+\left(u_z'^2\right)}{2}$ —the turbulence kinetic energy;

u_i' —velocity fluctuations;

P_k —a member of the turbulence energy generation;

ω —relative turbulence dissipation rate;

ν_T —turbulent viscosity.

— transfer equation of relative velocity of turbulence energy dissipation:

$$\frac{\partial \omega}{\partial t}+\overline{u_j}\frac{\partial \omega}{\partial x_j}=\alpha S^2-\beta\omega^2+\frac{\partial}{\partial x_i}\left[\left(\nu+\sigma_\omega\nu_T\right)\frac{\partial \omega}{\partial x_j}\right]+2\left(1-F_j\right)\sigma_{\omega 2}\frac{1}{\omega}\frac{\partial k}{\partial x_j}\frac{\partial \omega}{\partial x_j}$$

The Reynolds tensions in the dynamics equations are based on the Boussinesque hypothesis:

$$\rho u_i u_j=2\mu_T\left[\frac{1}{2}\left(\frac{\partial \overline{u_i}}{\partial x_j}+\frac{\partial \overline{u_j}}{\partial x_i}\right)-\frac{1}{3}\frac{\partial \overline{u_k}}{\partial x_k}\delta_{ij}\right]-\frac{2}{3}\rho k\delta_{ij},$$

where δ_{ij} —Kroneker symbol.

When using the empirical closing coefficients of these equations it is possible to obtain a numerical solution of the turbulent liquid flow in the calculation area.

The local resistance coefficient is determined by the formula

$$\Delta p=\zeta\rho\frac{V^2}{2}$$

The flow part of the recirculation paths was simulated with a mesh consisting of 2.7 million cells. In the core of the flow cell have a polyhedral shape, the solid walls of the pipe - prismatic. The design mesh for the recirculation paths is shown in Figure 3.

IOP Conf. Series: Materials Science and Engineering **779** (2020) 012039 doi:10.1088/1757-899X/779/1/012039

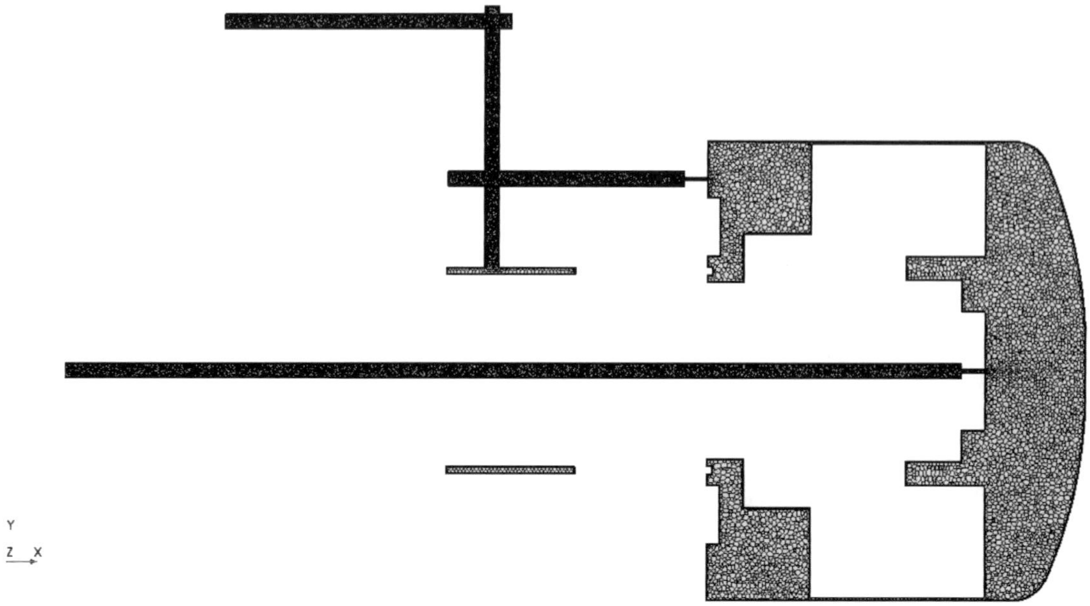

Fig.3. Mesh.

2. Results and analysis

Boundary conditions of calculation are outlet pressure (Po = 0 Pa) and stagnation at the inlet (wheel head H = 519341.4 Pa). In order to determine the dependence of the flow rate in the recirculation paths on the head, a calculation was made at different pump pressures. The dependence is shown in Figure 4.

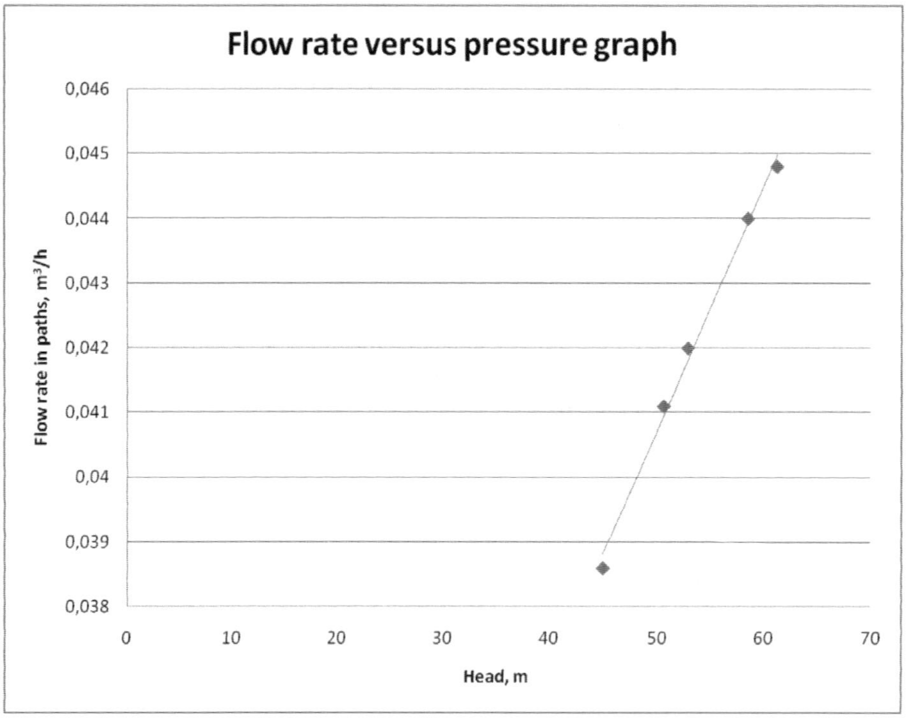

Fig. 4. Flow rate versus head graph.

The flow rate at nominal mode is Q= 0,042m^3/h.
Figures 5.6 show the scene of pressure and velocity distribution in the recirculation channels.

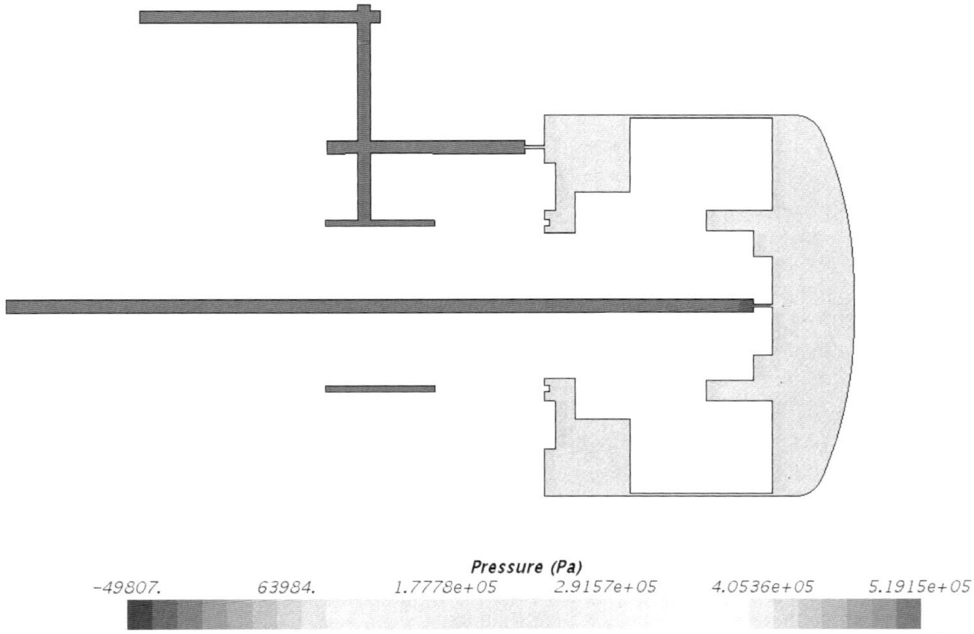

Fig.5. Pressure distribution scene in recirculation paths.

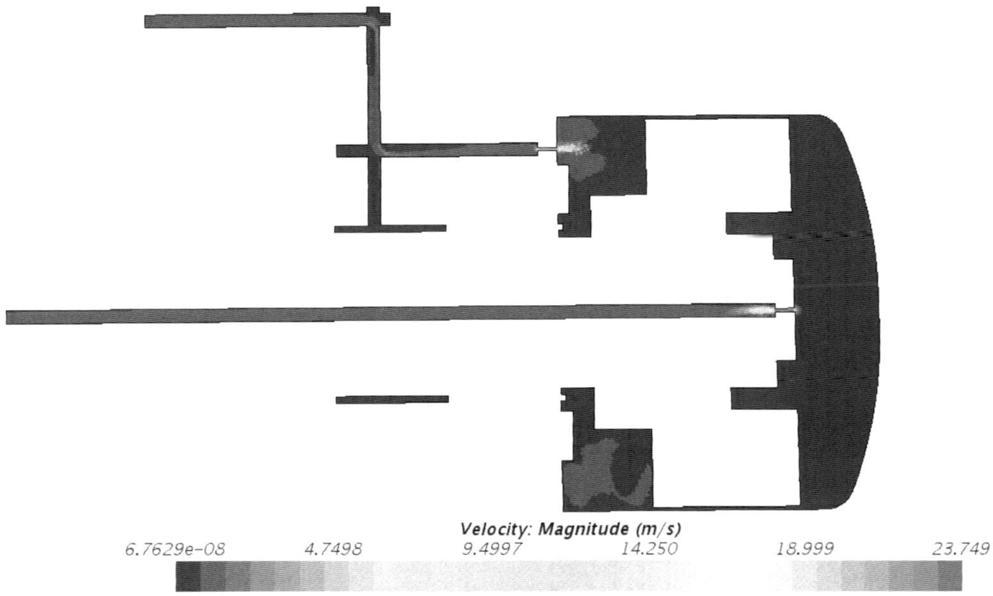

Fig.6. Velocity distribution scene in recirculation paths.

Figure 7 shows the scheme of recirculation paths, which is necessary for theoretical calculation.

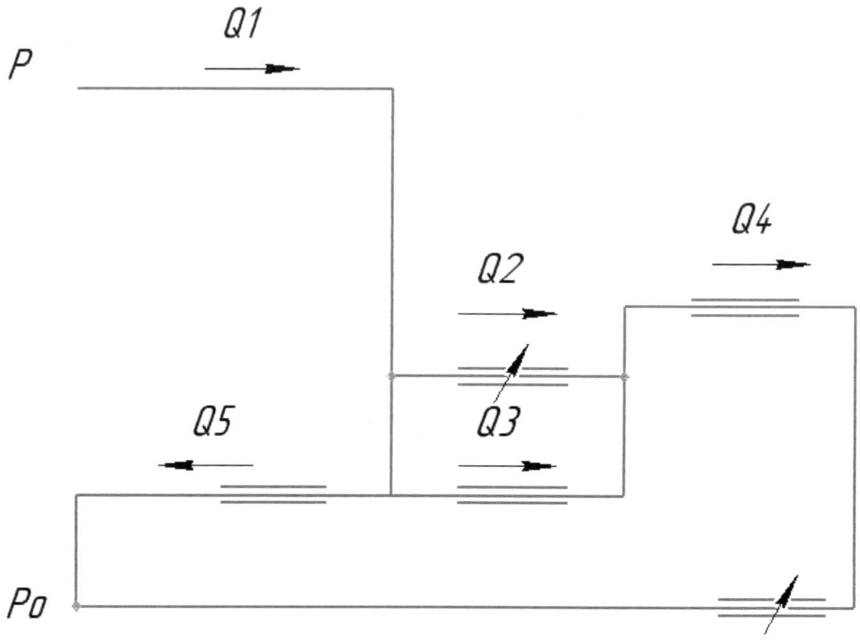

Fig.7. Scheme of recirculation paths

The system for calculating offlow rate.

$$\begin{cases} Q_1 = Q_5 + Q_4 \\ p - p_0 = k_p \cdot Q_5^2 \\ p_2 - p_0 = k_d \cdot Q_4^2 \\ p_1 - p_2 = k_m \cdot Q_4^2 \\ p - p_1 = k \cdot Q_4^2 \end{cases}$$

where k_p —resistance of bearings, k_m — resistance of magnet coupling, k_d —resistance of throttle, k — total throttle and bearing resistance.

As a result of the theoretical calculation, the flow rate through the channels isQ_1= 0,0804 m^3/h. From the above results it can be seen that the experimental values are 2 times less than the theoretical values. Thus, the real values of the volume efficiency will be higher than the calculated one.

3. Conclusions

According to the results of the conducted calculations, the dependence of the flow rate in the recirculation paths on the pump head was obtained, which allows to draw conclusions about its working capacity on different operating modes. A theoretical calculation was also carried out, the results values of which differ from the numerical modeling ones. This is explained by the fact that the theoretical method did not take into account the influence of flow rotation, friction of channels and changes in their area.

References.

[1] A Petrov, A Morozov, M Kuleshova. Investigation of hermetic low-speed pumps performance when operating on highly viscous liquids. IOP Conference Series: Materials Science and Engineering, Volume 589, Issue 1, 16 August 2019, 012030. DOI: 10.1088/1757-899X/589/1/012030

[2] Posch, S., Berger, E., Heimel, M., Hopfgartner, J., Almbauer, R., Schöllauf, P. Fluid Flow in the Oil Pumping System of a Hermetic Compressor. IOP Conference Series: Materials Science and Engineering, Volume 90, Issue 1, 30 July 2015, 012028. DOI: 10.1088/1757-899X/90/1/012028

[3] Posch, S., Hopfgartner, J., Berger, E., Zuber, B., Almbauer, R., Schöllauf, P. Determination of the oil distribution in a hermetic compressor using numerical simulation. IOP Conference Series: Materials Science and Engineering, Volume 232, Issue 1, 6 September 2017, 012009. DOI: 10.1088/1757-899X/232/1/012009

[4] Braga, V.M., Barbosa, J.R., Deschamps, C.J. Numerical investigation of refrigerant outgassing in the screw pump of a hermetic reciprocating compressor oil supply system. IOP Conference Series: Materials Science and Engineering, Volume 604, Issue 1, 3 September 2019, 011009. DOI: 10.1088/1757-899X/604/1/011009

[5] Mosshammer, M., Benigni, H., Jaberg, H., Krämer, S., Dahlke, C. Multistage canned motor pump — Getting deep insights with CFD — Simulation, optimization and validation. IOP Conference Series: Earth and Environmental Science, Volume 240, Issue 3, 28 March 2019, 032004. DOI: 10.1088/1755-1315/240/3/032004

[6] Martynyuk, A., Petrov, A., Kuleshova, M. Comparative analysis of the use of the spiral and bladed stators in a compact high-speed centrifugal pump with hydrodynamic modeling methods. IOP Conference Series: Materials Science and Engineering. Volume 589, Issue 1, 16 August 2019, 012029. DOI: 10.1088/1757-899X/589/1/012029

[7] Chaburko, P., Kossova, Z. Wet part hybrid optimization method of hermetic pump. IOP Conference Series: Materials Science and Engineering, Volume 492, Issue 1, 13 March 2019, 012011. DOI: 10.1088/1757-899X/492/1/012011

[8] Shibata, T., Fukushima, H., Segewa, K. Improvement of steam turbine stage efficiency by controlling rotor shroud leakage flows - Part I: Design concept and typical performance of a swirl breaker. Journal of Engineering for Gas Turbines and Power. Volume 141, Issue 4, 1 April 2019. DOI: 10.1115/1.4041650

[9] Yong, S. Analysis of Cooling Performance of a Marine Air-cooled Sliding Intermediate Bearing Based on Fluid-Structure Interaction. IOP Conference Series: Materials Science and Engineering, Volume 361, Issue 1, 4 June 2018, 012017. DOI: 10.1088/1757-899X/361/1/012017

[10] Ke, Z., Zinan, W., Qingyuan, W. The characteristics of studying high speed motorized spindle dynamic magnetic coupled. IOP Conference Series: Materials Science and Engineering, Volume 399, Issue 1, 2018, 012027. DOI: 10.1088/1757-899X/399/1/012027

[11] Li, Y., Liu, J., Li, L., Yang, D. Effect of environmental pressure on dynamic characteristics of aerodynamic gas bearing of boost turbo expander. IOP Conference Series: Materials Science and Engineering, Volume 382, Issue 3, 13 July 2018, 032018. DOI: 10.1088/1757-899X/382/3/032018

[12] Zhao, L.J., Nguyen, T.T., Zhang, B., Cui, X., Zhang, H. Study on lubrication performance of heavy duty bearing of shearer based on fluid-solid coupling. Journal of Physics: Conference Series, Volume 1168, Issue 2, 12 March 2019, 022092. DOI: 10.1088/1742-6596/1168/2/022092

[13] Javorova, J., Alexandrov, V. Effects of fluid inertia and bearing flexibility on the performance of finite length journal bearing. IOP Conference Series: Materials Science and Engineering, Volume 174, Issue 1, 1 March 2017, 012039. DOI: 10.1088/1757-899X/174/1/012039

[14] Fedorov, S.V., Bolotina, I.A., Strukov, Yu.A. Thermal effects during inertial stretching of high-gradient conductive rod with "frozen" magnetic field. Herald of the Bauman Moscow State Technical University, Series Natural Sciences. Issue 2, 2018, Pages 39–59. DOI: 10.18698/1812-3368-2018-2-39-59

[15] Isaev, N. The effect of design parameters of the closed type regenerative pump the energy characteristics. IOP Conference Series: Materials Science and Engineering, Volume 492, Issue 1, 13 March 2019, 012026. DOI: 10.1088/1757-899X/492/1/012026
[16] Protopopov, A., Vigovskij, V. Development of methods for high-speed centrifugal pump analysis. IOP Conference Series: Materials Science and Engineering, Volume 492, Issue 1, 13 March 2019, 012003. DOI: 10.1088/1757-899X/492/1/012003
[17] LOYTSYANSKIY,L. Mechanics of Fluid and Gas Textbook for HEIs. — 7th edition, Prosp. — Moscow: Drofa, 2003.
[18] Menter F. R., Kuntz M., Langtry R.B. Ten Years of Industrial Experience with the SSTTurbulence Model, 2003.
[19] Matyushenko, A.A., Garbaruk, A.V. Adjustment of the k-ω SST turbulence model for prediction of airfoil characteristics near stall. Journal of Physics: Conference Series, Volume 769, Issue 1, 23 November 2016, 012082. DOI: 10.1088/1742-6596/769/1/012082
[20] Driver D.M. Reynolds shear stress measurements in a separated boundary layer flow. AIAA91-1787, 1991

IOP Conf. Series: Materials Science and Engineering **779** (2020) 012040 doi:10.1088/1757-899X/779/1/012040

Development and application of vortex control valves in pneumahydraulic systems

A Uss[1,3], A Pugachuk[1,2] and A Chernyshov[1]

[1]Bauman Moscow State Technical University
[2]United Institute of High Temperatures of Russian Academy of Sciences, Moscow, Russia

[3]E-mail: uss25.ru@mail.ru

Annotation: The current state of the development and study of vortex jet devices that are used as control valves based on the working medium vortex flow principle is reviewed and analyzed in the article. The scientific and technical literature is reviewed. The article considers vortex valves without mechanically moving parts, as well as a semi-mechanical vortex valve. The principle of operation of a vortex jet device is described. The rationale of the use of a vortex jet device as an actuator in a valve for controlling the flow of a working medium, as well as in a control system of a sleeve-valve cylinder, is given. The classification of vortex jet devices by the feed type of the supply and control flows of the working medium is carried out for the first time.

Introduction. Actuality

Nowadays, devices designed to control the flow of the working medium (WM) are widely used in many areas of practical human activity. The working medium is to be understood as compressible and not compressible Newtonian viscous fluids. This class of devices is usually called block-and-regulating devices (BRD). BRD include valves, faucet, cutoff plates, butterfly dampers, pressure or flow WM regulators, etc. This paper will tell about vortex valve as one of the crucial elements in a pneumohydraulic system. Being an automatic device the valve includes such basic elements as actuating unit and drive component. Such type of valves refers to mechanical valves. The object under consideration is the regenerative stage with a pressure pipe and a separator.

Valves largely determine the reliability and faultless performance of the pneumohydraulic systems in which they operate. First of all this can be explained due to the presence of elements, that move at high speeds and collide with parts and components of the configuration. The moving elements of the valve are exposed to non-linear gas-static and gas-dynamic loads. The valve parts are in interaction with the flow of the working fluid with a high speed under the influence of which a partial and in certain cases complete destruction of the sealing element is possible. The use of mechanical valves in systems with contaminated, aggressive and/or high-temperature mediums is extremely difficult and commonly simply impossible.

Regulating and shut-off valves constitute a significant part of the controlling the multiphase flow equipment in modern industry, power plants, mining, in the pneumohydraulic sections of various mechanisms, etc. They can be used to maintain the given flow rates at a constant high resistance as well as to change the resistance and flow rate by moving or turning the element of the final adjustment of the system in accordance with a given program.

Content from this work may be used under the terms of the Creative Commons Attribution 3.0 licence. Any further distribution of this work must maintain attribution to the author(s) and the title of the work, journal citation and DOI.
Published under licence by IOP Publishing Ltd

IOP Conf. Series: Materials Science and Engineering **779** (2020) 012040 doi:10.1088/1757-899X/779/1/012040

It is possible to increase the reliability of the BRD by creating construction designs without moving elements. The operating principle of which is based solely on gas-dynamic effects such as the Coanda effect, turbulization of the WM flow as well as the use of the swirl flow of the WM. Devices that use the listed processes and effects to control the WM flow are called "vortex jet devices" and can be used as an actuation device. The classical scheme of the vortex jet device is shown in (Fig. 1a).

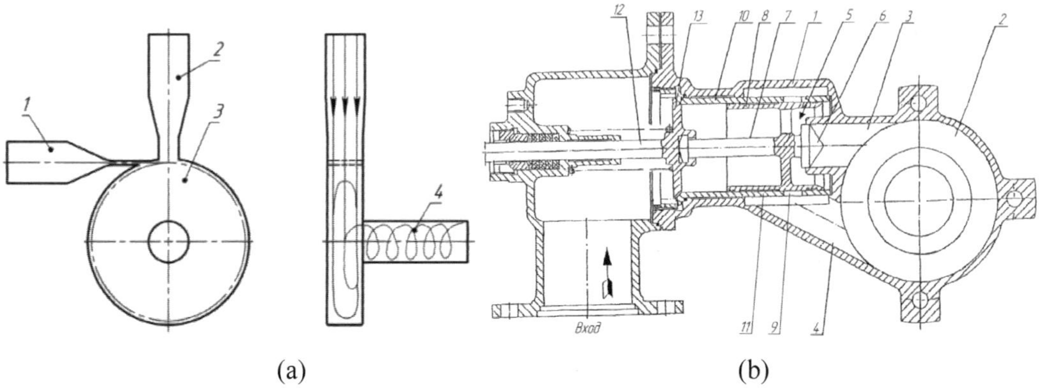

(a) (b)

Figure 1. (a) — vortex jet device switching circuit; (b) — application of a vortex jet device in a semi-mechanical vortex control valve.

The operating principle of the vortex jet device (hereinafter VJD) shown in Figure 1a is as follows: the power flow enters the channel 2, then into the vortex chamber 3, fills it and passes to the outlet pipe 4. If it is necessary to lower the pressure/flow rate in the outlet pipe 4 an auxiliary stream — control stream must be supplied to the control channel 1. As a result of the interaction between the supply and control jets, they merge so that the mass of the resulting stream represents the sum of the masses of the individual jets. The direction of the resulting flow is determined by the geometric sum of the number of motions of the colliding flows.

Bugaenko V.F. in [1] suggested to introduce the following notation: α — angle of deviation of the supply stream, β — angle between the axes of the supply and control channels $I_1 = F_1 \upsilon_1^2 Q$ — momentum of the force of the control flows, $I_0 = F_0 \upsilon_0^2 Q$— momentum of the power of the supply flow.

Then the equation describing the deviation of the resulting flow can be represented as

$$\alpha = \eta arctg \frac{(I_1 / I_0)\sin\beta}{1+(I_1 / I_0)\cos\beta}, \quad (1)$$

where η — deviation coefficient of the resulting jet from the original direction of the supply jet. It is known that for vortex controllers having channels with a circular bore $\eta \approx 0.8$ from the scientific and technical literature. Formula (1) is based on the assumption that the static pressure is constant in the field of interaction of the jets and that the velocity distribution over the cross section is uniform in the supply channels and in the resulting stream.

After the interaction of the supply and control flows the resulting flow under the influence of the Coanda effect adheres to the cylindrical wall of the vortex chamber. As a result of this, a vortex flow of the working medium is formed in the vortex chamber. The main principle of reducing the pressure in the outlet pipe is to create significant hydraulic resistance during the spiral movement of the working medium from the periphery of the vortex chamber to the center. The resistance is due to the vortex flow, which results in the formation of a field of centrifugal forces acting on the rotating mass of the working medium. Under the influence of centrifugal forces, the working medium is discarded to the periphery of the vortex chamber. As a result of this, a pressure drop is created: in the center of the vortex — reduced pressure; at the periphery of the vortex — increased pressure. At that moment, when the pressure at the periphery of the vortex chamber becomes equal to the pressure in the power channel, the flow from the power channel stops. This effect is called the effect of blocking the supply flow in the vortex jet device.

Based on the principle of operation of the VJD, we can conclude that the main purpose of the control flow is to reject the supply flow by 90 ° to form a vortex flow of the working medium in the vortex chamber of the VJD.

Analyzing the formula (1), we can conclude that to deviate the supply flow, the control flow should have a larger momentum of force compared to the momentum of the supply flow.

The deviation of the power flow is constructively provided as follows:

1. by making the design of the passage section of the control channel smaller than the passage section of the power channel, whereas it is necessary to increase the pressure in the control channel to increase the momentum of the control flow force;
2. by making the design of the control and supply channels with equal through sections, while the pressure and flow rate in the control channel remain equal to the pressure and flow rate in the power channel. In this configuration, the blocking effect is not possible, and the vortex jet device will act as a vortex choke. The considered concept was implemented in [2,3].

The relevance of the development of control devices using the principle of the vortex flow of the working medium is confirmed by many developments of similar products in various fields of technology. Let us consider the most significant developments that use the vortex valve in hydraulics.

The analysis of existing problems in hydraulic control equipment for multiphase transport pipelines of high pressure and flow shows that the development of a new valve capable of providing high hydraulic resistance (about 30-40 MPa and even more if necessary) is currently a serious task. All valves designed for high pressures are subjected to cavitation, severe erosion and, as a result, shortened device service life.

Throttling of the fluid flow at large pressure drops is usually accompanied by cavitation. The fluid passing through the throttle element of the control valve experiences changes in speed, since it enters a narrow constriction of the valve body (increase in speed) and then enters the expanding region of the valve body downstream after the throttle section (decrease in speed). These changes in velocity likewise lead to a change in the kinetic energy of the liquid molecules. To save energy in a moving fluid stream, any increase in kinetic energy due to an increase in velocity is accompanied by an additional decrease in potential energy, usually in the form of a liquid pressure. It indicates that the fluid pressure decreases at the point of maximum restriction in the valve, then again increases (or recovers) downstream after the throttling section. As long as the working medium is not a compressible liquid, and the pressure at the point of maximum narrowing is less than the saturated vapors pressure of this liquid at the flow temperature, the liquid will spontaneously boil. If the pressure is restored to a point above the vapor pressure of the liquid, the vapor will again condense into the liquid. This phenomenon is called cavitation. In case of cavitation appears close to the surface of the part, erosion may occur, what ultimately will lead to the leakage of the valve assembly, and subsequently to the destruction of the entire structure.

One of the technical solutions that can effectively deal with cavitation are valves based on the principle of the vortex flow of the working medium (Fig. 1b). The vortex valve has improved cavitation characteristics, is easy to manufacture. It can operate reliably in an almost unlimited range of pressure drops and high temperatures.

Vortex valves typically have at least one radial inlet channel and one or more tangential control channels. The supply pressure should be kept constant for optimal modulation of the flow rate at the outlet, and the control pressure should be 1.2-2.0 times higher than the supply pressure at the minimum flow rate.

New vortex valves in which the interaction of opposite tangential flows is used appeared to eliminate the above (Fig. 1b). The vortex movement allows creating a pressure gradient inside the vortex chamber, because of which the gas bubbles are displaced to the center of the vortex flow (at a distance from the structure wall). That excludes direct contact of pulsating and collapsing bubbles with the valve working surfaces, prevents the formation of erosion and thereby prolongs the service life devices.

The flow is not twisted and its resistance is minimal at the same flow rates through the tangential channels of the vortex chamber. When closing the supply pipe to one of the channels, the flow swirl increases, as well as the hydraulic resistance to the flow passing through the valve.

Constructively the valve comprises a housing 1 with a vortex chamber 2, inside which tangential channels 3 and 4 create opposite flows. The housing contains a sleeve 10 with via-holes 9 for the passage of the working fluid from the inlet chamber to the tangential channel 4. The regulating element 5 is installed in the housing coaxially to the channel 3. Element 5 consists of a cylindrical shell 8 and a central shaft 7 with a shaped head 6 interconnected by ribs 11. The rod of the regulatory element is connected to the drive shaft 12.

When the control element is moved by the drive shaft 12, the channels 9 begin to close, which ceases the equality of flows through channels 3 and 4 and creates a rotating flow in the vortex chamber, thereby increasing the hydraulic resistance of the valve. When all the supply channels are closed, the further movement of the regulating element brings its contour head into interaction with the entrance to the channel 3, further increasing the valve resistance. The required valve flow characteristic is ensured by designing via-holes in sleeve 10 and stem head 7.

Thereby, the operation of such a vortex valve does not require a control flow with a higher pressure level. The distribution element is installed on the pipeline through which not more than half the flow rate of the working fluid is supplied.

Vortex jet devices can be used to control the thermodynamic parameters of high-energy WM flows, which include high enthalpic WM flotation, as well as high-flow WM flows.

A vortex valve scheme with a distributed supply of feed and a concentrated supply of working medium control flows was proposed in [4] (Fig. 2). Constructively the valve consists of a test unit, a control chamber, two central bodies, two receiving nozzles. The vortex chamber of the valve is formed by the following surfaces. The end walls of the vortex chamber are formed by the end surfaces of the central bodies, the cylindrical wall of the vortex chamber is formed by the cylindrical surface of the control chamber. The vortex valve principle of operation is following. The working power medium enters the cavity of the supply flow, then it is distributively feed through the annular gap between the central body and the control chamber. The control flow is supplied to the control collector, then through the tangentially located control channels into the vortex chamber, where it interacts with the supply flow. As a result of this, a vortex flow of the working medium is formed in the vortex chamber of the valve. As a result of the formation of a vortex flow at the exit of the vortex chamber, the torch is defocused, thus the working medium moves along the cone, surrounds the "receiving nozzle" part and enters the discharge channel. In the absence of a supply stream, a vortex flow is not formed and the working medium when it flows out of the vortex chamber represents itself a flooded stream, thus entering the receiving nozzle and then into the outlet pipe.

The vortex valve developed in paper [4] is proposed to be used in the control system of the hydraulic cylinder. Conceptually the hydraulic cylinder control scheme (Fig. 3) consists of a pilot stage representing itself an electric drive with a flapper-nozzle, which provides flow control for two vortex valves I and II. These valves are part of the hydraulic cylinder control power stage. Power stage vortex valves are configured with a hydraulic cylinder.

The servo valve hydraulic circuit diagram is shown in Fig. 3. The supply flow from the flow driver doubles and enters the power channels of the vortex valves I and II. The pilot stage flapper-nozzle operates as follows. The control flow from the flow driver enters the working cavity of the damper nozzle. When the shutter moves in the direction 1, the left nozzle closes and the right nozzle opens, thereby the fluid enters the control channel of the vortex valve II, locking it, while the vortex valve I opens. Thus, the liquid with higher pressure enters the cavity 1 'of the hydraulic cylinder, the pressure drop acting on the piston induces a force acting in the direction from the cavity with higher pressure to the cavity with lower pressure, as a result of which the piston with the rod moves to the right. Fluid is forced from the right cavity of the hydraulic cylinder to the discharge channel of the vortex valve II. In a similar manner, the piston moves with the rod to the left.

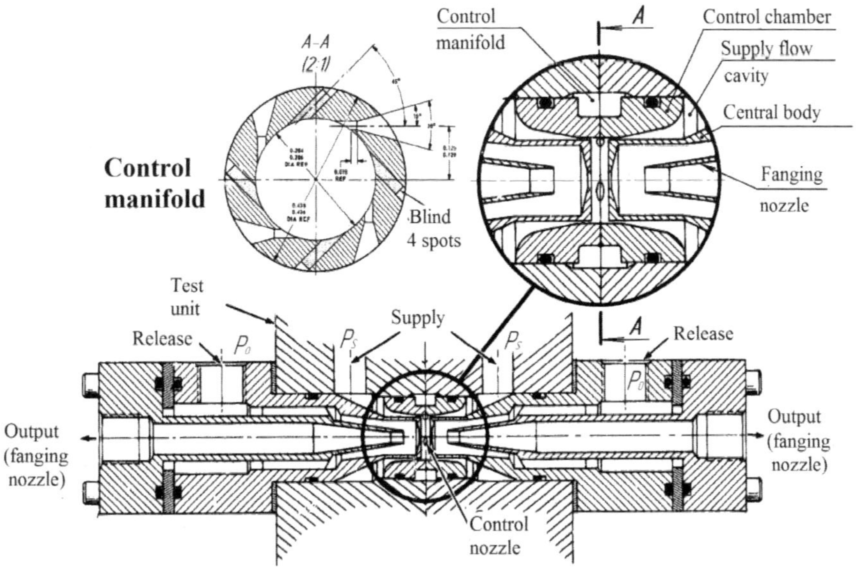

Figure 2. Vortex servo valve with distributed feed and concentrated feed of working medium control flows.

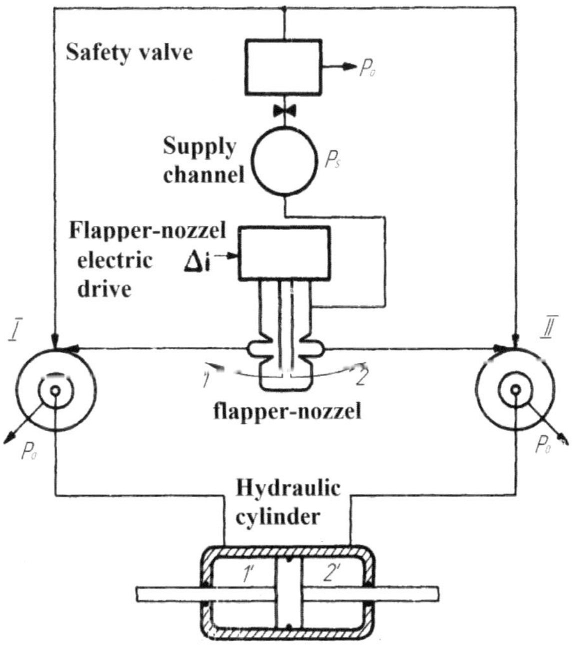

Figure 3. The scheme of the hydraulic cylinder control by means of vortex valves.

The design of a vortex valve with a concentrated feed of the supply and a distributed feed of the control flows of the working medium was proposed in [5] (Fig. 4).

A circuit diagram of the proposed vortex valve is shown in Figure 4. The device contains a vortex chamber 1 with tangential channels or holes for the control flow input and nozzle at the outlet of the

IOP Conf. Series: Materials Science and Engineering **779** (2020) 012040 doi:10.1088/1757-899X/779/1/012040

common cavity. The adjustable flow is supplied to the sleeve 2 and through a series of radial channels 4 enters the common cavity, where it interacts with the control flow before exiting the device. The diameter of the nozzle of the vortex chamber is designed large to approach the vortex control flow to the exit from the radial channels and rod 3 is installed to maintain nozzle cross section.

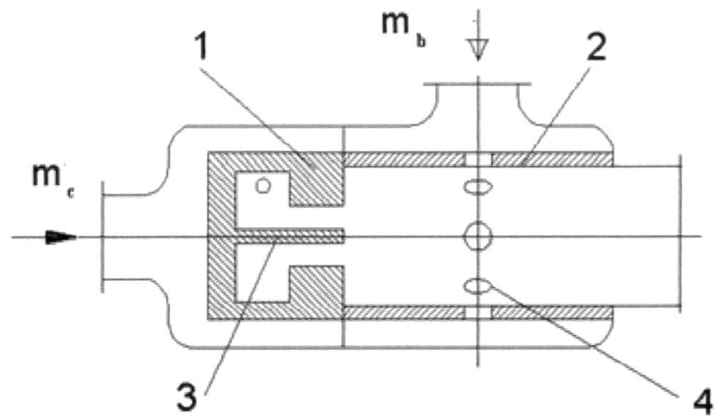

Figure 4. Scheme of a vortex valve with a concentrated supply and distributed feed of control flows of the working medium.

The reviewed vortex valves designs can be used to control the flow of a compressible fluid as well as to control the thermodynamic parameters of high-energy flows of WM, which include high-enthalpy flows of WM, and flows of WM with high pressure and flow-rate.

The author of this work proposed the design of a vortex valve [6] with a distributed feed of supply and control flows of the working medium (Fig. 4).

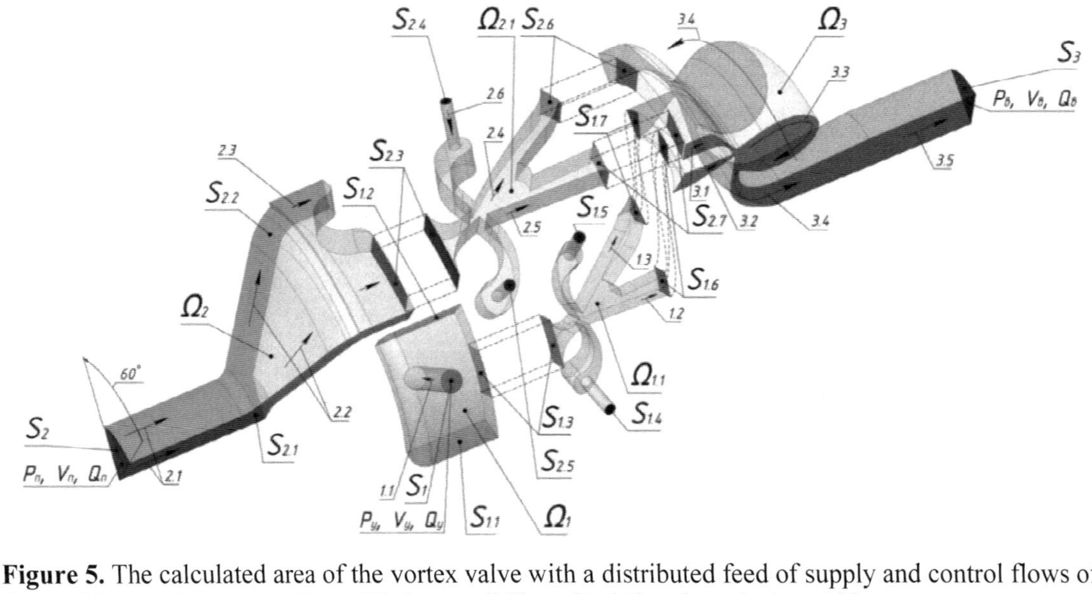

Figure 5. The calculated area of the vortex valve with a distributed feed of supply and control flows of the working medium, as well as with the possibility of twisting them (cutout 5/6).

The operating principle of the gas pressure regulator with a distributed feed of the supply control flow of the working gas and with the twist control of the supply and control flows, as well as the operating principle of the pressure regulator with the concentrated feed of control and supply flows, is based on the principle of interaction of the jets. However, in this design, the power supply 2.1 through section S_2 enters the supply pipe, then flowing around the central body enters the collector of six triggers $\Omega_{2.1}$. Arrows 2.1-2.2-2.3 indicate the gas flow in the supply channel. Trigger $\Omega_{2.1}$ works as follows: when the working fluid is supplied to section $S_{2.4}$ (under pressure equal to the supply pressure), flows 2.3 and 2.6 interact, and as a result, the resulting stream moves in the radial direction 2.5. The working medium (under pressure equal to the supply pressure) is supplied to section $S_{2.5}$ to deviate the feed stream in the tangential direction. As a consequence of this, the flows 2.3 and 2.5 interact and the resulting stream 2.4 moves in the tangential direction. Hereafter the working stream 3.1 is fed into a vortex toroidal chamber, where it interacts with the control flow 3.2. The control flow, as well as the supply flow, can be pre-twisted with the help of a collector of six triggers $\Omega_{1.1}$ on the control channel. The operating principle of the control collector is similar to the principle of the collector installed on the power supply. To feed the control flow 1.1 in the radial direction, the working fluid (under pressure equal to the control pressure) is fed into section $S_{1.5}$, to feed the control flow in the tangential direction; the working fluid is fed into section $S_{1.4}$. A complex spiral-like flow of the working fluid is formed in the interaction of the supply 3.1 and the control 3.2 flows of the working medium in the vortex toroidal chamber. As a result of it a pressure gradient is created in the vortex chamber: the minimum pressure at the center of the vortex and the maximum at the periphery of the vortex. The pressure of the working fluid at the periphery of the vortex if reached equal to the supply pressure, the decrease in the power consumption up to its closure appears.

The mathematical relations of the output parameters (pressure and flow rate at the device outlet) on the control parameters (pressure and flow rate in the control channel), as well as the influence of changes in the geometric parameters of the flow cavity (vortex chamber diameter D, vortex chamber width H, geometric parameters of the power and control channels their number and location) on the change in the value of thermodynamic parameters are described in [7–9]. The presented mathematical models were based on empirical dependences, in other words on the data of experimental studies. That practically can only be applied in a narrow range of thermodynamic parameters of WM and at the initial stage of calculation of similar structures.

Modern studies of the article's authors on the VJD: a patent review, a scientific and technical literature review, a design of vortex valves review, identifying of the advantages and disadvantages, a numerical study of gas-dynamic processes in the flow chamber of the VJD in distributed gas-dynamic parameters, an investigation of the effect of locking the feed channel as well as creating of a design, a series of experimental studies and verification of the mathematical model of the workflow are described in [10–13].

Classification of vortex valves according to the type of feed of the supply and control flows of the working medium.

As a review of the state of the issue on the development of vortex valve designs shows, currently there are many designs of the supply and control flows feed to the vortex chamber. That leads to a variety of vortex valve designs. Relative to it there is a need to classify structures according to the type of feed of the supply and control flows of the working environment.

By the type of feed of the working medium is meant the organization of the supply of the working medium to the vortex chamber of the vortex jet device. Concentrated feed - when the working medium is supplied through the channels in the form of nozzles.Distributed feed - when the working medium is fed through annular channels around the entire perimeter of the vortex chamber.

Vortex jet devices in general can be represented by the type of feed of the working medium in the form of a matrix of possible design options: the columns of which display channels (supply and control), and the line type of feed: concentrated and distributed. The numbers indicate the intersection of rows and columns.

Feed / Flow	Supply	Control
Concentrated	1	2
Distributed	3	4

- 1-2 with concentrated feed of the supply and control flow of the working medium. The number of power and control channels located around the perimeter of the vortex chamber can be different 1, 2 ... n supply and control channels. An example of such constructions can be found in [7,8];
- 3-2 with distributed feed and concentrated feed of the control flow of the working medium. There are known vortex jet devices [4], where the feed of the supply flow is organized in a distributed manner (the working medium flows around the central body and through the annular channel enters the vortex chamber), and the control flow is concentrated through the nozzles located around the perimeter of the vortex chamber.
- 1-4 with concentrated feed of the supply flow and distributed feed of the control flow [5].
- 3-4 with a distributed feed of supply and control flows of the working medium [6].

Conclusion

An analysis of the research conducted in this area and a review of existing technical solutions allows us to formulate the following conclusions:

- designs of control valves using the principle of the vortex flow of the working medium are considered. Their widespread application confirms the relevance of the development of such devices, and the legitimacy of their use to solve technical problems;
- analysis of technical solutions showed the advantage of vortex jet devices (high reliability, the ability to work with various working fluids, including working fluids with abrasive inclusions, low manufacturing costs, low operating costs, immunity to radiation and electromagnetic fields), which is why it is recommended to use a vortex jet device as an actuator in valves;
- the first classification of vortex jet devices according to the type of feed of the supply and control flows of the working medium;
- the article is a scientific source for further work on the development and study of vortex control valves [14–21].

References

[1] V. F. Bugaenko, Pneumatic automation of space rocket systems (Moscow, 1979), 168 p.
[2] On the development of a regulating valve design with improved cavitational characteristics Michael P. Levitsky, and Semyon P. Levitsky, 2006.
[3] CONTROL VORTEX VALVE, WO 02/50456, PCT/IL01/00666.
[4] Hydraulic Fluid Interaction Servovalve T.A. Phillips and A. Blatter 1968
[5] On a Decision of the Jet Flow Control Michael P. Levitsky, and I. M. Levitskaya
[6] A.Yu. Uss, N.V. Atamasov, A.V. Chernyshev. "Development of the Calculation Method and Designing of a Vortex Jet Device for Gas Flow Regulation Purposes", AIP Conference Proceedings, (2019); doi: 10.1063/1.5122078.
[7] L. A. Zalmanzon, Pneumatic element's theory (Nauka, Moscow,1969), 508 p.
[8] I. V. Lebedev, S. L. Treskunov and V. S. Yakovenko, Elements of jet automation (Moscow, 1973), pp. 289–314
[9] Works MSTU №244. Research and calculation of inkjet elements and chains of automatic control systems. 1977.
[10] A. Yu. Uss and A.V. Chernyshev, "The Development of the Vortex Gas Pressure Regulator", ProcediaEngineering ,(Moscow, 2016), vol.152. pp. 380–388
[11] A. Yu. Uss, A. V. Chernyshyov, and V.I. Krylov. "Development of Gas Pressure Vortex Regulator", AIP Conference Proceedings 1876, 020025 (2017); doi: 10.1063/1.4998845.
[12] O. V. Belova, A. A. Starodubcev and A. V. Chernyshev. "Calculation of the vortex gas pressure regulator", Engineering Journal: Science and Innovation, (Moscow,2013),vol. 5. URL: http://engbul.bmstu.ru/doc/740398.html
[13] O. V. Belova, A. A. Starodubcev and A. V. Chernyshev, "Vortex gas pressure regulator" in Engineering Magazine: Engineering Gazette, 2014, vol. 10.
URL: http://engjournal.ru/articles/760/760.pdf

[14] Y.A. Grishin, V.A. Zenkin and R.N.Khmelev, "Boundary conditions for numerical calculation of gas exchange in piston engines", Journal of Engineering Physics and Thermophysics (Moscow,2016), vol.90 No.4, pp.965–970
[15] A.I. Leont'ev, Y.A. Kuzma-Kichta and I.A. Popov, "Heat and mass transfer and hydrodynamics in swirling flows (review)", Thermal Engineering, (Moscow,2017), vol.64, No. 2, pp.111–126
[16] D.A.Zhidkov, V.A. Denisov and O.V. Kirikova, " Detailed consideration of the shock-wave (pulsation) concept of the Ranque-Hilsch vortex effect", Theoretical foundations of Chemical Engineering, (Moscow,2018) vol.52, No. 4, pp. 751–759
[17] X.Huang, W.Cheng, W. Zhong and X. Li, " Development of new pressure regulator with flowrate-amplification using vacuum ejector", Vacuum,(China,Zhenjiang,2017), vol.144, pp. 172–182.
[18] O.V. Mitrofanova, G.D. Podzorov and I.G. Pozdeeva, "Vortex structure of swirl flows", International journal of heat and mass transfer, (Moscow, 2013), vol.65, pp.225–234
[19] J. Chen, M. Veenstra, J. Purewal, B. Hobein and S. Papasavva, "Modeling a hydrogen pressure regulator in a fuel cell system with Joule-Thomson effect", International journal of hydrogen energy, No. 2, pp. 1272–1287
[20] T. Kato, K. Kawashima, T. Funaki, K. Tadano and T. Kagawa, "A new, high precision, quick response pressure regulator for active control of pneumatic vibration isolation table", Presicion engineering, 2010, vol.34, pp.43–48
[21] I.R. Chinyaev, A.V. Fominykh and E.A. Ilinykh, "The valve is a shutoff for the passive protection systems of pipelines", Procedia engineering, (Kurgan,2016), vol. 150, pp. 220–224

IOP Conf. Series: Materials Science and Engineering **779** (2020) 012041 doi:10.1088/1757-899X/779/1/012041

Design of an optimal hydraulic tank configuration

N Belov[1,2] and N Sosnovsky[1]

[1]Bauman Moscow State Technical University

[2]E-mail: ch5n@mail.ru

Abstract. The fluid behavior in the tank of the cutting machine hydrostatic actuator is considered. The hydrodynamic simulation technique of such processes is described. The velocity magnitudes, temperature distribution fields and fluid flow lines trajectories in the tank volume analyzed. The tank configuration changes that allow to an increase its operational efficiency as the part of the hydraulic actuator are proposed. Recommendations for selecting tank optimal geometric dimensions in accordance with the criterion for the most even fluid flow rates distribution in the tank entire volume are presented.

Introduction

The tank is as a rule a thin-wall reservoir in the hydraulic system. This is a container which main tasks are storage and maintain the hydraulic actuator oil properties. Its functions efficiency depends from many factors, such as the tank cavity geometric shape, material, surface coating, specific devices presence and much more. Cases are common when the tank design is not optimal in terms of the performed functions efficiency due to the calculation methods lack or the desire to oversimplify the tank design and manufacturing progress. Such phenomenon can be much often traced in units that have been manufactured long ago but still being operated. Possible problems anticipation methods at the design stage and timely measures for its elimination are considered in this study. When upgrading the hydraulic system of a previously used but well-proven unit in order to achieve for example higher performance there are often carried out some adjustments only to the system structure and no one for the hydraulic tank so the considering question is relevant at present time. Also the similar situation occurs in case the hydraulic stations development for various reasons. It is an important to take allowance for the tank significant impact on the overall hydrostatic actuator performance.

Similar researches have been previously conducted. Various tank design tusks can be solved using CDF (Computational Fluid Dynamics) methods which are discussed in sources [1]–[9]. Flow processes numerically investigated in tanks with and without internal additional solid baffles [1]–[3]. Issues related to the dissolved gases presence in the fluid have been studied [4, 5]. The operation fluid specifications dependence from tank configuration and main flow parameters are presented in studies [3], [6]–[8].

Automated machine hydrostatic actuator components

As an example stationary hydraulic system tank optimization of the 8V66A automated cutting machine is considered [10].

Hydrostatic actuator consists of following components: vane pump type G12-22 (nominal flow rate $Q = 18\ \mathrm{l/m}$); hydraulic motor; executive hydraulic cylinders (for vertical and horizontal jaw vice,

Content from this work may be used under the terms of the Creative Commons Attribution 3.0 licence. Any further distribution of this work must maintain attribution to the author(s) and the title of the work, journal citation and DOI.
Published under licence by IOP Publishing Ltd

IOP Conf. Series: Materials Science and Engineering **779** (2020) 012041 doi:10.1088/1757-899X/779/1/012041

circular saw lifting, work material lifting); control spool valves; pressure reducing and relief valves; throttles; fluid cooling radiators; pressure gauge and tank. The hydraulic actuator provides a quick supply, working and return headstock disk stoke, working material lifting above the bearing prism. Clamping and unclamping the work piece is carried out through a leaver system using a hydraulic cylinder.

E-20A industrial oil is used as the machine actuator's hydraulic fluid. It is a general-purpose oil without additives and has been used in hydrostatic actuators that operating indoor spaces. Main fluid specifications indicated in Table 1.

Table 1.E-20A oil main specifications.

Parameter name	Value according to GOST
Density at 20°C, kg/m^3, max	890
Density at 40°C, kg/m^3, max	859,5
Kinetic viscosity at 40°C, mm^2/s	25…35

The tank appearance one of the present hydrostatic actuator solutions is shown in Fig. 1.

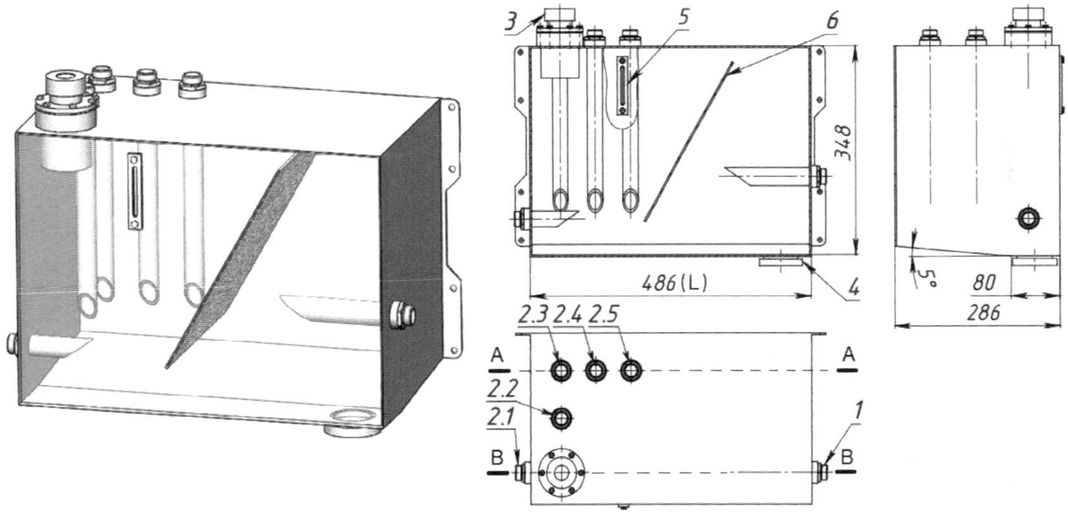

Figure 1. Tank overview and drawing

Tank parts are number-indicated: 1 — suction port; 2 — return ports (2.1 — general port of circular saw control hydraulic cylinders, 2.2 — relief valve port, 2.3 — hydraulic motor port, 2.4 — port of hydraulic cylinder horizontal vice, 2.5 — general port of vertical vice control hydraulic cylinders); 3 — tank filler opening that is combined with filler filter and breather; 4 — pollution collector bowl; fluid level indicator (32 l min, 40 l max); 6 — perforated baffle for the fluid deaeration. The tank nominal volume is 40 l (according to level indicator upper scale).

Methods

Following Favre averaged equations system describes viscous fluid hydrodynamics:

- continuity equation

$$\frac{\partial \overline{u}_i}{\partial x_i} = 0,$$

where $i = 1 \ldots 3$; $\overline{u}_i$ — Favre averaged stream wise$x$, y, z velocity components.

A constant density $\rho$is taken as the assumption. Thus, an incompressible Newton fluid is considered.

- momentum equation

$$\rho\left[\frac{\partial \overline{u}_i}{\partial t}+\overline{u}_j\frac{\partial \overline{u}_i}{\partial x_j}\right]=-\frac{\partial \overline{p}}{\partial x_i}+\frac{\partial}{\partial x_i}\left[T_{ij}^{'}-\tau u_i u_j\right],$$

where $j=1\ldots3$; $\overline{p}$ — mean static pressure; $T_{ij}^{'}=2\mu S_{ij}^{'}$ — molecular stress tensor; μ — dynamic viscosity; $S_{ij}^{'}=\frac{1}{2}\left[\frac{\partial \overline{u}_i}{\partial x_j}+\frac{\partial \overline{u}_j}{\partial x_i}\right]$ — strain rate pulsation tensor, $\tau u_i u_j$ — Reynolds-stress tensor.

- turbulent kinetic energy equation

$$\rho\frac{\partial k}{\partial t}+\rho\overline{u}_j\frac{\partial k}{\partial x_j}=\rho\tau_{ij}\frac{\partial \overline{u}_i}{\partial x_j}+\frac{\partial}{\partial x_i}\left[\left(\mu+\frac{\mu_T}{\sigma_{K1}}\right)\frac{\partial k}{\partial x_j}\right]-\beta_0^*\rho k\omega,$$

where $k=\frac{\left(u_x^{'}\right)^2+\left(u_y^{'}\right)^2+\left(u_z^{'}\right)^2}{2}$ — turbulent kinetic energy, $u_i^{'}$ — stream wise$x$, y, z pulsation velocity components; $\mu_T=\rho\frac{k}{\omega}$ — turbulent viscosity coefficient; ω — specific dissipation rate.

- specific dissipation rate equation

$$\rho\frac{\partial \omega}{\partial t}+\rho\overline{u}_j\frac{\partial \omega}{\partial x_j}=\alpha\frac{\omega}{k}\rho\tau_{ij}\frac{\partial \overline{u}_i}{\partial x_j}+\frac{\partial}{\partial x_i}\left[\left(\mu+\frac{\mu_T}{\sigma_{\omega1}}\right)\frac{\partial \omega}{\partial x_j}\right]-\beta_0\rho\omega^2$$

Model numerical constants values are: $\beta_0^*=0{,}09$, $\alpha=5/9$, $\beta_0=0{,}075$, $\sigma_{K1}=\sigma_{\omega1}=2$.

These equations represent the k - ω turbulence model.

The laminar flow calculation can also be implemented using described mathematical model. In this case the turbulent terms of the equations tend to zero and the result is correct. The similar mathematical model was previously used for simulation flow process in positive displacement pumps [11, 12] and hydrodynamic centrifugal pumps [13]–[17] and actual results were provided.

The hydrodynamic simulation is used because this approach allows study fluid behavior in the tank cavity during the actuator operation. The occurring fluid phenomena can be sufficiently predicted. The details of the hydrodynamic simulation researches were previously considered in studies [1, 2], [5]–[9]. The described above mathematical turbulence k - ω SST model is integrated in the STAR CCM + software package.

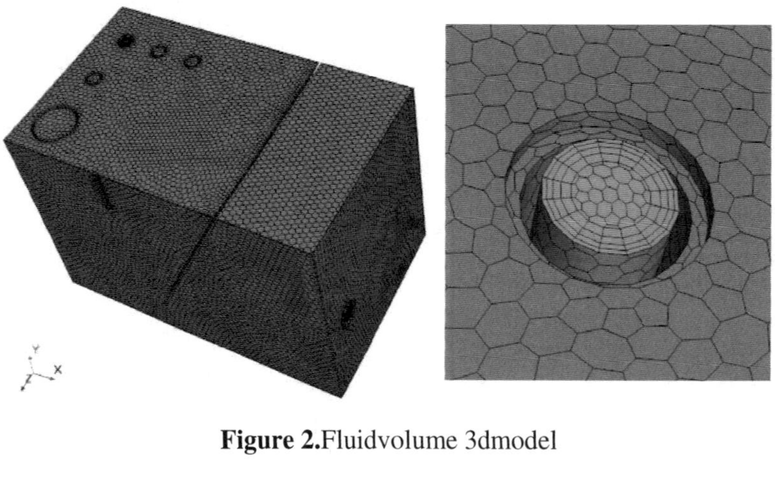

Figure 2. Fluidvolume 3dmodel

The 3dmodel of contained in the tank fluid is presented in Fig. 2.The model is divided into cells by a volumetric mesh. The cell size is selected in compliance with an acceptable relationship between the speed of the calculation process and the resulting accuracy. Nominal cell size is 6 mm. The boundary prismatic layer parameters are: cell size 50% of the nominal; the coefficient of prismatic layer extension is 1.3; prismatic layer count is 4. As an example, the prismatic layer in the suction pipe wall region is also shown in Fig. 2.

The simulation is assumed at nominal operating mode of the vane hydraulic-motor. In this case, the fluid enters the tank cavity through port 2.3 (Fig. 1). The fluid circulates in a closed circuit. Therefore the return and suction lines volume flow rates are equal. Entering fluid average temperature is T ≈ 45°C. Hydrostatic actuator pump is driven by an electric motor with a constant speed.

Boundary conditions: the velocity distribution at the pipes ports is described by a function in polar coordinate system

$$u = \begin{cases} U\left(1-(2r/d)^2\right), 0 \le r \le d \\ 0, r > d \end{cases}$$

where U — max velocity; r — radial coordinate axis; d — pipe diameter. Boundary conditions are presented in Table 2.

Table 2. Boundary conditions

U_{re}, m/s	U_{su}, m/s	d_{re}, m	d_{su}, m
1,59	1,44	$20 \cdot 10^{-3}$	$24 \cdot 10^{-3}$

The simulation is performed using following quantities:entering fluid average temperature; fluid (mineral oil) and medium carbon steel specific heat capacity, thermal conductance; fluid density and dynamic viscosity. Values are summarized in Table 3.

Table 3. Constant quantities

T_{in},°C	C_{fl},J/(kg·K)	C_{st},J/(kg·K)	K_{fl},W/(m·K)	K_{st},W/(m·K)	ρ,kg/ m^3	μ,Pa·s
45	1780	474	$9 \cdot 10^{-2}$	46	859,5	$25 \cdot 10^{-3}$

The most informative in fluid movement assessment terms are the scalar fields of the velocity distribution over sections A-A and B-B (Fig. 1), as well as flow lines in the fluid volume and the scalar heat propagation field. The simulation results are presented in Fig. 3and Fig. 4.

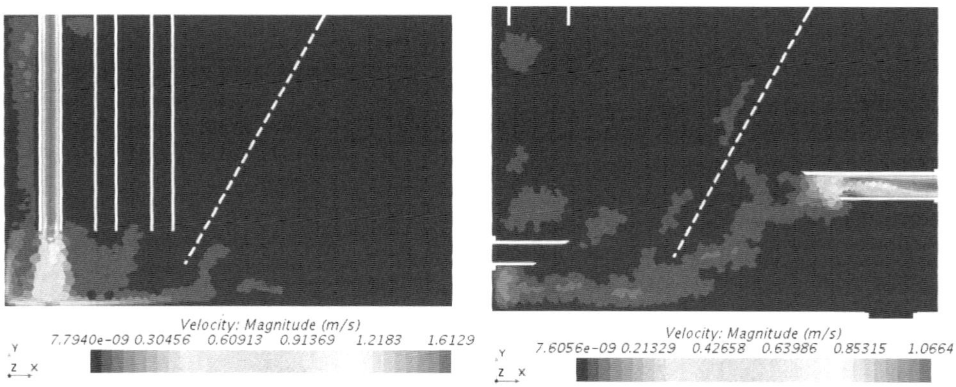

Figure3.Scalar fields of the velocity distribution over sections A-A (left) and B-B (right)

IOP Conf. Series: Materials Science and Engineering **779** (2020) 012041 doi:10.1088/1757-899X/779/1/012041

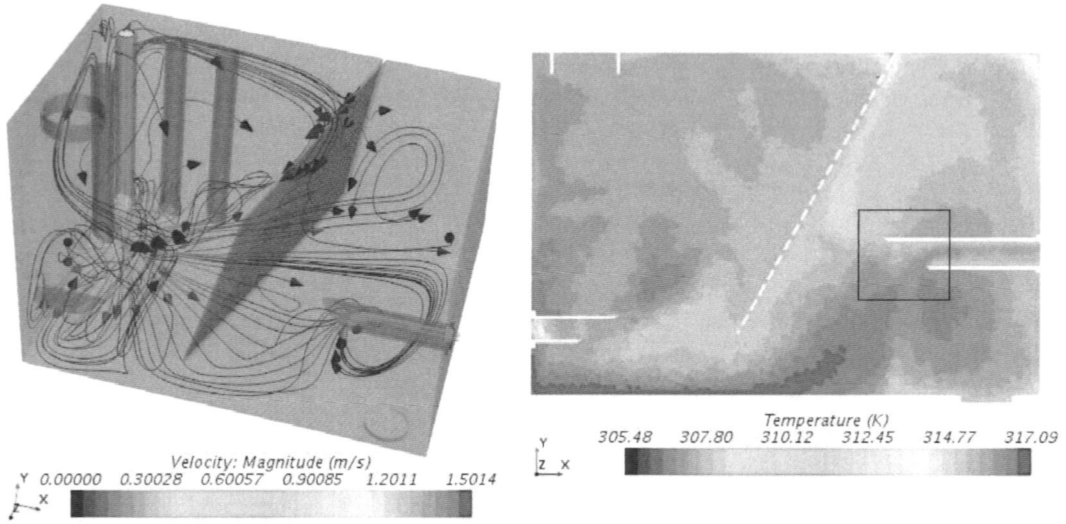

Figure 4. Flow lines (left) and scalar field of the temperature distribution over section B-B (right)

As can be seen from Fig. 3, the scalar fluid velocities distribution fields are heterogeneous: there are clearly defined regions with a more intense flow velocity and viceversa with practically zero velocity value. After entering the tank fluid immediate lyrushes to the suction pipe and only a small parts preads in the all over tank volume. Located mainly away from the suction and return line pipes fluid is less flow involved. It is necessary to study the flow lines distribution (Fig. 4) in order to unambiguous lye valuate the flow nature.

Entering the pump suction port fluid heat can be estimated as the average temperature calculated in cubic region volume near the suction pipe inlet (highlighted in a black square in Fig. 4) [18, 19].The a veragetemperatureis 315,3 K.

As the way to improve tank efficiency a nenhanced model with addition baffles which separating suction region is proposed. A tank 3dmodelis presented in Fig. 5 (1 — suction port region; 2 — inclined screen grain).

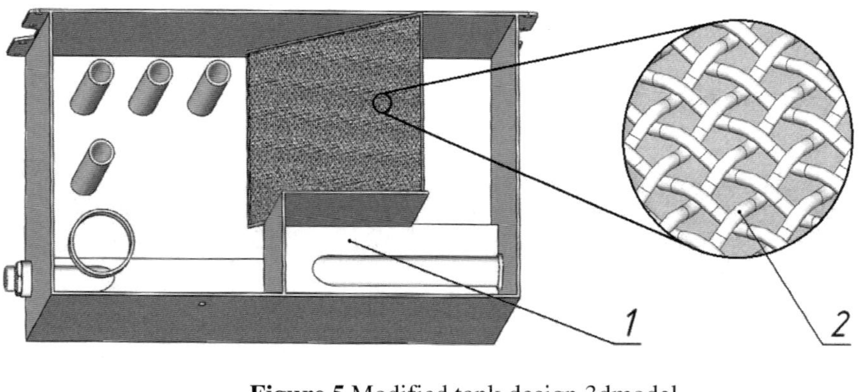

Figure 5. Modified tank design 3dmodel

The ports and functional elements (excepting deaerating inclined screen) location correspond to the initial drawing data. The presented configuration significantly increase fluid path without changing the tank volume. It is recommended to apply a metal grid that in structure is similar to the coarse filter's grid as an inclined screen. The screen should not provide excessive resistance to the fluid flow so grid opening dimensions has been selected taking this into account. The grid screen may also have as pecialcoating that increases it ability of air finely divided bubbles catching. The fluid flow action provides air bubbles gradually movement to the surface. Thus, it is possible ominimize the air ingress into the suction line. The fluid deaeration is increased. In particular cases at relatively high flow rate through the screen it makes sense put to use several grid layers.

Similar lyper formed (the same boundary conditions) tank's fluid flow simulation results are presented in Fig. 6 and Fig. 7.

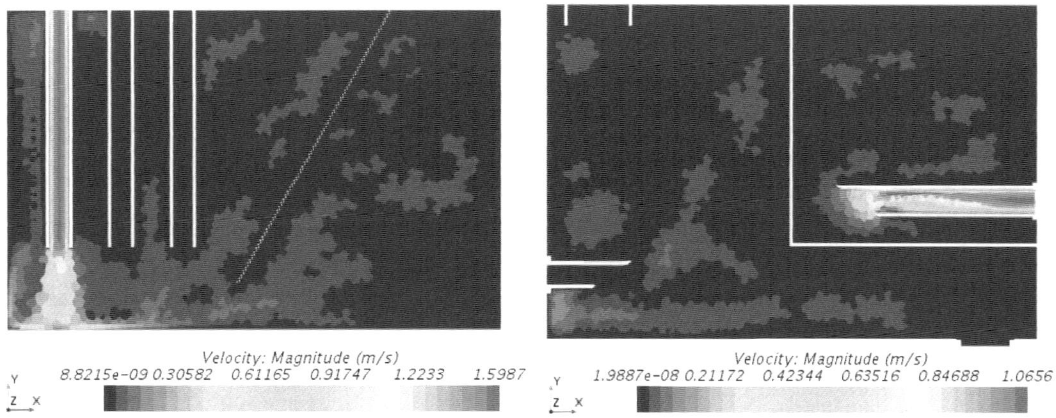

Figure 6. Scalar fields of the velocity distribution over sections A-A (left) and B-B (right)

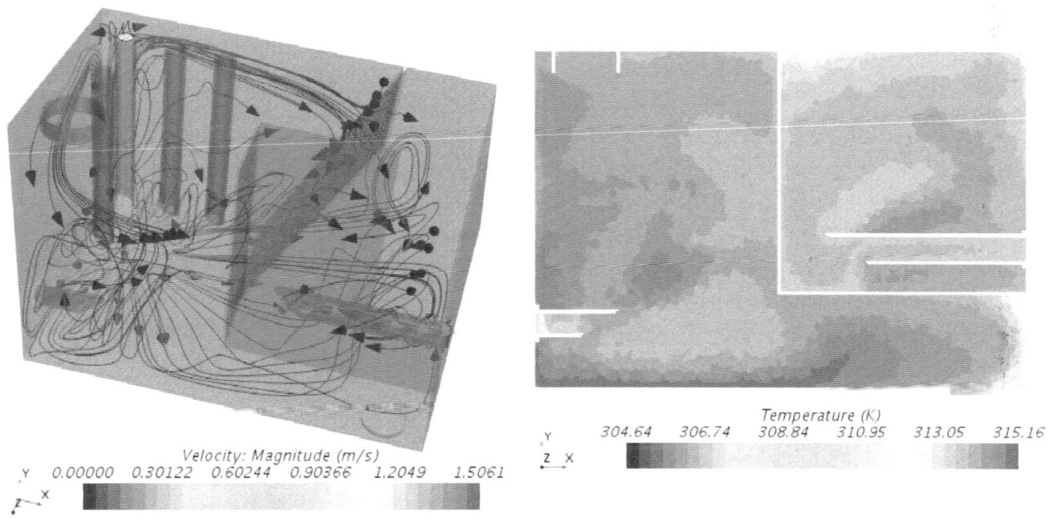

Figure 7. Flow lines (left) and scalar field of the temperature distribution over sectionB-B (right)

In this case before achieve the suction pipe the coming from the return pipe stream flows around additional solid baffles and the residence fluid in the tank time increases. The processes of oil cooling, settling contaminants to the bottom, rising air bubbles to the surface are more intense [9].

Proposed tank design contributes to a more uniform flow velocities distribution and flow lines in the tank volume which positively affects the cooling, deaerating and cleaning reservoir functions [4, 20]. The average fluid temperature is 313,6 K in suction pipe inlet region.

Optimal tank overall l dimensions selecting

Since described above tank design provides al on ger path for the fluid to pass from the return port to the suction port the tank dimensions may be changed to reduce its volume. As an example the large stoverall dimension is considered — the tank length (marked L dimension in Fig. 1). The task is the selection length value which a flow regime favorable from the point of most efficiency view isobserved. Leng this determined by optimization according to a criterion of the ensuring the most uniform velo city distribution in the fluid volume. The quadratic velocity difference is used to assess the velocity scalar field non-uniformity

$$V_r = \sum_{j=1}^{n} \left(V_j - V_{j+1}\right)^2,$$

where j — fluid volume points index; V_j — velocity value at a point without taking into account flow direction. Sixteen control points $n = 16$ were selected for optimization. Control points are set directly in the fluid volume with the exception suction and return tubes regions. A count of control points may be wider in order to increase accuracy. The results of five consecutive calculations $(i = 1 \ldots 5)$ corresponding to length L variation from 402 mm to 570 mm are presented as a schedule in Fig. 8.

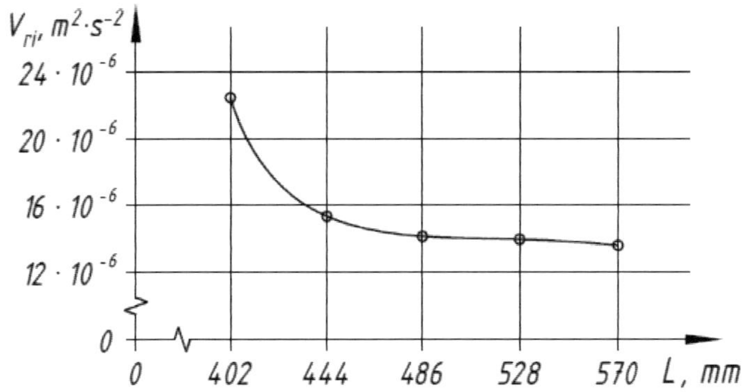

Figure 8. Velocity scalar field uniformity criterion V_{ri} dependence from L dimension value

Selection L value recommend dations have been done based on the obtained results. An intensive increase in unevenness begins with a starting from L ≈ 440 mm length decrease, the unevenness decrease with an increase in L: the dependence turns out to be similar to hyperbolic. It is not any sense to produce very long tank because that approach provides a relatively low unevenness reduction but its manufacture cost is greatly increased at the same time. In described case is recommended to choose L value from range of 440…480 mm.

Conclusions

Proposed tank design changes contribute to:
- reduction of the tank size in favor of a more efficient contained in it fluid use;
- better removal of contaminants and gas bubbles from the fluid;
- uniform use entire fluid volume and improving tank property heat removal.

The presented simulation procedure and the obtained during study results can be useful for hydrostatic actuator tank development. The main design features that affect the tank functions efficiency are noted.

Reference

[1] Tič, V. &Lovrec, D. 2012. Design of modern hydraulic tank using fluid flow simulation. International Journal of Simulation Modelling, Vol. 11, No. 2, pp. 77–88. https://doi.org/10.2507/IJSIMM11(2)2.202

[2] Singala, V., Bajaj, J., Awalgaonkara, N. &Tibdewalc, S. 2014. CFD Analysis of a kerosene fuel tank to reduce liquid sloshing, Procedia Engineering, Vol. 69, pp. 1365–1371. https://doi.org/10.1016/j.proeng.2014.03.130
[3] Zhang, Ch. 2015. Application of an improved semi-Lagrangian procedure to fully-nonlinear simulation of sloshing in non-wall sided tanks, Applied Ocean Research 51, 74–92, 2015. https://doi.org/10.1016/j.apor.2015.03.001
[4] Sakama, S., Tanaka, Y., Higashi, H., Goto, H. & Suzuki, R. 2014. Air Bubble Separation and Elimination from Working Fluids for Performance Improvement of Hydraulic Systems. Proc. IFPE 2014, March 4-8, 2014, Las Vegas, USA. 8 p. https://doi.org/10.13140/2.1.4927.4245
[5] Vollmer, T., Untch, J. Capabilities and Challenges of CDF Multiphase Simulation of Hydraulic Tanks. In: *Proceedings of the 8th FPNI Ph.D Symposium on Fluid Power*, Lappeenranta, Finland, June 11–13, 2014. https://doi.org/10.1115/FPNI2014-7849
[6] Wohlers, A., Backes, A. and Schönfeld, D. "An Approach to Optimize the Design of Hydraulic Reservoirs," in 10. InternationalesFluidtechnischesKolloquium, Dresden, 2016.
[7] Močilan, M., Žmindák, M., Pecháč, P. & Weis, P. 2017. CFD Simulation of Hydraulic Tank. ProcediaEngineering, Vol. 192, pp. 609–614. https://doi.org/10.1016/j.proeng.2017.06.105
[8] Ling Hou, Fangcheng Li, Chunliang Wu, A Numerical Study of Liquid Sloshing in Two-dimensional Tank under External Excitations, Journal of Marine Science Applications 11, 305–310, 2012. https://doi.org/10.1007/s11804-012-1137-y
[9] Tič, V.; Lovrec, D. Air-release and Solid Particles Sedimentation Processwithin a Hydraulic Reservoir. // TehničkiVjesnik- Technical Gazette. 20, 3(2013), pp. 407–412.ISSN 1848-6339
[10] Birykov, B.N. Hydraulic equipment for metal cutting machines. M.: Mashinostroenie, 1979. — 112 p.
[11] Salutagi, S., Creswick, M., Yuan, Q., Ding, H. Axial Piston Pump Performance Prediction using 3D CFD Simulation. EatonCorp. SimericsInc. 2015.
[12] RodionovL.V. Features of simulation the gear pump operation process hydrodynamics. Izvestiya Samar. scientific RAN Center. 2017. 19, No 4, part. 1, pp. 15–21.
[13] V Cheremushkin and APolyakov 2019 *IOP Conf. Ser.: Mater. Sci. Eng.***589** 012001
[14] P Chaburko et al 2019 *IOPConf. Ser.: Mater. Sci. Eng.***492** 012011
[15] V Tkachuk et al 2019 *IOP Conf. Ser.: Mater. Sci. Eng.***589** 012007
[16] V Lomakin et al 2019 *IOP Conf. Ser.: Mater. Sci. Eng.***492** 012012
[17] A Boyarshinova et al 2019 *IOP Conf. Ser.: Mater. Sci. Eng.***589** 012014
[18] Wu, C., Xu, C., Mao, X., Li, B., Hu, J. & Liu, Y. 2017. Heating analysis in constant-pressure hydraulic system based on energy analysis. IOP Conf. Series: Earth and Environmental Science, Vol. 100, p. 012147. https://doi.org/10.1088/1755-1315/100/1/012147
[19] Pimonov, G.G., Pimonov, I.G. Research and development of heat-control system for hydraulic actuator. Vestnik HNADU, issue 65–66, 2014.
[20] Rylyakin, E.G. Operation factors influence on the hydraulic actuator technical conditions changes/ Materialy X mezinarodnivedecko-prakticka conference «Modernivymozenostivedy—2014». —Dil 38. Technickevedy. // Praha. Publishing House «Education and Science». — 2014. — pp. 13–15.

IOP Conf. Series: Materials Science and Engineering **779** (2020) 012042 doi:10.1088/1757-899X/779/1/012042

Conceptual features for improving the flow part of the multiphase stages of ESP submersible plants for small and medium feeds for extracting stratal liquid with a high free gas content

A Trulev[1,4], M Kayuda[1], S Timushev[2] and V Lomakin[3]

[1]Rimera Group of companies, Moscow, Russian Federation
[2]Moscow Aviation Institute, Moscow, Russian Federation
[3]Bauman Moscow State Technical University

[4]E-mail: aleksey.trulev@rimera.com

Annotation. This article analyzes the operation numerical simulation for serial and multiphase stages of various manufacturers on water and gas-liquid mixtures (GLM). Are identified the local gas separation areas in serial stages, which can lead to the formation of gas plugs and disruption of the pump supply.
These factors include: the creation of a finely dispersed flow structure at the entrance of the stage, the preservation along the flow with small free gas bubbles diameters, the sustenance of the value for the allowed free gas content by circulation means of fluid part in the impeller, a defined relationship between the pressure gradient and the GLM flow rate. A mathematical model is selected, hydrodynamic calculations of structural options, optimization are carried out, a conceptual option is selected. Bench tests were carried out, confirming the high parameters of the product when working on GLM.

Introduction

There is a steady tendency in the world to increase the number of low-yield wells, which make up about 30-40% of the total well stock in the Russian Federation. In order to increase the oil recovery coefficient (ORC), the down hole pressure is reduced, respectively, the free gas content fand solids at the pump inlet increases.

Serial submersible pumping equipment for operating wells is often unable to operate in such difficult conditions.

It is valid to question about effective technology for operating low-production wells with high average production time [1], [2].

In oil wells with a high free gas content, an increase in the efficiency of electric centrifugal pumps (ECPs) is achieved by installing multiphase stages at the inlet. The task of the stages is to effectively disperse, grind the gas-liquid mixture, reduce the diameter of the free gas bubbles and compress, in order to partially transfer the gas from the free state to the dissolved one, to reduce the free gas content at the pump inlet. After passing through the pump, free gas is released in the tubing and due to the effect of the gas lift helps to deliver the stratal liquid to the surface.

Content from this work may be used under the terms of the Creative Commons Attribution 3.0 licence. Any further distribution of this work must maintain attribution to the author(s) and the title of the work, journal citation and DOI.
Published under licence by IOP Publishing Ltd

The efficiency of multiphase stages depends significantly on the dispersion of the gas-liquid structure of the pumped fluid, on the diameters of gas bubbles, water cut of the formation fluid, the presence of surface-active substances, pressure at the inlet and outlet of the pump [3, 4]. The average diameter of the gas phase bubbles in the stratal liquid at the pump inlet can be quite large: 300 — 1 mm or more. As a result of dispersion, it is in the range from 30–40 μm for non-watered oil, about 130 μm for watered oil [5], [6]–[29].

To increase the efficiency of the oil production process from low-production wells, Rimera Group has developed and offers complete equipment, mainly transferred to serial production. That is the conceptual design of low flow stages, gas separator, multiphase module, hydroprotection.

Multiphase stages can be installed directly at the section inlet or in a separate multiphase phase converter module at the pump inlet.

The topic is also of interest for the creation of small-sized gas-liquid compressors, for general industrial use, for small-sized aircraft engines of unmanned vehicles (drones). If the engine compressor runs on a mixture of gas and liquid fuel, the mass fraction of which reaches 10%, the density and, accordingly, the pressure of the gas-liquid fuel increases by two orders of magnitude, respectively, the diameters can be reduced by about ten times, and the quality of mixing fuel with gas can noticeably increase. Compression occurs according to a quasi-isothermal process, which can increase the efficiency of the installation.

Analysis of structural schemes.

The fig. 1 shows the multiphase stages main structural schemes of pumps for small and medium flows.

1. Series multiphase stages with axial stages, see Fig. 1a.
2. An analog of multiphase stages of one of foreign manufacturers, impellers of a half-open type without a cover disk, with holes in the drive disk made at the entrance to the impeller, see Fig. 1b.
3. New multiphase stages with fluid circulation in the impeller, see Fig. 1c.
4. New multiphase centrifugal-disk stages with fluid circulation in the impeller, differing from stages on the point 3 by the presence of a longitudinal cut in the middle part of the blade, this design is at the patenting stage.

Multiphase stages can be installed directly or in a separate module at the entrance to the section, while their operation on the GHS is essentially the same.

Mathematical model

In this article, we use the model of a multiphase incompressible fluid flow (ρ = const). Numerical simulation is based on solving discrete analogs of the basic hydrodynamic equations.

The calculation is carried out on the basis of a mathematical model of a divided multiphase flow. That is, for each phase, the equations of mass and momentum transfer are solved separately, but the pressure field is the same for all phases.

The mathematical model consists of a set of differential and algebraic equations:

1. The volume of the i-th phase in each calculated cell is calculated as:

$$Vi=\int \alpha i V dV$$

whereαi— is the concentration of the i-th phase in the cell.

The sum of the concentrations of all phases in the cell is one

$$\sum_{i6}^{n} \alpha_i = 1$$

2. The equation of conservation of mass (continuity equation):

$$\frac{\partial}{\partial t}\int_V \alpha_i \rho_i dV + \int_A \alpha_i \rho_i \overline{V_i} d\bar{a} = 0$$

whereρi—i-th phase density

$V\vec{i}$—i-th phase velocity (in the case of turbulent flow modeling by a RANS-type model averaged over time)

3. The equation for the change in the amount of motion:

$$\frac{\partial}{\partial t}\int\limits_V \alpha_i\rho_i dV + \int\limits_A \alpha_i\rho_i\left(\overrightarrow{V_i V_i}\right)d\vec{a} = -\int\limits_V \alpha_i \nabla p dV + +\int\limits_V \alpha_i\rho_i \vec{g} dV + \int\limits_A \left[\alpha_i\left(T_i + T_i^t\right)\right]d\vec{a} + \int\limits_V \overrightarrow{M_t} dV$$

where($V\vec{i}V\vec{i}$) — tensor product of the velocity vectors of the i-th phase.

p— pressure

g— mass intensity vector (in this case, the gravity force is 9.81 m / s2 and the inertial force due to the rotation of the computational domain)

Ti— molecular viscosity stress tensor

Tit— turbulent stress tensor

$\vec{M}$— vector of total intensity of interfacial interaction forces per unit volume, for a vector $M\vec{i}$fair equality: $\sum \overrightarrow{M_t} = 0$

Vector $M\vec{i}$characterizes all the forces that separate phases interact with each other.

$$M_i = \sum_{i \neq j}\left(\overrightarrow{F_{ij}^D} + \overrightarrow{F_{ij}^{VM}} + \overrightarrow{F_{ij}^L} + \overrightarrow{F_{ij}^{TD}} + \overrightarrow{F_{ij}^{WL}}\right)$$

Where

$FijD\vec{}$— resistance force, $FijVM\vec{}$ power of the virtual mass

$FijL\vec{}$— lifting force

$FijTD\vec{}$— turbulent dispersion force

$FijWL\vec{}$— turbulent dispersion force

Power of resistance

In the case of modeling the flow of continuous and dispersed media, the resistance force acting on the dispersed medium I from the side of the continuous medium j is equal to:

$$\overrightarrow{F_{ij}^D} = A_D \overrightarrow{V_r},$$

Where

AD— linearized drag coefficient

$Vr\vec{}$ =$Vj\vec{}$ −$Vi\vec{}$— relative velocity of one medium relative to another

$$A_D = C_D \frac{1}{2}\rho_C \left|\overrightarrow{V_r}\right|\left(\frac{a_{cd}}{4}\right),$$

Where

CD— drag coefficient

ρC— continuous phase density

acd— interaction area of the phases (in this case, the member$acd4$ is the projected area of the spherical particle on the plane).

The drag coefficient is based on the relation:

$$C_D = f_D C_{D\infty},$$

Where

$CD\infty$ — coefficient of resistance of a single spherical particle moving in an infinite flow

fD— coefficient taking into account the concentration of particles

$$C_{D\infty} = \frac{24}{R_d}\left(1+0.15R_d^{0.687}\right) npu R_d < 1000,$$

and

$$C_{D\infty} = 0.44 npu R_d > 1000$$

$$R_d = \frac{\rho_C |V_r| l}{\mu_C}$$

l— characteristic interaction length or bubble size
μC— dynamic viscosity coefficient of continuous medium
Coefficient $fD$is the ratio:

$$f_D = \alpha_C^{nD}$$

αC— continuous phase concentration
nD=−8.3 for spherical particle
Virtual mass
The inertia of the surrounding fluid affects the acceleration of the particle immersed in the fluid. This effect is modeled by adding mass to the dispersed particle.
The force from the virtual mass acting on phase I, moving rapidly with respect to phase j is found as:

$$\overrightarrow{F_{ij}^{VM}} = C_{VM}\rho_C\alpha_C\left(\overrightarrow{a_j} + \overrightarrow{a_i}\right)$$

$aj,i^{\rightarrow}$— acceleration of jth and ith phase
CVM=0,5 — virtual mass coefficient for a spherical particle
Lift force
In the case where the flow surrounding the dispersed particle is non-uniform or the particle is swirling, it experiences a force perpendicular to the relative velocity.

$$\overrightarrow{F_{ij}^{L}} = C_{Leff}\rho_C\alpha_d\left(\overrightarrow{V_r} \times \left(\nabla \times \overrightarrow{V_c}\right)\right)$$

$Vr^{\rightarrow}$— relative phase velocity
$Vc^{\rightarrow}$— continuous phase speed
αd— dispersed phase concentration
CL=0,25 — lift force coefficient
Turbulent dispersion force
Additional change in phase concentrations caused by flow turbulence is modeled as the force of turbulent dispersion

$$\overrightarrow{F_{ij}^{TD}} = A_D\overrightarrow{V_{TD}}$$

$FijTD^{\rightarrow}=ADVTD^{\rightarrow}$
A_D—drag force coefficient
$VTD^{\rightarrow}$—relatives lip speed

$$VTD^{\rightarrow}=DTD(\nabla\alpha d\alpha d-\nabla\alpha c\alpha c)$$

$D_{TD}=C0vct\sigma\alpha l$—turbulent diffusion tensor
C_0=1
vct— turbulent viscosity coefficient kinematic $\sigma\alpha$– turbulent Prandtl number

IOP Conf. Series: Materials Science and Engineering **779** (2020) 012042 doi:10.1088/1757-899X/779/1/012042

I— unit tensor

$$\sigma_\alpha = \sigma_0 \sqrt{1 + C_\beta \xi^2}\, \frac{1+\eta}{B+\eta}$$

σα=σ0√1+Cβξ21+ηb+η
σ0=1 — unmodified turbulent Prandtl number
Cβ=1,8 — correction factor
ξ— particle sliding velocity related to the speed of turbulent fluctuations
η— particle interaction time related to relaxation time
b— the ratio of the accelerations of the continuous / dispersed phase

$$b = \frac{1 + C_{VM}}{\frac{\rho_d}{\rho_c} + C_{VM}}$$

$$\eta = \frac{\tau_I}{\tau_R}$$

τI— characteristic time of interaction of a particle and a turbulent vortex
τR— particle relaxation time

$$\tau_I = \frac{\tau_T}{\sigma_0} \frac{1}{\sqrt{1 + C_\beta \xi^2}}$$

$$\tau_T = \frac{2}{3} C_\mu \frac{k_c}{\epsilon_c}$$

ϵc-continuous phase turbulence energy dissipation rate

$$\tau_R = \tau_D \left(1 + \frac{\rho_c}{\rho_d} C_{VM}\right)$$

$\tau_D = \frac{\rho_d d^2}{18\mu_c}$ —*characteristic time scale for a dispersed particle*

$$\xi = \frac{\left|\overrightarrow{V_r}\right|}{\sqrt{\frac{2}{3} k_c}}$$

kc— kinetic turbulence energy of the continuous phase
Wall strength
Being located near the solid wall, the gas bubble undergoes an asymmetrical action from the liquid. The force per unit volume that a gas bubble experiences is equal to:

$$\overrightarrow{F_{ij}^{VL}} = -C_{WL} \gamma_W \alpha_D \rho_c \frac{\left|\overrightarrow{V_r}\right|^2}{d_p} \vec{n}$$

Where
yw— distance from the wall
αD— dispersed phase concentration
ρC— dispersed phase density

dp— bubble diameter

CWL— coefficient is a function of the distance from the wall and decreasing with increasing distance

$\vec{n}$— unit normal to the wall at the point closest to the bubble

$Vr,\vec{\tau}$— tangent to the wall component of the relative velocity

Coefficient $CWL$is like:

$$C_{WL} = max\left(C_{w1} + \left(\frac{C_{w2}}{\gamma_w}\right)d_p, 0\right)$$

Coefficients are $CW1$=–0,01, $CW2$=0,05.

Thus the force caused by the influence of the wall disappears at a distance from the wall equal to the five diameters of the bubble.

The mathematical modeling results.

To carry out hydrodynamic calculations, models of the flow part of the serial stage and multiphase stages, which were presented above in the analysis of structural schemes, were built.

In fig. Figure 2 shows the results of numerical simulation of the operation of a serial stage on water. 2a and on the GLM, see Fig. 2b.

An analysis of the serial stage operation shows that there are two areas between the impeller disks and the upper (region 1) and lower (region 2) disks of adjacent guide vanes installed above and below the impeller in which there is no main stream flow. In these “stagnant” areas, conditions have been created for the separation of gas bubbles, enlargement, and the possibility of periodic exit of large gas bubbles beyond the outer diameter of the driving and driven disks into the flow part of the impeller. Since the pressure gradient, even with partial liquid separation during operation on the GLM, is quite high, see Fig. 2b for the separation of large bubbles from the exit to the input of the impeller with the possibility of the formation of a gas plug and a stall

For reliable operation of multiphase stages, areas in which separation and enlargement of free gas bubbles should be excluded.

A numerical simulation analysis of the stages using water and GLM helps to analyze the factors affecting the stages effective operation on the GLM without interruption of supply, on the possibility of increasing the permissible free gas content at the inlet to the centrifugal pump, with multiphase stages installed at it. Such factors include: creating a finely dispersed flow structure with small diameters of free gas bubbles at the entrance to the stage and maintaining the entire length of the flow, maintaining the value of the permissible content of free gas, for example, due to the circulation of a part of the liquid in the impeller, a certain ratio between the pressure gradient and GLM flow rate. The conceptual option is selected, see Fig. 3.

A new concept of the operation of compressor dispersing stages (CDS) is the circulation of a part of the liquid in the impeller. When working in each step (see Fig. 3) can distinguish the next points:

1. The GLM enters to the impeller, ejection by a jet of partially separated liquid,
2. Dispersion due to special grooves and a special blade design,
3. Compression in the flowing part of the impeller,
4. Partial gas separation at the impeller outlet
5. Gas with the main stream goes to the center and up to the next stage.
6. Part of the fluid, through the radial clearance, is directed to the entrance of the same impeller.

The design principle of operation allows to ensure the free gas content value in the impeller at the level that is necessary for reliable operation, without interruption of supply.

The CDS are crushed, the bubbles of the gas-liquid mixture are dispersed, and a homogeneous, finely dispersed flow structure is prepared, pre-compressed and fed to the main stages of the pump.

When the pump is working together with the GDS, the maximum volumetric free gas content at the inlet is allowed 15% higher compared to using a pump without GDS.

The Fig. Figure 4 shows the numerical simulation results of the multiphase stages assembly on water. The stages are sequentially installed:

1. A multiphase stages analogue of one of the foreign manufacturers, the impeller is half-open type without a cover disk, with holes in the drive disk made at the inlet, see Fig. 1b.
2. A new multiphase centrifugal-disk stage with a longitudinal cut in the middle of the blade.
3. Multiphase stage with a new operating concept, with fluid circulation in the impeller, see Fig. 1c

The stages differed only in the design of the impellers, the guiding devices were the same.

In Fig. 4a shows the velocity plots, in Fig. 4b pressure plots.

In Fig. Figure 5 shows the calculated values of efficiency (a), head (b) and mass flow (c). The numbers indicate the values for each of the successively set stages, the number 4 shows the integral values for the stages assembly.

An analysis of the data shows that the highest pressure and efficiency values have a stage with the circulation of part of the liquid inside the impeller. The flow rate through the impeller in stages with circulation is higher than the flow rate of traditional stages by the amount of flow from the output back to the input of this wheel. The step 3 has a higher mass flow through the wheel, since the pressure of this step is higher than the pressure of the centrifugal - disk stage, and accordingly, higher than the fluid flow through the gap between the guide sleeve and the driven wheel disk.

However, due to better dispersion, the centrifugal - disk stage will be able to work better in off-design modes.

Theoretically, the stages with the circulation of a part of the separated liquid can, in contrast to the known analogues, also work at a 100% free gas content and can be used as gas-liquid compressors. In known multiphase stages, at a certain content of free gas, coalescence processes begin, combining gas bubbles, which lead to a failure of the supply. The new design guarantees a certain amount of fluid circulates in the stages, ensuring reliable operation.

For calculation, the StarCCM software package was used. The previously constructed flow geometry was imported as a surface mesh.

The following parameters are accepted as a physical model:

• Multiphase interaction, the scale of the interaction length is 0. 1mm.
• The phases are water H2O and gas Air (air)
• Initial liquid / gas distribution 1/0
• The input pressure is set to a total pressure of 0 Pa and a liquid / gas ratio of 4/1
• A negative velocity value of 4.5 m / s is set at the fluid outlet boundary
• Atmospheric pressure is set at the gas outlet boundary

The following parameters were taken as the studied parameters: head, separation coefficient (fraction of water at the separator outlet) and power depending on the flow rate.

Results of bench tests.

The Fig. Figure 6 shows the results of bench testing of a section on industrial water with air, without surface active substances (SAS), with multiphase dispersing stages of various designs installed at the inlet:

1. Axial Stages, see Fig. 1a.
2. Stages with impellers of a half-open type, with holes in the drive disk, see Fig. 1b.
3. Stages with fluid circulation in the impeller, see Fig. 1c.

Analysis of the results shows the advantages of multiphase stages of a conceptual design.

Design features of a multiphase module.

In the universal module for pumps for small and medium flows, the module is filled only with multiphase centrifugal compressor dispersing stages. For low feed pumps, it is advisable to install an additional stage in the form of a labyrinth-screw pair for preliminary dispersion of large gas bubbles.

In Fig. 7 shows the design of a new multiphase module, which has been put into mass production for low-flow pumps.

Conclusions

The results of numerical modeling and bench tests showed that:

1. A new conceptual design of the multiphase stage has been developed.
2. The conceptual novelty of the design is the circulation of part of the liquid in the centrifugal impeller, due to this, the required ratio between the amount of free gas and liquid in the flowing part of the impeller is maintained, which ensures reliable operation of the stage. An issue has been resolved that limits the allowable free gas content in known similar steps.
3. The results of bench tests on GLM the characteristics of new stages exceed the level of the best serial products.
4. The conceptual ideas inherent in the design can serve as the basis for the development of multiphase stages for other sizes.

List of references:

[1] Yakimov S. B. on the prospects of using radially stabilized compression electro-centrifugal pumps to improve the efficiency of operation of wells in the AB group of Samotlorskoye field / / Territory of NEFTEGAZ. 2016. No. 6. Pp. 78-86.

[2] Kosilov D. A. Improving the efficiency of well management in the current macroeconomic environment/ Mechanized production / / Materials of the specialized section of the Rosneft conference. - M: 2015. - Pp. 8-11.

[3] Igrevsky L. V. Improving the efficiency of operation of submersible pump-ejector systems for oil production: dis. Cand. Techn. sciences'. Moscow, 2002. 216 PP.

[4] Dengaev A.V. Improving the efficiency of well operation with submersible centrifugal pumps when pumping gas-liquid mixtures: dis. Cand. Techn. sciences'. Moscow, 2005. 212 s

[5] Vasiliev Yu. N., Maksutov R. A., Bashkirov A. I. Experimental study of the structure of oil and gas flow in a fountain well. 1961. No. 4. Pp. 41-44.

[6] Lomakin V O, Chaburko P S, Kuleshova M S 2017 Multi-criteria Optimization of the Flow of a Centrifugal Pump on Energy and VibroacousticCharacteristics*Procedia Engineering* **176** pp 476–482

[7] Lomakin V O, Kuleshova M S, Bozh'eva S M 2016 Numerical Modeling of Liquid Flow in a Pump Station*Power Technology and Engineering* **5**pp 324–327

[8] Lomakin V O 2015 Proceedings of 2015 International Conference on Fluid Power and Mechatronics

[9] Lomakin V O, Kuleshova M S, Kraeva E A 2015 Fluid Flow in the Throttle Channel in the Presence of Cavitation*Procedia Engineering***106** pp 27–35

[10] Shargatov, V.A., Gorkunov, S.V., Il'ichev, A.T. Dynamics of front-like water evaporation phase transition interfaces (2019) Communications in Nonlinear Science and Numerical Simulation, 67, pp. 223–236.

[11] Arefyev, K.Y., Prokhorov, A.N., Saveliev, A.S. Study of the breakup of liquid droplets in the vortex wake behind pylon at high airspeeds (2018) Thermophysics and Aeromechanics, 25 (1), pp. 55–66.

[12] Borovsky B. I., Petrov V. I. high-Speed shovel pumps. Moscow, "Mechanical Engineering", 1975.

[13] Gouskov, A.M., Lomakin, V.O., Banin, E.P., Kuleshova, M.S. Minimization of Hemolysis and Improvement of the Hydrodynamic Efficiency of a Circulatory Support Pump by Optimizing the Pump Flowpath (2017) Biomedical Engineering, 51 (4), pp. 229–233.

[14] Arefyev, K.Y., Voronetsky, A.V., Suchkov, S.A., Ilchenko, M.A. Computational and experimental study of the two-phase mixing in gas-dynamic ignition system (2017) Thermophysics and Aeromechanics, 24 (2), pp. 225–237.
DOI: 10.1134/S086986431702007X

[15] Lomakin, V.O., Kuleshovav, M.S., Bozh'eva, S.M. Numerical Modeling of Liquid Flow in a Pump Station (2016) Power Technology and Engineering, 49 (5), pp. 324–327.

DOI: 10.1007/s10749-016-0623-9
[16] Arefyev, K.Y., Voronetsky, A.V. Modelling of the process of fragmentation and vaporization of non-reacting liquid droplets in high-enthalpy gas flows (2015) Thermophysics and Aeromechanics, 22 (5), pp. 585–596. DOI: 10.1134/S0869864315050078
[17] Belov, P.A., Kobets, L.P., Borodulin, A.S. Impregnation kinetics of fibers with liquids: Simulation within the generalization of Navier-Stokes equations (2014) Inorganic Materials: Applied Research, 5 (4), pp. 403–406. DOI: 10.1134/S2075113314040182
[18] Логинов В.Ф. Компрессионные насосы для нефтедобычи в осложненных условиях. // Oilandgasjournal. 2016, № 6. С. 54–56.
[19] D. C. Wilcox (1994). “Turbulence modeling for CFD” // DCW Industries, Inc., 460 pages.
[20] Aref'Ev, K.Yu., Voronetskii, A.V., Il'Chenko, M.A. Dynamic characteristics of a resonant gas-dynamic system for ignition of a fuel mixture (2013) Combustion, Explosion and Shock Waves, 49 (6), pp. 657-661.
[21] Trulev A.V. and others. Patent of the Russian Federation No. 2622578 "Multiphase stage of a submersible multistage centrifugal pump". Application date 13.09.2016.
[22] Trulev A.V. and others. Russian Federation Patent no. 2586801 "Submersible vane multiphase pump". Datapolitical 29.06.2015.
[23] Trulev A V 2015 New equipment of CJSC Rimera for wells with complicated operating conditions, Proceedings of the conference on mechanized production Oil and Gas Vertical (NeftegazovayaVertikal)19
[24] Trulev A V 2015 New equipment of OJSC Alnas for wells with complicated operatingconditions Engineering Practice (Inzhenernayapraktika) (12) pp 71–73
[25] Trulev A V, Sabirov A A, Sibirev S V, Verbitsky V S and Timushev S F 2017 Submersible ESP units with wide channels in the flow part for the production of formation fluid from marginal wells with a high content of mechanical impurities Engineering Practice (Inzhenernayapraktika) (1–2) pp 60–63
[26] Trulev A V 2017 Submersible ESP unit of new generation. Innovative solutions for combatting complicated conditions in case of the marginal field, Oil and Gas Journal Russia (4) pp 30-34
[27] Trulev A V 2017 New submersible centrifugal pumps with gray cast iron stages and between bearing design Engineering Practice (Inzhenernayapraktika) (5) pp 74–76
[28] Trulev A V and Grachev V V 2018 Bench tests of a submersible gas and mechanical impurities separator in model mixtures Gas Industry (Gazovayapromyshlennost') (8) pp 20–24
[29] A.V.Trulev, V.F.Loginov, S.I.Gorbunov, S.F.Timushev, A.A.Sabirov, V.O.Lomakin, evelopment and pilot operation of ESP systems of a conceptually new design for the operation of depleted wells with a high content of free gas and mechanical impurities. Composite volume the laureates of the International Competition of Scientific, Technical and Innovative Developments aimed at the development of the fuel and energy and extractive industries in 2019.

IOP Conf. Series: Materials Science and Engineering **779** (2020) 012043 doi:10.1088/1757-899X/779/1/012043

Analysis of force factors affecting valve fitting process at the saddle

S Semenov

Bauman Moscow State Technical University

E-mail: a1e10@rambler.ru

Annotation. The analysis results of the action of various force factors that slows the hydraulic valve when fitting on the saddle are presented. Manifestations of inertia forces in a thin layer of liquid located in the gap between the valve and the seat immediately before the closing of the gap are shown. The components of inertia forces are distinguished, which differ in the mechanism of occurrence. The degree of their influence on the fitting process in relation to the viscosity forces is shown.

Introduction

Valves are ubiquitous in hydraulic equipment. The valve assembly in volumetric pumps with valve distribution is one of the most important and, as a rule, one of the most problematic. Despite the fact that the pump mechanisms can be very complex [1], it is often the valves that determine the ultimate capabilities of such pumps. Irrationally designed or faulty valves cause increased noise, reduced volumetric efficiency, and reduced pump life. Valve assemblies are widely used in other types of machines and systems [2, 3, 4, 5].

Most of the problems are somehow related to the process of seating the valve on the saddle. In volumetric pumps working with liquid at high pressure, valve parts are made of metal or high-strength hard plastic. High strength ceramics are sometimes used. At the same time, at pressures of tens of MPa, metals become practically uncontested. The tightness of the closed valve in this case is achieved due to the accuracy of the closable surfaces and the density of their fit after closure.

At the moment of approach to the saddle, the valve moves with a finite speed, and if it is not slowed down, then when solid surfaces are closed at significant speeds, shock forces inevitably arise. If the valve and saddle are made of hard materials with high density, the shock pulse may be short, but very large in amplitude. This leads to significant mechanical stresses, which, acting cyclically, destroy the valve and seat.

The design of valve assemblies used in volumetric pumps is quite diverse. In particular, there are many design methods known to reduce impact forces during fitting. However, as the pressure and operating frequency of the pump increase, the choice of design solutions narrows. At a pressure approaching 100 MPa, the shut-off element becomes either spherical or takes the form of a cone or cylinder, while the material must be strong, rigid, and the movable element should also be as light as possible. As a result, it becomes difficult to add special elements to the structure that damp the valve during fitting, as well as use materials that effectively dissipate impact energy in the structure.

To a large extent, a rational choice of the design parameters of the valve assembly, in which the damping ability of the liquid located in the gap between the joined surfaces is effectively used, allows

Content from this work may be used under the terms of the Creative Commons Attribution 3.0 licence. Any further distribution of this work must maintain attribution to the author(s) and the title of the work, journal citation and DOI.
Published under licence by IOP Publishing Ltd

IOP Conf. Series: Materials Science and Engineering **779** (2020) 012043 doi:10.1088/1757-899X/779/1/012043

to solve the problem. For this, it is necessary to correctly understand how various force factors appear in a thin layer of liquid between the valves at the fitting moment.

Often, if not in most cases, engineers use generalized experimental data and recommendations set forth in [6] or similar approaches when designing valves of volumetric pumps. The prevailing point of view is that at the end of the stroke, the valve is stopped due to the viscous friction forces of the fluid displaced from the closing gap under the valve. Is this always the case? What is the scale and origin of the forces caused by hydrodynamic processes in the fluid displaced from under the valve? Answers to these questions will allow a more conscious approach to valve design and improve their performance.

Analysis

To obtain the most common and visual analysis results, we will adopt an extremely simplified design scheme and accept the following assumptions:

- It will be considered that all surfaces are flattened;
- the liquid layer in the closable gap is thin (its size is small compared to the width of the sealing girdle);
- It will be accepted the parabolic law of the velocities distribution over the thickness of the extruded layer, as with a steady laminar flow in a thin gap;
- It will be considered that the sealing girdle narrow, that is, its width is much smaller than the hole radius in the valve seat, which allows us to neglect the curvature of the girdle;
- The fluid pressure flow through the closing gap is not taken into account under the influence of the differential pressure on the gap;
- The fluid flow displaced from the gap is considered to be completely symmetrical relative to the middle of the girdle;
- Fluid compressibility is not considered.

As a result, the valve during the fitting process can be represented as a rectangular prism (bar), falling to the plane and displacing the liquid from the gap between its lower surface and this plane. Due to symmetry, we will consider the half-bar with respect to the sealing girdle of width moving along the vertical plane passing through the middle of the girdle. Accordingly, there is no flow through this plane. There is also no fluid leakage at the ends of the gap. The design scheme is shown in Fig. 1.

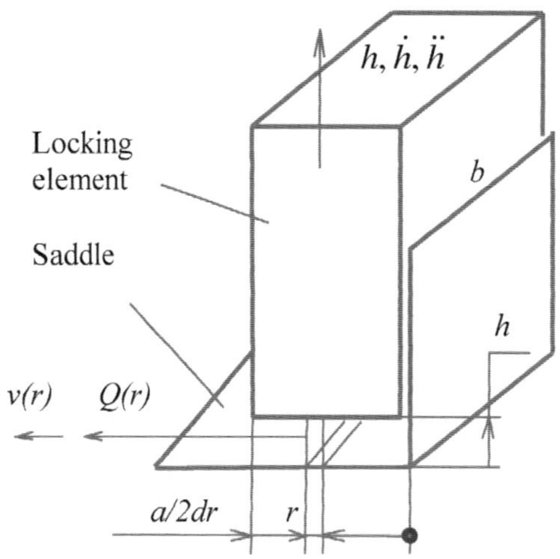

Fig. 1. The design scheme

The flow displaced from under the bar, depending on the distance from the vertical plane in the middle of the girdle is calculated as follows:

$$Q(r,\dot{h}) = -b\dot{h}r$$

Where$b$— length of the girdle;

r— Distance from the middle of the girdle;

$\dot{h}$—bar speed taking into account the sign (Fig. 1).

Then the average current velocity depending on the distance from the middle of the girdle, height and bar speed is calculated:

$$v_{cp}(r,h,\dot{h}) = -b\dot{h}r / bh = -\dot{h}r / h$$

Then the hydraulic friction losses in the gap in a section of length dr at a distance r from the middle of the girdle are calculated (having adopted, as a first approximation, the parabolic law of velocity distribution)

$$dp_{mp} = -\frac{k_{\phi э}12\mu}{bh^3}Q\,dr = \frac{k_{\phi э}12\mu}{h^3}\dot{h}r\,dr = \frac{k_m}{h^3}\dot{h}r\,dr$$

Where$k_{\phi э}$— correction factor that allows (if necessary) to take into account the deviation of the structure of the real flow in the gap from that adopted in the first approximation (at this stage it is taken equal to unity);

$k_т$— equivalent resistance coefficient;

$$k_m = k_{\phi э}12\mu$$

μ— Dynamic viscosity coefficient.

Acceleration of fluid in the gap (at average speed)

$$\frac{d}{dt}v(r,h,\dot{h}) = \frac{\partial v}{\partial r}\dot{r} + \frac{\partial v}{\partial h}\dot{h} + \frac{\partial v}{\partial \dot{h}}\ddot{h} = -\frac{\dot{h}}{h}\left(-\frac{\dot{h}r}{h}\right) + \frac{\dot{h}r}{h^2}\dot{h} - \frac{r}{h}\ddot{h}.$$

Therefore, when moving both up and down

$$\dot{v} = \frac{2r\dot{h}^2}{h^2} - \frac{r\ddot{h}}{h}$$

The pressure increment in the gap in the dr section is due to the inertia of the fluid, expressed as a function of the distance r from the middle of the girdle during accelerated movement of the valve and the fluid in the gap

$$dp_{ин} = -\rho\dot{v}dr = -\rho\left(\frac{2r\dot{h}^2}{h^2} - \frac{r\ddot{h}}{h}\right)dr = -\left(\frac{2\rho r\dot{h}^2}{h^2} - \frac{\rho r\ddot{h}}{h}\right)dr.$$

The total pressure increment in the dr section as a function of r

$$dp(r) = \left(-\frac{2\rho\dot{h}^2}{h^2} + \frac{\rho\ddot{h}}{h} + \frac{k_m\dot{h}}{h^3}\right)rdr.$$

After integration, we obtain the law of pressure distribution over the girdle width

$$p(r) = \frac{1}{2}\left(-\frac{2\rho\dot{h}^2}{h^2} + \frac{\rho\ddot{h}}{h} + \frac{k_m\dot{h}}{h^3}\right)(r^2 - \frac{a^2}{4}),$$

Where a— girdle width.

IOP Conf. Series: Materials Science and Engineering **779** (2020) 012043 doi:10.1088/1757-899X/779/1/012043

The maximum pressure in the gap (in the middle of the girdle)

$$p_{\max} = -\frac{a^2}{8}\left(-\frac{2\rho\dot{h}^2}{h^2} + \frac{\rho\ddot{h}}{h} + \frac{k_m\dot{h}}{h^3}\right) =$$

$$= \frac{\rho\dot{h}^2 a^2}{4h^2} - \frac{\rho\ddot{h}a^2}{8h} - \frac{k_m\dot{h}a^2}{8h^3}.$$

The physical meaning of the terms of the last expression can be interpreted as follows:

The first term is called the convective component of pressure. It is equally determined by two phenomena: firstly, it is, in fact, a manifestation of the velocity head of a fluid moving in the gap; secondly, this is a manifestation of the inertial pressure from the acceleration that occurs when the fluid moves from a region with a lower velocity near the middle of the girdle to a region with a higher velocity near its edge. It depends squarely on the speed of lowering the valve.

The second term is called the reduced pressure component. It is determined by the inertia forces caused by the acceleration component of the fluid, which arises due to the accelerated movement of the valve (the speed of the displaced fluid is kinematically related to the speed of the valve). This term linearly depends on the acceleration of the valve.

The third term is called the viscosity component of pressure. It linearly depends on the speed of movement of the locking element and is determined by the viscosity component of the hydraulic resistance of the gap.

The vertical force (composed of the forces acting on both half-belts) is obtained by integrating the pressure over the area of the belt

$$P = 2\int_0^{a/2} p(r)b\,dr = \frac{2b\rho a^3\dot{h}^2}{12h^2} - \frac{b\rho a^3\ddot{h}}{12h} - \frac{k_m b a^3\dot{h}}{12h^3}$$

Note that the inertia forces appear in the fluid squeezed out of the thin gap, appear even with uniform movement of the valve. In this case, the second term of the last expression is zeroed, but the first remains. Moreover, even with an insignificant viscosity of the liquid and, accordingly, the smallness of the third term, a force preventing the closure of the gap will still arise, and its value will be determined by the first term.

Experiment

To confirm the above calculations, as well as to determine the degree of influence of the listed physical effects on the valve fitting process, a computational and physical experiment was conducted.

A computational experiment was carried out by numerically solving the differential equation of motion of a spring-loaded valve, obtained on the basis of the above expressions

$$m\ddot{h} = \frac{2b\rho a^3\dot{h}^2}{12h^2} - \frac{b\rho a^3\ddot{h}}{12h} - \frac{k_m b a^3\dot{h}}{12h^3} - mg - k_{mp}\dot{h} - c(h + h_0)$$

or

$$(m + \frac{b\rho a^3}{12h})\ddot{h} + (\frac{k_m b a^3}{12h^3} + k_{mp})\dot{h} - \frac{2b\rho a^3}{12h^2}\dot{h}^2 + c(h + h_0) = -mg$$

Where h_0— initial spring preload;

c— spring stiffness;

$k_{\text{тр}}$— coefficient of viscous friction when the locking element moves along the guide.

In this equation, it is worth paying attention that an additional term is added to the mass of the valve, which expresses the reduced mass of the liquid squeezed out of the closable gap and increases as the valve lowers.

The initial data (parameters and initial conditions) adopted during the simulation correspond to the parameters of the experimental setup:

m=0,142 кг; a=0,006 м;
ρ=1000 кг/м^3; b=0,157 м;
$k_м$=12; $k_{тр}$=13,1 Н·с/м;
c=1226 Н/м; h_0= 0;
$h(0)$=0.01 м; $\dot{h}(0)$=0 м/с.

At the initial moment, the valve is raised above a flat surface to a height h (0) and accelerates downward under the action of a spring and gravity. As it approaches the supporting plane, it is slowed by the liquid layer (water) between the flat girdle at its lower end and the supporting flat surface.

Two cases were simulated. In the first case, both viscous friction forces and inertia forces occurring in the liquid layer displaced from the gap are taken into account. In the second case, only viscous friction forces are taken into account. In both cases, the maximum valve speed is approximately the same and is vmax = 0.57 m / s. During braking, the maximum calculated acceleration in the first case is 668 m / s^2, and in the second –3560 m / s^2. Accordingly, the maximum pressure (in the middle of the girdle) is in the first case 0.15 MPa, and in the second –0.93 MPa. The corresponding graphs are shown in Fig. 2 and 3.

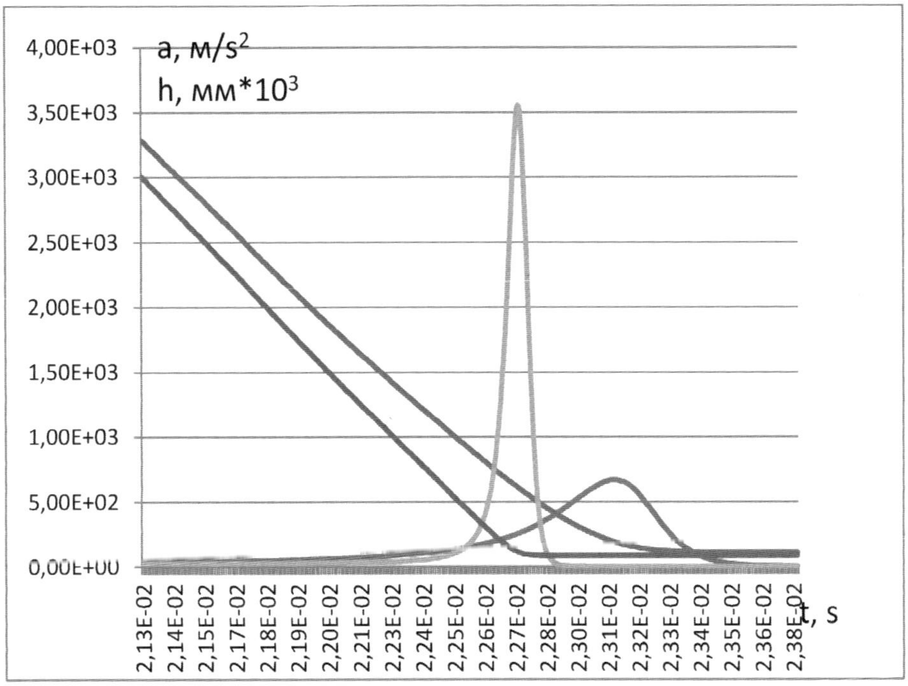

Fig. 2. The calculated dependences of the position (mm * 10^3) and valve acceleration (m / s^2) versus time (s) taking into account the inertia forces in the liquid (corresponding blue and red graphs) and without taking into account the inertia forces (green and violet graphs)

To conduct a physical experiment, the setup shown in Fig. 4 is used. It consists of a steel base with a flat upper surface, a steel guide cylinder with a cylindrical inner surface and a spring-loaded aluminum cup with a flat girdle on the lower surface, acting as a valve. The cup has the ability to freely move inside the guide cylinder but with a small gap. During the experiment, the entire structure is placed in a container of water. Grooves and holes are made in the cup, the guide cylinder and the base, which allow the water displaced from the closable gap to flow freely. To measure the cup acceleration at the fitting time, a shock accelerometer *PCB 350 C 04* is fixed in it.

IOP Conf. Series: Materials Science and Engineering **779** (2020) 012043 doi:10.1088/1757-899X/779/1/012043

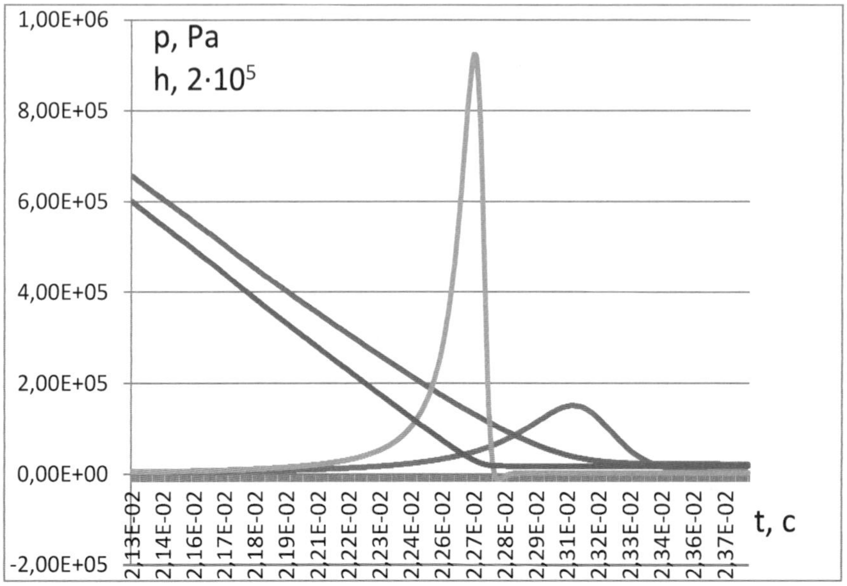

Fig. 3. The calculated dependences of the position (mm * $2 \cdot 10^5$) and the maximum (in the middle of the girdle) pressure of the displaced liquid (Pa) versus time (s) taking into account the inertial forces in the liquid (corresponding blue and red graphs) and without taking into account the inertia forces (green and purple graphics)

Setupparameter:	
– Cup weight	137 g;
– Spring weight	14 g;
– Spring stiffness	1230 N/м;
– Girdle outer diameter	56 мм;
– Girdle inner diameter	44 мм.
Parameters of shock accelerometer:	
– Sensitivity	0,1мV/(м/s²) (1,0 мV/*g*);
– Maximum impact acceleration	±49000 м/s² (±5000 *g*);
– Frequency range	(±1dB) from 0.4 to 10000 Hz
–Mass	5,4 g.

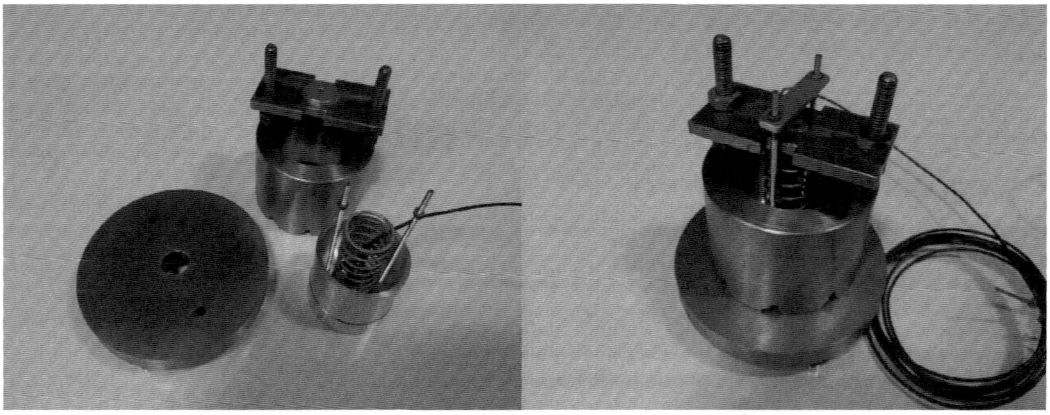

a) б)

Fig. 4. The experimental setup a) disassembled, b) assembled

IOP Conf. Series: Materials Science and Engineering **779** (2020) 012043 doi:10.1088/1757-899X/779/1/012043

The electric pulse generated by the shock accelerometer when the cup is located on the base in the presence of water in the gap and recorded by the oscilloscope is shown in Fig. 5.

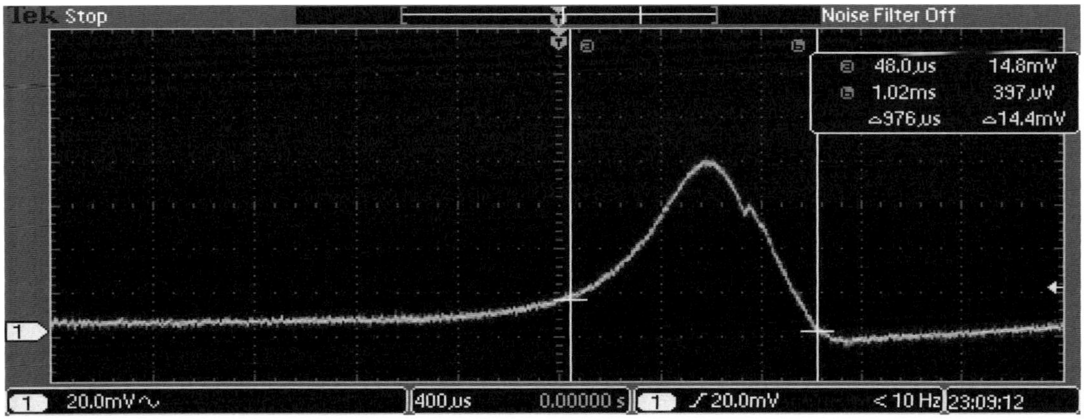

Fig. 5. Recording on an oscilloscope an acceleration pulse recorded by a shock oscilloscope during a physical experiment. Experiment graph of acceleration during the valve fitting layout taken in the experiment

Generalized data obtained as a result of processing data from computational and physical experiments:

For the physical experiment:

Maximum acceleration Impulse duration.

760м/s^2;**0,85**мs

For the computational experiment:

Maximum acceleration, duration, pressure:

– Option 1 **668**м/s^2; (-12%) **0,8**мs (-6,3%) 0,15 MPa;

– Option 2 3560 м/s^2; 0,93MPa.

Maximum speed: v_{max}=0,57м/c (approximately the same in both versions)

The data presented show a good approximation between theory and experiment, which allows us to use the above mathematical model of the valve fitting process in other studies. Figure 6 shows the calculated graph of the dependence of the largest (peak) pressure in the closing gap depending on the width of the girdle (other parameters remain the same). Reducing the width leads to a sharp, not proportional increase in peak pressure. So with a decrease in the width of the girdle to 0.5 mm, the calculated peak pressure reaches 370 MPa. Obviously, in this case, to increase the accuracy of the calculation, it is necessary to use a more complete mathematical model that takes into account the compressibility of the liquid. However, the goal of obtaining the above model was just to obtain a rational combination of valve assembly parameters at which pressures do not reach large values, and the working surfaces of the valve and saddle are durable.

In the same way, one can obtain the dependence of the peak pressure in the gap on the viscosity of the liquid (Fig. 7). The graph shows that this dependence is manifested much weaker.

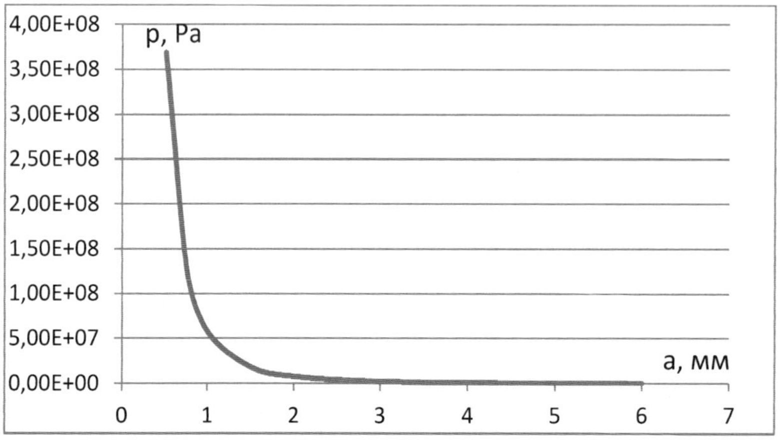

Fig. 6. Dependence of the largest (peak) pressure in the closing gap on the width of the girdle (other valve parameters are the same)

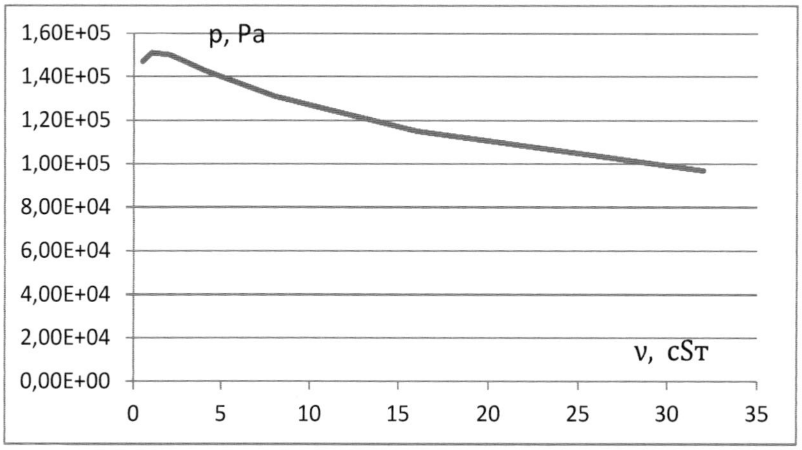

Fig. 7. The dependence of the liquid viscosity peak pressure in the gap

Conclusion

The results show that the inertia forces arising in a thin layer of fluid displaced from the closable gap between the closing valve and the saddle can be not only significant, but decisive in the process of valve seating on the seat, which requires consideration when designing volumetric pump valves.

References

[1] Semenov, S., Kulakov, D. Mathematical modeling of the mechanisms of volumetric hydraulic machines (2019) IOP Conference Series: Materials Science and Engineering, 492 (1), № 012042

[2] Rekach, F.V., Shambina, S.L., Sinichenko, E.K., Belousov, Y.V.Influence of the valve-pull-down equipment on the area of high pressure peaks in long pipelines(2019) International Journal of Recent Technology and Engineering, 8 (2 Special Issue 11), pp.

[3] Gouskov, A., Lomakin, V., Banin, E., Kuleshova, M., Chaburko, P.Investigation of the influence of centrifugal pump wet part geometry on hemolysis index(2019) IOP Conference Series: Materials Science and Engineering, 492 (1), № 012013, .

[4] Demikhov, K.E., Ochkov, A.A., Gordeeva, U.S.Experimental study of a vacuum fluid transportation system(2018) IOP Conference Series: Materials Science and Engineering, 387 (1), № 012015, .

[5] Naumov, V., Pogulyaev, Y., Baytimerov, R., Chizhov, D.New Diesel Engine Fuel Supply System Able to Control Pressure of Pre- and Post-Main Injections(2015) SAE Technical Papers, 2015-September, .
[6] Bastet.M., Rudnevs.S., Nekrasov.B. IDR. Hydraulics, hydraulic machines and hydraulic drives. Textbook for engineering universitics-2nd ed., pererab. - M.: mechanical engineering, 1982 — - 423 p.

IOP Conf. Series: Materials Science and Engineering **779** (2020) 012044 doi:10.1088/1757-899X/779/1/012044

Application of optimization algorithms to improve the vibroacoustic characteristics of pumps

V Lomakin[1,2], T Valiev[1] and P Chaburko[1]

[1]Bauman Moscow State Technical University

[2]E-mail: lomakin@bmstu.ru

Abstract: The article describes an attempt to optimize the flow part of the centrifugal pump by vibroacoustic characteristics, suggests a method of evaluation of these properties by methods of hydrodynamic modeling based on the analysis of force pulsations and moments acting on the rotor and stator of the flow part. The obtained data are decomposed into a spectrum by the Fourier method and presented in the form of dependence of sound pressure level on frequency. By results of optimization, the model with the best vibroacoustic characteristics within the limits of numerical modeling is found.

Introduction

In certain areas of application, the characteristics of the pumps may be subject to requirements for vibration and noise characteristics, for example, when it takes to design pumping units for the submarine fleet [1–3].

Improving the vibroacoustic characteristics of the pump (reducing the noise and vibration generated by a pump) is a non-trivial task for the engineer. In addition to the well-known recommendations for the design of such equipment (the choice of basic sizes, the selection of the number of impeller blades and guide diffusers) there is a possibility to improve the vibroacoustic properties by varying the values of a large number of geometric parameters of the flow part (angles of impeller blades, diffusors and return guide vanes, the shape of the blades, etc.), which affect to the noise and vibrations [4–8]. Thus, it is necessary to carry out optimization on the specified parameters. It should not be forgotten that it is often necessary to consider other criteria, such as efficiency, NPSH value, etc. Therefore, the following purpose was set in our work: to improve the vibroacoustic characteristics of the centrifugal pump flow part by means of optimization methods.

The following tasks had to be solved for this purpose:

- Select the optimization parameters and criteria;
- Choose the method of evaluation of vibroacoustic characteristics based on CFD methods;
- Perform optimization and compare the obtained results.

The object of the study in this work was chosen as the flow part of the four-stage hermetic pump, presented in Figure 1, for which the vibroacoustic characteristics are a priority.

Optimization method — an algorithm based on global and targeted search methods. Optimization parameters are presented in Table 1. First of all these are design parameters directly affecting the flow, namely: the angles of blades in the impeller and diffusor and return guide channels, the number of blades and their angular length, width and diameter of the wheel.

Content from this work may be used under the terms of the Creative Commons Attribution 3.0 licence. Any further distribution of this work must maintain attribution to the author(s) and the title of the work, journal citation and DOI.
Published under licence by IOP Publishing Ltd

IOP Conf. Series: Materials Science and Engineering **779** (2020) 012044 doi:10.1088/1757-899X/779/1/012044

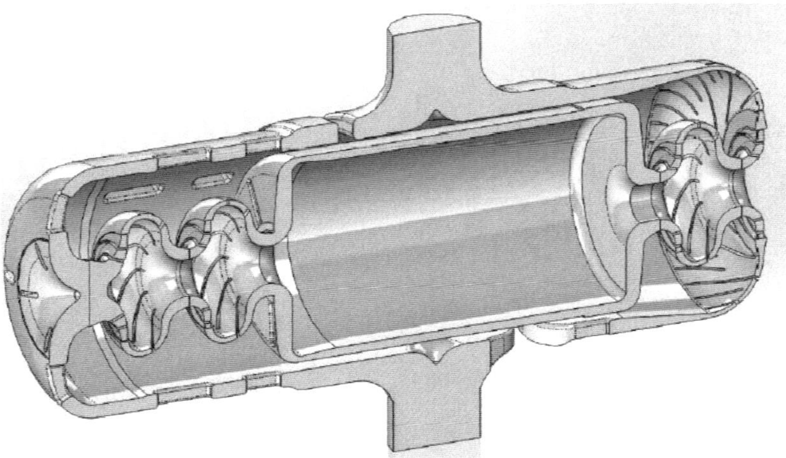

Figure 1. The flow part of the hermetic pump

The optimization criterion was the sum of sound pressure levels at 5 different frequency ranges. Also limitation on the pressure of one stage was introduced like a criterion.

Table 1. Parameters of optimization

Parameter	Value	
	Min	Max
Blade angle at the impeller output, deg	10	30
The angle of the blade at the entrance to the guide diffuser, deg	5	40
Output impeller width, mm	12	25
Diameter of impeller, mm	190	220
Blades of impeller	6	8
Blades of guide diffuser	8	12
Angular length of impeller blade, deg	80	140
Angular length of diffusor blade, deg	50	110

Methods

The main question in setting the task of optimizing the presented pump by its vibroacoustic characteristics is the completeness of the used mathematical model [9]. The direct method of calculation is the solution of the conjugate problem, i.e. the calculation of the flow field and the calculation of the housing wall deformation simultaneously. However, such statement of the problem requires long-term preparation of models and is demanding to computational resources, so in the process of optimization it is inexpedient to use it [10–13]. To compare the vibroacoustic characteristics of the flow part, it is most convenient to use a calculation model that includes only the flow part. At the same time, we will calculate the force influence of the flow on solid walls as integral parameters of the pressure field and evaluate the spectral characteristics of this influence.

In this work, we used the STAR CCM + software product specializing in hydrodynamic modeling as an applied software package.

IOP Conf. Series: Materials Science and Engineering **779** (2020) 012044 doi:10.1088/1757-899X/779/1/012044

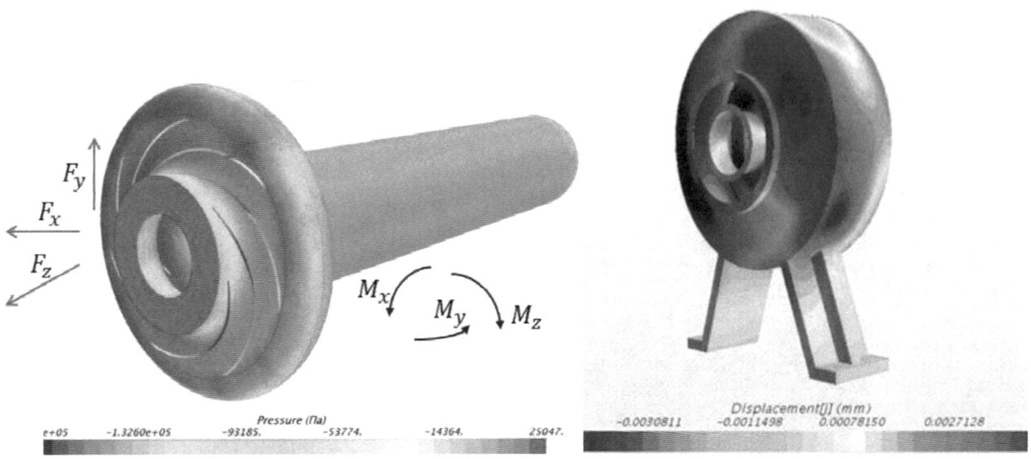

Figure 2. Methods of estimation of vibroacoustic characteristics at CFD modeling

The method of numerical modelling is based on solving discrete analogs of the basic hydrodynamic equations [14–15]. In the case of an incompressible fluid model (ρ = const), this is:

Mass conservation equation (continuity equation):

$$\frac{\partial u_j}{\partial x_j} = 0 \quad (1)$$

where u_j — the averaged value of the fluid velocity in the projection on the j axis (j = 1, 2, 3);

The equation of momentum conservation (Reynolds averaging):

$$\rho\left[\frac{\partial u_i}{\partial t} + u_i\frac{\partial u_i}{\partial x_j}\right] = -\frac{\partial p}{\partial x_i} + \frac{\partial}{\partial x_i}\left[T_{ij} - \rho\overline{u_i' u_j'}\right] \quad (2)$$

where u, p — averaged speed and pressure;

$$T_{ij} = 2\mu s_{ij} - \frac{2}{3}\frac{\partial u_j}{\partial x_j} - \textit{viscous stress tensor for incompressible fluid};$$

$$s_{ij} = \frac{1}{2}\left(\frac{\partial u_i}{\partial x_j} + \frac{\partial u_j}{\partial x_i}\right) - \textit{instant strain rate tensor};$$

$$\rho\overline{u_i' u_j'} - \textit{Reynolds stresses};$$

$$u_i' - \textit{ripple component of velocity}$$

$$\mu - \textit{the dynamic viscosity coefficient of the fluid}$$

The introduction of the Navier—Stokes equation, Reynolds-averaged, makes the system of equations non-closed, since additional unknown Reynolds stresses appear. A semiempirical k—ω SST model of turbulence has been used for solving the system in this task, which introduces the necessary additional equations: the transfer equations for the kinetic energy of turbulence and the relative dissipation rate of the energy:

$$\frac{\partial k}{\partial t} + u_j\frac{\partial k}{\partial x_j} = P_k - \beta^* k\omega + \frac{\partial}{\partial x_j}\left[\left(\mu + \sigma_k \nu_t\right)\frac{\partial k}{\partial x_j}\right]; (3)$$

$$\frac{\partial \omega}{\partial t}+u_j\frac{\partial \omega}{\partial x_j}=\alpha\frac{\omega}{k}P_k-\beta\omega^2+\frac{\partial}{\partial x_j}\left[\left(\nu+\sigma_\omega\nu_t\right)\frac{\partial \omega}{\partial x_j}\right]++2\cdot\left(1-F_i\right)\sigma_{\omega 2}\cdot\frac{1}{\omega}\frac{\partial k}{\partial x_i}\frac{\partial \omega}{\partial x_i}, \quad (4)$$

The following combination of boundary conditions was used to calculate the flow rate in the pump:

- Stagnation on inlet;
- Velocity on outlet

The simulation was carried out in an unsteady statement with a time step of 10^{-5} s and 10^{-4} s. The number of internal iterations per time step is 10.

The calculation grid has different topology. In the core of the cell flow are polyhedrons of different shapes and sizes. Cells near solid walls are polyhedral prisms stretched in the direction perpendicular to the wall. The number of prismatic layers is taken equal to 5 with a stretch coefficient of 1.3. The thickness relative to the base size of the polyhedral cell is 35%. The total number of cells was about 1 million elements. The calculation was carried out on one pump stage.

Results

Figure 3 shows the pressure field established by the calculation. The forces and moments in the unsteady statement and processing of the received data were taken by means of the tool of the frequency analysis built in a program complex. Fast Fourier Transform was used to isolate amplitudes and frequencies of the spectrum of forces and moments obtained.

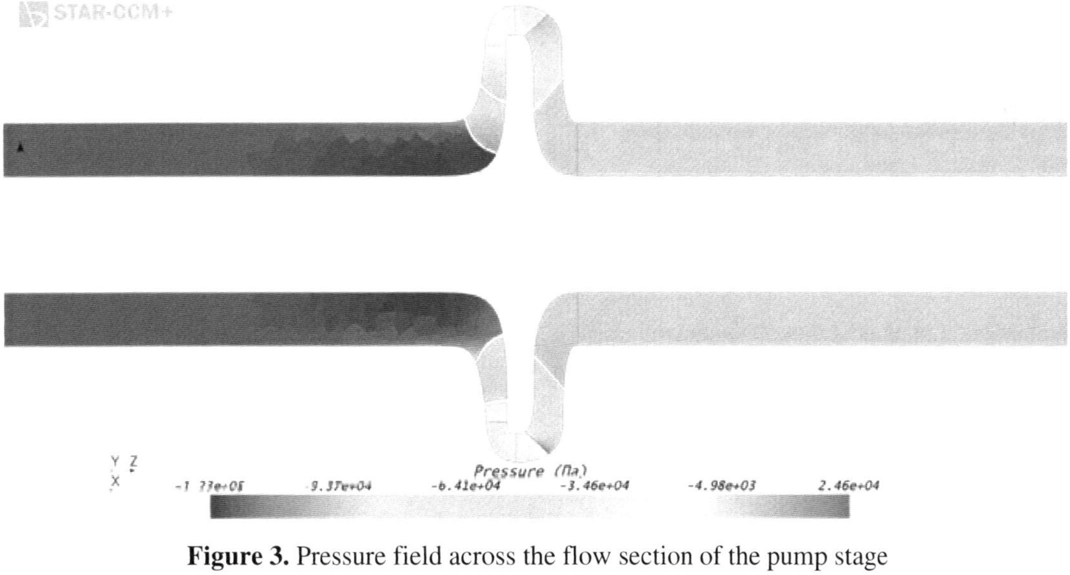

Figure 3. Pressure field across the flow section of the pump stage

According to the results of the numerical experiment, the forces and moments depend on time in the flow parts of the wheel and the guide apparatus. Figure 4 shows a graph of forces in three axes of coordinates acting on the impeller.

The graph below shows that these forces fluctuate with several frequencies. To determine the frequencies and amplitudes we used the Fourier transform, the essence of which is to decompose the function into elementary components — harmonic oscillations with different frequencies. As a result, we obtained the spectrum shown in Figure 5, where the level of sound pressure as an axis of ordinates, and the frequency as an axis of x. The reference pressure is 1 μPa for water. Green graph is the best model as a result of optimization, red and blue graphs are the results obtained in the calculation of models with available experimental data. As can be seen, the best result was shown by 106 model from the optimization, the amplitudes of oscillations were significantly lower than those of other models in the low-frequency part of the spectrum.

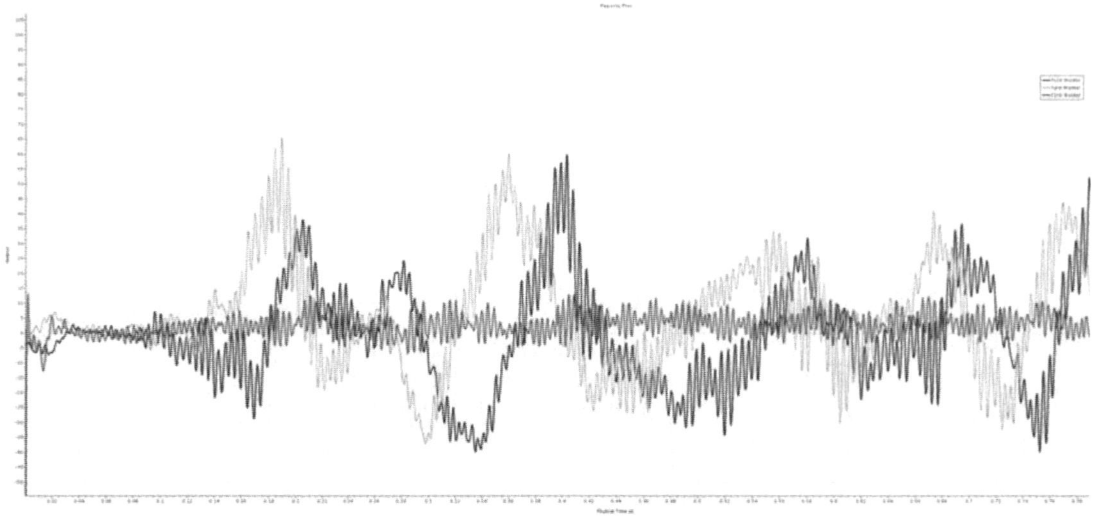

Figure 4. Graphs of forces on the impeller in three axes

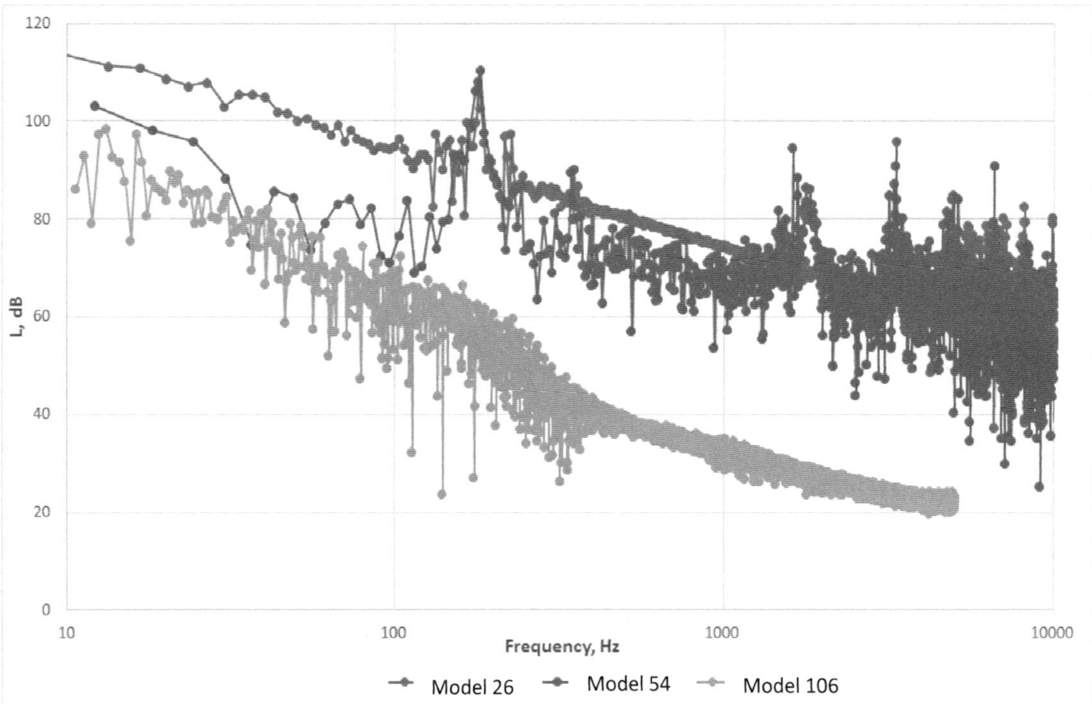

Figure 5. Sound pressure level depending on frequency

Conclusion

Thus, we can draw the following conclusions from the work done:

- With the help of optimization methods, the vibroacoustic properties of the flow part of the hermetic centrifugal pump have been improved;
- Numerical calculation of the flow part with calculation of the force influence of the flow on the solid walls and their subsequent decomposition into the amplitude-frequency spectrum allows us to evaluate the vibroacoustic characteristics.

References

[1] V Lomakin, P Chaburko, M Kuleshova. Multi-criteria Optimization of the Flow of a Centrifugal Pump on Energy and Vibroacoustic Characteristics. Procedia Engineering, 2017, 176: 476–482.

[2] Min'kov, L.L., Krokhina, A.V., Ducckc, J. Hydrodynamic mechanisms of the influence of injection on the classification characteristics of a hydrocyclone (2011) Journal of Engineering Physics and Thermophysics, 84 (4), pp. 807–819.

[3] Ionaitis, R.R., Chain, P.L. Principles for engineering active-passive hydrodynamic control-and-safety systems for nuclear reactors (2009) Atomic Energy, 106 (3), pp. 175–184.

[4] K.A. Morch. Dynamics of Cavitation bubbles and Cavitating liquids //Treatise on Materials Science and Technology, 1976. 16. P. 309–335.

[5] Kendrick H. Light. Development of cavitation erosion resistant advanced material system. Orono (Maine, USA) The University of Maine, 2005. 76 p.

[6] Choi, J.S.; Mclaughlin, D.K.; Thompson, D.E. Experiments on the unsteady flow field and noise generation in a centrifugal pump impeller. J. Sound Vib. 2003, 263, 493–514.

[7] Chu, S.; Dong, R.; Katz, J. Relationship between unsteady flow, pressure fluctuations, and noise in a centrifugal pump—Part A: Use of PDV data to compute the pressure field. J. Fluids Eng. 1995, 117, 24–29.

[8] Hou-lin Liu, Han-wei Dai, Jian Ding, Ming-gao Tan, Yong Wang, Hao-qin Huang. Numerical and experimental studies of hydraulic noise induced by surface dipole sources in a centrifugal pump. Journal of Hydrodynamics, 2016, 28(1): 43–51.

[9] Norton M. P., Karczub D. G. Fundamentals of noise and vibration analysis for engineers. Noise Control Engineering Journal, 2007, 55(2): 275–276.

[10] V Lomakin and O Bibik 2019 IOP Conf. Ser.: Mater. Sci. Eng. 492 012037

[11] T Valiev and APetrov 2019 IOP Conf. Ser.: Mater. Sci. Eng. 492 012038

[12] P Chaburko and Z Kossova 2019 IOP Conf. Ser.: Mater. Sci. Eng. 492 012011

[13] V Lomakin, P Chaburko, M Kuleshova. Multi-criteria Optimization of the Flow of a Centrifugal Pump on Energy and Vibroacoustic Characteristics. Procedia Engineering, 2017, 176: 476–482.

[14] A Petrov, A Lysenko, T Valiev, N Isaev. Optimal design of inlet device of multistage centrifugal pump. 2019 IOP Conf. Ser.: Mater. Sci. Eng. 492 012040.

[15] Patankar S 1984 Numerical methods for solving problems of heat transfer and fluid dynamics (Moscow: Energoatomizdat Publ.) 152 p.

IOP Conf. Series: Materials Science and Engineering **779** (2020) 012045 doi:10.1088/1757-899X/779/1/012045

Experimental study of cavitation resistance of restoring coatings

E Aksenova[1], V Lomakin[1,2] and V Cheremushkin[1]

[1]Bauman Moscow State Technical University
2nd Baumanskaya st., 5, Moscow, 105005 Russian Federation

[2]E-mail: lomakin@bmstu.ru

Annotation. The article presents the results of an experimental study of the resistance of various metallic and nonmetallic coatings to cavitation failure. The method of accelerated cavitation resistance tests using an ultrasonic emitter is described. Comparative characteristics of the resistance of materials are given.

Introduction

The cavitation phenomenon has been known for a long time and is quite common. As a rule, its formation is negative, since the pair forms various structural elements on the surface (impellers of pumps, hydraulic turbines, narrow sections in cooling systems), which leads to their gradual and sometimes very intense wear [1].

Despite the popularity and prevalence, this phenomenon is difficult to mathematically describe. There are semi-empirical models in packages of numerical hydrodynamic modeling [2, 3]. It should be noted that such models require the determination of a number of empirical coefficients, for example, by comparing the results of modeling and experiment [4]. Sometimes cavitation models are supplemented by erosion models that allow one to assess the degree of surface wear [5].

Due to the imperfection of the mathematical apparatus for calculating cavitation, one of the main methods for assessing its intensity and influence is experiment [6, 7, 8]. Two main approaches can be distinguished in the experimental study of cavitation at the moment. The first is to directly test the device (some hydraulic system) in which cavitation wear occurs. This method will allow to determine the localization of cavitation spots, but not its intensity, since it takes a lot of hours to obtain significant cavitation erosion.

The second approach involves the application of accelerated cavitation wear methods, for example, cavitating jet and ultrasonic [9], which is described in [10].

This approach allows you to determine the resistance of a particular material to the effects of cavitation, if it is already known where exactly it takes place.

In this article, we considered the question of determining the relative cavitation resistance of various coatings applied using different technologies. Usually it is a question of determining the relative cavitation resistance [11, 12, 13], since it is difficult to draw a correspondence between the results of accelerated tests and wear in a real design.

This question is relevant at present, since many real structures where cavitation occurs can no longer be fundamentally changed to completely eliminate it. However, the use of cavitation-resistant coatings can significantly extend the service life of the units, bringing significant benefits.

Content from this work may be used under the terms of the Creative Commons Attribution 3.0 licence. Any further distribution of this work must maintain attribution to the author(s) and the title of the work, journal citation and DOI.
Published under licence by IOP Publishing Ltd

IOP Conf. Series: Materials Science and Engineering **779** (2020) 012045 doi:10.1088/1757-899X/779/1/012045

Description of the laboratory setup and selected materials

For testing, metallic and non-metallic coatings were selected. For electric arc surfacing, four different coating materials were selected. Their properties are showning the table below:

Table 1. Mechanical properties of welding wires

№ s/p	Name	σ_f, MPa	σ_t, MPa	KCV for 10°C, J/sm2	KCV for -20°C, J/sm2	KCV for -45°C, J/sm2	T
1	4	640	840	-	69	56	-
2	3	600	840	100	-	-	-
3	1	-	-	-	-	-	380 HB
4	2	-	-	-	-	-	540 HB

These welding wires are characterized by high strength properties or high surface hardness. Wire 3 is used for the manufacture of hydropower equipment operating in conditions of cavitation wear. However, the achievement of the parameters indicated for it is ensured only after a long heat treatment, which is impossible in most cases of repair of hydraulic turbines, therefore, during laboratory tests, the sample coated with this material was also not subjected to heat treatment.

Wire 4 is a flux-cored wire, the deposited metal of which is characterized by high strength and plastic properties in combination with good corrosion resistance. The main field of application is the production of technological equipment for the pulp and paper industry and offshore platforms for the processing and transportation of oil and gas.

Wires 1 and 2 are solid wires and are recommended for surfacing products subject to intense mechanical wear. The deposited metal is characterized by high hardness due to the presence in the composition of a large amount of carbon (C) and chromium (Cr). For electrospark surfacing the following materials were selected: 6 — rhenium, 7 — molybdenum, 8 — VK8, 9 — nickel. For high strength properties, alloy 8 was selected for laboratory tests. The main mechanical properties are shown in table 2.

Table 2. Mechanical properties of the alloy

№ s/p	Name	$\sigma_т$, MPa	$\sigma_в$, MPa	a_HJ/sm^2	Hardness
1	8	-	1666	35	87,5 HRC

Belzona coatings were chosen as a non-metallic coating.

10 — Two-component polyurethane resin for coating metal and rubber parts of equipment that protects against cavitation and abrasive wear.

11 — Two-component high-temperature epoxy coating to protect equipment operating in continuous immersion at temperatures up to 95 ° C from erosion and corrosion.

12 — A two-component epoxy coating specially designed to increase the efficiency of pumps, pipelines, valves and other hydraulic pump equipment, which at the same time provides long-term protection against erosion and corrosion.

Since there are no domestic standards containing the installation diagram and methods for testing cavitation samples using the ultrasonic method, the American standard ASTM G32-10 "Standard test method for cavitation erosion using vibratory apparatus" was taken as a basis

The installation scheme for testing samples taken from the mentioned standard is shown in Figure 1. The installation includes a high-frequency generator, a magnetostrictive ultrasonic emitter, a metal extension cord — a waveguide, a container with distilled water for immersing the sample, and a thermal stabilization system.

IOP Conf. Series: Materials Science and Engineering **779** (2020) 012045 doi:10.1088/1757-899X/779/1/012045

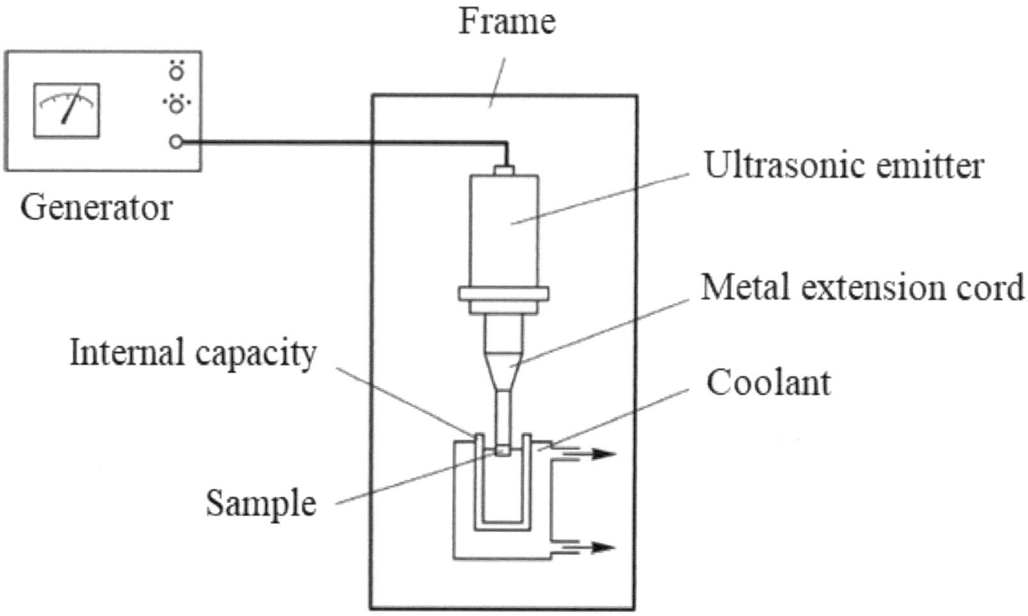

Figure 1. Installation diagram for testing samples for cavitation erosion by the ultrasonic method

Figure 2 shows an ultrasonic emitter and a high-frequency generator vertically mounted on a frame. A metal extension cord (horn) is rigidly attached to the end face of the output core of the emitter (on which ultrasonic vibrations are generated), to the end of which, in turn, a test specimen made of various materials is rigidly attached to the thread (test specimen). In this case, ultrasonic vibrations are transmitted from the core to the extension cord and from it to the sample due to the fact that their end metal surfaces are tightly adjacent to each other.

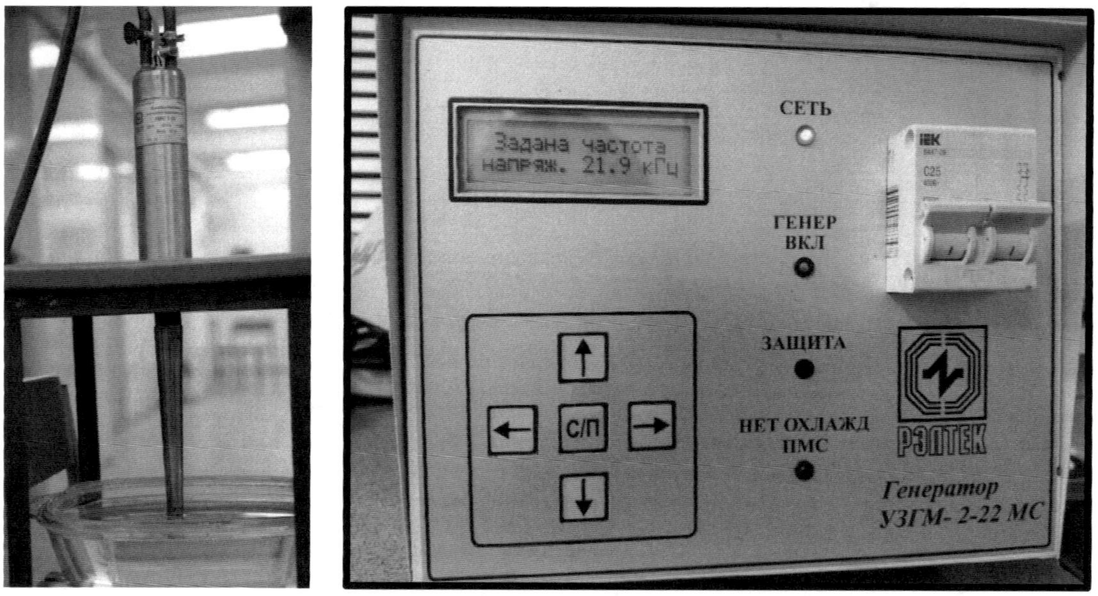

Figure 2. Photos of the test bench for samples created at the Department of E-10 MSTU. N.E. Bauman (the emitter on the stand on the left, the ultrasonic oscillation generator on the right)

IOP Conf. Series: Materials Science and Engineering **779** (2020) 012045 doi:10.1088/1757-899X/779/1/012045

In order to maintain a constant water temperature during the tests, a thermal stabilization system was mounted on the test bench based on the digital thermostat STH0024UG-v3 with a remote sensor DS18B20. This module has all the necessary functionality and technical characteristics for the task. It also compares favorably with its analogues with a small measurement error (± 0.5 °C) and a customizable hysteresis from 0.1 °C.

The degree of cavitation wear was controlled by the gramometric method using an analytical balance (accuracy — +/– 0.1 mg). The tests were carried out in pure water with a temperature of plus 25 ± 0.5 ° C using the high-frequency vibration of the sample with a frequency of 20 ± 0.5 kHz. The oscillation amplitude of the sample was (according to the documentation for the emitter) 50 nm, the radiated acoustic power (measured on an ultrasonic oscillator) ranged from 0.9 to 1 kW.

Test results.

The test results are the obtained comparative dependences of the weight loss of the samples on time, as well as the rate of weight loss, presented in the form of the graphs below. In addition to weighing, visual inspection of the end surfaces of the samples was also carried out.

Sample 1

Sample 2

Sample 3

Sample 4

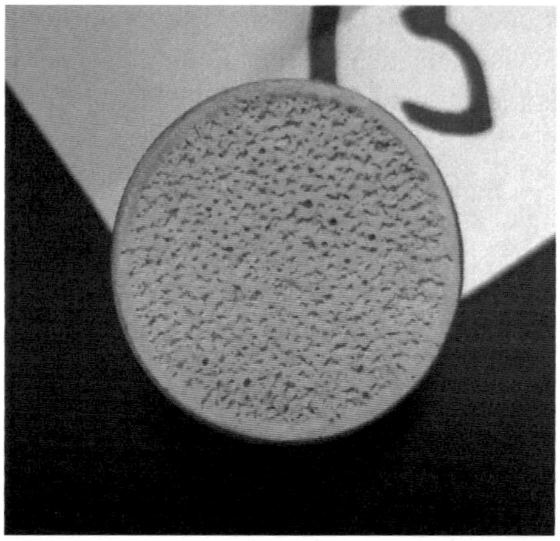

Sample10

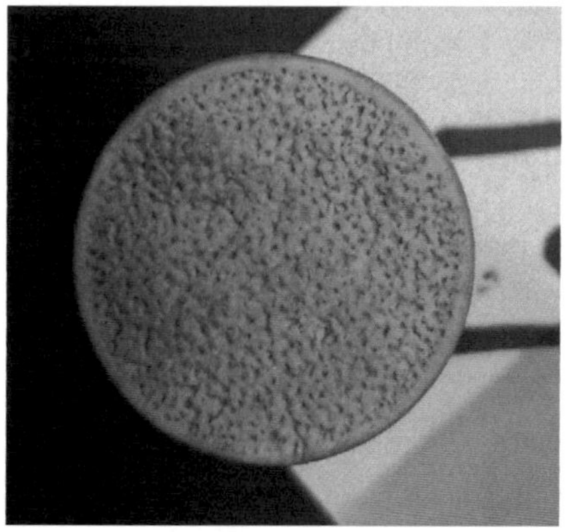

Sample11

Sample12

12X18H10T

06X12H3Д

The test results are presented in the plots below.

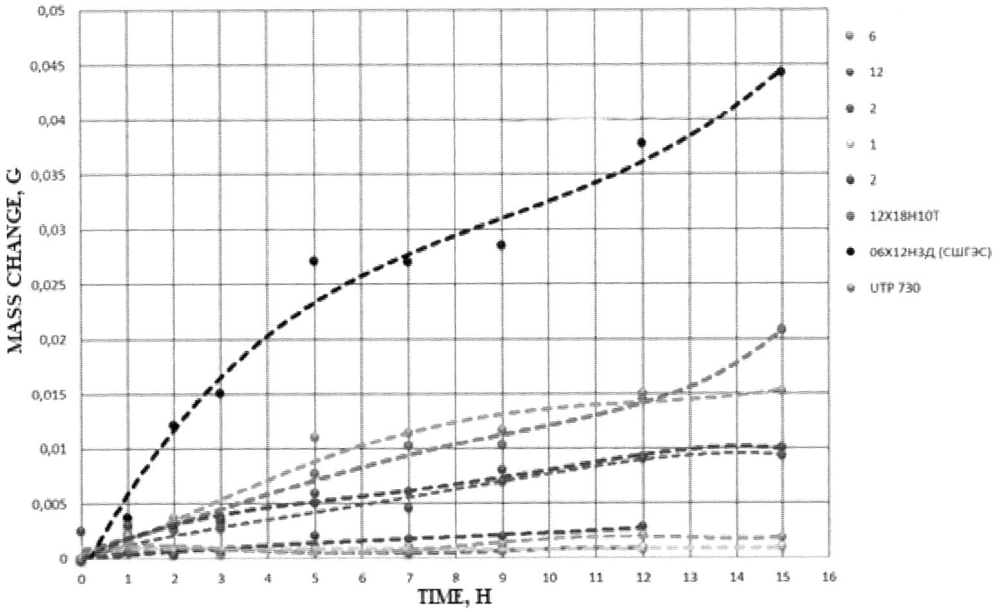

Plot 1. Comparative dependence of weight loss on time of metal samples

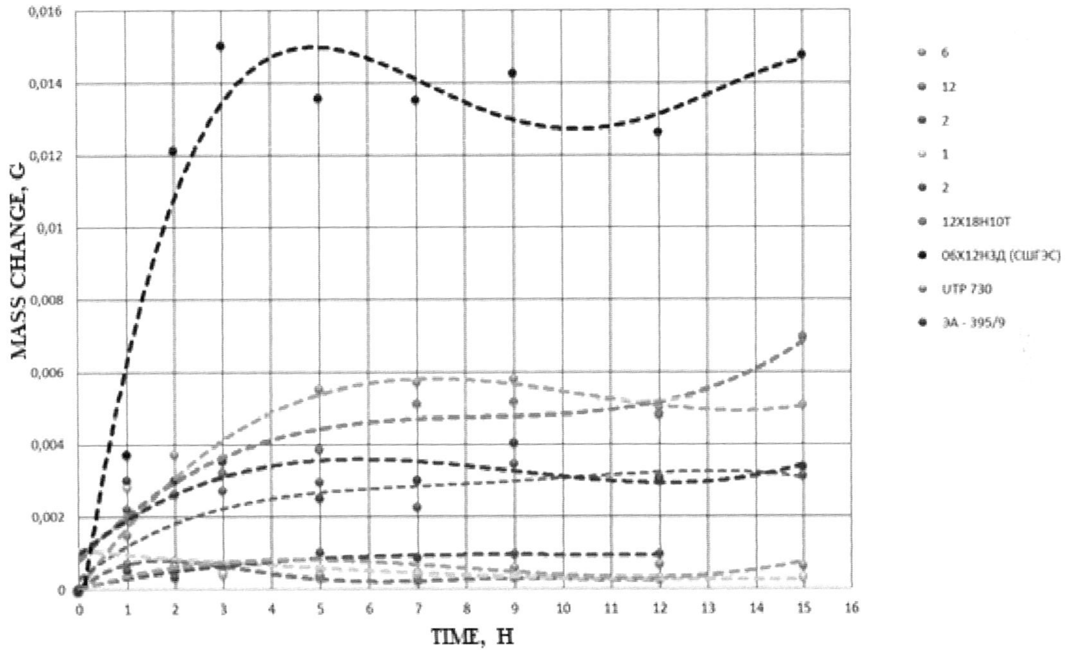

Plot 2. Comparative dependences of weight loss on time of non-metallic samples

Conclusion

Based on the results of the tests, graphs were constructed for a relative comparison of the cavitation resistance of coating materials.

As a result of comparative tests, polymer coatings showed a result no worse than the best metal coatings

IOP Conf. Series: Materials Science and Engineering **779** (2020) 012045 doi:10.1088/1757-899X/779/1/012045

List of references.

[1] I. Pirsol, Cavitation, Mills and Boon Ltd., London (1972).

[2] Lomakin V. Investigation of two-phase flow in axial-centrifugal impeller by hydrodynamic modeling methods. Proceedings of 2015 International Conference on Fluid Power and Mechatronics, FPM 2015, No 7337302, pp 1204–1206. DOI: 10.1109/FPM.2015.7337302.

[3] Lomakin V., Kulesova M., Kraeva E. Fluid Flow in the Throttle Channel in the Presence of Cavitation. Procedia Engineering, Vol 106, pp 27–35, 2015. DOI: 10.1016/j.proeng.2015.06.005

[4] Lomakin V., Bibik O. Numerical prediction of the gas content effect on the cavitation characteristics of the pump using the simplified Rayleigh-Plesset equation. IOP Conference Series: Materials Science and Engineering, Vol 492, Issue 1, 2019. DOI: 10.1088/1757-899X/492/1/012037.

[5] Timushev, S., Knyazev, V., Panaiotti, S., Soldatov, V., Rohatgi, U.Numerical procedure for assessment of centrifugal pump cavitation erosion. ASME International Mechanical Engineering Congress and Exposition, Proceedings, Vol 14, pp 7–78, 2009. DOI: 10.1115/IMECE2008-67177

[6] Potoceanu, N., Nedeloni, M. Chirus, D., Florea, D. Cavitation erosion behavior of the antifriction alloy YSn83. International Symposium on Metallography, Metallography 2013, Vol 782, pp 257–262, 2014. DOI: 10.4028/www.scientific.net/MSF.782.257.

[7] Petrov, A., Saatashvili, G., Averianov, A., Bilenko, I. Study of the corrosion rate of the paths walls of aluminum heat exchangers in the ethylene glycol antifreezes environment under developed aeration of power fluid. IOP Conference Series: Materials Science and Engineering, Vol 589, Issue 1, No 012031, 2019. DOI: 10.1088/1757-899X/589/1/012031.

[8] Li, Z., Han, J., Lu, J., Zhou, J.,Chen, J. Vibratory cavitation erosion behavior of AISI 304 stainless steel in water at elevated temperatures. Wear, Vol 321, pp 33–37, 2014. DOI: 10.1016/j.wear.2014.09.012

[9] Hattori, S., Maekawa, N.,Kuwahara, M.Comparison of cavitation erosion between cavitating jet and vibratory methods specified in ASTM standard. Nihon Kikai Gakkai Ronbunshu, A Hen/Transactions of the Japan Society of Mechanical Engineers, Part A, Vol 67, Issue 655,pp 470–475, 2001.

[10] ASTM G32-10 «Standard test method for cavitation erosion using vibratory apparatus» Pub.Dec. 1, 2010.

[11] Li, Z., Han, J., Lu, J., Chen, J.Cavitation erosion behavior of Hastelloy C-276 nickel-based alloy. Journal of Alloys and Compounds, Vol 619, pp 754-759, 2015. DOI: 10.1016/j.jallcom.2014.08.248.

[12] Court, S., Corni, I.,Symonds, N.Cavitation erosion performance of Steel, Ceramics, Carbide, and Victrex peek materials. Materials Performance and Characterization, Vol 7, Issue 5, 2018. DOI: 10.1520/MPC20180027.

[13] A Protopopov and D Bondareva 2019 IOP Conf. Ser.: Mater. Sci. Eng. 492 012002

IOP Conf. Series: Materials Science and Engineering **779** (2020) 012046 doi:10.1088/1757-899X/779/1/012046

Optimization of parameters of the pressure regulator with variable characteristic

I Kolodin

Bauman Moscow State Technical University, Moscow

E-mail: kolodin@student.bmstu.ru

Abstract. When designing a variable capacity pressure regulator, it is necessary to select parameters that meet certain criteria. In this article optimization of regulator parameters by LPτ search method is considered. The mathematical model of the regulator is made, a possible set of quality criteria is given, dependent criteria are established and the majority of approximate effective points are found. Setting of criteria restrictions is performed in the dialog mode. The methods described in the article allow to obtain an optimal design of the pressure regulator with variable characteristic.

Introduction

The use of pumps with a pressure regulator is a widespread way to increase the energy efficiency of hydraulic systems. Improving the performance of hydraulic systems is an urgent task [1–4]. In many hydraulic systems operating at constant pressure, there is a need to change the pressure itself. There are circuits with a variable characteristic in which, at a constant load, a change in the characteristic causes a transition to a new static state with a new pressure value.

Choosing the parameters of a pressure regulator is a complex multi-criteria task [5–9]. Choosing a set of quality criteria is also not an easy task. It is necessary to determine a set of quality criteria, which, on the one hand, reflects as much information as possible about the operation of the system, and on the other hand, is minimal for solving the optimization problem.

There are various methods for solving optimization problems [10–13]. We will optimize the parameters of a pressure regulator with a variable characteristic by the LPτ search method, which is widely used and underlies many other methods [14–19].

The mathematical model of the regulator.

The schematic diagram of the controller is presented in Fig. 1.

The principle of operation and the mathematical description of the equilibrium of this regulator is given in [20]. Consider the dynamics equations of the regulator.

Spool movement equation:

$$m_1 \cdot \frac{d^2}{dt^2} x + k_{f1} \cdot \frac{d}{dt} x + c_1 (x_0 + x) + p_{ext} \cdot A_{cf} = p_d \cdot A_{sf} \quad (1)$$

Where m_1 — reduced mass of the spool,

Content from this work may be used under the terms of the Creative Commons Attribution 3.0 licence. Any further distribution of this work must maintain attribution to the author(s) and the title of the work, journal citation and DOI.
Published under licence by IOP Publishing Ltd

t— Time
x — Spool offset from neutral,
k_{f1} — reduced damping coefficient of the spool,
c_1 — The stiffness of the spring acting on the spool,
x_0 — initial spring preload,
p_{ext} —external pressure
A_{cf} —theareaaffectedbyexternalpressure,
p_d —is the discharge pressure
A_{sf} —The area of the end face of the spool.

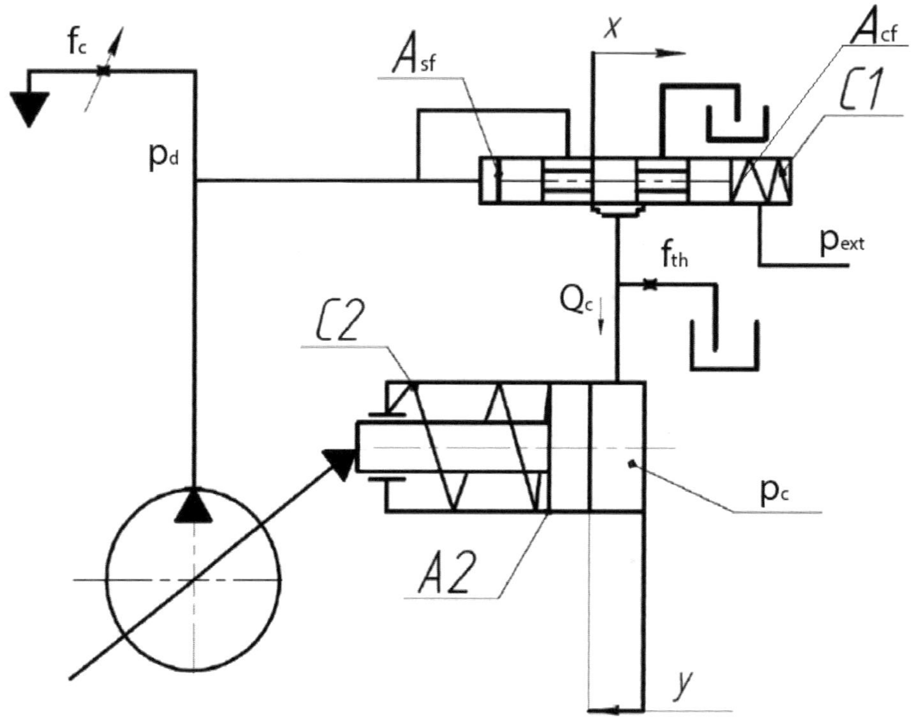

Fig. 1. Schematic diagram of the controller.

Equation of piston movement:

$$m_2 \cdot \frac{d^2}{dt^2} y + k_{f2} \cdot \frac{d}{dt} y + c_2 \cdot y = p_c \cdot A_2 \quad (2)$$

Where m_2 — the reduced mass of the piston,
y — The coordinate of the piston position,
k_{f2} — The reduced damping coefficient of the piston,
c_2 — The stiffness of the control piston spring,
p_c — The control pressure
A_2 — The area of the end face of the control piston.
The equation of the balances of the costs of the pump - throttle simulating the load:

$$Q_d + Q_{com1} = Q \quad (3)$$

Where Q_d — flow rate on the throttle simulating the load,
Q_{com1} — Flow rate for fluid compression in the discharge line,
Q — pump feed.
We replace the input quantities with the expressions for their calculation.

$$\mu \cdot f_d \cdot \sqrt{\frac{2 \cdot p_d}{\rho}} + \frac{V_{lin}}{B}\frac{dp_d}{dt} = Q_{max} - k_c y \quad (4)$$

Where μ — flow rate
f_d — flow area of the throttle simulating the load,
ρ — Fluid density,
V_{lin} —discharge line volume (assumed constant),
B — Reduced volume modulus of elasticity of the working fluid,
Q_{max} — Maximum pump flow.

$$k_c = \frac{Q_{max}}{y_{max}} \quad (5)$$

Where y_{max} — maximum movement of the control piston.
The equation of the balance of expenses:

$$Q_{fs} = Q_{com2} + Q_{ds} + Q_c + Q_{th} \quad (6)$$

Where Q_{fs} — flow rate through the filling gap,
Q_{com2} — Fluid compression rate in the control cavity,
Q_{ds} — Flow rate through the drain slot
Q_c — control flow,
Q_{th} — flow on the throttle.
Replace the incoming values with expressions to calculate them.

$$\mu b X_f \sqrt{\frac{2}{\rho}} \cdot \sqrt{p_d - p_c} = \frac{V_c}{B}\frac{dp_y}{dt} + \mu b X_d \sqrt{\frac{2}{\rho}} \cdot \sqrt{p_c} + A_2 \frac{dy}{dt} + \mu \cdot f_{th} \cdot \sqrt{\frac{2 \cdot p_c}{\rho}} \quad (7)$$

Where b — the total width of the dispenser windows,
V_c — Volume of the control cavity (assumed constant),
f_{th} — Throttle bore area,
The following spool geometry model is adopted:

$$X_f = -2R + \sqrt{(\delta + 2R)^2 + 0,25 \cdot \left(S + 2R + x + |S + 2R + x|\right)^2} \quad (8)$$

$$X_d = -2R + \sqrt{(\delta + 2R)^2 + 0,25 \cdot \left(S + 2R - x + |S + 2R - x|\right)^2} \quad (9)$$

Where R — spool edge radius
δ — Spool radial clearance,
S—spool edge overlap.

Mathematical Modeling Results

A program was written in the Python programming language that allows the entire optimization process to be performed for this device. It includes an LPτ sequence generator, subroutines that accept a list of variable parameters and return a list of criteria corresponding to it, a dialog algorithm for specifying criteria constraints, and a generator of a set of approximately effective points. To calculate transients was used the scipy.integrate.solve_ivp function from the scipy library, which allows to find a solution to systems of ordinary differential equations. The integration method BDF is used (an implicit multi-step method of variable order (from 1 to 5) based on the backward differentiation formula). To work with data arrays, the numpy library is used.

To determine the optimal parameters of the controller, the following variable parameters were identified:

$a_1 = A_2$—end face of the control piston,

$a_2 = c_2$—control piston spring stiffness,

$a_3 = A_{sf}$ —Spool end face,

$a_4 = c_1$ — The stiffness of the spring acting on the spool,

$a_5 = f_{th}$ —throttle bore area,

$a_6 = b$ — The total width of the dispenser windows,

$a_7 = S$ — overlapping spool edges,

$a_8 = k_c$ — Coefficient of influence of piston movement on pump flow,

$a_9 = k_{f1}$ — reduced damping coefficient of the spool,

$a_{10} = k_{f2}$ —Reduced piston damping coefficient.

The parameters were varied using an LPτ sequence generator based on an arithmetic algorithm. At the output of the generator, we have an LPτ sequence in a unit cube, which is then recalculated for any given parallelepiped P. The accepted number of points in the sequence is N=256.

The value calculation A_{cf}, x_0 is made separately for each selected point in the space of variable parameters so that the static characteristics set has the same value of the regulation start pressure, which is the initial value for design.

Variable parameters are sought in the multidimensional parallelepiped P, which is determined by the inequalities:

$$a_i^* \le a_i \le a_i^{**} \; (i = 1, 2 \ldots 10) \quad (10)$$

Where a_i^* — the restriction on the ith criterion from below,

a_i^{**} —restriction on the ith criterion from above.

Before solving the optimization problem, a_i^0 — the values of the parameters corresponding to some prototype. And found $\Phi_\vartheta^0 \; (\vartheta = 1, 2 \ldots 16)$ — values of the quality criteria for the prototype.

Tab. 1. Values of parametric restrictions on the values of the parameters of the prototype

i	1	2	3	4	5	6	7	8	9	10
a_i^* / a_i^0	0,56	0,67	0,56	0,90	0,88	0,83	0,80	0,82	0,61	0,75
a_i^{**} / a_i^0	1,27	3,06	1,59	1,10	1,32	1,17	1,50	1,36	1,46	1,25

The following quality criteria were assigned as optimality criteria for the regulator.

Group I (criteria related to the static characteristic).

$$\Phi_1 = \frac{1}{k_{Qp1}}$$

Where k_{Qp1}—gain flow pressure at a flow rate of $\frac{Q_{max}}{2}$ and at maximum p_{ext}.

$$\Phi_2 = Q_{fs01}$$

Where Q_{fs01} — flow rate through the filling gap at the flow rate point $\frac{Q_{max}}{2}$ and at maximum p_{ext}.

$$\Phi_3 = \frac{1}{k_{Qp2}}$$

Where k_{Qp2}—the pressure gain of the flow rate at the point with the flow rate $\frac{Q_{max}}{2}$ and at maximum p_{ext}.

$$\Phi_4 = Q_{fs02}$$

Where Q_{fs02} — flow rate through the filling gap at the flow rate point $\frac{Q_{max}}{2}$ and at maximum p_{ext}.

Group II (criteria related to the dynamics of the regulator).

For each selected point in the space of variable parameters, the following system responses are constructed $p_d(t)$.

1) System response to step increase p_{ext}. Initially, the system is in equilibrium at a point with a flow rate of $\frac{Q_{max}}{2}$ with a minimal p_{ext}. The magnitude of the step effect at the input is chosen such that the change in p_d is $\frac{1}{3}$ of the original value.

2) System response to step increase p_{ext}. Initially, the system is in equilibrium at a point with a flow rate of $\frac{Q_{max}}{2}$ with a maximal p_{ext}. The magnitude of the step effect at the input is chosen such that the change in p_d is $\frac{1}{3}$ of the original value.

3) System response to step increase f_d. Initially, the system is in equilibrium at a point with a flow rate of 0 with a minimal p_{ext}. The magnitude of the step effect at the input is chosen such that the flow rate varies from 0 to $\frac{Q_{max}}{2}$.

4) System response to step increase f_d. Initially, the system is in equilibrium at a point with a flow rate of 0 with a minimal p_{ext}. The magnitude of the step effect at the input is chosen such that the flow rate varies from Q_{max} to $\frac{Q_{max}}{2}$.

IOP Conf. Series: Materials Science and Engineering **779** (2020) 012046 doi:10.1088/1757-899X/779/1/012046

5) System response to step increase f_d. Initially, the system is in equilibrium at a point with a flow rate of 0 with a maximal p_{ext}. The magnitude of the step effect at the input is chosen such that the flow rate varies from 0 to $\frac{Q_{max}}{2}$.

6) System response to step increase f_d. Initially, the system is in equilibrium at a point with a flow rate of 0 with a maximal p_{ext}. The magnitude of the step effect at the input is chosen such that the flow rate varies from Q_{max} to $\frac{Q_{max}}{2}$.

A stepwise action is defined by an exponential sigmoid. Operating points, if necessary, are found by the half-division method. For each i-th transient, the transient time t_i and the amount of overshoot are determined by $\Delta h_i \ (i = 1, 2 \ldots 6)$.

Then

$$\Phi_{2i+3} = t_i, (11)$$

$$\Phi_{2i+4} = \Delta h_i . (12)$$

Total we have 16 quality criteria.

After the values of the quality criteria were obtained for each selected point from the space of variable criteria, the correlation coefficients between the criteria were found. The exclusion of dependent criteria was carried out as follows: criteria that were desired to be maintained $(\Phi_1, \Phi_2, \Phi_{15}, \Phi_{16})$, those criteria that had a correlation coefficient with these criteria greater than 0.9 were excluded from consideration. As a result, the set of criteria considered has decreased $(\Phi_1, \Phi_2, \Phi_7, \Phi_{15}, \Phi_{16})$. Test tables were compiled for each of these criteria, based on which criteria limitations were set. Taking into account the criteria constraints, a set of admissible points D was formed, and the criteria constraints were entered in the interactive mode with the subsequent verification of the set D for emptiness.

From the set D, points are selected that form the set of approximately effective ones. An analysis of the table containing the parameters and quality criteria of approximately effective points made it possible to select the final set of parameters.

Tab. 2. Values of the selected parameters relative to the values of the parameters of the prototype.

$\frac{a_1}{a_1^0}$	$\frac{a_2}{a_2^0}$	$\frac{a_3}{a_3^0}$	$\frac{a_4}{a_4^0}$	$\frac{a_5}{a_5^0}$	$\frac{a_6}{a_6^0}$	$\frac{a_7}{a_7^0}$	$\frac{a_8}{a_8^0}$	$\frac{a_9}{a_9^0}$	$\frac{a_{10}}{a_{10}^0}$
1,230	1,313	0,916	0,900	0,948	0,868	1,257	1,253	1,287	0,881

Tab. 3. Values of the quality criteria of the selected option relative to the values of the quality criteria of the prototype.

$\frac{\Phi_1}{\Phi_1^0}$	$\frac{\Phi_2}{\Phi_2^0}$	$\frac{\Phi_3}{\Phi_3^0}$	$\frac{\Phi_4}{\Phi_4^0}$	$\frac{\Phi_5}{\Phi_5^0}$	$\frac{\Phi_6}{\Phi_6^0}$	$\frac{\Phi_7}{\Phi_7^0}$	$\frac{\Phi_8}{\Phi_8^0}$
0,999	0,875	0,997	0,875	1,009	0	0,857	0,773
$\frac{\Phi_9}{\Phi_9^0}$	$\frac{\Phi_{10}}{\Phi_{10}^0}$	$\frac{\Phi_{11}}{\Phi_{11}^0}$	$\frac{\Phi_{12}}{\Phi_{12}^0}$	$\frac{\Phi_{13}}{\Phi_{13}^0}$	$\frac{\Phi_{14}}{\Phi_{14}^0}$	$\frac{\Phi_{15}}{\Phi_{15}^0}$	$\frac{\Phi_{16}}{\Phi_{16}^0}$
1,453	0	1,067	0	0,758	0,168	1,194	0,034

Conclusions.
A lot of approximately effective points are obtained. The point selected from this set, according to the results of expert evaluation and the selected quality criteria, is the best compared to the point corresponding to the original prototype, and meets the specified requirements. The same calculation can be performed for a larger number of points. In thc presented model, a number of assumptions are made, in particular, it was assumed that the moment on the inclined disk changes slightly.

Reference
[1] Semenov, S., Kulakov, D. Mathematical modeling of the mechanisms of volumetric hydraulic machines (2019) IOP Conference Series: Materials Science and Engineering, 492 (1), № 012042,
[2] Semenov, S.E. Mathematical modeling of the electro-hydraulic actuation systems of the machines with tree-like kinematic structure (2015) Proceedings of 2015 International Conference on Fluid Power and Mechatronics, FPM 2015, № 7337184, pp. 583–592.
[3] Kulakov, D.B., Semenov, S.E., Kulakov, B.B., Shcherbachev, P.V., Tarasov, O.I. Hydraulic Bipedal Robots Locomotion Mathematical Modeling (2015) Procedia Engineering, 106, pp. 62–70.
[4] Kulakov, D., Kulakov, B. Study of the dynamics of an electro-hydraulic servo drive physical model on FESTO learning system (2019) IOP Conference Series:
[5] Fulbright, N.J., Van De Ven, J.D. Dynamic response of pressure compensated variable displacement linkage pump (2018) BATH/ASME 2018 Symposium on Fluid Power and Motion Control, FPMC 2018, .
[6] Henke, R.W. UNDERSTANDING PRESSURE-COMPENSATED PUMPS. (1985) Hydraulics and Pneumatics, 28 (11), pp. 80–83.
[7] Baz, A. OPTIMIZATION OF THE DYNAMICS OF PRESSURE-COMPENSATED AXIAL PISTON PUMPS. (1983) Journal of Fluid Control, 15 (2), pp. 64–81.
[8] Jenkins, R.P., Ivantysynova, M. Investigation of instability of a pressure compensated vane pump (2016) 9th FPNI Ph.D. Symposium on Fluid Power, FPNI 2016
[9] Pery, Arie EXPLANATION OF LOW FREQUENCY PEAK IN PRESSURE COMPENSATED PUMPS DISCHARGE PRESSURE FREQUENCY SPECTRUM. (1984) Proceedings of the National Conference on Fluid Power, Annual Meeting, 38, pp. 119–125.
[10] Rao, R.V. Rao algorithms: Three metaphor-less simple algorithms for solving optimization problems (2020) International Journal of Industrial Engineering Computations, 11 (1), pp. 107–130.
[11] Li, J., Lu, C., Zhang, H. A modified SSDP method for nonlinear semidefinie programming (2020) Journal of Computational Analysis and Applications, 28 (5), pp. 836–847.
[12] Meneses, A.A.D.M., da Silva, P.V., Nast, F.N., Araujo, L.M., Schirru, R. Application of Cuckoo Search algorithm to Loading Pattern Optimization problems (2020) Annals of Nuclear Energy, 139, № 107214, .
[13] Wang, Y., Chen, X. Hybrid quantum particle swarm optimization algorithm and its application (2020) Science China Information Sciences, 63 (5), № 159201, .
[14] Jordanov, I., Georgieva, A. Neural network learning With global heuristic search (2007) IEEE Transactions on Neural Networks, 18 (3), pp. 937–942.
[15] Antonova, G.M. Analysis of real-time-systems by means of simulation and optimization in LPτ -search with averaging (2017) Horizons in Computer Science Research, 15, pp. 69–100.
[16] Tynchenko, V.V., Pavlenko, A.A., Bukhtoyarov, V.V., Tikhonenko, D.V., Tynchenko, S.V., Tsvettsykh, A.V. Formation of initial point of initialization methods for optimization algorithms (2019) Journal of Physics: Conference Series, 1353 (1), № 012115, .
[17] Pavlenko, A.A., Kukartsev, V.V., Tynchenko, V.S., Mikhalev, A.S., Chzhan, E.A., Lozitskaya, E.V. Optimal parameters selection of the genetic algorithm for global optimization (2019) Journal of Physics: Conference Series, 1353 (1), № 012105, .

IOP Conf. Series: Materials Science and Engineering **779** (2020) 012046 doi:10.1088/1757-899X/779/1/012046

[18] Stupina, A.A., Pavlenko, A.A., Tikhonenko, D.V., Tynchenko, V.V., Bukhtoyarov, V.V., Tynchenko, S.V. Study of the point scattering uniform algorithms in R40 space (2019) Journal of Physics: Conference Series, 1353 (1), № 012112, .
[19] Antonova, G.M., Makarov, V.V. Recognition of nonlinear functions for estimating region of efficiency in LP τ-search with averaging (2015) Pattern Recognition and Image Analysis, 25 (2), pp. 193–200.
[20] Kolodin, I., Ryabinin, M. Mathematical representation of pressure regulator with variable characteristic (2019) IOP Conference Series: Materials Science and Engineering, 589 (1), № 012018.

IOP Conf. Series: Materials Science and Engineering **779** (2020) 012047 doi:10.1088/1757-899X/779/1/012047

Hydrodynamic modeling of cavitation in a multistage centrifugal pump during its operation in the constant feed mode with a change in the rotor speed of the pump

M Kasatkin[1] and A Petrov[1,2]
[1]Bauman Moscow State Technical University

[2]E-mail: alex_i_petrov@mail.ru

Abstract: Hydrodynamic modeling of cavitation for a multi-section high-speed centrifugal pump operating at a constant supply is carried out. The cavitation characteristics of the pump for various pump operating modes are determined taking into account changes in the rotor speed. The operation parameters of the mobile pumping system for these modes are calculated. It has been established that improving the cavitation qualities of a pump by reducing the speed of the pump rotor with a constant flow rate is not a productive method.

Introduction

In modern mobile deployable pumping systems, in particular, in oil, fuel or water transportation systems, retaining installations are used to provide the backwater necessary for the cavitation-free operation of the main pumps. Such installations include pumps with a small allowable cavitation margin, as they are required to collect the pumped liquid from various containers in which the liquid level may be below the level of the suction pipe of the pump. The use of a retaining installation is a standard solution for the operation of mobile pumping systems for various liquids, but in this case it is required to operate 2 types of equipment: a linear pumping station and a retaining installation [1]–[4].

In this situation, it becomes urgent to reduce the range of equipment used: to make possible to operate a linear station as a retaining station. It is considered the option of reducing the pump rotor speed of a linear pumping station with a diesel engine and a multistage centrifugal pump at the same value of the liquid flow rate in the system pipeline (by regulating the operation of subsequent stations) to reduce the cavitation stock of the pump and, therefore, the operation of such a station as a booster.

The difficulty in solving the problem lies in the fact that when the speed changes while the pump supply remains constant (due to the operation of other series-connected stations), its operating mode leaves the zone of optimal operating modes. Consequently, the cavitation characteristics of the pump, increasing due to a decrease in the rotor speed, are simultaneously deteriorating due to changes in the conditions of entry into the impeller in off-design modes. Hydrodynamic modeling of a two-phase flow in the first stage of the pump should answer the question of how these two factors together affect the suction capacity of the pump.

Method

Theoretical part.

Hydrodynamic modeling is based on the following equations [5]–[9]:

Content from this work may be used under the terms of the Creative Commons Attribution 3.0 licence. Any further distribution of this work must maintain attribution to the author(s) and the title of the work, journal citation and DOI.
Published under licence by IOP Publishing Ltd

- Fluid continuity equation:

$$\frac{\partial \overline{u_x}}{\partial x}+\frac{\partial \overline{u_y}}{\partial y}+\frac{\partial \overline{u_z}}{\partial z}=0,$$

Where $\overline{u_i}$ —time-averaged projections of fluid velocities on the corresponding axes.

- equation of change in momentum averaged over time:

$$\rho\left[\frac{\partial \overline{u_i}}{\partial t}+\overline{u_j}\frac{\partial \overline{u_i}}{\partial x_j}\right]=-\frac{\partial \overline{p}}{\partial x_i}+\frac{\partial}{\partial x_i}\left[T_{ij}^{(v)}-\rho u_i u_j\right],$$

Where $\overline{w_i}$ — average relative speed:

$\vec{w}=\vec{V}-\vec{u}$ — Ratio of relative, absolute and portable speed;

$\overline{p}$ — Averaged pressure;

$\tilde{T}_{ij}^{(v)}=2\mu\tilde{s}_{ij}$ — Viscous stress tensor for incompressible fluid;

$\tilde{s}_{ij}=\frac{1}{2}\left[\frac{\partial \overline{w_i}}{\partial x_j}+\frac{\partial \overline{w_j}}{\partial x_i}\right]$ —deformation rate tensor;

$\rho w_i w_j$ — Reynolds stresses.

Due to the presence of Reynolds stresses, the system of equations cannot be closed. In this case, the k-ω SST turbulence model is used to close the system. A feature of this model is the combination of the advantages of both k-ω and k-ε models: the k-ε model is used in the flow core, and the k-ω model in the near-wall region.

Two equations of transport of turbulence parameters are the basis of this model:

- equation of kinetic energy transfer of turbulence:

$$\frac{\partial k}{\partial t}+\overline{u_j}\frac{\partial k}{\partial x_j}=P_k-\beta^* k\omega+\frac{\partial}{\partial x_i}\left[\left(\nu+\sigma_k\nu_T\right)\frac{\partial k}{\partial x_j}\right],$$

Where $k=\frac{\left(u_x'^2\right)+\left(u_y'^2\right)+\left(u_z'^2\right)}{2}$ —kinetic energy of turbulence;

u_i' —Velocity fluctuations;

P_k — Factor of turbulence energy generation;

ω — Relative turbulence dissipation rate;

ν_T — Turbulent viscosity.

- Equation of transfer of relative velocity of turbulence energy dissipation:

$$\frac{\partial \omega}{\partial t}+\overline{u_j}\frac{\partial \omega}{\partial x_j}=\alpha S^2-\beta\omega^2+\frac{\partial}{\partial x_i}\left[\left(\nu+\sigma_\omega\nu_T\right)\frac{\partial \omega}{\partial x_j}\right]+2\left(1-F_j\right)\sigma_{\omega 2}\frac{1}{\omega}\frac{\partial k}{\partial x_j}\frac{\partial \omega}{\partial x_j},$$

The Boussines q hypothesis allows us to calculate Reynolds stresses in the equations of dynamics:

$$\rho u_i u_j=2\mu_T\left[\frac{1}{2}\left(\frac{\partial \overline{u_i}}{\partial x_j}+\frac{\partial \overline{u_j}}{\partial x_i}\right)-\frac{1}{3}\frac{\partial \overline{u_k}}{\partial x_k}\delta_{ij}\right]-\frac{2}{3}\rho k\delta_{ij},$$

Where δ_{ij} —Kronecker symbol.

Using empirical closure coefficients of these equations allows a numerical solution of the turbulent fluid flow in the computational domain.

The following equation is required to calculate the pump head:

$$H = \frac{p_2}{\rho g} + \frac{V_2^2}{2g} - \frac{p_1}{\rho g} - \frac{V_1^2}{2\rho},$$

Where p_2 — pressure at the outlet of the first stage guide vane;

p_1 — pump inlet pressure;

V_2 — Average speed at the exit of the first stage guide vane;

V_1 — Average speed at the pump inlet.

In the numerical model, the rotor speed and pump feed are set.

Turbulent fluid flow prevails in the flow part of the pump.

For the region of fluid flow that rotates with the rotor, the equation of change in momentum is written as follows:

$$\rho\left[\frac{\partial \overline{w_i}}{\partial t} + \overline{u_j}\frac{\partial \overline{w_i}}{\partial x_j}\right] = -\frac{\partial \overline{p}}{\partial x_i} + \frac{\partial}{\partial x_i}\left[T_{ij}^{(v)} - \rho w_i w_j\right] + F_u + F_\kappa,$$

Where $F_u = \rho\omega^2 r$ —the tension forces of inertia;

$F_\kappa = 2\omega w_{xy}$ —Coriolis force tension.

The calculation of the pump parameters was carried out in the following sequence: first, the calculation was carried out in a stationary setting, and after the relative convergence of the parameters — non-stationary. This approach can significantly reduce the time of hydrodynamic calculation.

To compare the cavitation characteristics of the pump in different operating modes, the cavitation speed coefficient given by S.S. Rudnev, is calculated by the formula:

$$C_k = \frac{5{,}62 n\sqrt{Q}}{\Delta h^{\frac{3}{4}}}$$

Practical part.

The 3D model of the calculated area is imported in the Parasolid (*) format.x b). The first (most cavitation prone) stage of a multistage centrifugal pump, which includes three subregions (regions): inlet, impeller, and guide vane, is taken as the calculation area. In this case, the pump inlet was applied in a special form, previously optimized for such multistage pumps [10].

This simulation does not take into account fluid flows between regions and flows in the space between the casing and the impeller disks. Such assumptions greatly simplify the calculation, but slightly reduce the accuracy of the calculation.

Figure 1 shows the imported model of the computational domain.

All boundaries are walls, except for the entrance to the pump (Inlet) and exit from the stage (Outlet). The types of boundaries for them are presented in table 1.

For the calculation, the following grid construction parameters are used (Table 2).

IOP Conf. Series: Materials Science and Engineering **779** (2020) 012047 doi:10.1088/1757-899X/779/1/012047

Figure 1. Imported three-dimensional model of the calculated area of the flow part of the first stage of the pump.

Table 1. Correspondence of borders with their types.

Border name	New type
Outlet	Input speed
Inlet	Input stagnation

Table 2. The main parameters of grid construction.

Parameter	Property	Value
Base	Value	3,5 мм
Prismatic layer thickness> Relative size	Absolute	1,5 мм
Surface Size> Relative minimum size	Percentage of base	25
Surface Size> Relative Desired Value	Percentage of base	100
Prismatic Stretching	Value	1.3
The number of prismatic layers	Value	5

In the region of the impeller, the values of the fluid flow rates reach their maximum mark, which causes the largest velocity gradients in this region. This circumstance significantly affects the calculation, therefore, the grid of this area needs to be crushed for a more accurate calculation.

The mesh parameters of the impeller region are presented in table 3.

In total, the regions were divided into 546083 cells, of which 231494 cells in the region of the impeller, 246832 cells in the guide vane and 67757 cells in the inlet.

Figure 2 shows the cross section of the areas of the impeller and the guide apparatus.

IOP Conf. Series: Materials Science and Engineering **779** (2020) 012047 doi:10.1088/1757-899X/779/1/012047

Table 3. Parameters of the impeller construction grid

Parameter	Property	Value
Surface Size> Relative minimum size	Percentage of base	20
Surface Size> Relative Desired Value	Percentageofbase	50

Figure 2. The cross section of the impeller with the guide apparatus of the grid model of the calculation area

The following models were used in the calculation:

• Time-non-stationary task;
• working environment — Euler multiphase;
• Euler multiphase model — volume of fluid (VOF);
• Solver — divided;
• Flow regime — turbulent;
• k-epsilon turbulence model. The phases selected are water and steam. The value of the saturation pressure of the water selected value p = 2338 Pa, corresponding to 20 ° C.

Results

In the process of hydrodynamic modeling, scenes of the distribution of the vapor phase in the computational domain are created for clarity. Figure 3 shows the distribution of the vapor phase during non-cavitation operation of the pump, and Figure 4 — in the cavitation mode.

IOP Conf. Series: Materials Science and Engineering **779** (2020) 012047 doi:10.1088/1757-899X/779/1/012047

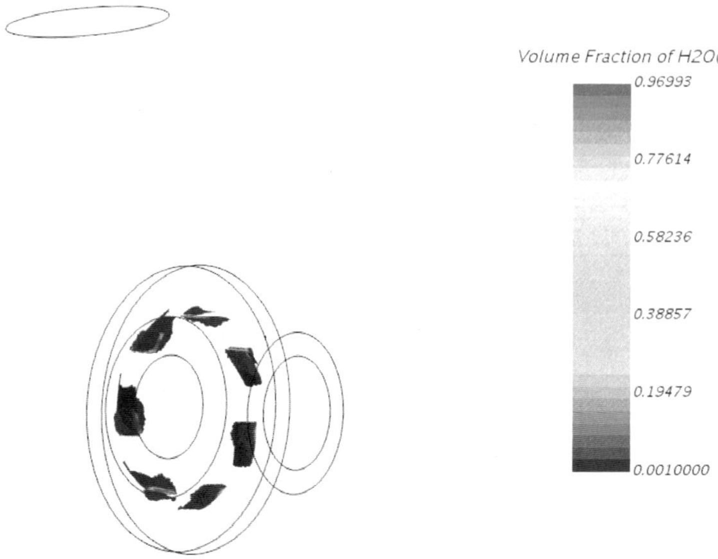

Figure 3. Distribution of the vapor phase in the non-cavitation operation of the pump.

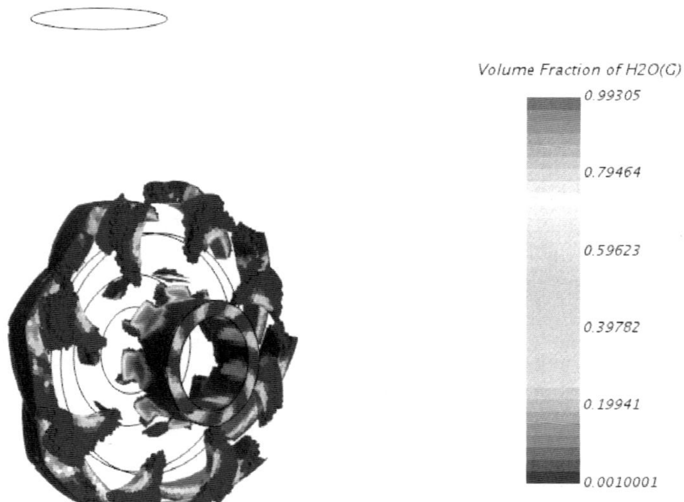

Figure 4. Distribution of the vapor phase in the cavitation mode.

The first basic operating mode of the pump of this hydrodynamic simulation is the point in the optimal operating mode of the pump: the pump rotor speed n = 7000 rpm at a flow rate in the system is $Q = 100\frac{м^3}{h}$. As a result, it is obtained a graph of the partial cavitation characteristics of the pump with these parameters (Fig. 5). Cavitation margin value for the second critical pump mode is $\Delta h = 13{,}3 м$.

IOP Conf. Series: Materials Science and Engineering **779** (2020) 012047 doi:10.1088/1757-899X/779/1/012047

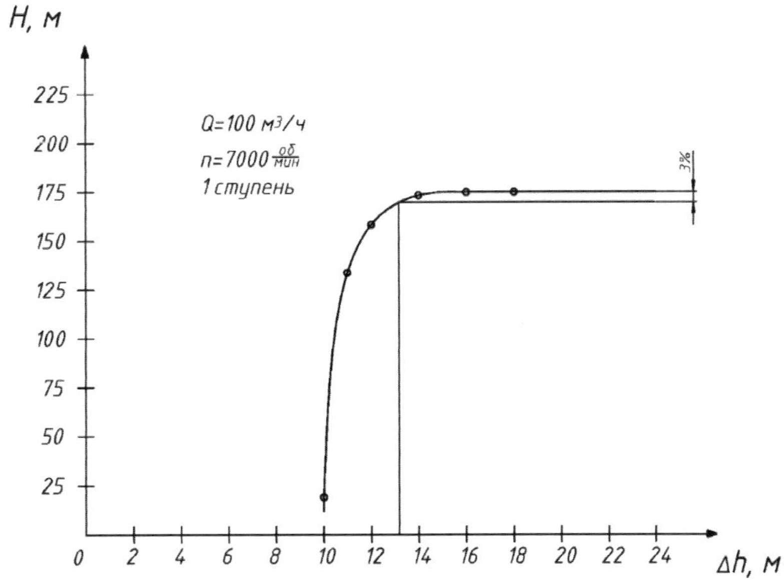

Figure 5. Cavitation characteristic of the pump at $n = 7000\frac{об}{мин}$

The second operating point has the following parameters: pump rotor speed is $n = 5500\,rpm$ at a flow rate of $Q = 100\frac{м^3}{h}$. The graph of the particular cavitation characteristic is shown in Figure 6. The value of the cavitation reserve is $\Delta h = 8,2\,м$.

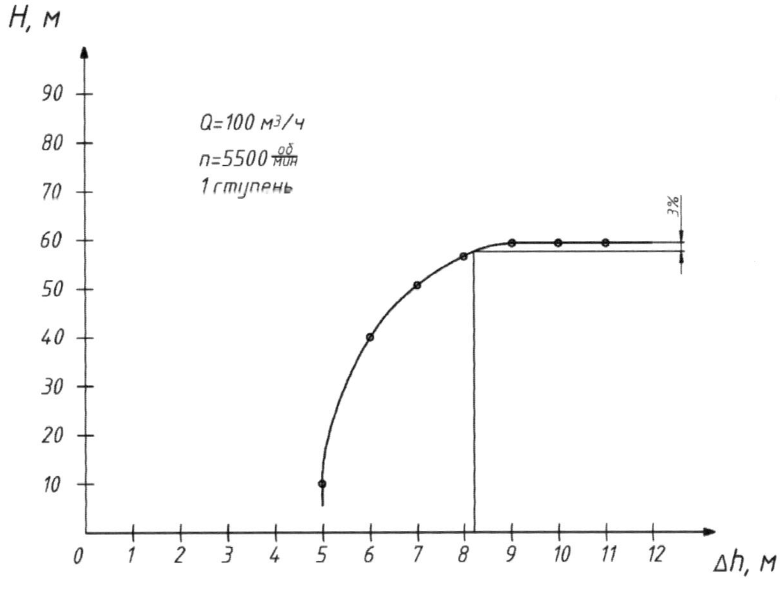

Figure 6. Cavitation characteristic of the pump at $n = 5500\frac{об}{мин}$

IOP Conf. Series: Materials Science and Engineering **779** (2020) 012047 doi:10.1088/1757-899X/779/1/012047

The third working point with the minimum engine rotor speed after the gear ratio of the multiplier has the following parameters: pump rotor speed is $n = 4000\,rpm$ at a flow rate of $Q = 100\frac{м^3}{h}$. The graph of the particular cavitation characteristic is shown in Figure 7. The value of the cavitation reserve is $\Delta h = 8,7\,м$.

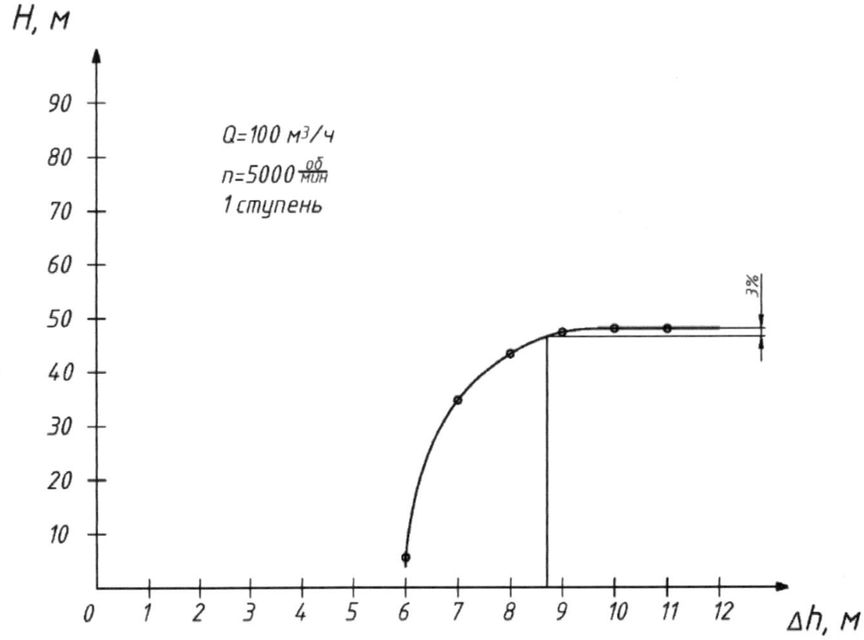

Figure 7. Cavitation characteristic of the pump at $n = 4000\,rpm$

The results of calculating the cavitation characteristics of the pump, as well as the calculated cavitation coefficient of speed are presented in table 4.

Table 4. Cavitation characteristics of the pump in different operating modes.

Name	Flow rate, cubic m / h	Head, m	Cavitation reserve, m	Cavitational speed coefficient
At $n = 5000\,rpm$		48	8,7	928
At $n = 5500\,rpm$	100	60	8,2	1067
At $n = 7000\,rpm$		175	13,3	945

Figure 8 shows a graphical display of the tabular data.

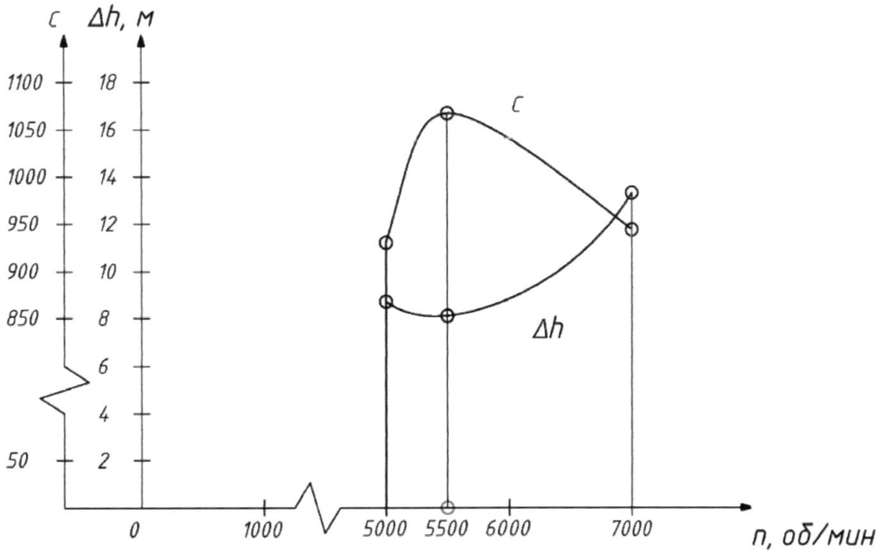

Figure 8. Graphs of the dependences of the cavitation stock and cavitation coefficient of speed on the rotational speed of the pump rotor.

Conclusion

The results obtained show that a decrease in the rotor speed of the pump rotor with a constant feed value can somewhat reduce the allowable cavitation reserve of the pump. But at the same time, the pressure value of the pump drops sharply, since the pump is operating in a non-optimal mode.

Also, on the graph, it can be noticed an inverse relationship: when a certain value of the rotor speed of the pump is reached, the cavitation characteristics begin to deteriorate. This phenomenon is due to the fact that the fluid flow rate to a large extent begins to differ from the rotational speed of the impeller, thereby the impeller no longer transfers energy to the flow, but instead takes it. A flow throttling phenomenon occurs in the flow part, in which cavitation is actively developing, which is the reason for the deterioration of cavitation characteristics.

Therefore, for a pump of this type, the proposed method for increasing cavitation characteristics is in principle permissible, but only with a slight, no more than 20%, decrease in the rotor speed of the pump. With a further decrease in cavitation characteristics again begin to deteriorate.

List of references

[1] A Protopopov and D Bondareva 2019 *IOP Conf. Ser.: Mater. Sci. Eng.* 492 012002
[2] A Protopopov and V Vigovskij 2019 *IOP Conf. Ser.: Mater. Sci. Eng.* 492 012003
[3] V Lomakin et al 2019 *IOP Conf. Ser.: Mater. Sci. Eng.* 492 012012
[4] N Egorkina and APetrov 2019 *IOP Conf. Ser.: Mater. Sci. Eng.* 492 012015
[5] N Isaev 2019 IOP *Conf. Ser.: Mater. Sci. Eng.* 492 012026
[6] A Protopopov and C Jakovich 2019 *IOP Conf. Ser.: Mater. Sci. Eng.* 492 012034
[7] A Petrov et al 2019 *IOP Conf. Ser.: Mater. Sci. Eng.* 492 01203
[8] T Valiev and APetrov 2019 *IOP Conf. Ser.: Mater. Sci. Eng.* 492 012038
[9] V Cheremushkin and V Lomakin 2019 *IOP Conf. Ser.: Mater. Sci. Eng.* 492 012039
[10] *Petrov A., Lysenko A., Valiev T., Isaev N.* Optimal design of inlet device of multistage centrifugal pump. IOP Conference Series: Materials Science and Engineering 2019. — Vol. 492, Issue 1. — Art. no 012040; DOI: 10.1088/1757-899X/492/1/012040.

IOP Conf. Series: Materials Science and Engineering **779** (2020) 012048 doi:10.1088/1757-899X/779/1/012048

Mathematical modeling of a centrifugal pump with a spiral tap of simplified geometry with an open and closed wheel

A Averyanov[1, 2] and A Protopopov[1]

[1]Bauman Moscow State Technical University

[2]E-mail:dronaver98@mail.ru

Annotation. Abstract: the characteristics of a pump with open and closed impellers are compared. The paper considers a model of a centrifugal pump for General industrial purposes. The simulation was carried out in the STAR-CCM + package. A graphical comparison of the influence of different pump design sizes on the parameters is made and the most suitable model is identified.

Introduction

In modern science, there are two large classes of hydraulic machines: volumetric and dynamic. Differ in the principle of action. Volumetric hydraulic machines have an overlapping flow part, that is, each volumetric machine has its own working volume. Dynamic hydraulic machines have a non-overlapping flow part. The most common are the latter. In this article we will talk about such a pump, belonging to the type of blade.

Vane pumps are the most common class of machines used in almost all sectors of the national economy. Questions related to the calculation of centrifugal pumps and their individual parts occupy a lot of space in the literature [1]–[5]. Vane pumps belonging to the class of dynamic pumps have become more widespread. The enormous pace of development of the national economy of the USSR, the creation of new industries have set specific tasks for pumping. It was necessary to perform a large amount of work on the development, research and development of serial production of an extensive range of new types and designs of pumps, the parameters of which differed significantly from the parameters previously produced [6].

Vane pumps are capable of pumping large volumes of liquid in a relatively short time. However, they do not create such high pressures as can create bulk. The efficiency of vane pumps varies in the region of 75-80%.

An important role in the operation of the pump is played by the size of the supply pipe. As practice shows, the show itself is better than a narrow inlet of the device, rather than wider. As we know from the theory of hydraulics, in narrower spaces the fluid velocity is greater and the pressure is less. This often leads to cavitation-the phenomenon of local transition of liquid into steam due to local pressure drop. Cavitation is a harmful phenomenon for pumps. Gas bubbles are surrounded by high pressure and collapse near the surface of the wheel or body, gradually destroying it. Therefore, in practice, often try to make the size of the inlet pipe large. In this article, the first part of the study is devoted to the construction of the pump performance characteristics depending on the size of the supply pipe.

In some cases, pumps with an open (or semi-open) impeller may be used. In particular, it can be waste-mass pumps (type cm), pumps for chemical reagents, pumps for pumping water containing

Content from this work may be used under the terms of the Creative Commons Attribution 3.0 licence. Any further distribution of this work must maintain attribution to the author(s) and the title of the work, journal citation and DOI.
Published under licence by IOP Publishing Ltd

small impurities (1-3 mm), sludge water containing large particles (up to 20-30 mm), water containing long-fiber inclusions, petroleum products, liquids with a high content of abrasive, and so on.

Pumps with an open impeller have a lower efficiency, but are less prone to clogging, because the pumped medium has more space for passing the flow part. In this article, the second part of the study is devoted to the study of a pump with an open impeller with different gaps between the side of the blades and the body wall.

Figure 1.3D model of the flow part of the pump with a closed impeller

In the first part of the study, a characteristic of the pump was constructed, the flow part of which was designed in accordance with the systematics [7]. Further, the diameter of the inlet pipe was changed to 1, 2 millimeters in each direction and characteristics were also built. The best one has been chosen.

Figure 2.3D model of the flow part of the pump with an open impeller

In the second part of the study, three characteristics were constructed corresponding to three different sizes of the gap between the side of the blades and the wall of the housing at the pump with an open impeller. The best characteristic is selected.

Methods

In General, in vector form, the Navier-Stokes equation for a liquid has the form:

$$\frac{\partial \vec{v}}{\partial t} + (\vec{v} \cdot \nabla)\vec{v} = -\frac{1}{\rho}\nabla p + \nu\Delta\vec{v} + \vec{f}$$

where$t$ is time, ∇ is the Nabla operator, Δ is the Laplace vector operator, ρ is density, p is pressure, ν is kinematic viscosity, $\vec{v}$ is the velocity vector field, $\vec{f}$ is the mass force vector field.

The introduction of the Navier-Stokes equation averaged by Reynolds leaves the system of equations open, as additional unknown Reynolds stresses appear. To solve this system, we used the semi-empirical $k-\omega SST$ model of turbulence, which introduces the necessary additional equations: the equations of transfer of the kinetic energy of turbulence and the relative rate of dissipation of this energy:

$$\frac{\partial k}{\partial t} + U_j\frac{\partial k}{\partial x_j} = P_k - \beta k\omega + \frac{\partial}{\partial x_j}\left[(v + \sigma_k v_T)\cdot\frac{\partial k}{\partial x_j}\right]$$

$$\frac{\partial \omega}{\partial t} + U_j\frac{\partial \omega}{\partial x_j} = \alpha \cdot S^2 - \beta\cdot\omega^2 + \frac{\partial}{\partial x_j}\left[(v + \sigma_\omega v_T)\cdot\frac{\partial \omega}{\partial x_k}\right] + 2\cdot(1 - F_i)\cdot\sigma_{\omega 2}\cdot\frac{1}{\omega}\cdot\frac{\partial k}{\partial x_j}\cdot\frac{\partial \omega}{\partial x_j}$$

U, P — averaged speed and pressure;

$\overline{S_{ij}} = \frac{1}{2}\left[\frac{\partial u_i}{\partial x_j} + \frac{\partial u_j}{\partial x_i}\right]$ —instantaneous strain rate tensor;

The continuity equation also holds:

$$\frac{\partial \rho}{\partial t} + div\,\rho v = 0$$

The 3D models specified in the introduction were created using the CFTurbo application packages and in SolidWorks. This method is widely presented in the literature [8]–[14].

To simulate the flow, the model was loaded into the star-CCM + package. The following physical models were used: liquid, constant density, turbulent flow, K-ω turbulence model, non-stationary implicit calculation. At the entrance, the speed was set according to the required flow rate, at the exit-a pressure equal to zero.

To build the grid, we used: a surface grid generator, a polyhedral cell generator (base size 0.8 mm), a prismatic layer generator (5 layers, stretching 1.4, thickness-40% of the base size).

The calculation was carried out first in a stationary mode, in which the reference coordinate system, tied to the wheel, rotates at an angular speed equal to the frequency of rotation of the impeller. After the stationary calculation went to a constant mode (or steady-state fluctuations), the calculation continued in a non-stationary mode, in which the wheel itself rotates. By repeatedly modeling the flow, a family of points was obtained at different fluid flow rates. They formed the basis of the performance characteristics.

According to the same principle, the operating characteristics of the pump were built with other diameters of the inlet pipe. In the second part of the study, several performance characteristics were obtained in a similar way.

IOP Conf. Series: Materials Science and Engineering **779** (2020) 012048 doi:10.1088/1757-899X/779/1/012048

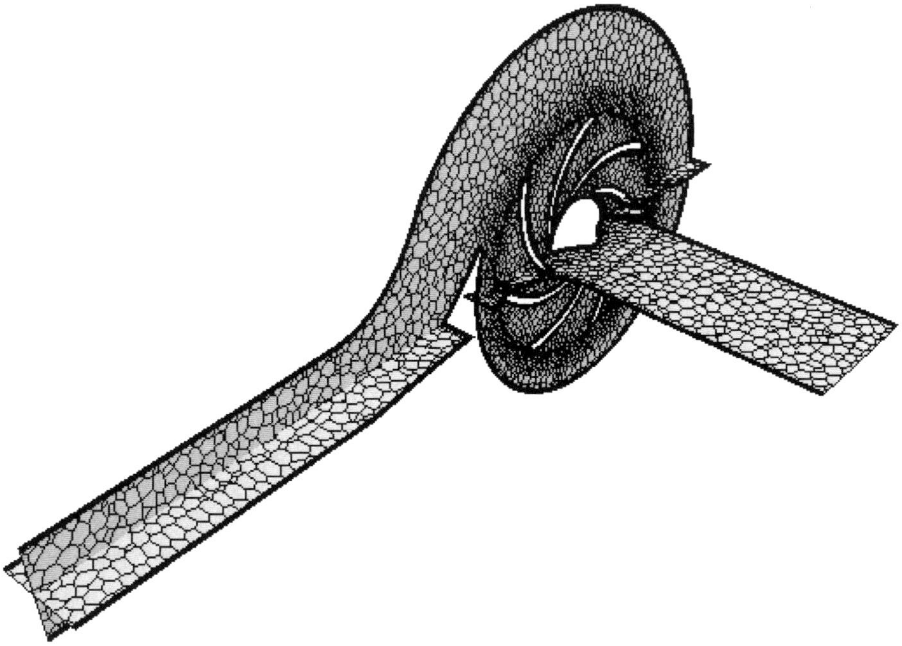

Figure 3. Calculated cross-section grid of the flow section

Results

In the process of modeling, we obtained pictures of the distribution of speeds and pressures in the flow part of the pump.

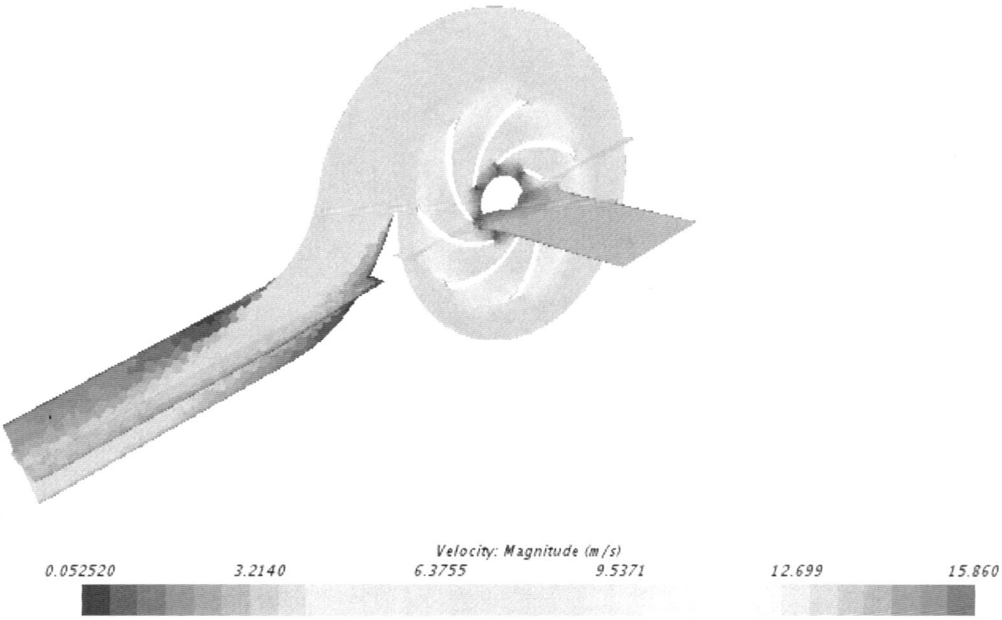

Figure 4. Speed distribution in the flow part of the pump with a closed wheel. Consumption 12 l/min

IOP Conf. Series: Materials Science and Engineering **779** (2020) 012048 doi:10.1088/1757-899X/779/1/012048

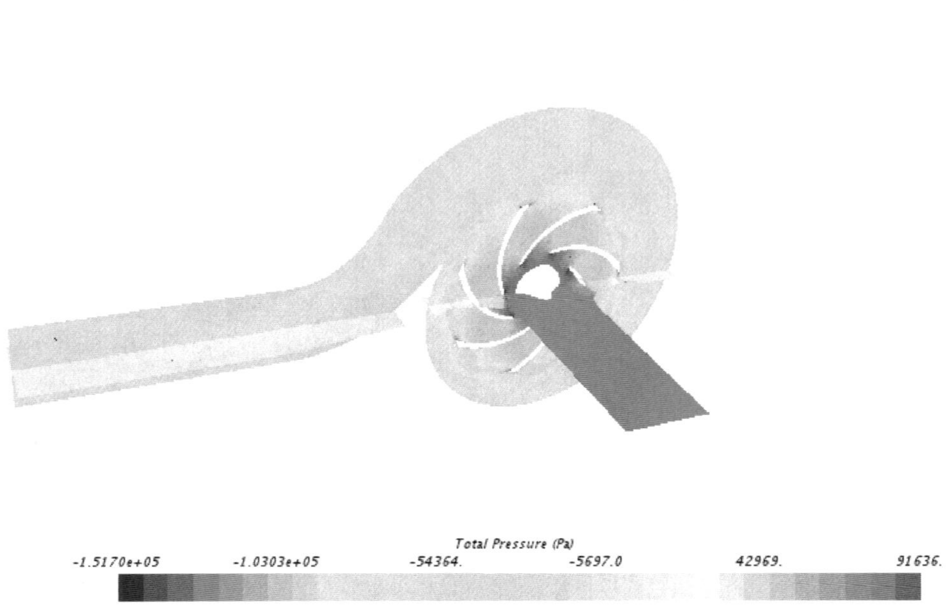

Figure 5. Pressure distribution in the flow part of the pump with the closed wheel. Consumption 12 *l / min*

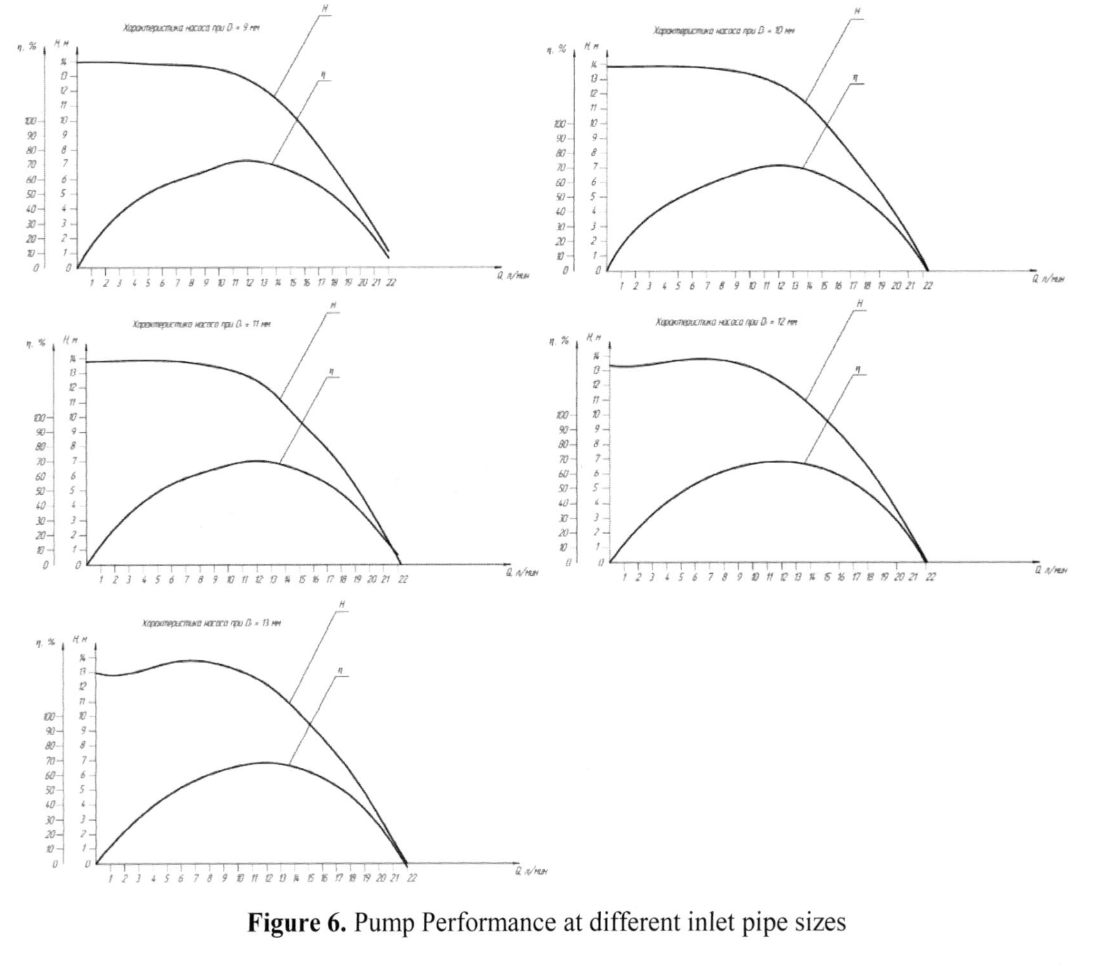

Figure 6. Pump Performance at different inlet pipe sizes

In the second part of the study, the approach was the same except that the diameter of the supply pipe did not change, but the gap between the sides of the blades and the body wall. For the calculation, a model with a diameter of the inlet pipe $D_1 = 9$ mm was taken.

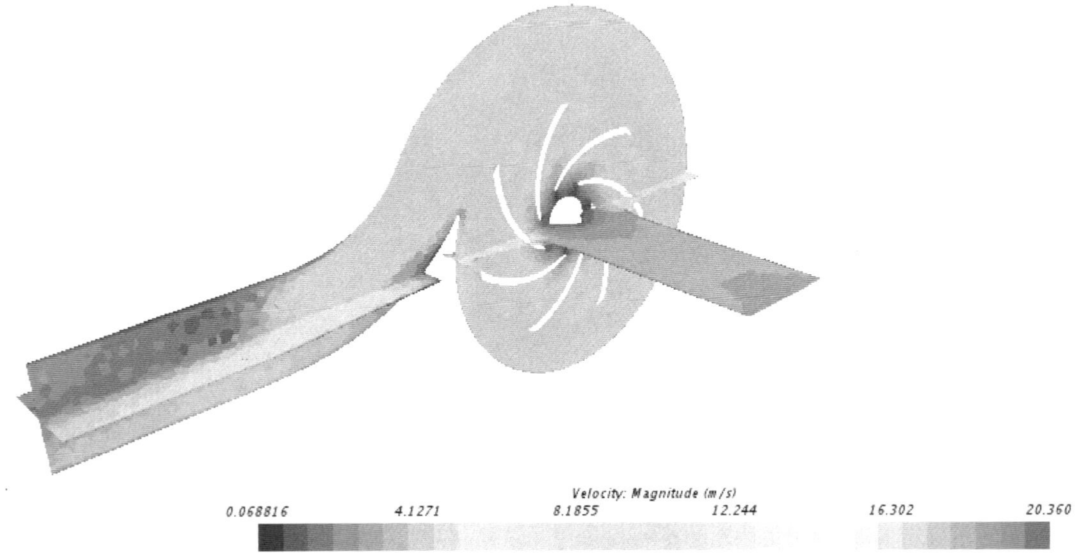

Figure 7. Speed distribution in the flow part of the pump with an open wheel. Consumption 12 *l / min*

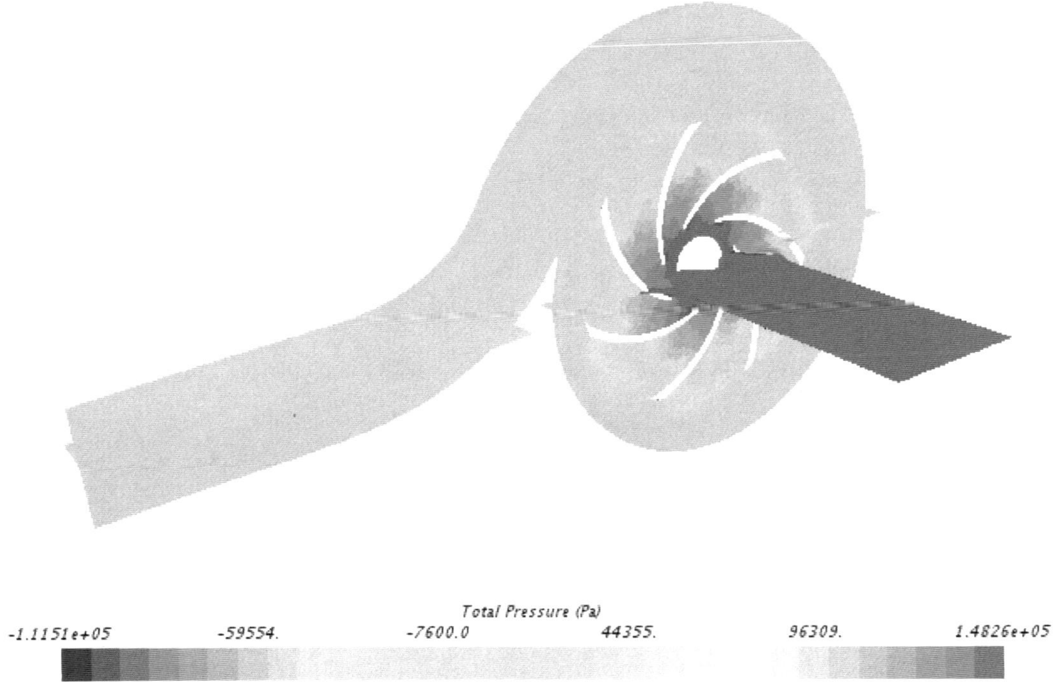

Figure 8. Pressure distribution in the flow part of the pump with an open wheel. Consumption 12 *l / min*

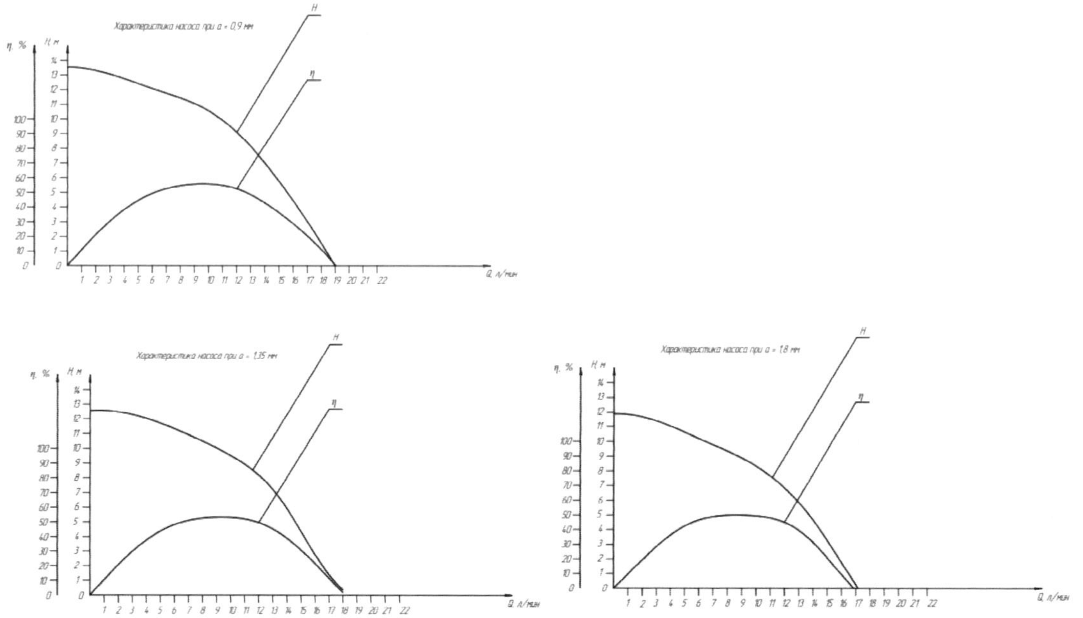

Figure 9. Operating characteristics of an open-wheel pump with different front clearance sizes

Summary

The results of the study confirmed the assumption that higher efficiencies are pumps with closed impeller, because there is no additional loss on the vortex formation because of the spread of liquid in the space between the side blades and the wall of the housing; then directed straight from the center to the periphery of the wheel well and wraps around the disks.

You can also see in the pressure distribution scenes that there is a negative pressure. This is due to the fact that the outlet pressure is assumed to be zero. This is physically incorrect, but the calculation will not change if you set any other pressure value. The pressure drop is important, not the absolute pressure value.

List of references

[1] Irfan, M., Alwadie, A., &Glowacz, A. (2019). Development of a new method of electrical diagnostics for the analysis of faults of centrifugal pumps. Applied Sciences (Switzerland), 9 (23) doi:10.3390 / app9235093

[2] Wang, L., Lu, J., Liao, W., Zhao, Y., & Wang, W. (2019). Numerical simulation of vortex characteristics of tip leakage in a semi-open centrifugal pump. Applied Sciences (Switzerland), 9 (23) doi:10.3390 / app9235244

[3] Guo, C., Wang, J., &Gao, M. (2019). Numerical study of the distribution and evolutionary characteristics of the acoustic field in the time domain of a centrifugal pump based on the theory of Powell's vortex sound. Applied Sciences (Switzerland), 9 (23) doi:10.3390 / app9235018

[4] Rapur, J. S., & Tiwari, R. (2019). Experimental diagnostics of faults for known and invisible modes of operation of centrifugal pumps using analysis based on MSVM and WPT. Measurement: journal of the International measurement Confederation, 147 doi: 10.1016 / j.measurement.2019.07.037

[5] Guo, S., Gao, M., Wang, J., Shi, Y., & He, S. (2019). The influence of the blade exit angle on the characteristics of the distribution of the acoustic field of a centrifugal pump is based on the theory of Powell's vortex sound. Applied Acoustics, 155, 297–308. doi: 10.1016/j. apacoust.2019.05.031

IOP Conf. Series: Materials Science and Engineering **779** (2020) 012048 doi:10.1088/1757-899X/779/1/012048

[6] Mikhailov a. K., Malyushenko V. V. Blade pumps. Theory, calculation and construction. M. "Mechanical Engineering" 1977.
[7] Rudnev S. S., Matveev I. V. Methodical manual on course design of blade pumps. M. Ed. N. E. Bauman, 1975.
[8] P Chaburko and Z Kossova 2019 VGD Conf. Serial.: Mother. Sci. Ang. 492 012011
[9] inLomakin et al. 2019 VGD Conf. Serial.: Mother. Sci. Ang. 492 012012
[10] A Gouskov et al 2019 IOP Conf. Serial.: Mother. Sci. Ang. 492 012013
[11] N Yegorkina and a Petrov 2019 IOP Conf. Serial.: Mother. Sci. Ang. 492 012015
[12] K Dobrokhodov and a Petrov 2019 IOP Conf. Serial.: Mother. Sci. Ang. 492 012016
[13] N Isaev 2019 IOP Conf. Serial.: Mother. Sci. Ang. 492 012026
[14] M. Saprykina and V. Lomakin 2019 IOP Conf. Serial.: Mother. Sci. Ang. 589 012017

Investigation of the influence of the angle of the installation of the blades of the impeller of the regenerative pump on its energy characteristics by the methods of hydrodynamic modeling

A Petrov[1,2], N Isaev[1] and S Radchenko[1]

[1]Bauman Moscow State Technical University

[2]E-mail: alex_i_petrov@mail.ru

Annotation. This paper focuses on the study of the influence of the vortex stage impeller blades angle installation in a centrifugal vortex pump. The mathematical model used in the numerical modeling is described. The 3 variants models of working regenerative impeller with different installation angles are considered. The paper shows the influence of the angle of the blades on the energy characteristics. The evaluation criteria were pressure, hydraulic efficiency and manufacturability. The results obtained during hydrodynamic modeling are presented in the form of velocity distribution vector fields. This paper will be of interest to specialists in the field of designing vortex hydraulic machines.

Introduction

The working part of a regenerative pump is the impeller [1] with radial or inclined blades. (Picture 1)

In practice closed regenerative pumps with impellers having an installation angle of 90 ° are most often found due to the ease of manufacture. But in the literature [1] mathematically is described the fluid flow process in pumps with impellers, which have a sickle-shaped cross-section in the blades. Changing the installation angle leads to an increase in pressure, this is due to the principle of operation of the regenerative pump.

However, practical studies of the influence of the angle of the impeller vanes in the regenerative pump are not given due attention.

Methods

The numerical simulation method is based on the solution of discrete analogues of the basic equations of hydrodynamics [2, 3, 4]. In the case of an incompressible fluid model (ρ = const), this is:

The mass conservation equation (continuity equation):

$$\frac{\partial u_j}{\partial x_j} = 0;$$

Content from this work may be used under the terms of the Creative Commons Attribution 3.0 licence. Any further distribution of this work must maintain attribution to the author(s) and the title of the work, journal citation and DOI.
Published under licence by IOP Publishing Ltd

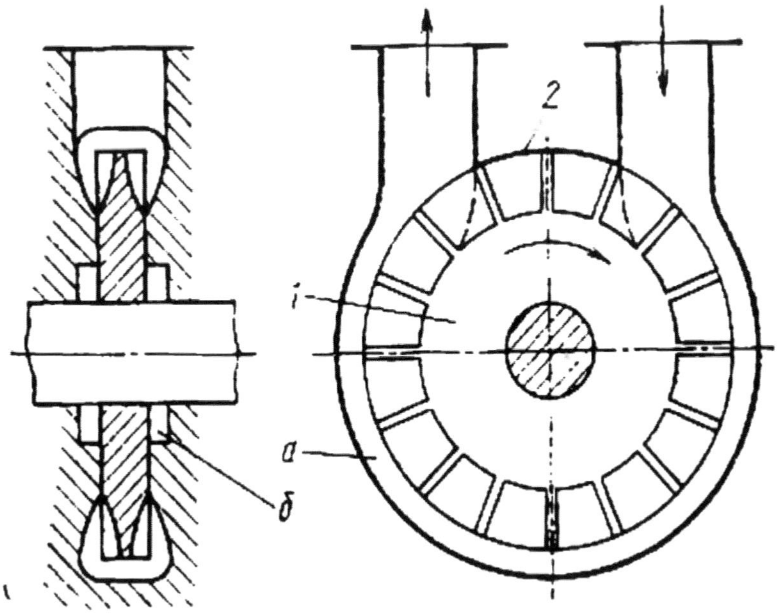

Figure 1. Diagram of a closed regenerative pump

Where u_j — average value of fluid velocity in projection onto the j-th axis ($j = 1,2,3$);

Equation of conservation of momentum (averaging by Reynolds):

$$\rho\left[\frac{\partial U_i}{\partial t}+U_i\frac{\partial U_i}{\partial x_j}\right]=\frac{-\partial P}{\partial x_i}+\frac{\partial}{\partial x_i}\left[\overset{\cdot(v)}{T_{ij}}-\rho u_i u_j\right];$$

Where U, P — averaged speed and pressure;

$\overset{\cdot(v)}{T_{ij}} = 2\mu \dot{S}_{ij}$ — Viscous stress tensor for incompressible fluid;

$\dot{S}_{ij}=\frac{1}{2}\left[\frac{\partial \tilde{u}_i}{\partial x_j}+\frac{\partial \tilde{u}_j}{\partial x_i}\right]$ —instant strain rate tensor;

$\rho u_i u_j$ — Reynolds stresses.

The introduction of the Navier-Stokes equation averaged by Reynolds makes the system of equations open, as additional unknown Reynolds stresses appear. To solve this system in this problem, a semi-empirical k-ω SST turbulence model was used, which introduces the necessary additional equations: the equations of transfer of the kinetic energy of turbulence and the relative dissipation rate of this energy:

$$\frac{\partial k}{\partial t}+U_j\frac{\partial k}{\partial x_j}=P_k-\beta k\omega+\frac{\partial}{\partial x_j}\left[\left(v+\sigma_k v_T\right)\cdot\frac{\partial k}{\partial x_j}\right]$$

$$\frac{\partial \omega}{\partial t}+U_j\frac{\partial \omega}{\partial x_j}=\alpha\cdot S^2-\beta\cdot\omega^2+\frac{\partial}{\partial x_j}\left[\left(v+\sigma_\omega v_T\right)\cdot\frac{\partial \omega}{\partial x_j}\right]+2\cdot\left(1-F_l\right)\cdot\sigma_{\omega 2}\cdot\frac{1}{\omega}\cdot\frac{\partial k}{\partial x_j}\cdot\frac{\partial \omega}{\partial x_j}$$

The flow part of the vortex stage of the pump was designed for the following parameters:

$H = 40м$ — Head;

IOP Conf. Series: Materials Science and Engineering **779** (2020) 012049 doi:10.1088/1757-899X/779/1/012049

$Q = 32\frac{м^3}{h}$ — Flow Rate;

$n = 3000rpm$ — Rotation speed.

The determining factors are the pressure (N), efficiency (η), and the manufacturability of the impeller [4, 5].

We study 3 variants of vortex wheels with installation angles of 45°, 60°, and 90 ° (Figure 2)

Figure 2. Regenerative impeller with an installation angle of 60°

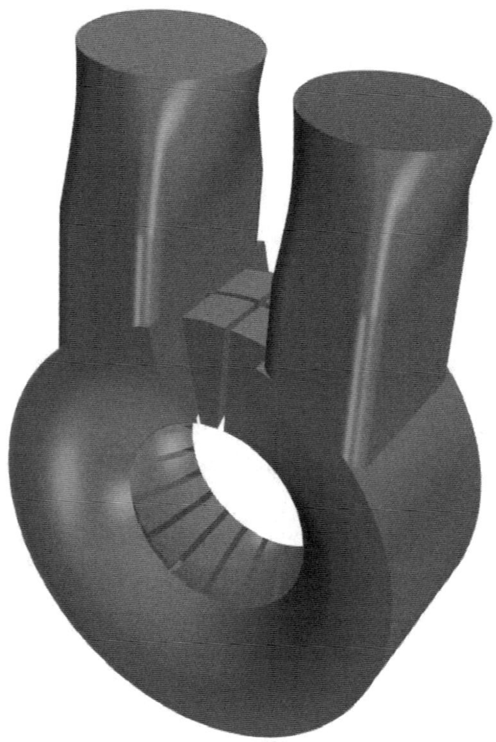

Figure 3. Calculation model

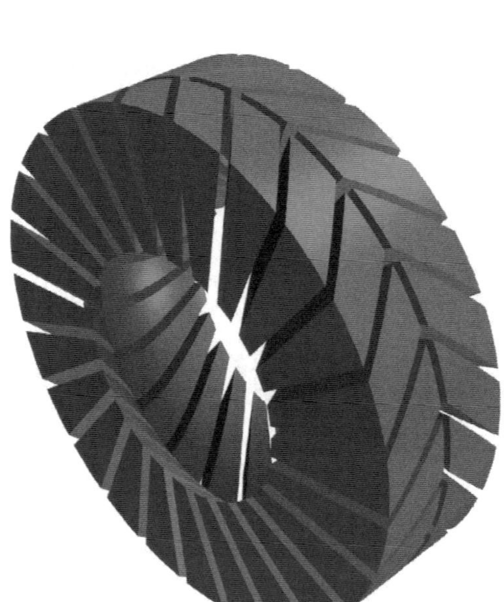

Figure 4. Design model of an impeller with an installation angle of 45°

IOP Conf. Series: Materials Science and Engineering **779** (2020) 012049 doi:10.1088/1757-899X/779/1/012049

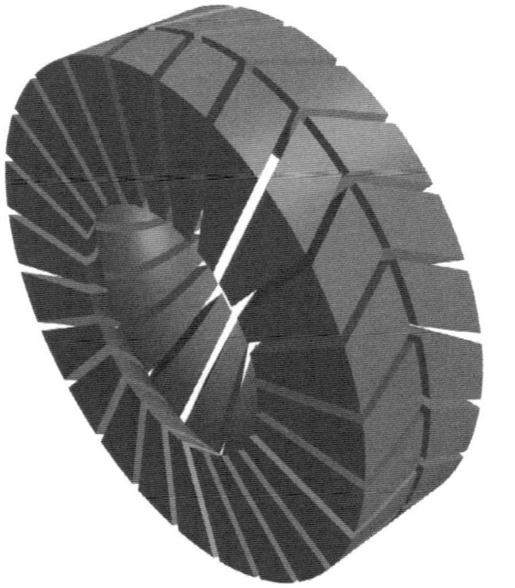

Figure 5. Design model of a wheel with an installation angle of 60º

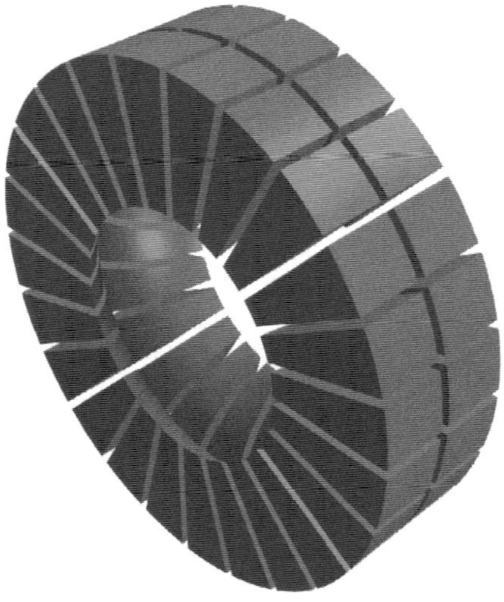

Figure 6. Design model of a wheel with an installation angle of 90º

To minimize the errors associated with numerical calculation, all models were calculated with the same parameters of the computational grid and with the same boundary conditions [6–10]. In the core of the flow cell, they have a multifaceted shape, at the solid walls of the pipe they have a prismatic shape. The computational grid for all models consisted of 600 thousand nodes (Figure 7).

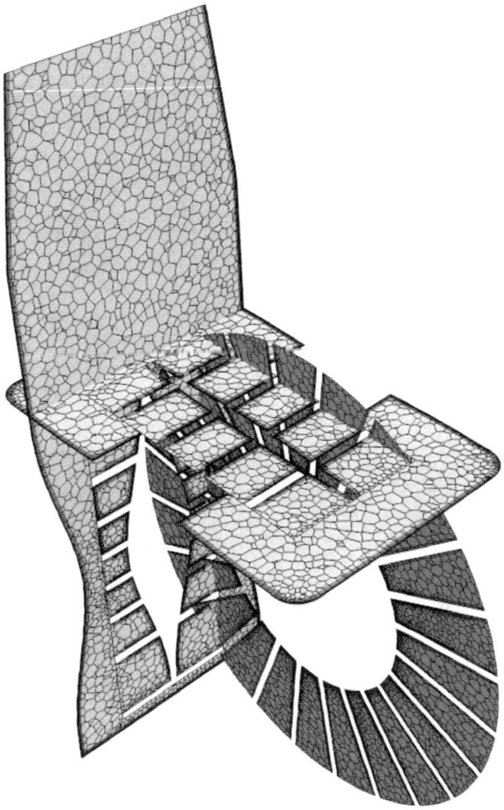

Figure 7. Grid of the computational model

IOP Conf. Series: Materials Science and Engineering **779** (2020) 012049 doi:10.1088/1757-899X/779/1/012049

As the boundary conditions were chosen: the total fluid pressure at the inlet (p = 0 Pa) and the fluid flow rate at the outlet. Time step equals to $t = 1e - 5c$.

Simulation results

The simulation results of 3 variants of the impellers are presented in figures 8-10 and are summarized in table 1.

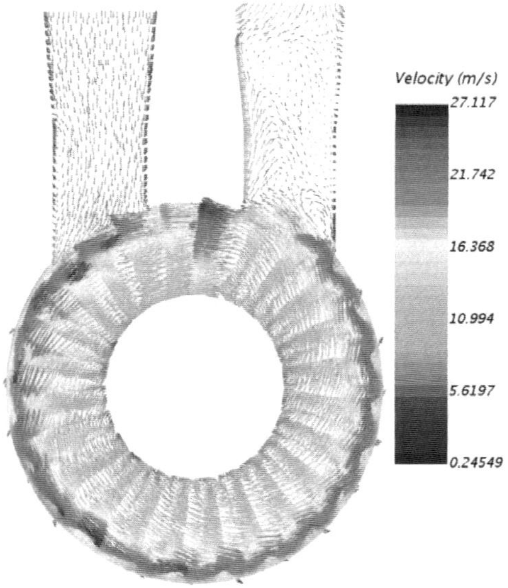

Figure 8. Vector field of velocity distribution in the flow part of the impeller with an installation angle of 45°

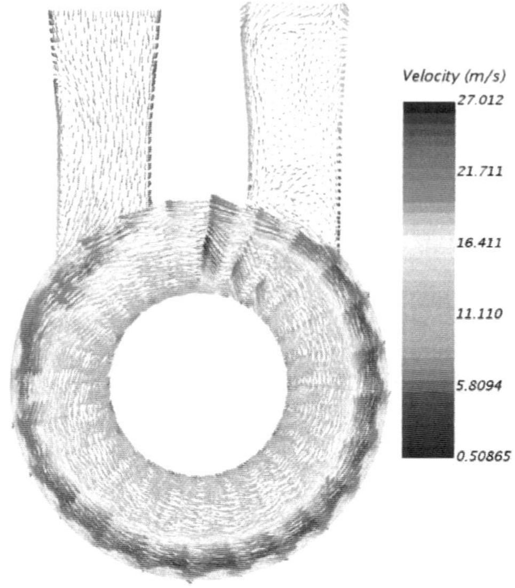

Figure 9. Vector field of velocity distribution in the flow part of the impeller with an installation angle of 60°

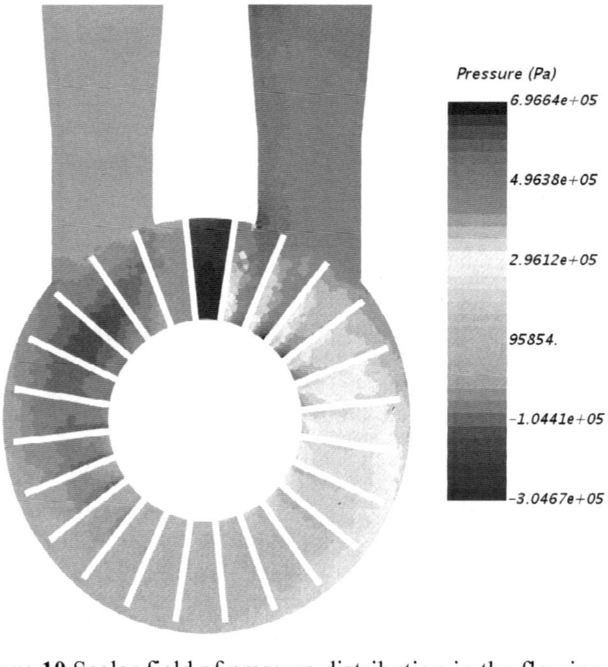

Figure 10. Scalar field of pressure distribution in the flowing part of the impeller with an installation angle of 90°

Table 1. Simulation results

Models	Installation angle, °	Head, m	Efficiency, %	Manufacturability
1	45	54,8	50	–
2	60	65,6	54	–
3	90	51,4	49	+

The graphs show a comparison of the data obtained as a result of numerical simulation with the data obtained as a result of mathematical calculation [1].

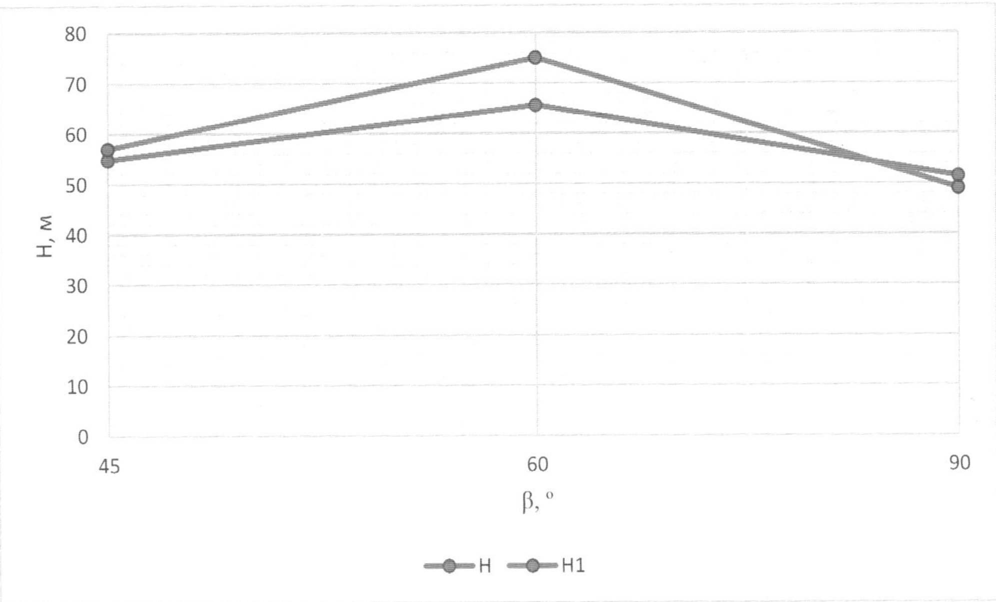

Figure 11. Head dependence on the blade angle

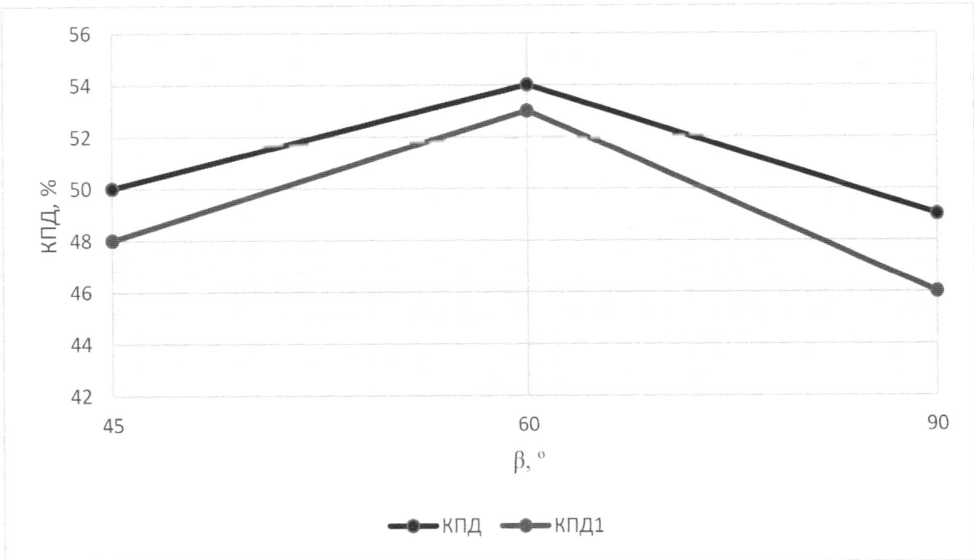

Figure 12. The efficiency dependence on the installation angle of the blades

Discussion

IOP Conf. Series: Materials Science and Engineering **779** (2020) 012049 doi:10.1088/1757-899X/779/1/012049

1. As a result of numerical hydrodynamic modeling, it was determined that 3 impellers have the best energy performance, however, it is much more complicated to manufacture than the option 1;
2. Using this technique, in the future it is possible to study the impellers with other installation angles, since option 3 is probably not optimal according to the criterion of energy efficiency.
3. To determine the flow part with the best energy characteristics, it is necessary to use multi-parameter optimization.
4. Additional experimental studies of the flow parts of vortex pumps are needed to confirm the results of numerical simulation.

List of references

[1] Baibakov O. V. Vortex hydraulic machines. - Moscow: Mashinostroenie, 1981.
[2] A Protopopov and D Bondareva 2019 *IOP Conf. Ser.: Mater. Sci. Eng.* **492** 012002
[3] N Isaev 2019 *IOP Conf. Ser.: Mater. Sci. Eng.* **492** 012026
[4] N Isaev et al 2019 *IOP Conf. Ser.: Mater. Sci. Eng.* **589** 012009
[5] N Isaev. The effect of design parameters of the closed type regenerative pump the energy characteristics. IOP Conference Series Materials Science and Engineering 492:012026. March 2019
[6] Yoo I S, Park M R and Chung M K 2005 Improved momentum exchange theory for incompressible regenerative turbomachinesProceedings of the Institution of Mechanical Engineers. Part A: Journal of Power and Energy, 219 (7) Article ID A09704 pp 567–581.
[7] Song J W, Engeda A and Chung M K 2003 A modified theory for the flow mechanism in a regenerative flow pump Proceedings of the Institution of Mechanical Engineers, Part A: Journal of Power and Energy 217 (3) pp 311–322.
[8] Sixsmith H and Altmann H 1977 A regenerative compressor Journal of Engineering for Industry 99(3) pp 637–647.
[9] V Cheremushkin and V Lomakin 2019 *IOP Conf. Ser.: Mater. Sci. Eng.* **492** 012039
[10] A Protopopov et al 2019 *IOP Conf. Ser.: Mater. Sci. Eng.* **589** 012006

IOP Conf. Series: Materials Science and Engineering **779** (2020) 012050 doi:10.1088/1757-899X/779/1/012050

Investigation of the flow part of a high-speed vane pump of a hydraulic unit using numerical hydrodynamic modeling

M Derevianko[1,2] and M Ryabinin[1]

[1]Bauman Moscow State Technical University

[2]E-mail: max.derew@yandex.ru

Annotation. On this paper was studied the flow part of a high-speed vane pump of a hydraulic unit for pumping fuel of a passenger transport aircraft. To solve the task was used a numerical method for hydrodynamic modeling. It is described the used mathematical model. The obtained pressure and torque graphs on the shaft are given, depending on the feed of the centrifugal pump. The obtained data can be used for further design of the hydraulic unit.

Introduction

The hydraulic unit operation shown in Figure 1 is as follows: the working fluid under pressure, passing through the speed controller (1), enters the discharge cavity of the axial-piston hydraulic motor with an inclined disk (2), as a result of which the cylinder block begins to rotate together with pistons that perform reciprocating motion relative to the cylinder block. As a result of the rotational movement of the block relative to the stationary distributor, the cylinders of hydraulic motors are connected periodically with the suction cavity, then with the discharge cavity.

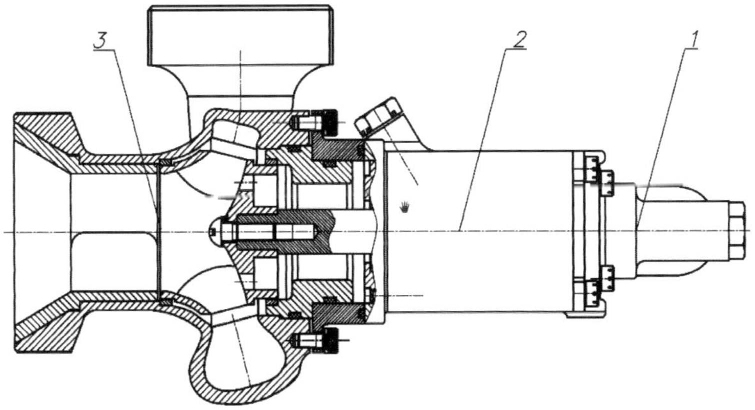

Fig. 1. Diagram of a hydraulic unit for pumping fuel of a passenger transport aircraft

The moment from the pressure forces from the cylinder block is transmitted via a spline connection to the shaft, and then through the spline connection it is transmitted to the impeller of a centrifugal pump (3). The pump casing is completely filled with liquid, and when the impeller rotates, the liquid that is in the channels of the impeller (between its vanes) will be displaced from its center to the

Content from this work may be used under the terms of the Creative Commons Attribution 3.0 licence. Any further distribution of this work must maintain attribution to the author(s) and the title of the work, journal citation and DOI.
Published under licence by IOP Publishing Ltd

IOP Conf. Series: Materials Science and Engineering **779** (2020) 012050 doi:10.1088/1757-899X/779/1/012050

periphery under the action of the arising forces. This will lead to the fact that a vacuum is created at the entrance to the impeller, and pressure increases on the periphery. With increasing pressure, fluid from the pump enters the pressure pipe. And under the action of a vacuum in the suction entrance, the liquid enters the pump thought the suction pipe from the feeder tank. Therefore, there is a continuous supply of fluid by a centrifugal pump from the suction to the pressure pipe [1]–[9].

The Centrifugal Pump Specifications are: Q_{nomcp}=4,5 l/s — nominal flow rate, n_{nomcp}=6650 rpm — nominal speed, H_{nomcp}=11 m — nominal head, pumped liquid-aviation kerosene TC-1.

To design the axial-piston hydraulic motor with inclined disk to act as a drive for a vane pump and to study the dynamic characteristics of a hydraulic unit, it is necessary to know the characteristics of the vane centrifugal pump. To obtain more accurate results, it is advisable to use hydrodynamic modeling methods. [10]–[11].

Flow mathematical model

This paper uses a turbulent flow model for an incompressible fluid (ρ=const). The numerical modeling is based on the solution of discrete analogues of the basic equations of hydrodynamics. The calculation is based on a mathematical model of a divided turbulent flow. The developed pump is used in the fuel transfer system of a passenger transport aircraft. It pumps aviation kerosene TS-1. The numerical hydrodynamic modeling was performed in the StarCCM + software package [12]–[15].

The mathematical model consists of a combination of differential and algebraic equations:

1)the mass conservation equation for an incompressible fluid (continuity equation): [16]–[20]

$$div(v)=0 \quad (1)$$

2)Equations of change in momentum (Navier-Stokes equations averaged by Reynolds): [16]–[20]

$$\rho\left[\frac{\partial \overline{u_i}}{\partial t}+\overline{u_j}\cdot\frac{\partial \overline{u_i}}{\partial x_j}\right]=-\frac{\partial \overline{p}}{\partial x_i}+\frac{\partial}{\partial x_i}\left[T_{ij}^{(v)}-\rho\cdot\left(\overline{u_i^{'}\cdot u_j^{'}}\right)\right] \quad (2)$$

Where $T_{ij}^{(v)}=2\mu S_{ij}$ *is the viscocity stress tensor for incompressible fluid*

$$S_{ij}=\frac{1}{2}\left[\frac{\partial u_i}{\partial x_j}+\frac{\partial u_j}{\partial x_i}\right]-\textit{is the strain rate tensor}$$

This system of equations is not closed. For its closure was used the $k-\omega SST3$ — semiempirical two-equation turbulence model.

3)The equation of kinetic energy transfer of turbulence:[2]–[9]

$$\frac{\partial K}{\partial t}+\overline{u_j}\cdot\frac{\partial K}{\partial x_j}=P_K-\beta^{*}\cdot K\cdot\omega+\frac{\partial}{\partial x_i}\left[\left(v+\sigma_K\cdot v_T\right)\cdot\frac{\partial K}{\partial x_j}\right] \quad (3)$$

4)The transport equation for the turbulence kinetic energy relative dissipation rate:[9]–[15]

$$\frac{\partial \omega}{\partial t}+\overline{u_j}\cdot\frac{\partial \omega}{\partial x_j}=\alpha\cdot S^2-\beta\cdot\omega^2+\frac{\partial}{\partial x_i}\left[\left(v+\sigma_{\omega 1}\cdot v_T\right)\cdot\frac{\partial \omega}{\partial x_j}\right]+2\cdot\left(1-F_j\right)\cdot\sigma_{\omega 2}\cdot\frac{1}{\omega}\frac{\partial K}{\partial x_j}\frac{\partial \omega}{\partial x_j} \quad (4)$$

Construction of the predicted characteristic of Hteor (Q) using the jet theory:

Theoretical head for a finite number of blades:[1]

$$H_T=\frac{\omega}{g}\left(R_2^2 y\omega-\frac{R_2 Q}{2\pi R_{u2} b_2 \psi_2 tg\beta_2}\right) \quad (5)$$

IOP Conf. Series: Materials Science and Engineering **779** (2020) 012050 doi:10.1088/1757-899X/779/1/012050

Fig. 2. Flow 3D-model for the centrifugal pump

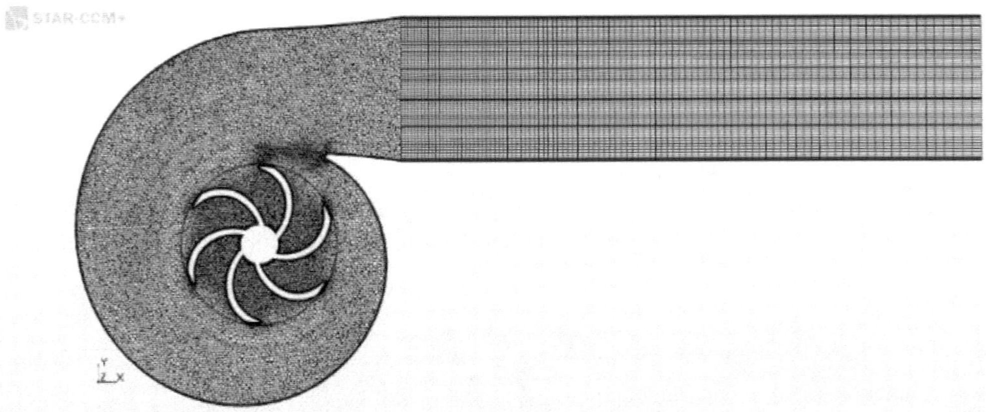

Fig. 3 Extruder grid

Outlet beam:

$$H_T = \frac{Q\omega}{A_s g} \quad (6)$$

Differential pressure on the gap seal:[1]

$$\Delta H = H_{Topt} - \eta_h H_{Topt} \quad (7)$$

Hydraulic loss:[2]

$$\Delta H = h_{fr}(Q) = H_T(1-\eta_h)\bar{Q}^2 = \Delta H \bar{Q}^2 \quad (8)$$

Where

$$\bar{Q} = \frac{Q}{Q_{opt}} = \frac{Q}{4{,}5 \cdot 10^{-3}} \quad (9)$$

IOP Conf. Series: Materials Science and Engineering **779** (2020) 012050 doi:10.1088/1757-899X/779/1/012050

We find the predicted characteristic by subtracting from the theoretical pressure head the hydraulic losses:(10)

$$Hteor(Q) = H_T - \mathrm{h}_{fr}(Q)$$

When constructing the predicted characteristics, an assumption was made about the shockless entry of the working fluid into the impeller. [2]

The magnitude of the flow Ql through the impeller seal near to the optimal regime is taken constant. This allows you to find for each point the corresponding value of the pump flow Qp = Q-Ql and build its predicted pressure characteristic Hteor (Q) at a given rotational speed of the impeller. [2]

Calculation results

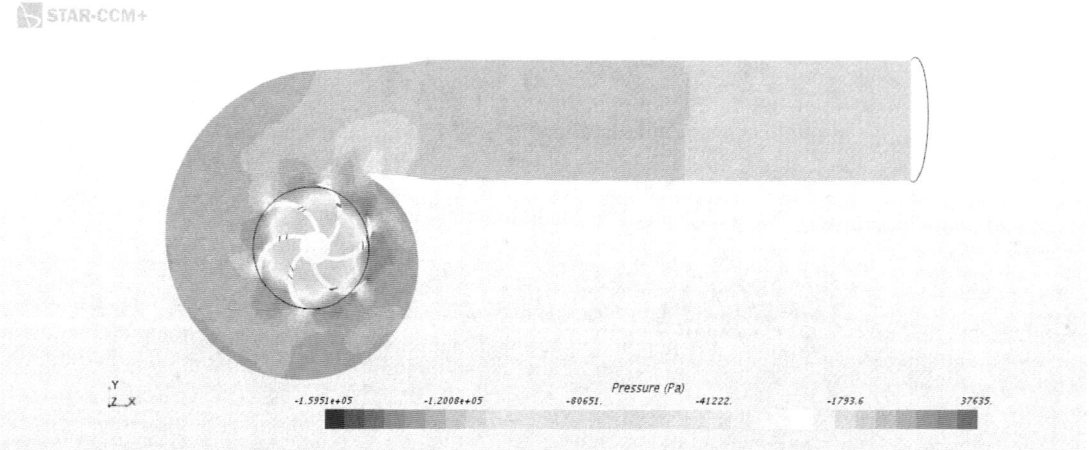

Fig. 4. Scalar field of the fluid pressure in the pump flow part at optimal operation

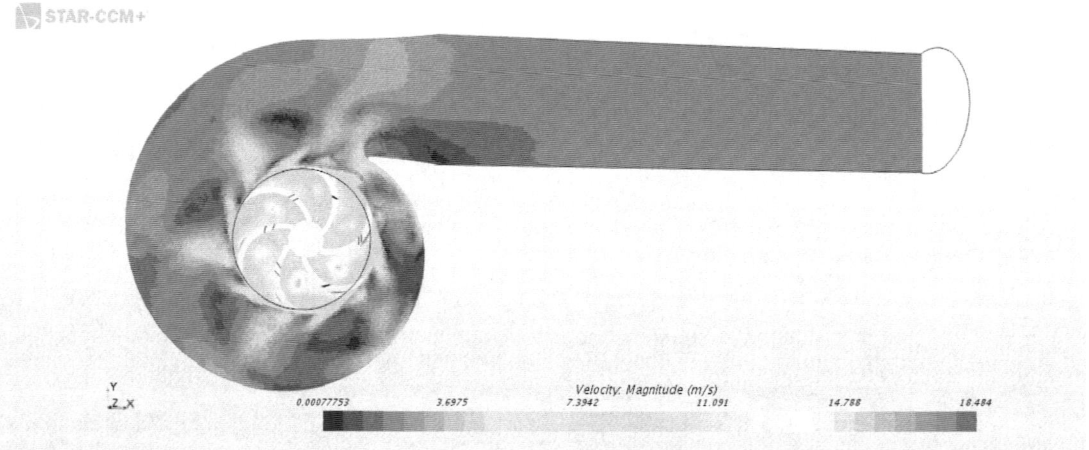

Fig. 5. Scalar field of the fluid velocity in the pump flow part at optimal operation

The calculation points were obtained from the STARCCM + software package, according to which the graphs of the desired dependencies were plotted in Excel, shown at Fig.6 and 7.

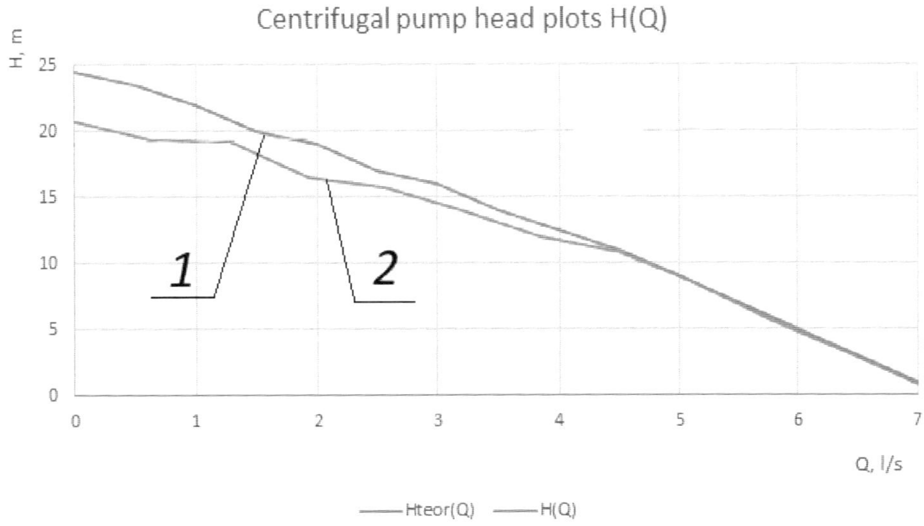

Fig. 6. Graph Head vs pump flow

Predictive characteristic of the Hteor (Q) vane pump, obtained using the jet theory;

1. Characteristic H (Q) of a vane pump obtained by hydrodynamic modeling in the STARCCM + software package.

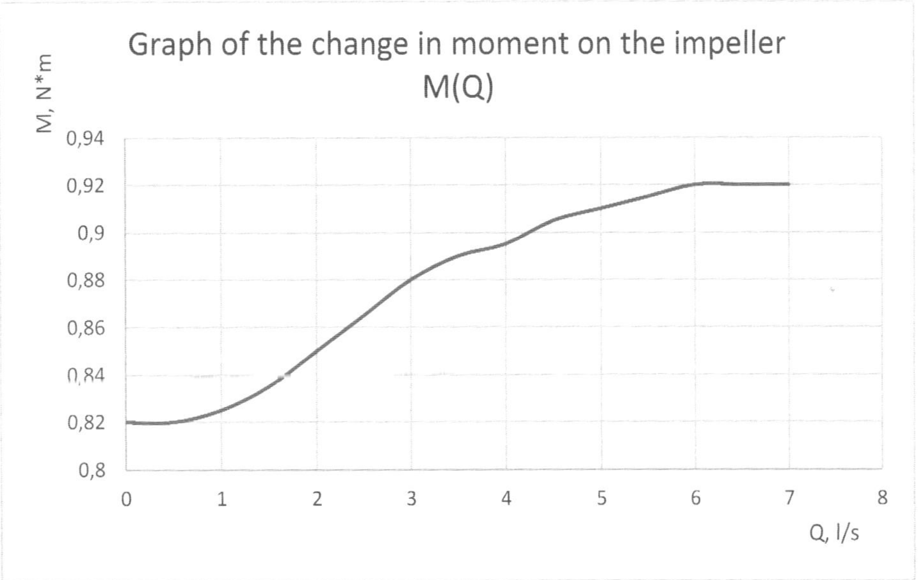

Fig. 7. Graph of the impeller torque based on pump flow

Conclusions

The analysis showed the effectiveness of this mathematical model. The calculation showed that the characteristic of the impeller pump H (Q) near the optimal operation of the hydraulic unit calculated using hydrodynamic modeling tools has a minimum discrepancy with the predicted characteristic Hteor (Q), calculated using the jet theory. Also were obtained the results for designing an axial piston hydraulic motor with an inclined disk as a vane pump drive and the dynamic characteristics of a hydraulic unit.

References

[1] Methodical manual on course design of vane pumps / Rudnev S. S., Matveev I. V.; Bauman higher technical school: Moscow, 1975. - 78 p.
[2] Centrifugal and axial pumps: theory, design and application / Stepanov L. I., Polyakovsky V. I.: Publishing house: State scientific and technical publishing house of machine-building literature, 1960.-468 PP.
[3] Ageev S., Druzhinin E., ESP with increased gas content at the inlet. Drilling and oil, 2003.
[4] Lomakin, V.O., Kuleshovav, M.S., Bozh'eva, S.M. Numerical Modeling of Liquid Flow in a Pump Station (2016) Power Technology and Engineering, 49 (5), pp. 324–327. DOI: 10.1007/s10749-016-0623-9
[5] Lomakin V.O. Investigationoftwo-phaseflowinaxial-centrifugalimpellerby hydrodynamic modeling methodsProceedings of 2015 International Conference on Fluid Power and Mechatronics, FPM 2015 no 7337302 pp 1204–6. DOI: 10.1109/FPM.2015.7337302
[6] Lomakin V O, Petrov AIandKuleshova M S2014Investigationof a two-phase flowinan axialcentrifugalwheelbymethodsofhydrodynamicmodelingScienceandeducation: electronic scientific edition of Bauman MSTU (9) pp 45–64 (in Russian)
[7] Brennen C.E. Fundamentals of Multiphase Flows. Cambridge: Cambridge University Press, 2005. 410 p.
[8] Wilcox D.C. Turbulence Modeling for CFD. 3rd ed. DCW Industries, 2006. 515 c.
[9] Ding H., Visser F.C., Jiang Y., Furmanczyk M. Demonstration and Validation of a 3D CFD Simulation Tool Predicting Pump Performance and Cavitation for Industrial Applications // Journal of Fluids Engineering. 2011. Vol. 133, no. 1. P. 011101-011101-1 DOI: 10.1115/1.4003196
[10] Lomakin V.O. Numerical simulation of flow parts of pump models and verification of simulation results by comparison of obtained values with experimental data. Naukaiobrazovanie MGTU im. N.E. Baumana = Science and Education of the Bauman MSTU, 2012, no. 5, pp. 52–62. DOI: 10.7463/0512.0356070 (in Russian).
[11] Lomakin V.O., Petrov A.I. Verification of the calculation results using hydrodynamic modeling package STAR-CCM + for flow channel of the centrifugal pump AX 50-32-200. Izvestiyavysshikhuchebnykhzavedeniy. Mashinostroenie [Proceedings of Higher Educational Institutions. MachineBuilding], 2012, no. S, pp. 6–9 (inRuss.).
[12] ZhuJ,ZhuH,Zhang Jand ZhangH-Q2019Anumericalstudyonflowpatternsinsidean electrical submersible pump(ESP) and comparisonwithvisualization experiments Journal of Petroleum Science and Engineering 173 pp 339–350 DOI: 10.1016/j.petrol.2018.10.038
[13] TimushevSF,KnyazevV A,PanaiottiSS,SoldatovVA andRohatgiUS2009 Numerical procedureforassessmentofcentrifugalpumpcavitationerosionASMEInternational Mechanical Engineering Congress and Exposition, Proceedings 14 pp 73–78 DOI: 10.1115/IMECE2008-67177
[14] Arefyev, K.Y., Voronetsky, A.V., Suchkov, S.A., Ilchenko, M.A. Computational and experimental study of the two-phase mixing in gas-dynamic ignition system (2017) Thermophysics and Aeromechanics, 24 (2), pp. 225–237. DOI: 10.1134/S086986431702007X
[15] Lozovetskii, V.V., Pelevin, F.V. Resistance of spherical fills in the case of flow of single- and two-phase media (2009) Journal of Engineering Physics and Thermophysics, 82 (2), pp. 277–282. DOI: 10.1007/s10891-009-0194-9
[16] Ivanovsky V.N., Sabirov A.A., Donskoy U.A. The effect of dissolved carbon dioxide on carbonate precipitation during oil production using electric centrifugal pumps. Oil industry, 2010
[17] Bellarya S.A.I., Husain A., Samadc A., KanaiaR.A.. Performance Optimization of Centrifugal Pump for Crude Oil Delivery. TheJournalofEngineeringResearch (TJER), Vol. 15, No. 1 (2018) 88–101

IOP Conf. Series: Materials Science and Engineering **779** (2020) 012050 doi:10.1088/1757-899X/779/1/012050

[18] Antonio C. Bannwart. Experimental Study of Centrifugal Pump Handling Viscous Fluid and Two-Phase Flow. SPE production&operations 30(02), 2013
[19] Larry Bachus, Angel Custodio. Know and Understand Centrifugal Pumps. ElsevierInc., 2003, 270 p.
[20] Gülich J.F. Centrifugal Pumps. Springer-Verlag Berlin Heidelberg, 2010, 964 p.

IOP Conf. Series: Materials Science and Engineering **779** (2020) 012051 doi:10.1088/1757-899X/779/1/012051

Investigation of the influence of wear in impeller seals on the axial force in double suction pumps

N Isaev[1,2], G Budaev[1], D Danilov[1] and K Dobrokhodov[1]

[1]Bauman Moscow State Technical University

[2]E-mail: isaev.nikita@bmstu.ru

Abstract. This article is devoted to the investigation of axial force on the rotor of the double suction centrifugal pump at unequal wear of impeller seals. The annular seal of the double suction centrifugal pump is considered as an object of the study. The distribution of the fields of velocities and pressure in the seals of the pump are obtained by numerical modeling. The leakage rate resulting from the simulation is compared to the existing formulas.

1. Introduction.

Double suction pumps [1,2,3] are horizontal, centrifugal, single-stage pumps with double suction impeller. The characteristics of these pumps are characterized by high efficiency and high permissible suction height (low value of allowable NPSH).

The pump housing of this type has an axial connector. The pressure and suction pipes of this type of the pumps are located at the bottom of the housing. This construction allows the D-type pumps to be assembled and disassembled without disconnecting the piping and the motor. The rotor bearings are antifriction bearings.

The pumped liquid can be water and other liquids with similar chemical parameters, the temperature of the liquid should not exceed 85 ° C, and the viscosity should not exceed 36 cSt.

The investigation was carried out for a pump with the following parameters [4](Table 1).

Table 1.

Parameter	Head, m	Flow rate, m^3/h	Rotational speed, rpm	Power, kW	Gap of the impeller seal mm
Value	50	315	2900	56	0,2

Double suction pumps (Fig. *1*) are designed to have no axial force during normal working conditions, but the impeller seals may wear out unequally during operation [5,6]. As a result of wear of the gap, the leakage through one of the seals changes, which in it's turn leads to asymmetry of the pressure distribution on different sides of the impeller and the appearance of unpredictable axial force, which leads to increased loads on the bearings and the reduction of service life.

Therefore, it is necessary to investigate the effects of wear on the annular seal in order to be able to predict the axial forces to prevent problems with the unit.

Content from this work may be used under the terms of the Creative Commons Attribution 3.0 licence. Any further distribution of this work must maintain attribution to the author(s) and the title of the work, journal citation and DOI.
Published under licence by IOP Publishing Ltd

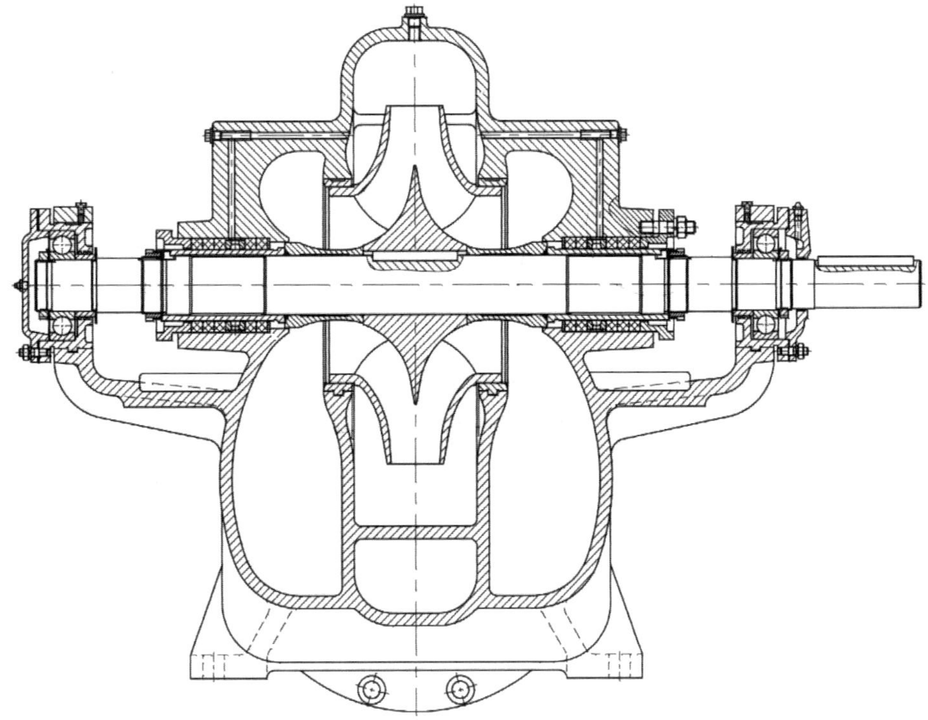

Figure1. Double suction pump.

2. Mathematical model and methods.

This paper uses the model of multiphase flow of incompressible liquid (ρ=const). Numerical modeling is based on solving discrete analogues of basic equations of hydrodynamics [7,8,9,10,11,12]:

Mass conservation equation (continuity equation)

$$\frac{\partial \tilde{u}_j}{\partial x_j} = 0,$$

where $\tilde{u}_j$ — average value of liquid velocity in the projection on the j-axis (j=1,2,3);

Equation of motion conservation (Reynolds averaging):

$$\rho\left[\frac{\partial U_i}{\partial t} + U_j\frac{\partial U_i}{\partial x_j}\right] = -\frac{\partial P}{\partial x_i} + \frac{\partial}{\partial x_i}\left[T_{ij}^{(v)} - \rho u_i u_j\right];$$

where U, P — averaged speed and pressure;

$\tilde{T}_{ij}^{(v)} = 2\mu\tilde{s}_{ij}$—viscous stress tensor for incompressible liquid;

$\tilde{s}_{ij} = \frac{1}{2}\left[\frac{\partial \tilde{u}_i}{\partial x_j} + \frac{\partial \tilde{u}_j}{\partial x_i}\right]$—viscous stress tensor for incompressible liquid;

$\rho u_i u_j$ — Reynolds stresses.

The Reynolds stresses are modeled via the k-ω SST turbulence model, which proved its validity in the calculation of dynamic pumps

The secondary flows of the pump unit was simulated on a grid consisting of ~6 million cells. Cells are multi-faceted at the core of the flow and prismatic near solid walls. Volume mesh for the gaps is shown in Fig. 2.

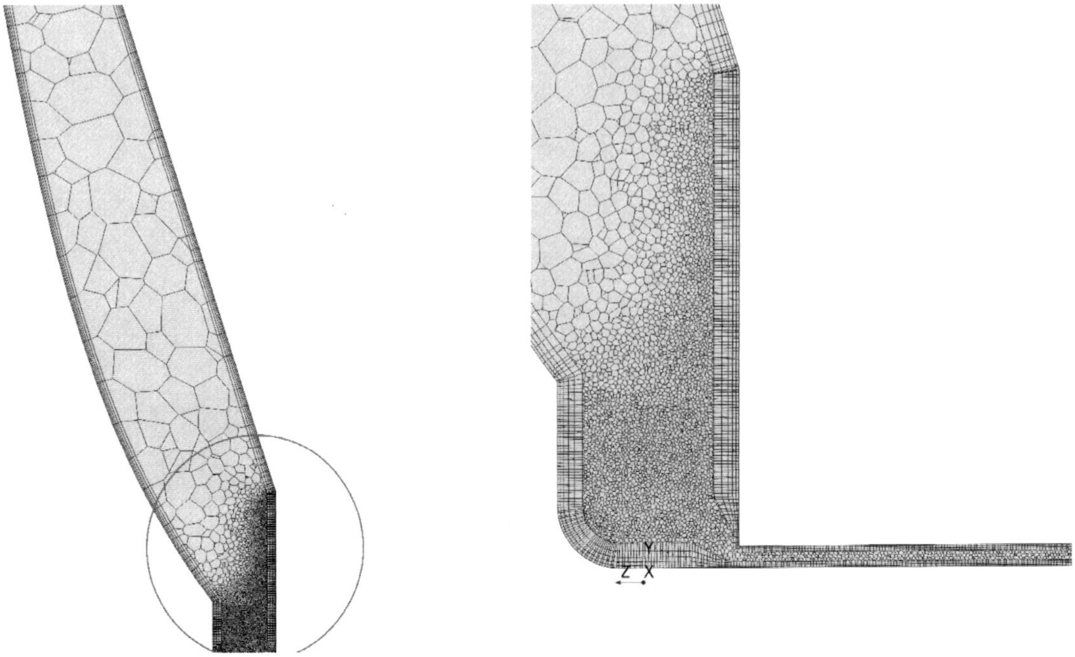

Figure 2. Volume mesh (The research area is enlarged).

Analytical calculation of the impeller seal flow rate is used to check the accuracy of the numerical simulation results, since the axial force depends on the leakage rate:

$$Q_L = \mu \pi D_S \delta \sqrt{2 g H_S};$$

where:

δ – *gap width*;

D_S – *impeller seal diameter*;

μ – *discharge coefficient*;

H_S – *pressure drop on the seal*;

$$\mu = \frac{1}{\sqrt{\zeta_{in} + \lambda_S \dfrac{c}{2\delta} + \zeta_{out}}};$$

where:

ζ_{in}, ζ_{out} – *seal inlet and outlet resistance coefficients*;

c – *gap length*;

λ_S – *gap friction coefficient*;

$$H_S = H_T - \frac{V_{2U}^2}{2g} - \frac{U_2^2}{8g}\left(1 - \frac{D_S^2}{D_2^2}\right);$$

where:

H_T – *theoretical head*;

IOP Conf. Series: Materials Science and Engineering **779** (2020) 012051 doi:10.1088/1757-899X/779/1/012051

V_{2U} *– circumferential component of the absolute flow rate at the impeller outlet;*
U_2 *– circumferential velocity at the impeller outlet;*
D_2 *– outlet diameter;*

3. Results and analysis.

For the purpose of numerical modeling, models of flow gaps were created together with impeller seals. The segment model was used to reduce the calculation time (Fig. *3*).

Figure 3. Impeller cavities segment.

When simulating, the gap of the impeller seal varied from 0.1 mm to 0.4 mm with a increment of 0.05 mm. The following boundary conditions are specified for the simulation: total inlet pressure — 0.55 MPa and outlet pressure — 0 MPa. To visualize the flow pattern of the gap appropriate scenes were created, Fig. 4,Fig. 5 show the scalar pressure distribution scene and velocity distribution vector field for the annular seal width $\delta = 0,2 mm$. Leakage flow rate and force generated by the fluid flow in the gap are obtained.

IOP Conf. Series: Materials Science and Engineering 779 (2020) 012051 doi:10.1088/1757-899X/779/1/012051

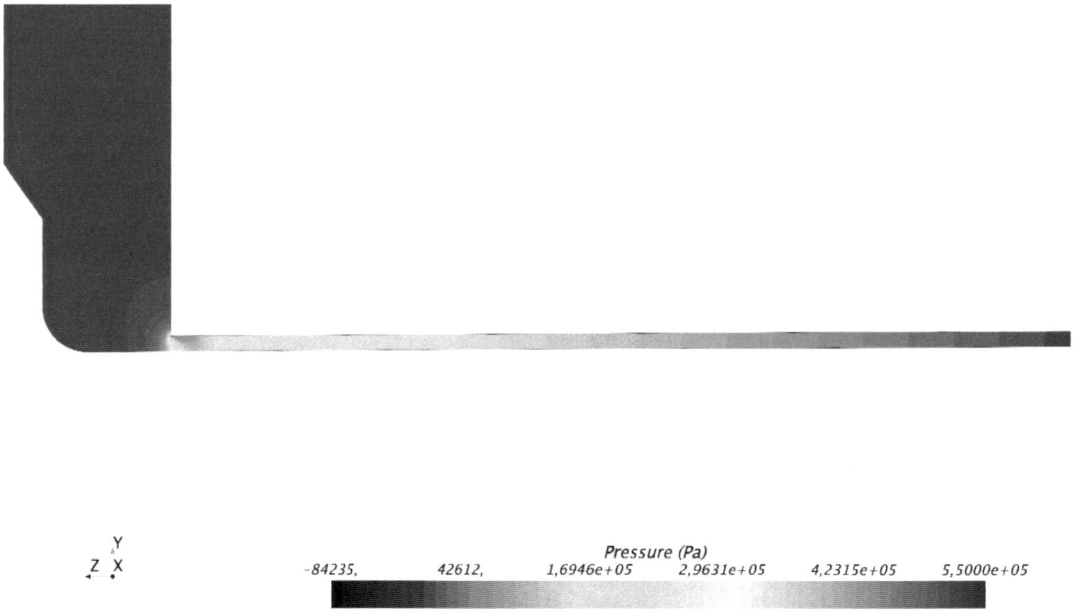

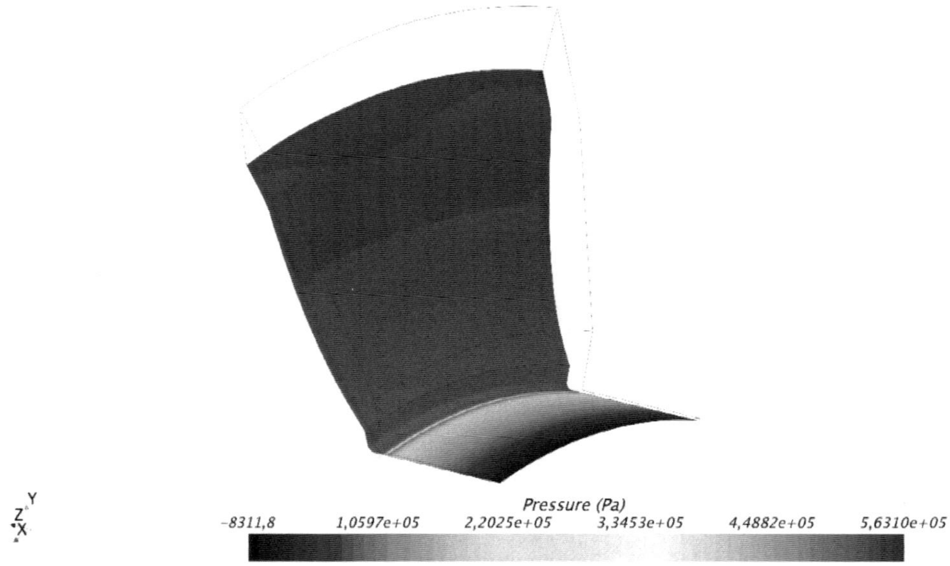

Figure 4. Scalar pressure distribution scene.

Figure 5. Vector velocity distribution field.

The data obtained during the analytical and numerical calculations are presented in Fig. *6*.As a result of comparison [13] it may be concluded that the values of hydrodynamic modelling, are appropriate, the difference in 15% a between the results is explained by incorrect behavior of a current theory when the gap width is considerably more than normal values. Additional experimental research is required to refine the theory.

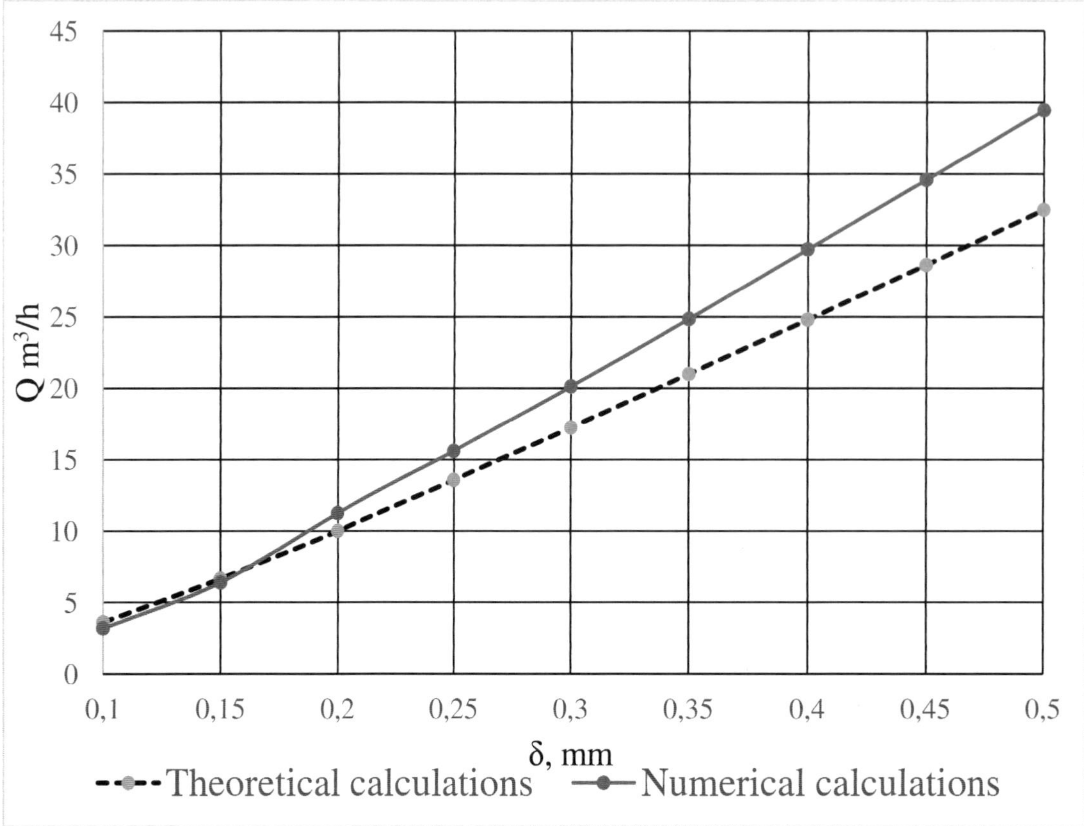

Figure 6. Leakage flow rate.

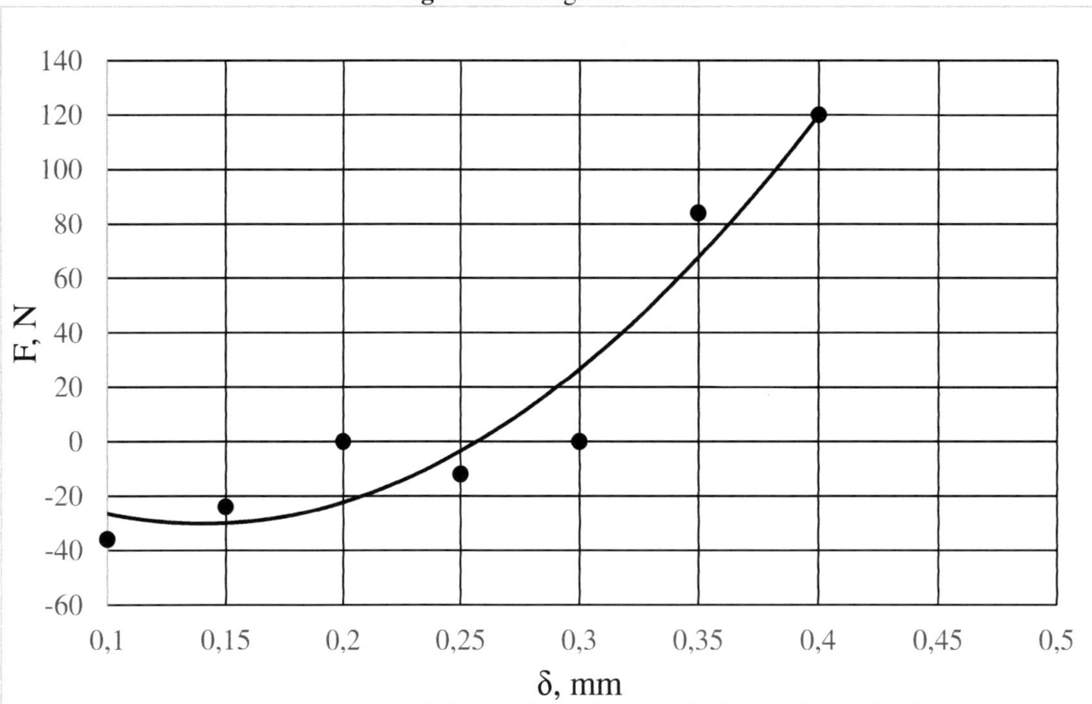

Figure 7. Axial force

An analysis of the dependence of the axial force [14], acting on the impeller on the width of the gap (Fig.7) was carried out. Axial force is the difference between two forces acting on the rotor from different sides. It occurs as a result of pressure difference on the outer covering discs of the impeller. Therefore, we calculate it as a difference of force at a normal gap ($\delta = 0,2mm$) and the force obtained in the calculation of the model with the current thickness of the gap.

$$F_a = F(\delta_i) - F(\delta = 0,2mm)$$

$$\delta_i = 0,1 + 0,05i; i = 0\ldots6$$

4. Conclusions.

According to the results of the performed investigation, the dependence of leakage flow rate and axial force on the gap of the seal was obtained, which makes it possible to make a conclusion about the change of forces acting on the impeller depending on the wear of the seal.

As the gap increases, the flow rate increases and the value of the force acting on the cavity side of the pump rotor increases.

With an increase in the gap difference of the impeller seals, the asymmetry of the loads on the impeller driven discs appears and leads to the appearance, and subsequently to an increase in the axial force.

References.

[1] Hatano, S., Kang, D., Kagawa, S., Nohmi, M., & Yokota, K. (2014). Study of cavitation instabilities in double-suction centrifugal pump. International Journal of Fluid Machinery and Systems, 7(3), 94–100.
[2] Song, Y., & Luo, Y. (2019). Influence of impeller vane arrangement on efficiency performance and pressure fluctuations of a double-suction centrifugal pump. Paper presented at the IOP Conference Series: Earth and Environmental Science, , 240(3)
[3] Maeda, T., Tsuru, W., Kang, D., Tsuneda, T., Kagawa, S., An, B., . . . Yokota, K. (2019). The suppression of cavitation surge in a double-suction centrifugal pump by using branch-type and slit-type accumulations installed at a pump outlet pipe. Paper presented at the IOP Conference Series: Earth and Environmental Science, , 240(3)
[4] A Protopopov and C Jakovich 2019 IOP Conf. Ser.: Mater. Sci. Eng. 492 012034
[5] Wu, T., & Andres, L. S. (2019). Pump grooved seals: A computational fluid dynamics approach to improve bulk-flow model predictions. Journal of Engineering for Gas Turbines and Power, 141(10)
[6] Untaroiu, A., Morgan, N., Hayrapetian, V., & Schiavello, B. (2017). Dynamic response analysis of balance drum labyrinth seal groove geometries optimized for minimum leakage. Journal of Vibration and Acoustics, Transactions of the ASME, 139(2)
[7] A Protopopov and V Vigovskij 2019 IOP Conf. Ser.: Mater. Sci. Eng. 492 012003
[8] Majidi, K. (2005). Numerical study of unsteady flow in a centrifugal pump. Journal of Turbomachinery, 127(2), 363–371.
[9] P Chaburko and Z Kossova 2019 IOP Conf. Ser.: Mater. Sci. Eng. 492 012011
[10] Lee, G. -.,& Bae, J. -. (2019). Comparative study for the prediction of cavitating flow inside a square-edged orifice using different commercial CFD software. Paper presented at the E3S Web of Conferences, , 128
[11] T Valiev and APetrov 2019 IOP Conf. Ser.: Mater. Sci. Eng. 492 012038
[12] Patil, A., Sundar, S., Delgado, A., &Gamboa, J. (2019). CFD based evaluation of conventional electrical submersible pump for high-speed application. Journal of Petroleum Science and Engineering, 182
[13] Zhang, W., Jiang, Q., Bois, G., Li, H., Liu, X., Yuan, S., &Heng, Y. (2019). Experimental and numerical analysis on flow characteristics in a double helix screw pump. Energies, 12(18)
[14] A Protopopov et al 2019 IOP Conf. Ser.: Mater. Sci. Eng. 589 012006

IOP Conf. Series: Materials Science and Engineering **779** (2020) 012052 doi:10.1088/1757-899X/779/1/012052

Assessment of the level of influence of various geometrical parameters of the centrifugal pump guide vane on its energy characteristics

A Petrov[1, 2], A Lysenko[1], E Konkeev[1] and D Konshin[1]

[1]Bauman Moscow State Technical University

[2]E-mail: alex_i_petrov@mail.ru

Abstract: The article is devoted to the research of the influence of geometrical parameters of the inlet section of the multistage pump radial channel guide device on its hydraulic efficiency. The obtained values of hydraulic efficiency, considered depending on the geometrical parameters of the guide device, are given. The efficiency parameters were obtained via hydrodynamic modeling. Recommendations on the optimal geometric parameters are offered and compared with already known semi-empirical formulas.

Introduction

One of the most important steps in the design of a centrifugal pump is the design of a diversion device. Radial channel guide devices are mainly used in multistage pumps with a single stage specific speed of $n_s < 120$, as well as in some single-stage pumps with $n_s = 200 - 500$. The guide device geometry is one of the most important factors determining the operating parameters and efficiency of the pump. [1]

Thus, it is necessary to take into account the influence of the basic geometric parameters on the overall efficiency of the pump flow part.

The proposed research can be carried out as experimentally, obtaining regularities by empirical methods, as well as by methods of hydrodynamic modeling [2, 3]. In this paper an attempt is made to optimize the parameters of the guide device by using the methods of hydrodynamic modeling, and a comparison with the available empirical data is made [4]–[8].

The object under consideration is the impeller of the centrifugal multistage pump and radial channel guide device.

Figure 1 represents the 3D-model of the impeller with the guide device mounted on a shaft.

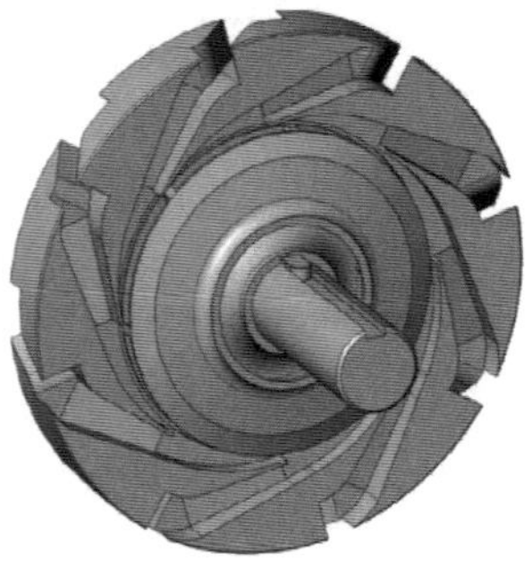

Figure 1. The guide device with the impeller mounted on a shaft

Content from this work may be used under the terms of the Creative Commons Attribution 3.0 licence. Any further distribution of this work must maintain attribution to the author(s) and the title of the work, journal citation and DOI.
Published under licence by IOP Publishing Ltd

Methods

The guide device consists of 4 main sections: spiral section, diffuser, transition channel and return channel. In the calculation given in [1], the idea of spiral capacity was used to determine the parameters of the spiral section:

$$\int_{r_3}^{r} \frac{b}{r} dr = k_{sp} \cdot \frac{Q}{K_3} = A_s,$$

where b is the width of the spiral; k_{sp} is the coefficient that depends on the specific speed; r_3 is the radius of the initial section of the spiral; K_3 is the moment of flow velocity at the inlet of the spiral; A_s is the expected throughput capacity.

The objective of this study was to analyze the influence of the volute throughput capacity and tongue angle on the operating parameters and efficiency of the pump. Also of special interest is the comparison of the obtained optimal parameters with the parameters of the guide device, designed on the basis of known empirical dependencies.

The spiral section is based on the principle that the throughput capacity of the section in it increases linearly with the increase of the angle. As the cross section of the spiral channel in the guide device considered has a rectangular shape, it was mathematically obtained that the shape of the curve providing the required increase in throughput capacity is a logarithmic spiral.

$$b = const;$$

$$A_s = b \cdot \int_{r_3}^{r} \frac{1}{r} dr = b \cdot \ln(r) + c$$

$$r = a \cdot e^{\frac{A_s}{b}}$$

This spiral should provide a steady increase in throughput capacity, but its shape does not take into account the influence of blade thickness in the target cross-section of the spiral. In order to provide the calculated throughput capacity in the target cross-section, the upper edge of the spiral moves upwards from the initial target cross-section to which the spiral is built.

The parameterized model of the initial section of the guide device was designed for the study.

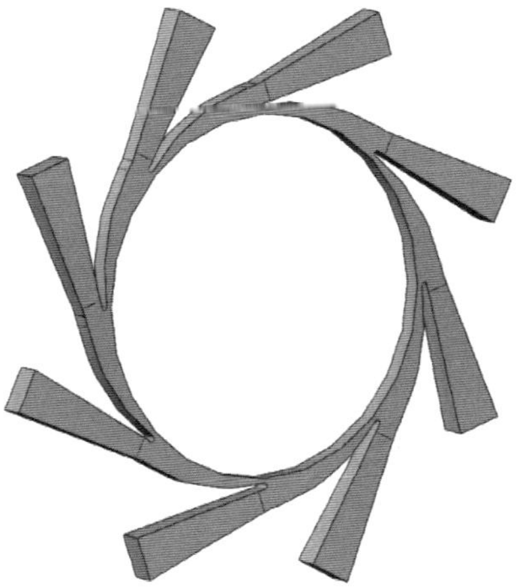

Figure 2. The parameterized model of the initial section of the guide device

The parameterized model is designed using the following method: mathematically precise logarithmic spiral is drawn on the basis of the defined radius r_3, the angle step $t = \frac{2\pi}{z}$, where the

z — number of channels of the guide device and the spiral endpoint that can be varied in height during the study.

The model provides the ability to vary the height of h_3 from h_{3init} to h_{3calc}', where:

h_{3init} — initial calculated height of the channel cross-section without taking into account the existence of blade;

h_{3calc}' —calculated height of the channel cross-section taking into account the existence of blade ($d_b = 3\ mm$)

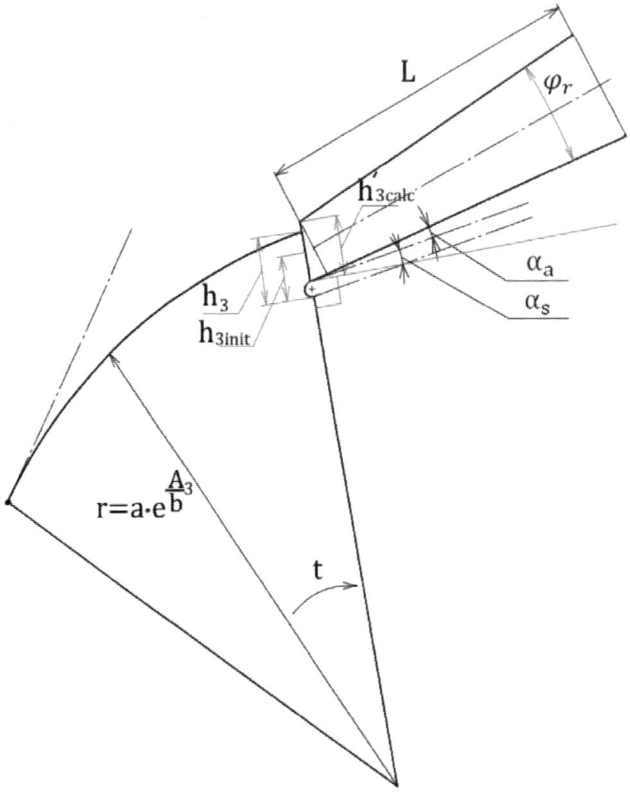

Figure 3. The method of designing the parameterized model

This is done by introducing the k_{sp} coefficient, which allows to vary the throughput capacity of the spiral section.

The diffuser is based on the strictly calculated input height and has remained constant in this study. Diffuser parameters were chosen to be optimal based on empirically developed relations, the angles of diffuser in radial and axial direction: $\varphi_r = 9°, \varphi_b = 4°$, diffuser coefficient 3,5.

Due to the difference in the helix final height and the initial diffuser height, the parameterized model automatically performs splicing of the helix based on the incircle tangent to the splicing lines.

It is possible to define the spiral throughput coefficient in the model, which allows to vary the difference between the initial and calculated cross-sections and the actual spiral endpoint.

$$A_{s1} = k_{sp} \cdot A_s,$$

where k is a variable coefficient; A_{s1} — is the resulting helix throughput capacity.

Another investigated parameter of the model is the initial angle of the blade, the minimum value of which is determined by the angle of tangent to the helix at its starting point.

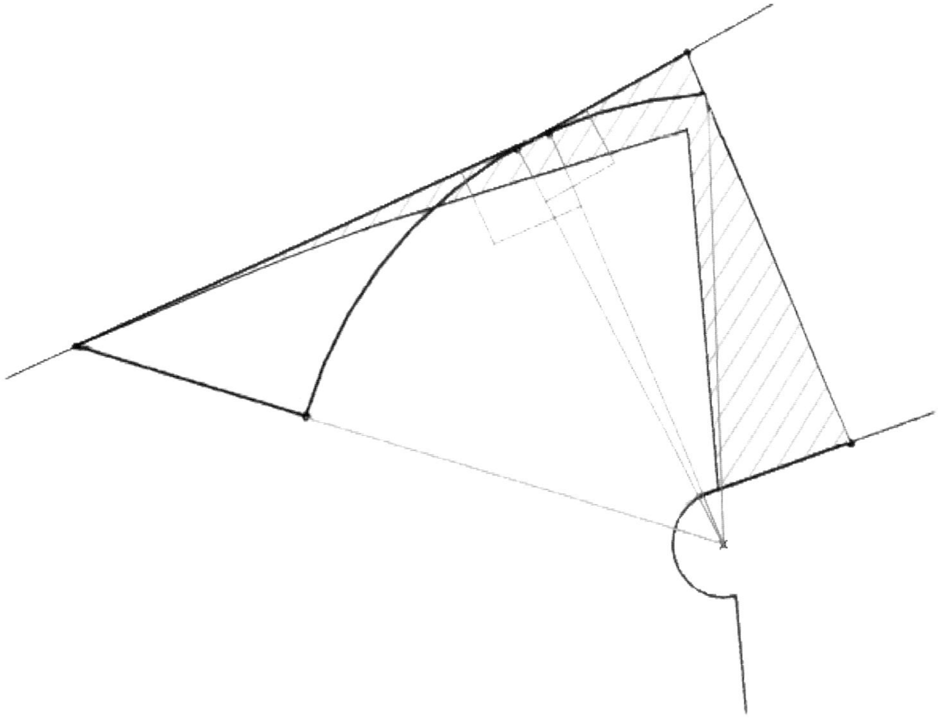

Figure 4. The method of splicing of the helix

Thus, this parameterized model was used to study the influence of the described parameters of the geometry of the input part of the guide device on its energy characteristics, especially the hydraulic efficiency. In order to capture the most comprehensive combination of possible combinations of parameters, 81 models were built, where the installation angle varied from 7 degrees to 25,6 degrees, and the capacity factor varied within (0.83..1.17).

Also the multistage pump impeller model was used for the study, the parameters of which have not been changed.

$$D_2 = 186mm,\ b_2 = 9mm,\ \beta_{2b} = 24°$$

Figure 5. The multistage pump impeller

The opportunities of modern hydrodynamic modeling software were used to measure hydraulic efficiency of the different guide devices. The numerical modeling method is based on solving discrete analogues of basic equations of hydrodynamics [9,10]. For incompressible fluid $(\rho = const)$ these are as follows:

Mass conservation equation (continuity equation)

$$\frac{\partial \overline{u_j}}{\partial x_j} = 0,$$

where $\overline{u_j}$ — the projection of fluid velocity averaged value on the j-axis (j=1,2,3);

The equation of conservation of momentum (Reynolds averaging):

$$\rho\left[\frac{\partial U_i}{\partial t} + U_i \frac{\partial U_i}{\partial x_j}\right] = -\frac{\partial P}{\partial x_i} + \frac{\partial}{\partial x_i}\left[\overline{T_{ij}^{(v)}} - \rho \overline{u_i u_j}\right], \text{ where}$$

U, P — averaged velocity and pressure;

$\overline{T_{ij}^{(v)}} = 2\mu \overline{S_{ij}}$ — viscous stress tensor for incompressible fluid;

$\overline{S_{ij}} = \frac{1}{2}\left[\frac{\partial u_i}{\partial x_j} + \frac{\partial u_j}{\partial x_i}\right]$ — instantaneous strain rate tensor;

$\rho u_i u_j$ — Reynolds stresses.

The introduction of the Reynolds averaged Navier — Stokes equation leaves the system of equations unclosed, since additional unknown Reynolds stresses appear. To solve this system, a semi-empirical *k-ω SST* turbulence model was used. It introduces the necessary additional equations: the equations for the transfer of the turbulence kinetic energy and the relative rate of dissipation of this energy:

$$\frac{\partial k}{\partial t} + U_j \frac{\partial k}{\partial x_j} = P_k - \beta k \omega + \frac{\partial}{\partial x_j}\left[\left(v + \sigma_k v_T\right) \cdot \frac{\partial k}{\partial x_j}\right]$$

$$\frac{\partial \omega}{\partial t} + U_j \frac{\partial \omega}{\partial x_j} = \alpha \cdot S^2 - \beta \cdot \omega^2 + \frac{\partial}{\partial x_j}\left[\left(v + \sigma_\omega v_T\right) \cdot \frac{\partial \omega}{\partial x_k}\right] + 2 \cdot \left(1 - F_i\right) \cdot \sigma_{\omega 2} \cdot \frac{1}{\omega} \cdot \frac{\partial k}{\partial x_j} \cdot \frac{\partial \omega}{\partial x_j}$$

Figure 6. Mesh

Each design flow part was simulated on a mesh of approximately 380,000 cells. The cells were built multi-faceted in the flow core and prismatic near the solid walls. The design mesh for one of the designs is represented in Figure 6.

Results

The field of distribution of pressures and velocities in the flow part for the best in efficiency model and for the model designed on the basis of known empirical dependencies is shown in Figures 7–10. The resulting hydraulic efficiency values depending on the investigated parameters of the guide device geometry are shown in Table 1.

Table 1. Hydraulic efficiency values depending on variable parameters

		Additional angle α_a, °								
Capacity factor K_{sp}	Spiral angle α_s, °	0.1	2	4	6	8	10	12	14	16
0.83	6.9	83.92	84.45	84.82	84.59	84.74	84.68	84.98	84.70	84.47
0.8725	7.2	84.36	84.58	84.84	84.76	84.72	84.81	84.90	84.71	84.46
0.915	7.5	84.52	84.66	84.73	84.71	84.80	84.96	84.71	84.58	84.51
0.9575	7.9	84.67	84.53	84.89	84.89	84.99	84.69	84.75	84.60	84.51
1	8.2	84.65	84.86	84.72	84.96	85.09	85.04	85.16	84.57	84.14
1.0425	8.6	84.58	84.80	84.86	84.97	84.84	85.15	84.70	84.63	84.32
1.085	8.9	84.61	84.80	84.66	84.75	84.65	84.98	84.37	84.57	84.17
1.1275	9.3	84.42	84.66	84.42	84.50	84.32	84.53	84.26	83.74	83.61
1.17	9.6	84.33	84.44	84.37	84.38	84.31	84.19	84.06	83.60	83.44

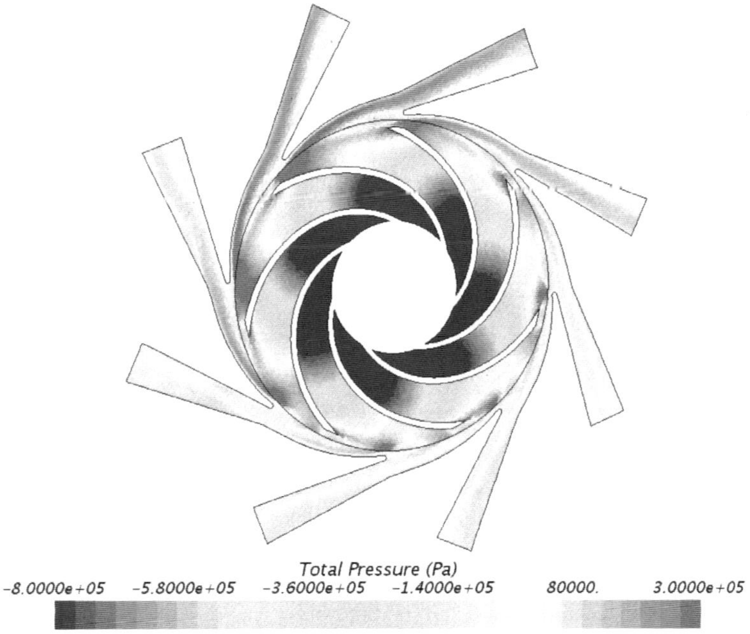

Figure 7. The field of distribution of pressure in the flow part for the best in efficiency model

IOP Conf. Series: Materials Science and Engineering **779** (2020) 012052 doi:10.1088/1757-899X/779/1/012052

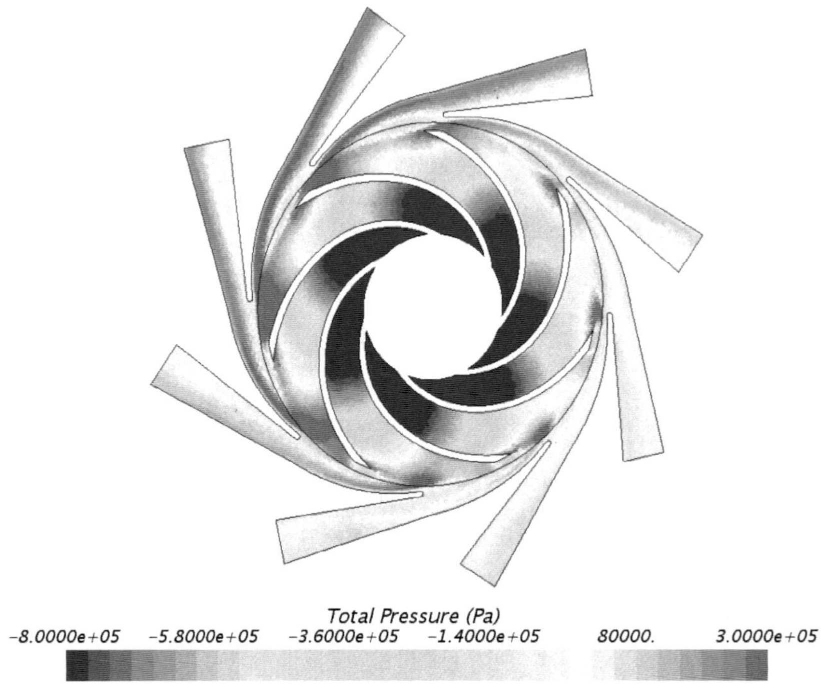

Figure 8. The field of distribution of pressure in the flow part for the model designed on the basis of known empirical dependencies.

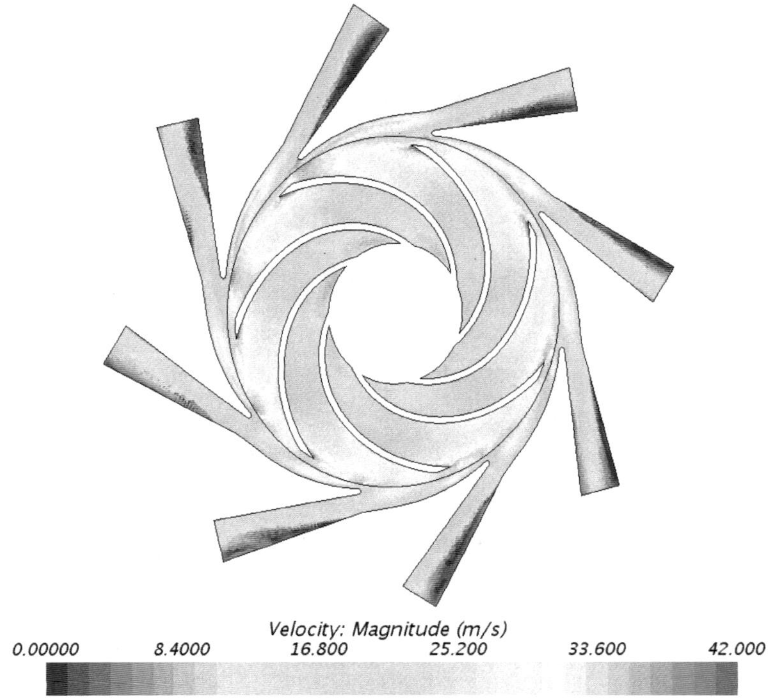

Figure 9. The field of distribution of velocity in the flow part for the best in efficiency model

IOP Conf. Series: Materials Science and Engineering **779** (2020) 012052 doi:10.1088/1757-899X/779/1/012052

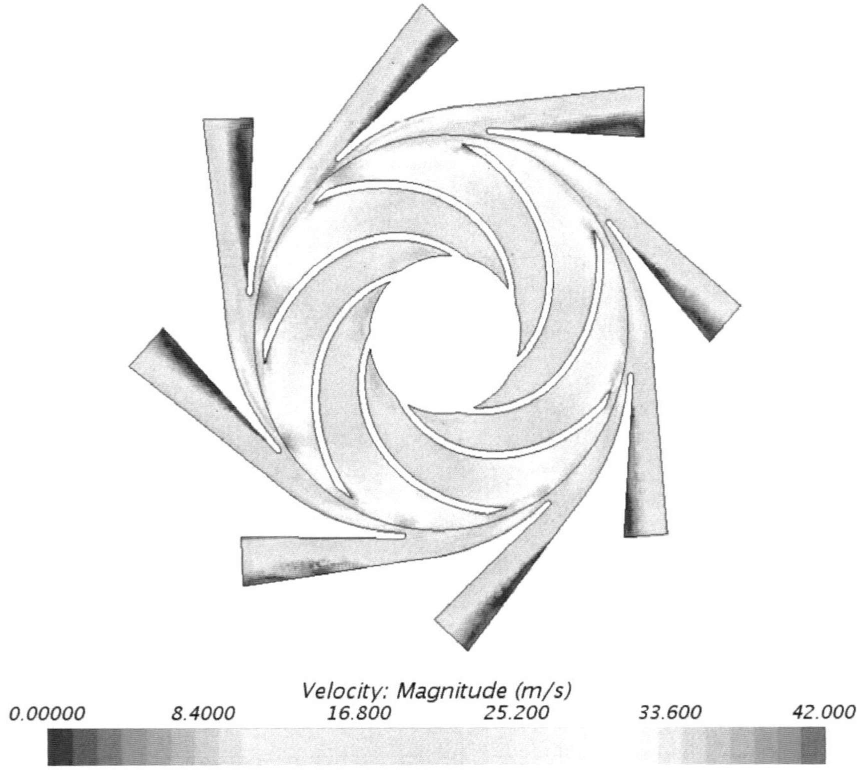

Figure 10. The field of distribution of pressure in the flow part for the model designed on the basis of known empirical dependencies.

Discussion

The analysis of the results of the hydrodynamic modeling has shown that in the case of the study of the influence of the initial angle of the blade, the model of the guide device with the angle deviating from the angle of shockless inflow $\alpha_{a-sh} = 10,2°$ has the best efficiency (calculated angle of shockless inflow $\alpha_{sh} = 10°$;*the angle deviating from the angle of shockless inflow*α_{a-sh}; *the absolute angle is* $\alpha = \alpha_u + \alpha_s = \alpha_{sh} + \alpha_{a-sh}$). This result deviates from the recommended by the empirical method, where $\alpha_{a-sh} = 2..3°$. [11]

In the case of the study of helix capacity influence, it was proved that the exactly calculated throughput capacity is the most optimal option for the design of guiding devices.

Conclusion

1. The determining factor in terms of losses in the guide apparatus is the flow in the diffuser, which is not necessarily associated with the shockless inflow condition.

2. The optimization of the guide apparatus must be carried out in a comprehensive way, taking into account both the parameters of the spiral part and the diffuser.

3. Due to the optimization it has been achieved to increase the hydraulic efficiency of the guide device from 84.33% to 85.16%. It is expected that a more complete optimization of the guide device will result in an even better result.

IOP Conf. Series: Materials Science and Engineering **779** (2020) 012052 doi:10.1088/1757-899X/779/1/012052

References

[1] K Abramov *2019 IOP Conf. Ser.: Mater. Sci. Eng.***589** 012013
[2] E Morozova et al *2019 IOP Conf. Ser.: Mater. Sci. Eng.***589** 012008
[3] N Isaev et al *2019 IOP Conf. Ser.: Mater. Sci. Eng.***589** 012009
[4] A Protopopov and V Vigovskij*2019 IOP Conf. Ser.: Mater. Sci. Eng.***492** 012003
[5] P Chaburko and Z Kossova*2019 IOP Conf. Ser.: Mater. Sci. Eng.***492** 012011
[6] N Isaev*2019 IOP Conf. Ser.: Mater. Sci. Eng.***492** 012026
[7] C Jakovich*2019 IOP Conf. Ser.: Mater. Sci. Eng.***492** 012034
[8] A Protopopov et al *2019 IOP Conf. Ser.: Mater. Sci. Eng.***589** 012005
[9] Zharkovsky A.,2003. Mathematical modeling of working processes in centrifugal pumps of low and medium speed for the solution of problems of the automated designing *.Dissertation and the abstract on HAC RF 05.04.13.*
[10] Loytsyanskiy L.,2003. Mechanics of Fluid and Gas Textbook for HEIs. *—7th edition, Prosp. — Moscow: Drofa.*
[11] Malushenko V V and Mikhailov A K 1981 Energy pumps: Reference book (Moscow: EnergoizdatPubl) 2000

Complex optimization of vaned stators of high-speed centrifugal pumps for thermal stabilization systems

A Kochetov[1], A Petrov[1,2] and M Kuleshova[1]
[1]Bauman Moscow State Technical University

[2]E-mail: alexeypetrov@bmstu.ru

Abstract. The scope of the pumps for thermal stabilization systems, the main elements of such pumps, their configurations and basic requirements for pumps of this type are considered. In order to increase efficiency, an optimization of the pump vaned stator is carried out. Such simulation results as efficiency and pressure values for each variant are obtained. The data received is analyzed and appropriate conclusion on the efficiency of optimization approach is made.

Introduction

High-speed centrifugal pumps are widely used in thermal stabilization systems in aerospace industry, for computing hardware, as well as in various ground-based installations and radar equipment.

The thermal stabilization system includes a circulation pump (to ensure the reliability of the system two pumps are used in it, one of which acts as a backup), a liquid heat exchanger, a pump control and monitoring unit, and system status sensors (temperature, pressure, etc.). Ethylene glycol coolants (C-40, C-65 and its analogues) are used as a working fluid in the system.

The pump includes an impeller, a vaned stator, a spiral volute with a diffuser and a power end, including a sealing element — pumps with a magnetic coupling and pumps with a lined motor [1] - [3], friction bearings [4], etc. are used. The introduction of the vaned stator to the pump design allows to decrease the radial force on the rotor, which is caused by the violation of velocities and pressure distribution around the circumference of the impeller. Also, by replacing the vaned stator with another one within the same pump, it is possible to change the characteristics of the pump, which in the case of an additional application, for example, frequency regulation of the pump [5] - [8], allows you to get the optimum working point in a wide range of characteristics, using, in fact, one and the same pump. Based on the results of optimization, it is recommended to perform full-scale tests using elements manufactured with additive technologies [9] - [12].

An indisputable fact is that optimization is an important part of pump designing process, especially in terms of impeller geometry. The most successful variants meet all the requirements and allow to achieve the best efficiency. However, the efficiency of the pump is also affected by losses in the stators and volutes. In this article, the optimization of the vaned stator and the expediency of its application is considered. One of the goals of this study is to determine the optimal parameters of a centrifugal pump vaned stator for heat stabilization systems. The parameters of the stator must comply with the requirements for pumps of this type, such as:

Content from this work may be used under the terms of the Creative Commons Attribution 3.0 licence. Any further distribution of this work must maintain attribution to the author(s) and the title of the work, journal citation and DOI.
Published under licence by IOP Publishing Ltd

IOP Conf. Series: Materials Science and Engineering **779** (2020) 012054 doi:10.1088/1757-899X/779/1/012054

1. Minimum pump weight;
2. Minimum pump size;
3. High efficiency.

The optimization is based on parametrized pump geometry and CFD-simulation of the flow inside it. The work flow might bedescribed as follows:

1. 3-D model of the pump is prepared;
2. Geometry is then transferred to the CFD-software;
3. Pump flows imulationis run;
4. Obtained results are analyzed.

Methods

The following geometrical parameters of the impeller were considered:

- Number of stator blades (nBl), impeller geometry is not changing and consists of 6 blades;
- Stator blade angle at the inlet (Beta1);
- Stator blade angle at the outlet (Beta2);
- The spacing between impeller outlet and blade leading edge (Inlet);
- The spacing between stator outlet and blade trailing edge(Outlet).

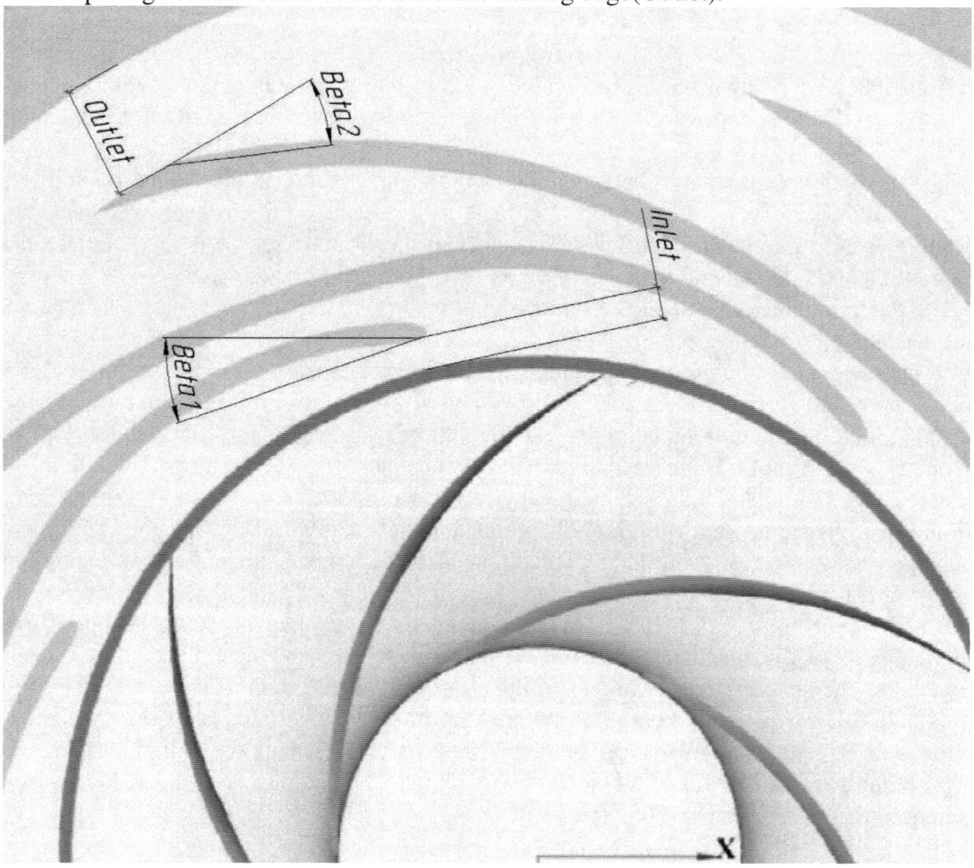

Fig. 1. Optimization variables

The range of optimization parameters values is chosen based on there equirements described above.

The simulation is based on mathematical models describing turbulent flow sin pump sand providing their characteristics with out actual field tests. To achieve these values of the turbulent flow it is required to solve a set of Reynolds-averaged Navier-Stokes equations[13]–[20].

Reynolds-averaged Navier-Stokese equation is as follows:

$$\rho(\frac{\partial U_i}{\partial t}+U_j\frac{\partial U_i}{\partial x_j})=-\frac{\partial P}{\partial x_i}+\frac{\partial}{\partial x_i}(T_{ij}^{(v)}-\rho\langle u_i u_j\rangle),\ \text{where}(1)$$

U_i, P—averaged velocity and pressure,

$T_{ij}^{(v)}$ — viscous stress tens or forin compressible fluid,

$\rho\langle u_i u_j\rangle$—Reynolds stresses.

Mass conservation equationis as follows:

$$(\frac{\partial \bar{u}_i}{\partial t}+\bar{u}_j\frac{\partial \bar{u}_i}{\partial x_j})=-\frac{1}{\bar{p}}\frac{\partial \bar{p}}{\partial x_j}+\vartheta\frac{\partial^2 \bar{u}_i}{\partial x_j^2}\ (2)$$

Introduction of the Reynolds-averaged Navier-Stokes equations makes the system of equationsnon-closed, since there are additional unknown Reynolds stress. To solve this system the semi-empiricalk-ωSS Tturbulence model isused. This model includes two additional equations.

Turbulence kinetic energy transport equationis as follows:

$$\frac{\partial k}{\partial t}+U_j\frac{\partial k}{\partial x_j}=P_k-\beta^* k\omega+\frac{\partial}{\partial x_j}\left[(v+\sigma_k v_T)\frac{\partial k}{\partial x_j}\right],\ \text{where}(3)$$

ω —specific dissipation rate,

k — turbulent kinetic energy,

v_T —turbulent viscosity,

P_k —turbulence energy generation.

Specific dissipation rate transferis described as follows:

$$\frac{\partial \omega}{\partial t}+U_j\frac{\partial \omega}{\partial x_j}=\alpha^* S^2-\beta\omega^2+\frac{\partial}{\partial x_i}\left[(v+\sigma_k v_T)\frac{\partial k}{\partial x_j}\right]+2(1-F_1)\sigma_{\omega 2}\frac{1}{\omega}\frac{\partial k}{\partial x_i}\frac{\partial \omega}{\partial x_i}\ (4)$$

Reynolds stresses are determine dusing the Boussinesq hypothesis

$$\rho\langle u_i u_j\rangle=2\mu_T\left[\frac{1}{2}\left(\frac{\partial U_i}{\partial x_j}+\frac{\partial u_j}{\partial x_i}\right)-\frac{1}{3}\frac{\partial U_k}{\partial x_k}\delta_{ij}\right]-\frac{2}{3}\rho k\delta_{ij},\ \text{where}(5)$$

δ_{ij} —Kronecker symbol.

AsaCFD-response pump head (H) and efficiency (EFF) are obtained.

Optimization work flow is fully automated: the geometry is changed by the optimization program and is then transferred to the CFD-code, the results are analyzed by the program as well. The decision on geometry changes is made by hybrid adaptive SHERPA algorithm. This method is a combination of several mathematical algorithms, which are used simultaneously and adapt the m selves to the case under consideration. The results are assessed by performance (E) which is the ratio of design under consideration efficiency (η_i) to initial design efficiency (η_0).

IOP Conf. Series: Materials Science and Engineering **779** (2020) 012054 doi:10.1088/1757-899X/779/1/012054

$$E=\frac{\eta_i}{\eta_0} \quad (6)$$

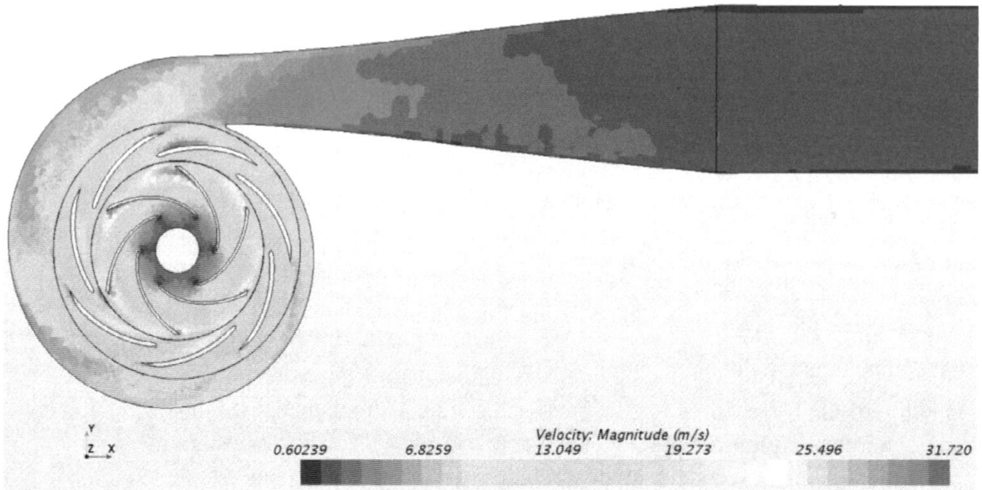

Fig.2. Velocity field distribution for the optimized design of the 3dset-up

Results

There searchin cludes 4 different set-ups for optimization and comparison, all of the minclude the same values of pump head and volumetric flow rate:

- Head(H)~45m;
- Volumetric flow rate(Q)=4.5m^3/h.

These setups differin rotation rate values:

1. Rotation rate(n)=3000(rpm);
2. Rotation rate(n)=8000(rpm);
3. Rotation rate(n)=20000(rpm);
4. Rotation rate(n)=24000(rpm).

125 variants for set-up1are considered. The results of top-5areshowninTable1.

Table 1. Optimization results for the1stset-up

Design variant	E	EFF	H	beta1	beta2	nBl	inlet	outlet
124	1.886	58.690	46.199	0.0677	0.25	7	0.0556	0.825
79	1.857	57.781	43.800	0.0633	0.24	7	0.055	0.839
71	1.857	57.780	43.800	0.0633	0.51	7	0.055	0.839
47	1.857	57.779	43.800	0.0633	0.2	7	0.055	0.839
48	1.696	52.778	51.448	0.0988	0.205	7	0.0526	0.849

Theresultsobtainedshowthatinitialdesignisnotefficientenough,andoptimizationhelpstoimprovethepu mpefficiency.

Then 100 variants for set-up 2 are considered, top-5 are showing in Table 2.

Based on the results of two first set-ups it might be concluded that optimization is an effective tool, that helps improve machine efficiency in a wide range of its operational conditions.

73 variants for the third setup are considered, the best variants are showing Table 3.

Table 2. Optimization results for the 2[d]set-up

Design variant	E	EFF	H	beta1	beta2	nBl	inlet	outlet
43	1.267	60.092	53.726	0.240	0.805	7	0.063	0.722
89	1.245	59.038	54.055	0.083	0.818	5	0.090	0.800
33	1.235	58.561	54.675	0.064	0.805	5	0.063	0.821
77	1.231	58.396	54.665	0.064	0.851	5	0.059	0.722
35	1.225	58.083	53.693	0.096	1.000	5	0.072	0.737

Table 3. Optimization results for the 3[d]set-up

Design variant	E	EFF	H	beta1	beta2	nBl	inlet	outlet
73	1.027	72.024	48.713	0.162	0.306	7	0.067	0.719
72	1.025	71.902	48.268	0.162	0.205	7	0.068	0.719
39	1.025	71.902	48.268	0.162	0.306	7	0.068	0.719
50	1.025	71.880	48.687	0.162	0.306	7	0.070	0.719
63	1.023	71.779	47.836	0.143	0.306	7	0.068	0.719

IOP Conf. Series: Materials Science and Engineering **779** (2020) 012054 doi:10.1088/1757-899X/779/1/012054

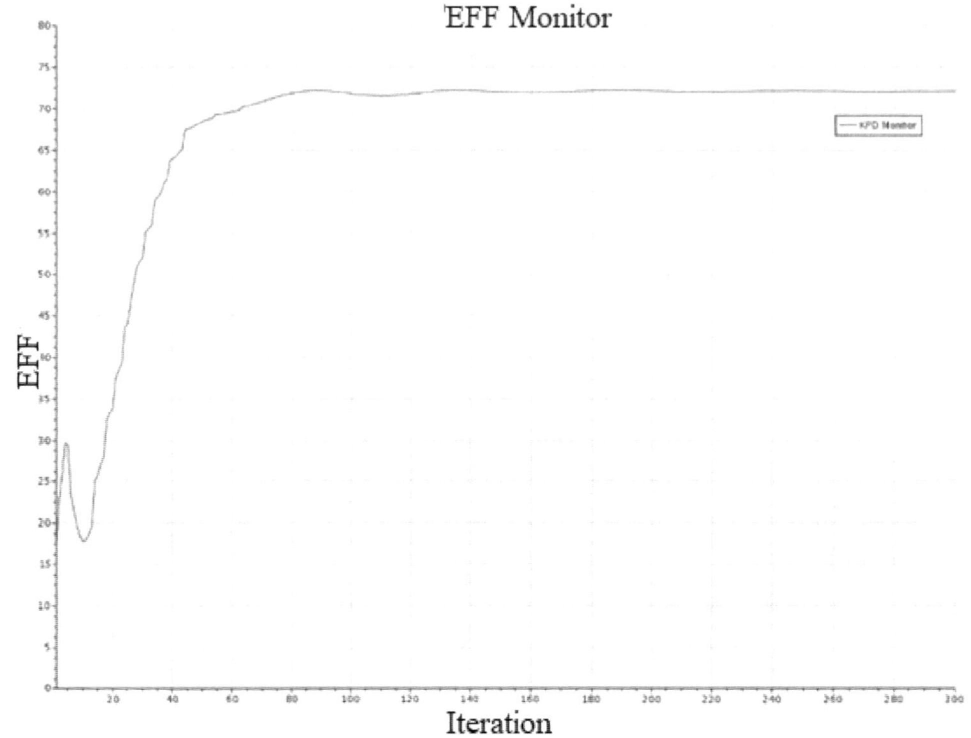

Fig.3. Efficiency for the best variant of the 3d set-up

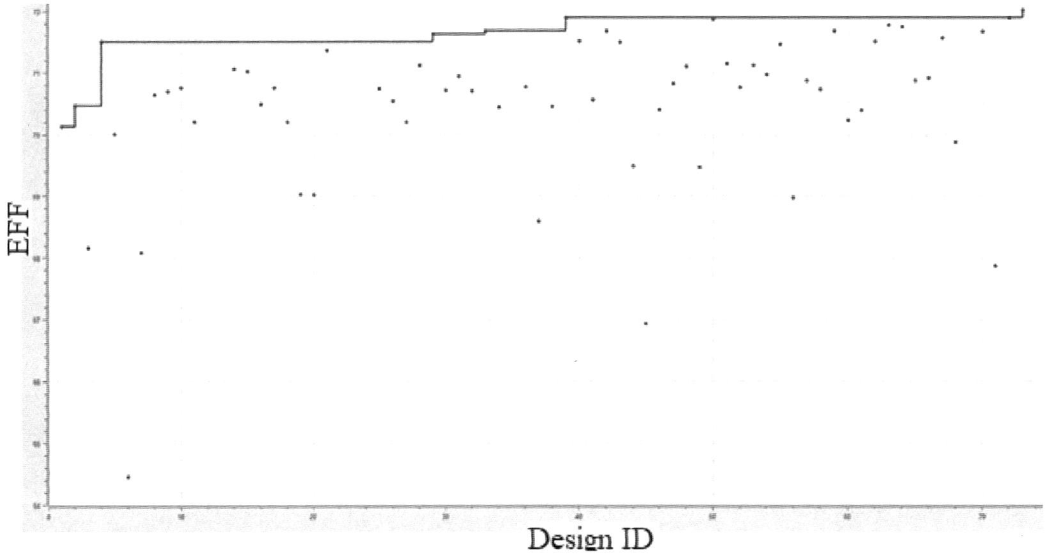

Fig.4. The change of efficiency for the 3dset-up during optimization

For the forth set-up 78 variant sare considered, the best of the mare showing Table4.

Table 4. Optimization results for the 4thset-up

Design variant	E	EFF	H	beta1	beta2	nBl	inlet	outlet

65	1.087	75.125	47.384	0.242	0.188	5	0.072	0.682
62	1.087	75.124	47.269	0.242	0.385	5	0.072	0.682
71	1.086	75.090	47.821	0.242	0.385	7	0.072	0.682
61	1.085	75.007	47.192	0.242	0.188	5	0.072	0.656
18	1.085	74.991	47.176	0.242	0.385	5	0.072	0.656

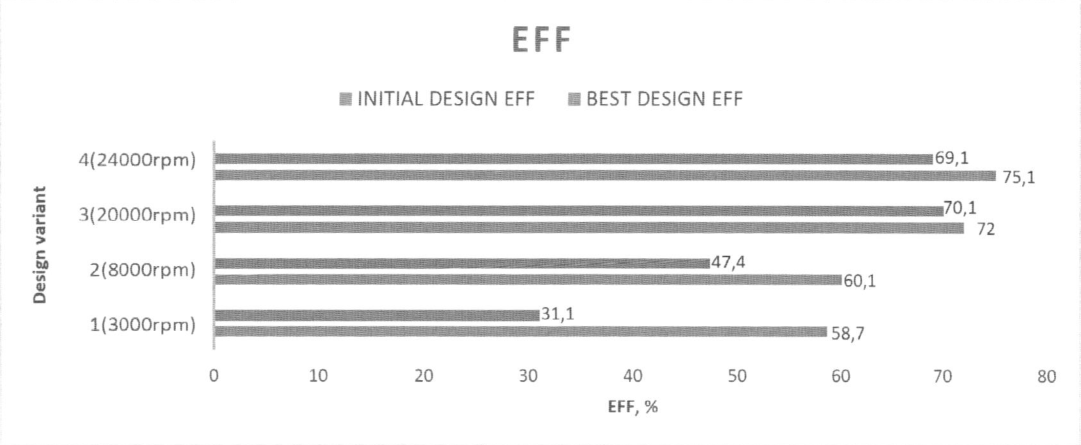

Fig.5. Comparis on of efficiency for the best configurations of different set-upswith initial values

The analysis of the results obtained shows that optimization is an effective tool helping achieve high efficiency for the pumps of this type only by changing vaned stator geometry, corresponding methods need to be integrated in the design process of such pumps and additional geometry changes of other pump elements should be also considered

Discussions

1.Achieved results show that optimization improve efficiency values of the pump, e.g. for the first set-up the best design efficiency is 58.7% when initial design value is only 31.1 % (for the second the yare 60.1% and 47.4% respectively, for the third set-up — 72.0 % and 70.1 % respectively and for the forth — 75. 1 % and 69.1 % respectively);

2.The comparison (Fig.5) shows that it is preferable to choose pumps with hig her rotation speed, but for such cases the irreliability should also be considered;

3.The results obtained call for additional investigations including optimization of all pump elements geometries, such as impeller, volute and diffuser.

References:

[1] Chaburko, P., Kossova, Z. Wet part hybrid optimization method of hermetic pump(2019) IOP Conference Series: Materials Science and Engineering, 492 (1), № 012011

[2] Guo,B.,Zhang,M.,Jin,F.Use of Magnetic Clutch to Improve Performance of Sucker Rod Pumps (2002) SPEWestern Regional/AAPG Pacific Section Joint Meeting, pp.541–552.

[3] Qiu, S.-J. ,Zhu,J.-H. Magnetic Drive Pump Selection in Small Flow rate and High Head Design Conditions (2018) Petrochemical Equipment,47(2),pp.20–24.

[4] Mahner, M., Bauer, M., Lehn, A., Schweizer, B. An experimental investigation on the influence of an assembly preload on the hysteresis, the drag torque, the lift-off speed and the thermal behavior of three-pad air foil journal bearings(2019) Tribology International, 137, pp. 113–126.

[5] Ni, J., Peng, L., Chen, G.Nonlinear PID synchro control on pattern drawing machine with four cylinders. (2011) Proceedings of 2011 International Conference on Modelling, Identification and Control, ICMIC 2011 5973725, pp. 325-330
[6] Zadorozhnaya, N.M., Lisitsyn, A.N.Automatic Tuner for PID Controllers with Elements of Artificial Intelligence. (2018) IOP Conference Series: Earth and Environmental Science 194(2),022046 DOI:10.1088/1755-1315/194/2/022046
[7] Bogdanova, Y., Guskov, A.Synergetic synthesis of control laws for left ventricular assist device rotor on magnetic suspension. (2016) Proceedings of 2016 International Conference "Stability and Oscillations of Nonlinear Control Systems" (Pyatnitskiy's Conference), STAB 2016 7541168. DOI: 10.1109/STAB.2016.7541168
[8] Bushuev A.Yu., Ivanov M.Yu., Korotaev D.V. Minimization of amis match time of movement of actuators of a throttle synchronization system.(2018) Journal of Physics: Conference Series 1141(1), 012090DOI:10.1088/1742-6596/1141/1/012090
[9] Azarov, A.V., Antonov, F.K., Golubev, M.V., Khaziev, A.R., Ushanov, S.A.Composite 3D printing for the small size unmanned aerial vehicle structure(2019) Composites Part B: Engineering, 169, pp. 157–163.
[10] Sanders P., Young A.J., Qin Y., Fancey K.S., Reithofer M.R., Guillet-Nicolas R., Kleitz F., Pamme N., Chin J.M. Stereolithographic 3D printing of extrinsic cally self-healing composites (2019) Scientific Reports, 9(1), №388,.
[11] Li,J., Rudykh. S. T Unable microstructure transformations and auxeticbehaviorin 3D-printed multiphase composites: The role of inclusion distribution (2019) Composites Part B: Engineering, 172, pp.352–362.
[12] Lengauer, W., Duretek, I., Fürst, M., Schwarz, V., Gonzalez-Gutierrez, J., Schuschnigg, S., Kukla, C., Kitzmantel, M., Neubauer, E., Lieberwirth, C., Morrison, V.Fabrication and properties of extrusion-based 3D-printed hardmetal and cermet components(2019) International Journal of Refractory Metals and Hard Materials, 82, pp. 141–149.
[13] Lomakin, V.O. Investigation of two-phase flow in axial-centrifugal impeller by hydrodynamic modeling methods(2015) Proceedings of 2015 International Conference on Fluid Power and Mechatronics, FPM 2015, № 7337302, pp. 1204–1206.
[14] Lomakin, V., Cheremushkin. V., Chaburko. P. Investigation of vortex and hysteresis effectsin the inlet device of a centrifugal pump (2018) 2018 Global Fluid Power Society PhD Symposium, GFPS2018, №8472374
[15] A Gouskov et al 2019 *IOP Conf. Ser.: Mater. Sci. Eng.* **492** 012013
[16] A Protopopov and V Vigovskij 2019 *IOP Conf. Ser.: Mater. Sci. Eng.* **492** 012003
[17] Sun-Sheng, Y., Singh, P., Zhang, H. Flow investigations of reverse running volute pumps with backward vanes in comparison to forward type turbine vanes (2019) Proceedings of the Institution of Mechanical Engineers, Part A: Journal of Power and Energy, 233 (1), pp. 111–131.
[18] Katalazhnova, I. Correlation and regression method of centrifugal pump geometry optimization (2019) Lecture Notesin Mechanical Engineering,(9783319956299),pp.1839–1845.
[19] Guleren, K.M. Automatic optimization of a centrifugal pump based on impeller–diffuser interaction (2018) Proceedings of the Institution of Mechanical Engineers, Part A: Journal of Power and Energy, 232(8), pp.1004–1018.
[20] Siddique, M.H., Afzal, A.,Samad, A.Design Optimization of the Centrifugal Pumps via Low Fidelity Models (2018) Mathematical Problems in Engineering, 2018, №3987594.

IOP Conf. Series: Materials Science and Engineering **779** (2020) 012055 doi:10.1088/1757-899X/779/1/012055

Development and research of a borehole centrifugal pump stage

V Cheremushkin[1,3], V Lomakin[1], N Kalin[1] and A Trulev[2]

[1]Bauman Moscow State Technical University
[2]AO «Rimera»

[3]E-mail:vcheremushkin@bmstu.ru

Abstract. This article is devoted to the model of a borehole centrifugal pump developed at the Department of Hydraulics of Bauman Moscow State Technical University. A universal layout has been developed for testing the stages of pumps of this type in a certain range of parameters. The article describes the mathematical model used. A comparison of numerical and field experiments is given. A problem has been identified when comparing their results.

1 Introduction

Currently, there is a huge scope of borehole pumps. For example, in the oil industry, in the agricultural, in everyday life. There are a lot of variations of such pumps and each one needs its own approach for testing.

The main number of such pumps are used in everyday life, for pumping water from wells. To date, no work has been done on the numerical study of this type of pump with small dimensions.

This work is associated with developments underway at the Department of Hydraulics of Bauman Moscow State Technical University. On the basis of the Department, a variant of the borehole centrifugal pump stage was developed. To obtain a wet part suitable for the given parameters, optimization was carried out using numerical hydrodynamic modeling.

Currently, two main areas of optimization of borehole centrifugal pumps are CFD modeling and improvement of gas-liquid centrifugal separators [1]. There is also a way to improve energy efficiency by improving control algorithms [2].

To conduct full-scale tests of the developed pump, a universal prototype was designed and produced, on which various flow parts can be tested with a fairly wide range of parameters and sizes. Figure 1 shows the design of such a layout.

The design includes universal units for supplying and discharging liquids, which also contain bearings and mechanical seals. The shaft has a margin in length, and the spacers in the form of a honed pipe between the elbows are interchangeable. Thus, the design of the layout allows one to test the stages of different radial and axial dimensions. Limitations in radial dimensions are 100 mm in diameter, since the inner diameter of the honed pipe is 100 mm. The axial limitation is only the length of this pipe and the length of the shaft. If necessary, the shaft can be made in a different length.

3D models of the impeller and the guide vanes were created (Figure 2), and subsequently made using 3D printing. The material is polyamide.

Content from this work may be used under the terms of the Creative Commons Attribution 3.0 licence. Any further distribution of this work must maintain attribution to the author(s) and the title of the work, journal citation and DOI.
Published under licence by IOP Publishing Ltd

IOP Conf. Series: Materials Science and Engineering **779** (2020) 012055 doi:10.1088/1757-899X/779/1/012055

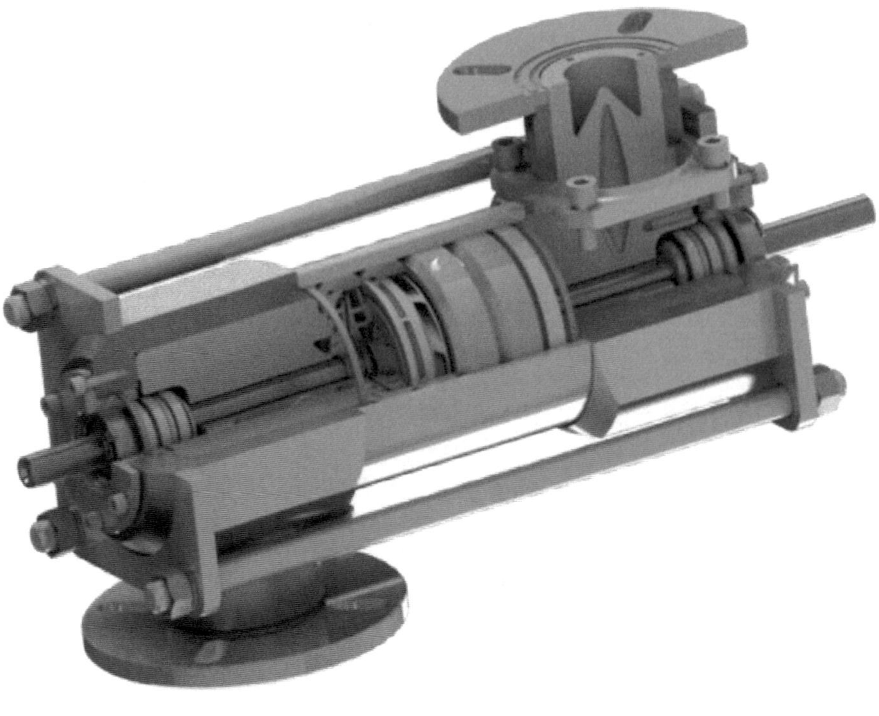

Figure 1. Stages test layout

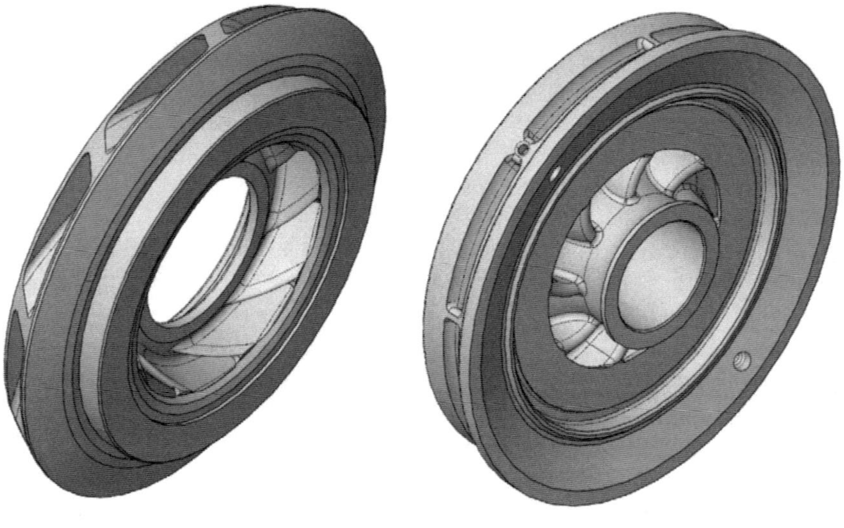

Figure 2.3D models of the impeller and guide vanes

The impeller is equipped with a bronze bushing (Figure 3) for precise alignment along the shaft, and the guide vanes is equipped with a metal sleeve (Figure 4) for centering in a honed tube.

Figure 3. Printed impeller with glued bronze bushing

Figure 4. Printed guide vanes glued to a metal sleeve

2 Problem statement and mathematical model.

The task was to develop stages for a borehole pump with certain parameters at a shaft speed of 2910 rpm.

To obtain a stage suitable for these parameters, the method of numerical hydrodynamic modeling was used to calculate the flow in the impeller and guide vanes.

The method of numerical simulation is based on the solution of discrete analogues of the basic equations of hydrodynamics [3, 4, 5]. In the case of an incompressible fluid model (ρ=const), this is:

Mass conservation equation (continuity equation)

$$\frac{\partial \tilde{u}_j}{\partial x_j} = 0,$$

where $\tilde{u}_j$ is the averaged value of the fluid velocity in the projection onto the jth axis (j=1,2,3);

Momentum conservation equation (averaged by Reynolds):

$$\rho\left[\frac{\partial U_i}{\partial t} + U_j\frac{\partial U_i}{\partial x_j}\right] = -\frac{\partial P}{\partial x_i} + \frac{\partial}{\partial x_i}\left[T_{ij}^{(v)} - \rho\langle u_i u_j\rangle\right];$$

where U, P — averaged speed and pressure;

$\tilde{T}_{ij}^{(v)} = 2\mu\tilde{s}_{ij}$ —viscous stress tensor for an incompressible fluid;

$\tilde{s}_{ij} = \frac{1}{2}\left[\frac{\partial \tilde{u}_i}{\partial x_j} + \frac{\partial \tilde{u}_j}{\partial x_i}\right]$ is the instantaneous strain rate tensor;

$\rho\langle u_i u_j\rangle$ — Reynolds stresses.

The introduction of the Navier-Stokes equations averaged by Reynolds makes the system of equations not closed, as additional unknowns appear - Reynolds stresses. To close the system, the semi-empirical k-ω SST turbulence model is used, which introduces the necessary additional equations: the equations of transfer of turbulence kinetic energy and the relative dissipation rate of this energy [6, 7, 8, 9]:

$$\frac{\partial k}{\partial t} + U_j\frac{\partial k}{\partial x_j} = P_k - \beta^* k\omega + \frac{\partial}{\partial x_j}\left[(\nu + \sigma_k\nu_T)\cdot\frac{\partial k}{\partial x_j}\right]$$

$$\frac{\partial \omega}{\partial t} + U_j\frac{\partial \omega}{\partial x_j} = \alpha\cdot S^2 - \beta\cdot\omega^2 + \frac{\partial}{\partial x_j}\left[(\nu + \sigma_\omega\nu_T)\cdot\frac{\partial \omega}{\partial x_j}\right] + 2\cdot(1 - F_1)\cdot\sigma_{\omega 2}\cdot\frac{1}{\omega}\cdot\frac{\partial k}{\partial x_i}\cdot\frac{\partial \omega}{\partial x_i}$$

The roughness function f is used to take into account the wall roughness:

$$f = \begin{cases} 1 & \text{если } R^+ < R^+_{smooth} \\ \left[B\cdot\left(\dfrac{R^+ - R^+_{smooth}}{R^+_{rough} - R^+_{smooth}}\right)\right]^a & \text{если } R^+_{smooth} < R^+ < R^+_{rough}; \\ B + CR^+ & \text{если } R^+ > R^+_{rough} \end{cases}$$

$$R^+ = \frac{ru^*}{\nu}, \text{где}$$

r— equivalent roughness value,

ν – kinematic viscosity,

u^*—characteristic velocity near the wall, determined depending on the turbulence model used.

$$a = sin\left[\frac{\pi}{2}\,\frac{\log\left(R^+ / R^+_{smooth}\right)}{\log\left(R^+_{rough} / R^+_{smooth}\right)}\right];$$

R^+_{smooth} —roughness value corresponding to smooth walls;

R^+_{rough} – R + value for roughness corresponding to a rough wall.

Then the logarithmic velocity profile is determined according to the expression:

$$u^+ = \frac{1}{K} ln\left(\frac{E}{f} y^+\right), \text{where}$$

$$y^+ = \frac{yu^*}{\nu};$$

y— distance from the wall;
K and *E*— constants.

The wet part of the borehole pump stage was modeled on a grid consisting of 739 thousand cells. In the core of the flow stream, the cells have a multifaceted shape; at the solid walls of the pipe, they are prismatic [10, 11]. The calculation grid for the stage is shown in the Figure 5.

Figure 5. Calculation grid

3 Optimization results

To obtain the best characteristics, suitable for the given requirements, 512 different models of the wet part of the stage were selected. Variable parameters and the range of their changes are presented in table 1.

Table1. Variableparameters

Variableparameter	Parameterchangerange
Outletbladeangle,β_2, °	20-35
Wrap angle of impeller blade, φ, °	90-130
Wrap angle of guide vanes channel, φ_s°	80-110
Impeller outlet width,b_2, mm	5-6

The optimization results are presented in table 2.

Table2. Optimization results

Model no.	Q/Qnom	H/Hnom	Efficiency, %
0	1	1.11	68.49
1	1	1.08	68.48
2	1	1.15	68.01
3	1	1.09	68.92
4	1	1.12	65.84
5	1	1.12	68.24
…	…	…	…
112	1	1.12	68.41
…	…	…	…
507	1	1.11	68.63
508	1	1.06	65.75
509	1	1.10	67.90
510	1	1.09	68.14
511	1	1.10	66.73
512	1	0	0

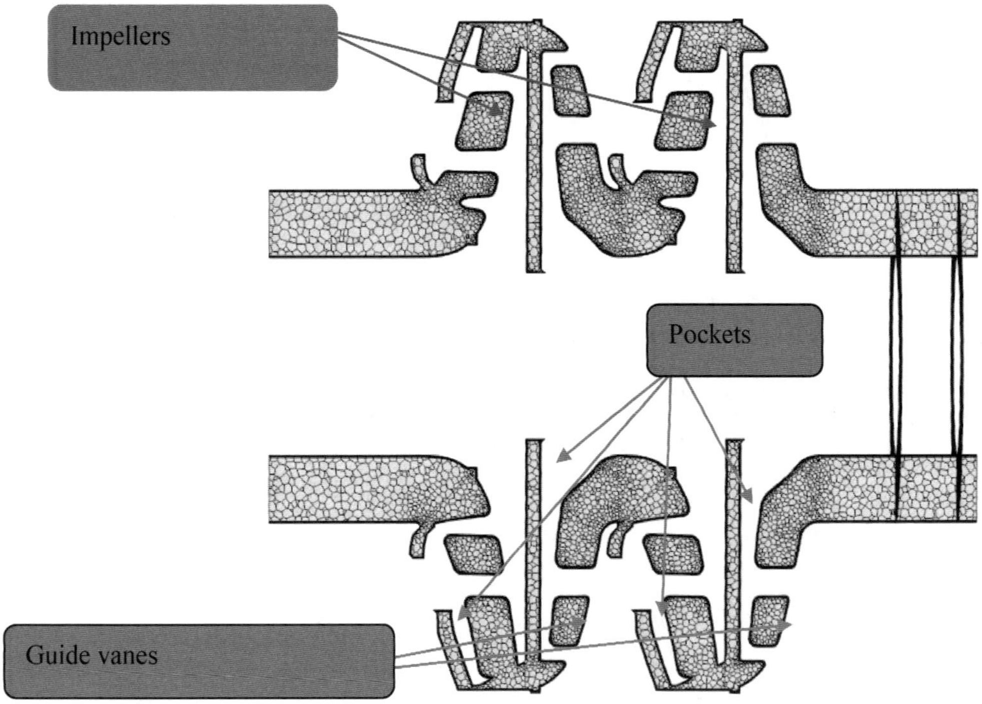

Figure 6. Calculation grid

The best model is No. 112. For it, a hydrodynamic numerical calculation was made in the unsteady mode of two stages, taking into account the pockets. In this calculation, the roughness of the walls, the mathematical model of which is described above, was taken into account. The calculation grid of the wet part is shown in the figure 6. It consists of 3290000 cells

The figure below shows scenes of the distribution of reference values in cross sections, such as velocities, pressures, and turbulent kinetic energy.

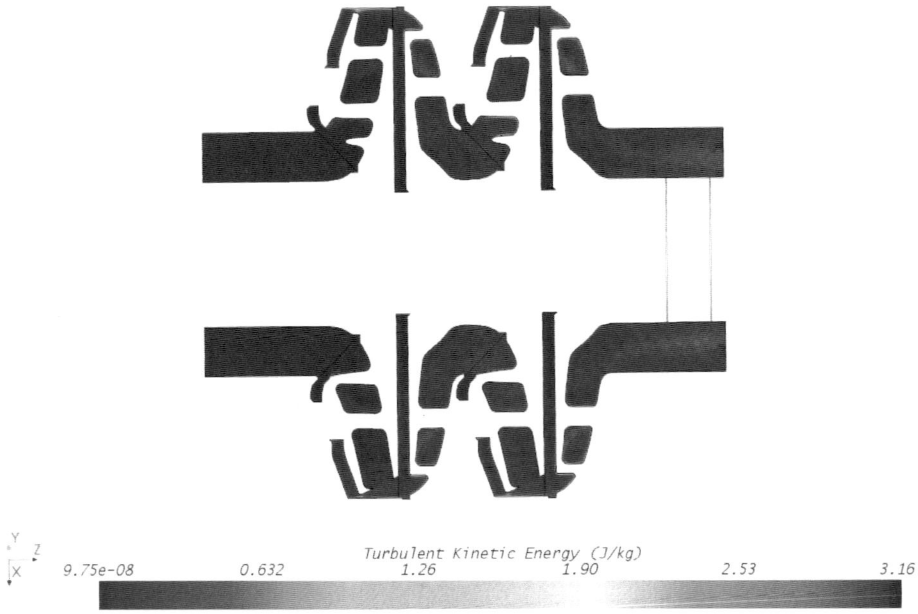

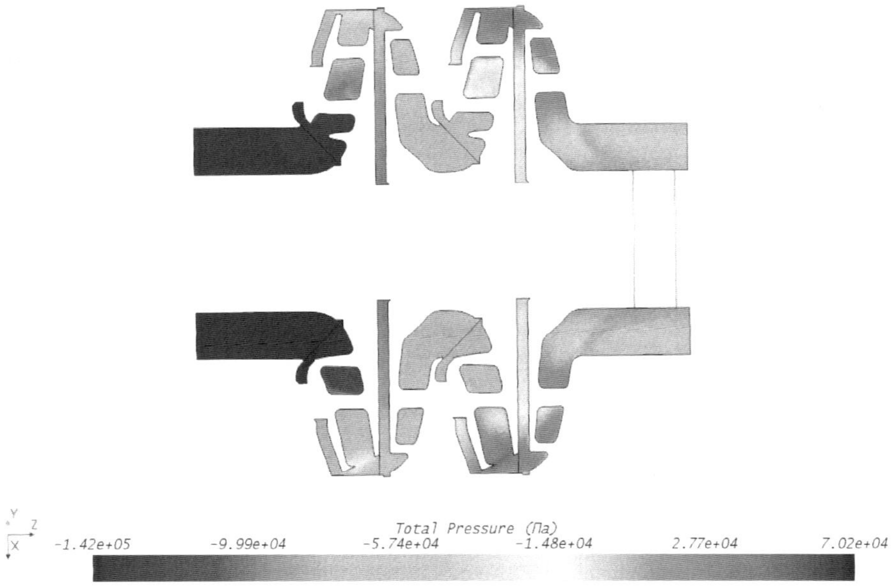

IOP Conf. Series: Materials Science and Engineering **779** (2020) 012055 doi:10.1088/1757-899X/779/1/012055

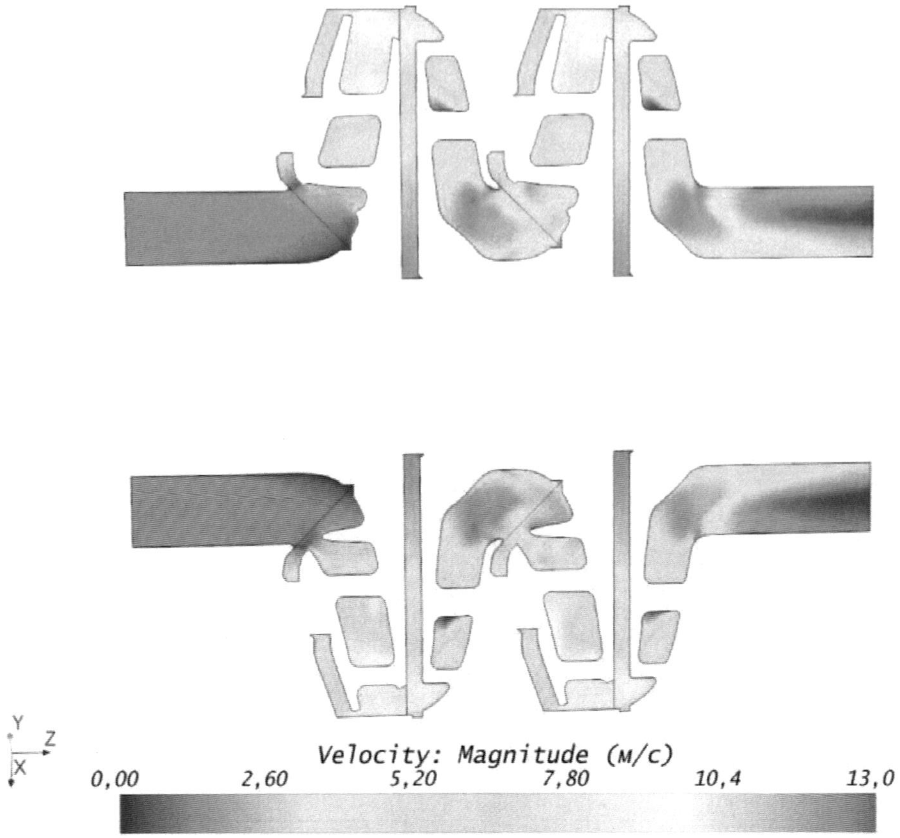

Figure 7. Scalar distribution scenes

The table shows the main parameters of the resulting optimized wet part after modeling of two stages, taking into account the roughness of the walls and the flow of fluid particles in the pockets.

Table 3

Model no.	Q/Qnom	H/Hnom	Efficiency,%
112	1	0.96	59.3

4 Field tests on the developed layout

To determine the hydraulic characteristics of the profiled wet part, its bench tests were carried out on a universal prototype (according to GOST 16504-81). The figure shows the layout mounted on the stand, as well as a general view of the test bench.

Tests were conducted for two and four stages. The resulting graphs are presented in figures 9 and 10.

Efficiency is given minus the dry friction moment — 0.45 N*m

IOP Conf. Series: Materials Science and Engineering **779** (2020) 012055 doi:10.1088/1757-899X/779/1/012055

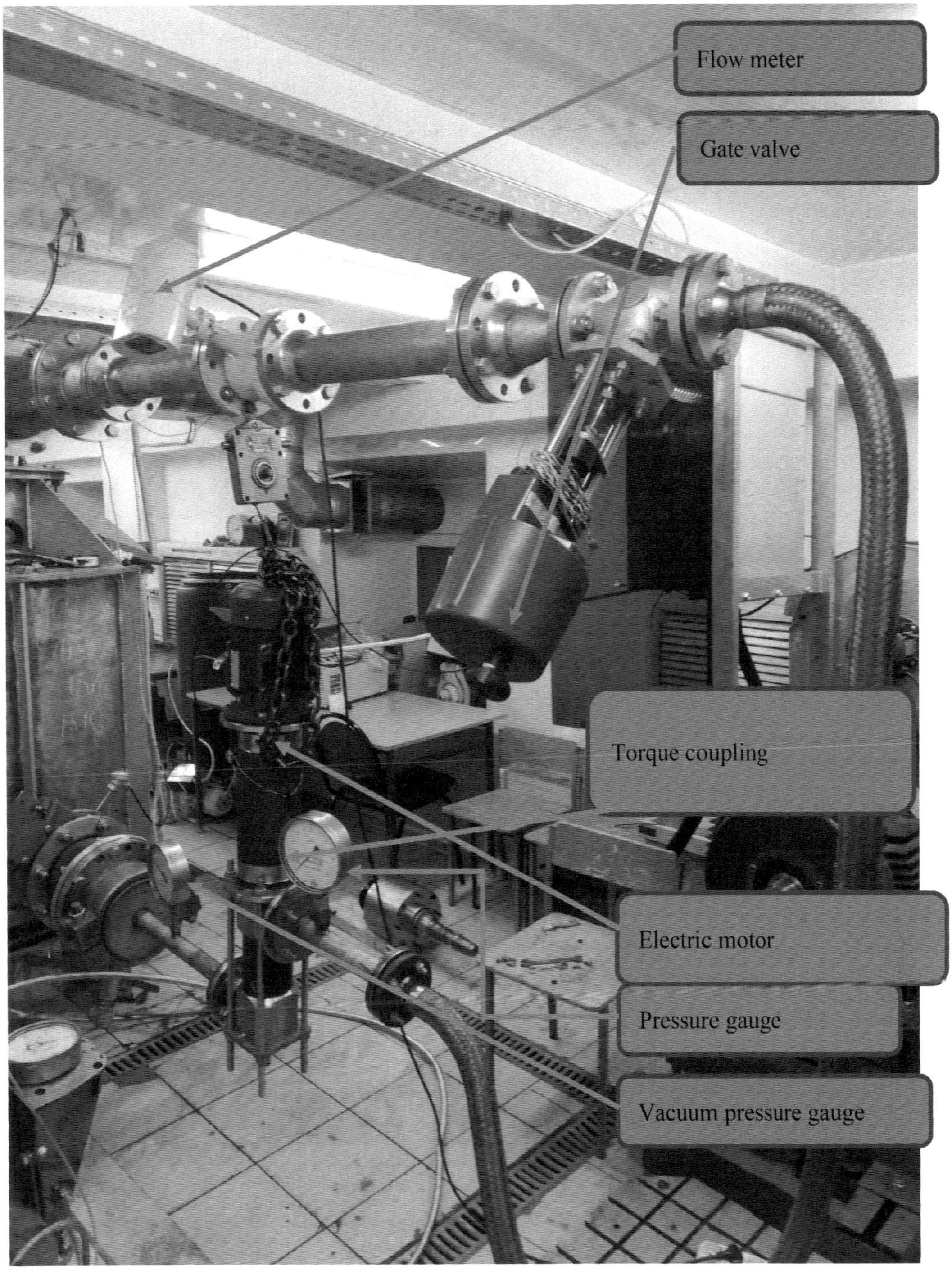

Figure 8. Test bench

IOP Conf. Series: Materials Science and Engineering **779** (2020) 012055 doi:10.1088/1757-899X/779/1/012055

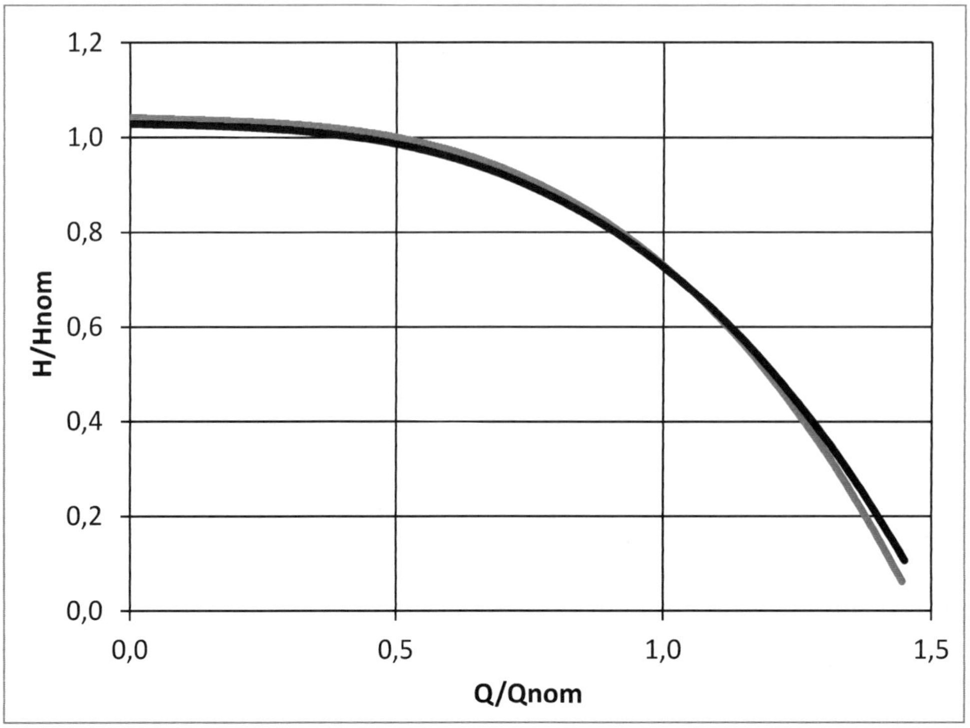

Figure 9. Pressure characteristics for 2 and 4 stages, rated to nominal

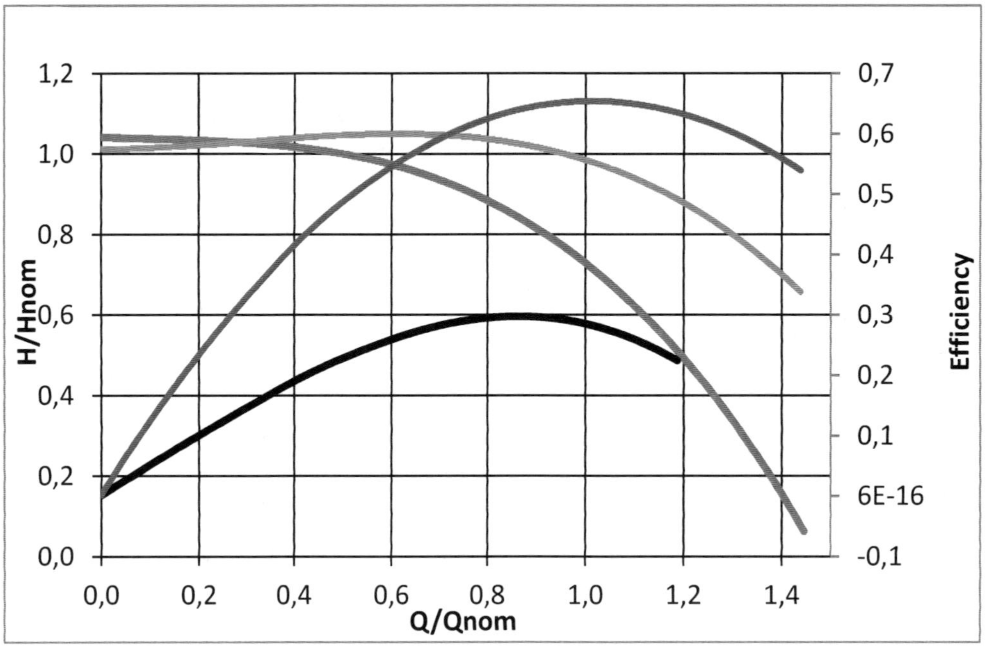

Figure 10. Pressure characteristics and efficiency obtained during field tests and by the method of hydrodynamic modeling. The black graph is the efficiency of the full-scale mock-up, the blue graph is the efficiency obtained using CFD modeling. The red characteristic is the pressure of the full-scale mock-up, the green characteristic is the pressure obtained using CFD modeling.

Table 4 compares the results of numerical calculation and experiment.

Table 4. Comparison of experimental and calculated data

Data	Head, m	Efficiency, %
Experiment	21,1	28,6
Calculation, 1 stage (conversion to 4 stages)	30,4	64,9
Calculation, 2 stages (conversion to 4 stages)	30,2	64,8
Calculation, 2 stages, taking into account the roughness (conversion to 4 stages)	28,4	59,5

5 Conclusion on the work performed

As a result of the work done by means of numerical hydrodynamic modeling, the wet part of the borehole centrifugal pump stage was obtained, which corresponds to the given parameters. To verify the conformity of the results of numerical calculation and field testing, a universal layout was designed. As a result, the efficiency obtained in the calculation numerically and in full-scale test differs by almost 2 times. A significant difference in pressure was also obtained, which can be seen in the comparative characteristics. The assumption was made that the error arose due to the fact that the roughness was not taken into account in the simulation. This assumption was verified by adding a roughness to the mathematical function, which affects the velocity distribution near solid walls. However, even taking into account the roughness, no correspondence was obtained between the numerical and field tests. Further research in this area is necessary to explain the results.

References.

[1] Budaev, G, Danilov, D, Kuznechov, A, Lomakin, V, Cheremushkin, V. Research of centrifugal gas-liquid separator // IOP Conference Series: Materials Science and Engineering, Vol.589 (1), № 012035. DOI: 10.1088/1757-899X/589/1/012035

[2] Brunman, V.E., Vataev, A.S., Volkov, A.N., Larin, M.Y., Matsko, O.N., Petkova, A.P., Plotnikov, D.G. Improving the energy efficiency of borehole pumps.(2017) Russian Engineering Research, 37 (7), pp. 579–580. DOI: 10.3103/S1068798X17070061

[3] Cheremushkin, V., Polyakov, A. Optimization of the output device of a disk pump for high viscous fluid // IOP Conference Series: Materials Science and Engineering, Vol.589 (1), № 012001. DOI: 10.1088/1757-899X/589/1/012001

[4] Lomakin, V.O., Chaburko, P.S., Kuleshova, M.S. Multi-criteria Optimization of the Flow of a Cen-trifugal Pump on Energy and Vibroacoustic Characteristics // Procedia Engineering. Vol. 176, 2017, pp 476–482. DOI: 10.1016/j.proeng 2017.02.347.

[5] Lomakin, V.O. Investigation of two-phase flow in axial-centrifugal impeller by hydrodynamic mod-eling methods. Proceedings of 2015 International Conference on Fluid Power and Mechatronics, FPM 2015, 2015, № 7337302, Pp 1204–1206.
DOI: 10.1109/FPM.2015.7337302

[6] Loitsianskii, L.G. Mechanics of liquisd and gases / Loitsianskii, L.G. -Moscow: Drofa, 2003. — 840p.

[7] Gouskov, A.M., Sorokin, F.D., Banin, E.P. Simulation of an Inlet Structure of an Implantable Axial Blood Pump. Biomedical Engineering, Vol 50, 2016, Pp 15-19. DOI: 10.1007/s10527-016-9578-2 13. Gouskov, A.M., Lomakin, V.O., Banin, E.P., Kuleshova, M.S. Minimization of Hemolysis and Im-provement of the Hydrodynamic Efficiency of a Circulatory Support

IOP Conf. Series: Materials Science and Engineering **779** (2020) 012055 doi:10.1088/1757-899X/779/1/012055

Pump by Optimizing the Pump Flowpath. Biomedical Engineering, Vol 51, 2017, Pp 229–233. DOI: 10.1007/s10527-017-9720-9

[8] Baturin, O., Popov, G., Kolmakova, D., Zubanov, V., Novikova, J., Korneeva, A. Optimization of Powerful Two-stage Screw Centrifugal Pump. 2nd International Conference on Mechanical, System and Control Engineering, ICMSC 2018; Moscow; Vol 220, 2018. DOI: 10.1051/matecconf/201822003009

[9] Tesch, K.Email Author, Kaczorowska-Ditrich, K. Discrete-continuous optimisation of an axial flow blood pump. 23rd Fluid Mechanics Conference, FMC 2018; Zawiercie; Poland. Vol 1101, 2018. DOI: 10.1088/1742-6596/1101/1/012044

[10] Boyarshinova, A., Lomakin, V., Petrov, A. Comparison of various simulation methods of a two-phase flow in a multiphase pump (2019).IOP Conference Series: Materials Science and Engineering, 589 (1), № 012014. DOI: 10.1088/1757-899X/589/1/012014

[11] Śmierciew, K.Email Author, Gagan, J., Butrymowicz, D. Application of numerical modelling for de-sign and improvement of performance of gas ejector. Applied Thermal Engineering. 2019, Pp 85-93. DOI: 10.1016/j.applthermaleng.2018.12.030

IOP Conf. Series: Materials Science and Engineering **779** (2020) 012056 doi:10.1088/1757-899X/779/1/012056

Investigation of the influence of geometric parameters of a multi-nozzle jet pump on its energy characteristics

V Cheremushkin[1,2] and A Polyakov[1]

[1]Bauman Moscow State Technical University

[2]E-mail: vcheremushkin@bmstu.ru

Abstract. In this paper the optimization process of a jet pump developed according to a multi-nozzle scheme and designed to pump a mixture of water and soil is considered. Its flow part was designed, some measures were taken to accelerate the calculation process, including a method of setting the boundary conditions. Criteria and optimization parameters, the mathematical model used are given in the text. According to the calculation results, the best models and some dependencies of the criteria on the variated parameters are shown. The flow patterns in the flow part of the jet pump are presented.

Introduction

A jet pump is a device for pumping liquid due to the work performed by the flow of a liquid. Figure 1 shows a schematic diagram of a jet pump.

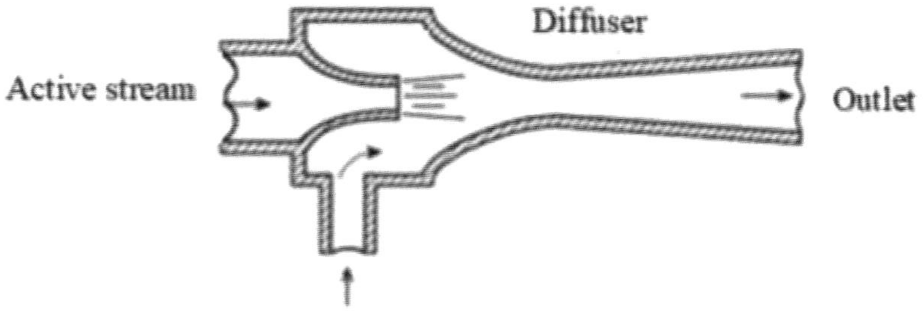

Figure1. Schematic diagram of a jet pump

Regardless of the design of the jet pump, it always directly transfers energy from a flow of a liquid to the flow of the ejected liquid. The first stream is active, and the second one is passive [1].

The necessary elements for the implementation of the jet pump workflow are the supply nozzles of the active and passive flows, the mixing chamber and the outlet diffuser.

Jet pumps are widespread in many areas of industry. Such qualities as high reliability associated with the absence of moving elements, the ability to work with aggressive liquids containing gas and solid inclusions greatly facilitated that. The resource of jet pumps is limited only by the properties of the materials that were used for its manufacture.

This article discusses the issue of optimizing the energy characteristics of a jet pump developed according to a multi-nozzle scheme with a peripheral disposition of active flow nozzles.

Content from this work may be used under the terms of the Creative Commons Attribution 3.0 licence. Any further distribution of this work must maintain attribution to the author(s) and the title of the work, journal citation and DOI.
Published under licence by IOP Publishing Ltd

IOP Conf. Series: Materials Science and Engineering **779** (2020) 012056 doi:10.1088/1757-899X/779/1/012056

Such a constructive decision was made according to the tasks that were faced in the process of development. A feature of the pumped medium is the presence of solid inclusions. These factors make it impossible to rely on existing works on the optimization of classical jet pumps [2–4].

Mathematical Model and Methods

The task of designing the diverting device is solved by means of CFD (computational fluid dynamics). The implementation of the numerical simulation method is based on solving the basic hydrodynamic equations discrete analogs [5–8]. In case of an incompressible fluid model, when the density is constant, these are:

Continuity Equation

$$\frac{\partial \tilde{u}_j}{\partial x_j} = 0,$$

where $\tilde{u}_j$ — j axis projection of an averaged value of the fluid velocity (j=1,2,3);

The momentum conservation equation (Reynolds averaged):

$$\rho\left[\frac{\partial \mathrm{U_i}}{\partial t} + \mathrm{U_j}\frac{\partial \mathrm{U_i}}{\partial x_j}\right] = -\frac{\partial \mathrm{P}}{\partial x_i} + \frac{\partial}{\partial x_i}\left[\mathrm{T}_{ij}^{(v)} - \rho u_i u_j\right];$$

where U, P — averaged velocity and pressure;

$\tilde{\mathrm{T}}_{ij}^{(v)} = 2\mu\tilde{s}_{ij}$ — viscous stress tensor for incompressible fluid;

$\tilde{s}_{ij} = \frac{1}{2}\left[\frac{\partial \tilde{u}_i}{\partial x_j} + \frac{\partial \tilde{u}_j}{\partial x_i}\right]$ — instant strain rate tensor;

$\rho u_i u_j$ — Reynolds stresses.

The introduction of the Reynolds averaged Navier-Stokes equation makes the equations system not closed, as additional unknowns, Reynolds stresses, appear.

To solve this system of equation, the k-ω SST turbulence model was used. It introduces the necessary additional equations: the transport equations for the turbulent kinetic energy and the relative dissipation rate of this energy:

$$\frac{\partial k}{\partial t} + U_j\frac{\partial k}{\partial x_j} = P_k - \beta^* k\omega + \frac{\partial}{\partial x_j}\left[\left(\nu + \sigma_k \nu_T\right)\cdot\frac{\partial k}{\partial x_j}\right]$$

$$\frac{\partial \omega}{\partial t} + U_j\frac{\partial \omega}{\partial x_j} = \alpha\cdot S^2 - \beta\cdot\omega^2 + \frac{\partial}{\partial x_j}\left[\left(\nu + \sigma_\omega \nu_T\right)\cdot\frac{\partial \omega}{\partial x_j}\right] + 2\cdot\left(1 - F_1\right)\cdot\sigma_{\omega 2}\cdot\frac{1}{\omega}\cdot\frac{\partial k}{\partial x_i}\cdot\frac{\partial \omega}{\partial x_i}$$

When searching for the values of the optimal geometric parameters of the elements of the jet pump, a combination of stochastic and local (directional) search was used [9].

Figure 2 shows the computational grid used in hydro modeling. The grid has polyhedral cells inside the flow in the flow core and prismatic near lid walls [10].

IOP Conf. Series: Materials Science and Engineering **779** (2020) 012056 doi:10.1088/1757-899X/779/1/012056

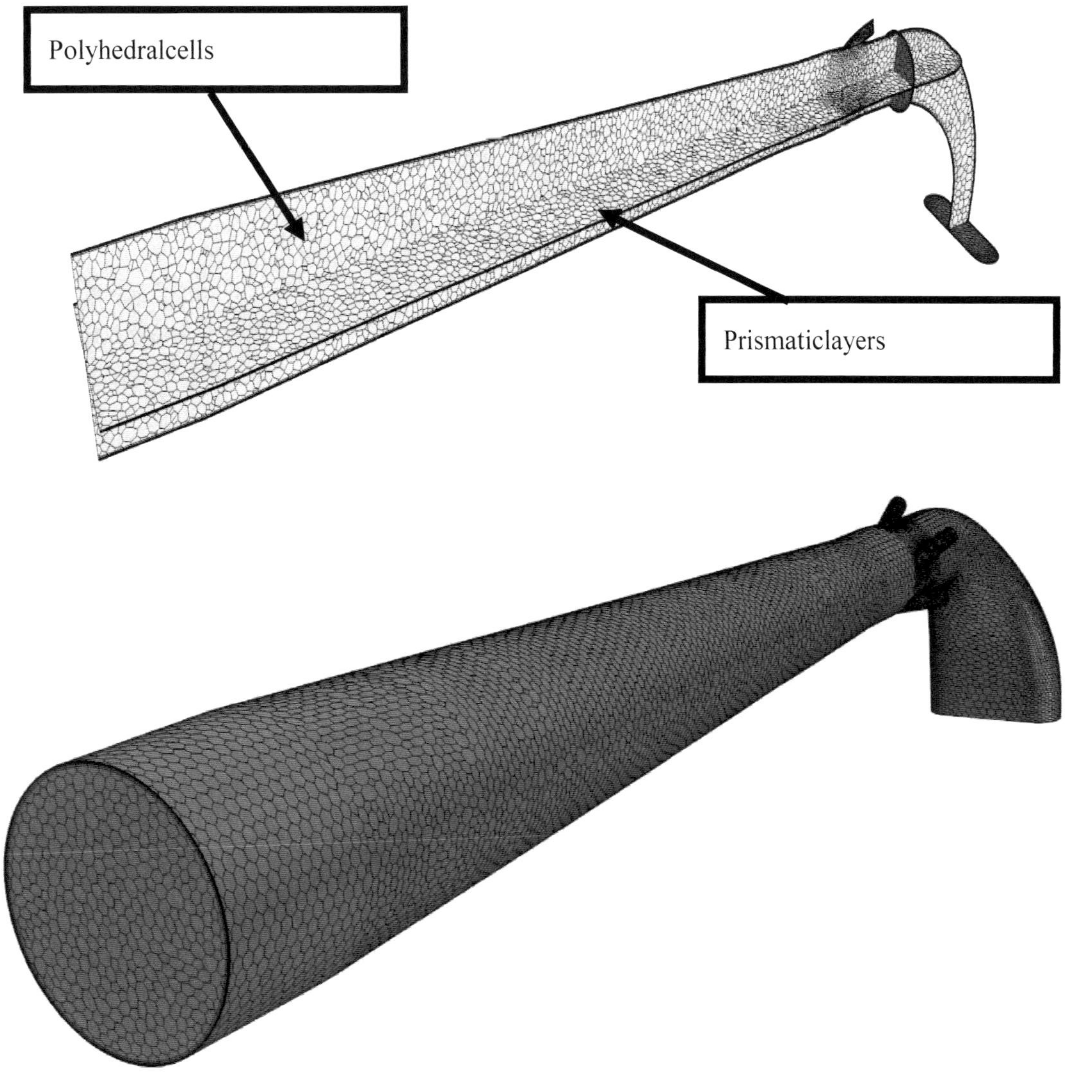

Figure2. Calculation grid

Description of the investigated model

To study the influence of the geometric parameters of the elements of the jet pump, a parameterized 3D model was made by methods of solid modeling during the development (Fig. 3).

Table 1 shows the geometric parameters that where changed during the optimization process and their ranges.

IOP Conf. Series: Materials Science and Engineering **779** (2020) 012056 doi:10.1088/1757-899X/779/1/012056

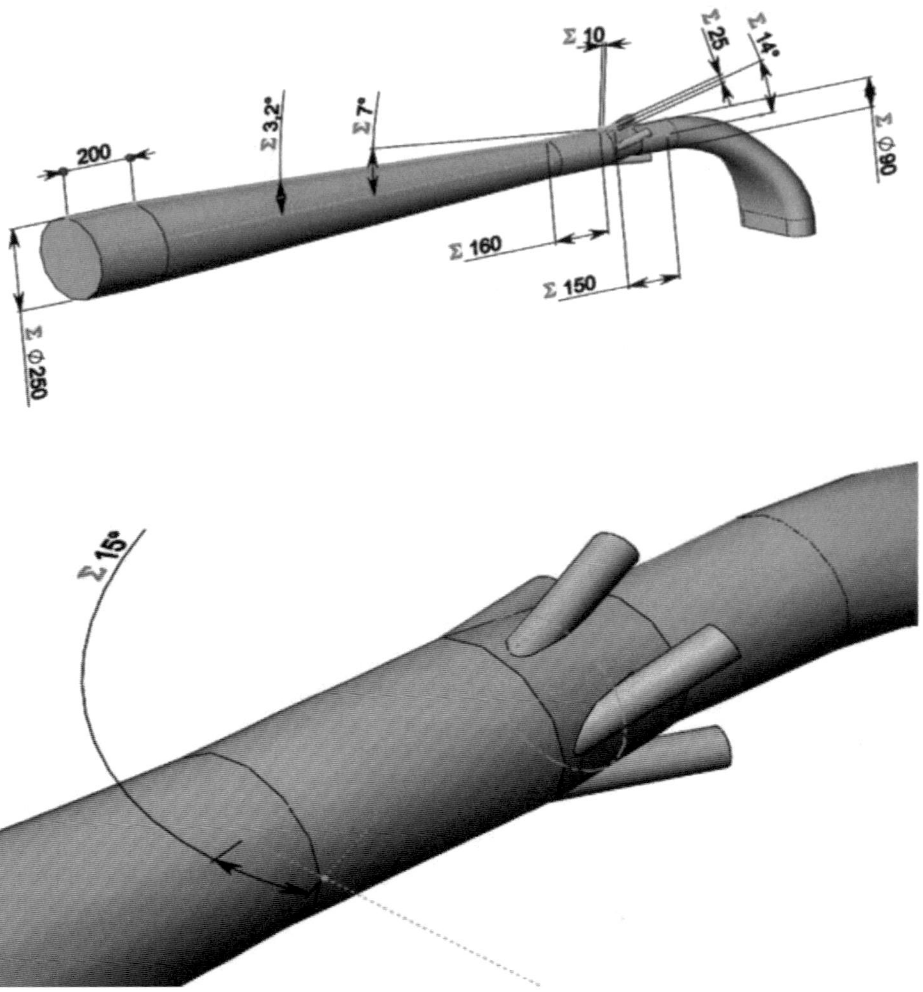

Figure3. Parameterized 3D model

Table1. Geometric parameters variation ranges

№	Parameter Name	Minimum	Maximum
1	Diameter of passive flow supply, **D1** mm	75	100
2	Active camera input opening angle, **fi3** degrees	5	45
3	Diameter of nozzles, **d** mm	23	27
4	The length of the stabilization section of the passive flow, **l1** mm	100	200
5	Mixing chamber length, **l3** mm	150	300
6	The size of the input chamber of the active stream,**delta** degrees	2	10
7	The angle of the nozzle to the axis, **fiax** degrees	10	20
8	Circumferential angle of nozzle inclination, **fir** degrees	0	30
9	Diffuser angle, **fi** degrees	1	5

In order to exclude the suction section from the optimization process, i.e. vertical pipe, it is advisable to calculate it separately. Thus, its drag coefficient can be determined and used to set the boundary condition. This approach can significantly reduce the volume of the computational domain to be optimized, and the calculation time respectively.

Figure 4 shows a 3D model of the flow part (fluid) of the suction pipe. In addition to the inner flow part of the pipe, a part of the inlet volume is included as a cube for precise calculation the inlet losses.

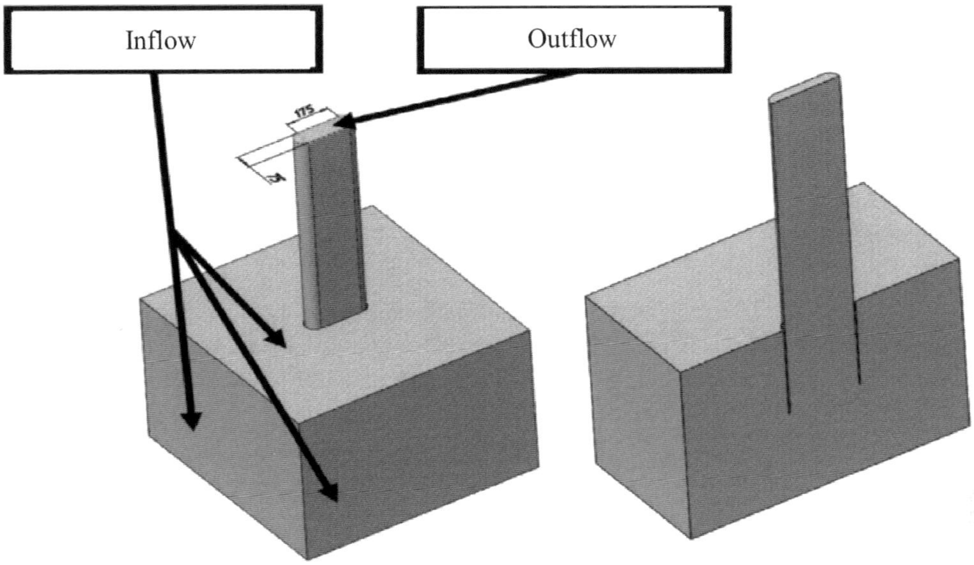

Figure 4. Flow part of the inlet area

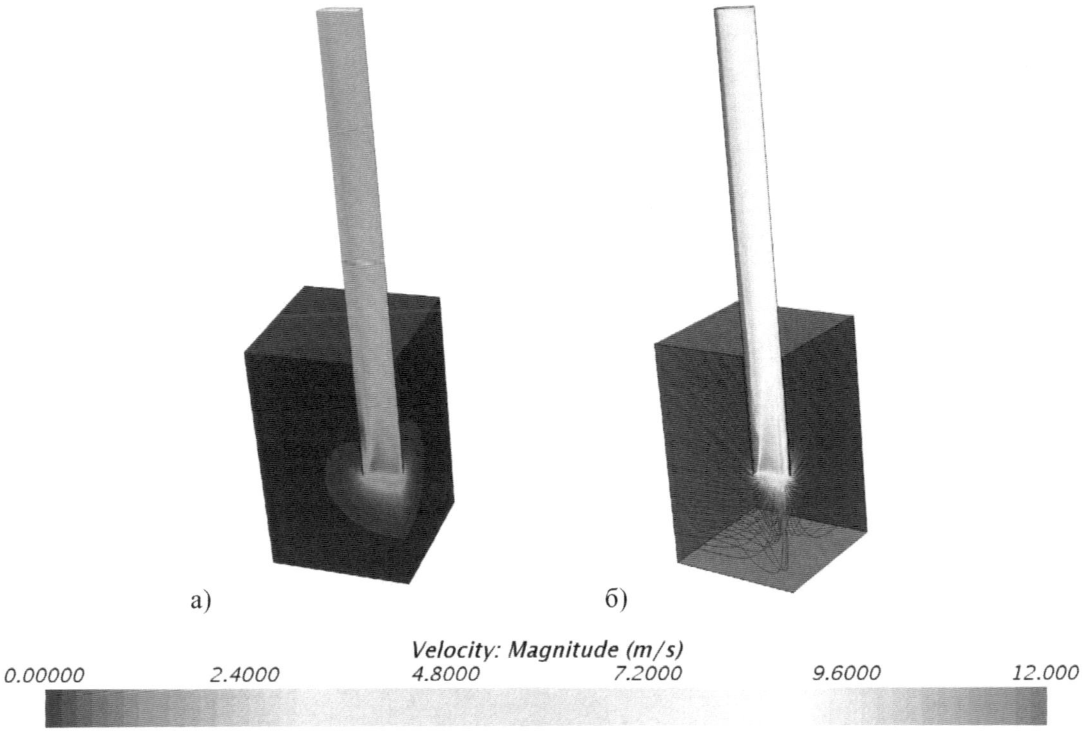

Figure 5. Distribution of velocity amplitude: a) along the planes of symmetry; b) along stream lines

The design of the calculation grid for the suction pipe is shown in Figure 4. The cells have a polyhedral shape in the flow core and prismatic near solid walls, because a finer mesh is required when computing the boundary layer.

To reduce the volume of the computational domain, the property of its geometric symmetry was used, i.e. in the process of modeling its quarter was used.

Figure 5 shows the character of the flow in the computational domain – velocity distributionalong streamlines and symmetry planes.

Losses are determined according to the formula:

$$\Delta h = (P_2 - P_1) / \rho g,$$

where Δh is the value of losses, m;

P_1 and P_2 — average values of total pressure at the input and output boundaries, respectively, Pa;

ρ—fluid density, kg/m^3;

g —acceleration of gravity, m/s^2.

Operation criteria and boundary conditions

The search for the optimal solution within the study was carried out by changing the previously given geometric parameters according to the following criteria:

1. Minimization of active flow power N_p;
2. The aspiration of the input velocity to the target value:V.

The maximum value of the power of the active stream is determined by its flow rate and pressure:

$$N_{pMAX} = P \cdot Q$$

Figure 6 shows the flow part used in the calculations with the mentioned boundary conditions.

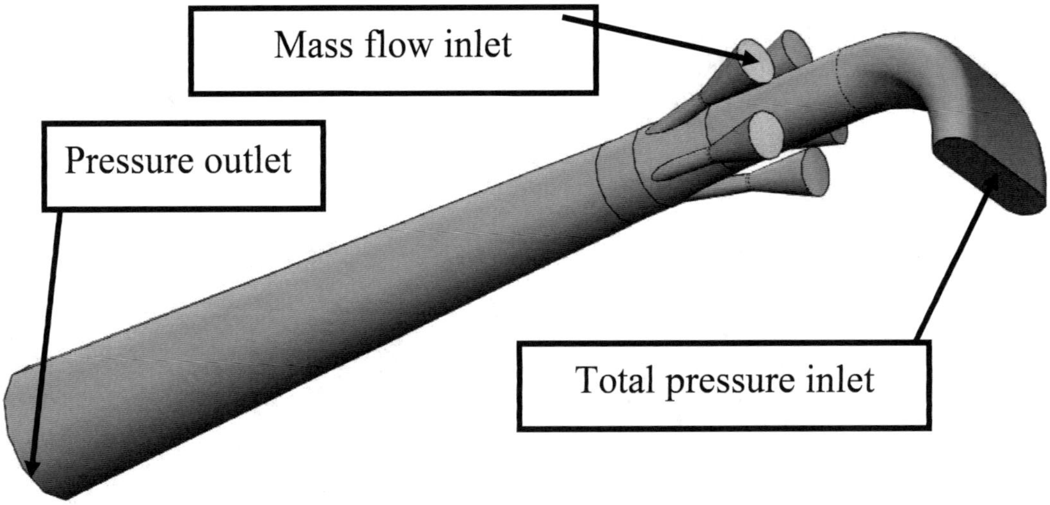

Figure 6. Boundary conditions

The total pressure of the passive flow at the inlet is determined according to the resistance characteristic of the suction pipe as follows:

$$H_{in} = \alpha \cdot V_{in}^2,$$

IOP Conf. Series: Materials Science and Engineering **779** (2020) 012056 doi:10.1088/1757-899X/779/1/012056

where:

α —a coefficient equal to the ratio of the magnitude of losses in the suction pipe to the flow rate of the liquid in it.

V_{in} — input passive flow rate.

Study results and analysis

Table 2 shows the 15 best models obtained as a result of the optimization. Power values are given in fractions of the maximum permissible, and the speed values are given in fractions of the minimum desired.

Table 2. Optimization results

№	КПД	Np	Vin	d	D1	fi_ax	fi	fi_r	l1	l3	D	delta	fi_3
1	0,111	0,907	0,849	25,7	75	12,2	4,6	13,4	111	166,5	285	6	22,7
2	0,109	0,914	0,847	25,7	75	12,2	4,6	13,4	111	250,5	285	6	22,7
3	0,210	0,919	1,055	25,2	85,8	13,9	3,2	13,4	155	150	285	6,3	8,3
4	0,210	0,921	1,055	25,2	85,8	13,9	3,2	13,4	155	160,5	285	6,3	8,3
5	0,081	0,922	0,769	26,7	85,2	13,2	2,1	1,4	191	250,5	285	5,1	25,3
6	0,219	0,927	1,071	25,2	86,8	13,3	3,2	14	155	160,5	285	6,3	7,9
7	0,076	0,932	0,752	26,8	84	12,8	1,6	2,6	112	172,5	285	5,3	28,9
8	0,087	0,935	0,790	26,6	84,6	13,1	1,9	2,2	185	250,5	285	5,3	22,9
9	0,089	0,935	0,796	26,6	86,2	13,2	1,8	2,4	187	150	285	5,7	23,5
10	0,220	0,940	1,079	25,2	87,4	13,8	3,2	14	165	160,5	285	6,3	8,5
11	0,220	0,942	1,080	25,2	87,4	13,8	3,2	14	155	160,5	285	6,3	8,5
12	0,208	0,943	1,060	25,1	85,8	13,9	3,3	13,4	157	162	285	6,5	8,3
13	0,194	0,951	1,039	25,2	85,8	13,9	3,2	13,4	155	243	285	6,3	8,3
14	0,223	0,954	1,089	25,1	87,4	13,8	3,3	14	157	162	285	6,5	8,5
15	0,232	0,954	1,103	25,2	88,8	13,8	3,2	14	155	160,5	285	6,3	8,5

Figures 7 and 8 show the dependences of the efficiency and speed in the suction pipe on the opening angle of the diffuser, the diameter of the nozzle, and the diameter of the mixing chamber.

The efficiency of the jet pump in this case is determined as follows:

$$\eta = \frac{N_2}{N_1};$$

Where:

N_2—passive flow power,

N_1—active flow power.

The choice of the useful work is explained by the main task of the pump — suction of the liquid-soil mixture.

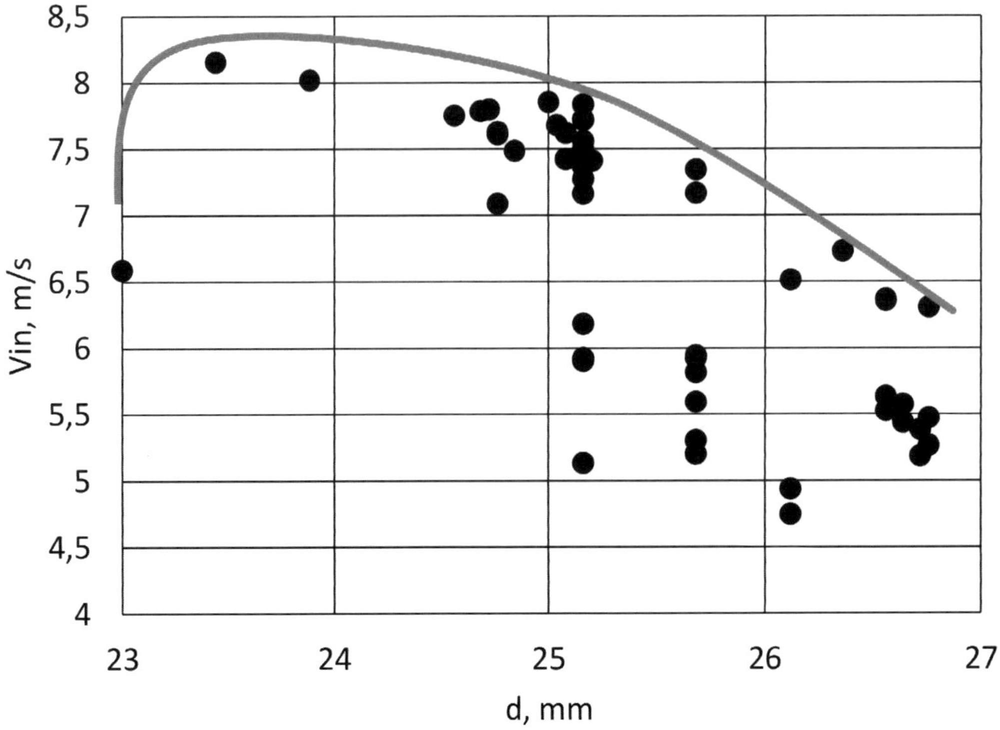
8,5
8
7,5
7
6,5
6
5,5
5
4,5
4
Vin, m/s
23
24
25
26
27
d, mm

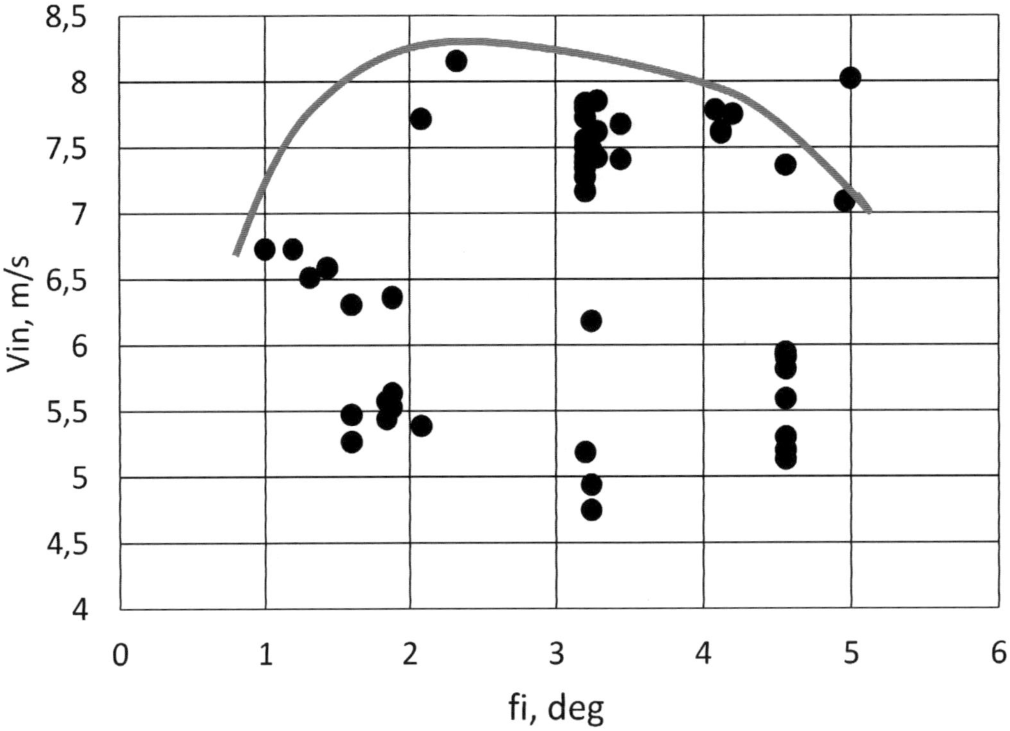
8,5
8
7,5
7
6,5
6
5,5
5
4,5
4
Vin, m/s
0
1
2
3
4
5
6
fi, deg

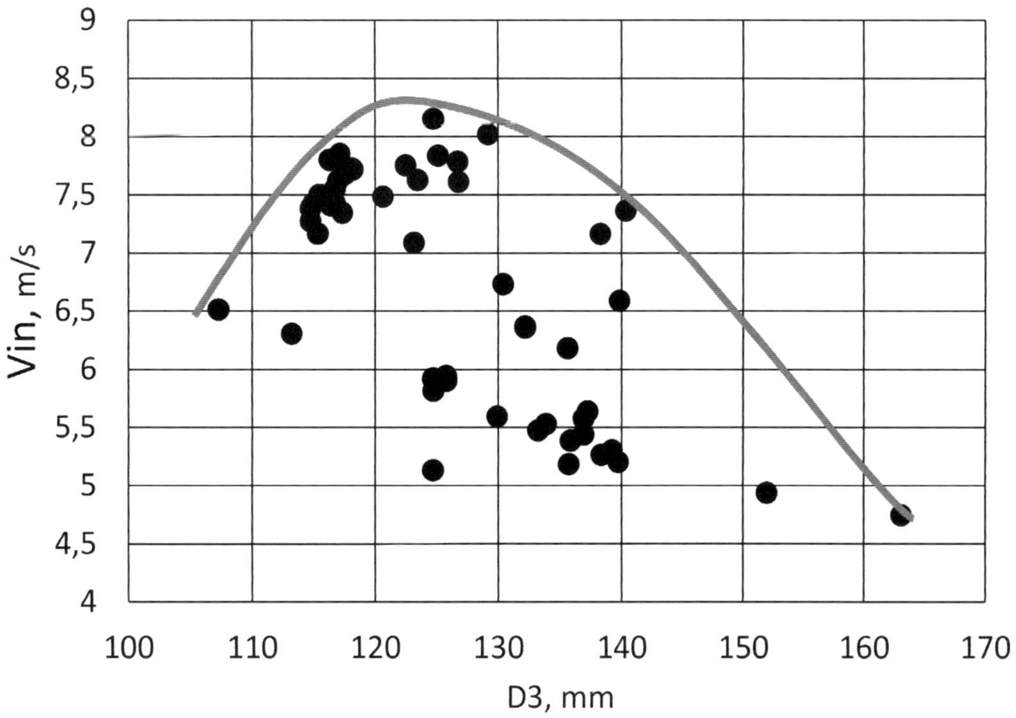

Figure 7. Dependences of speed in the inlet pipe on its geometric parameters

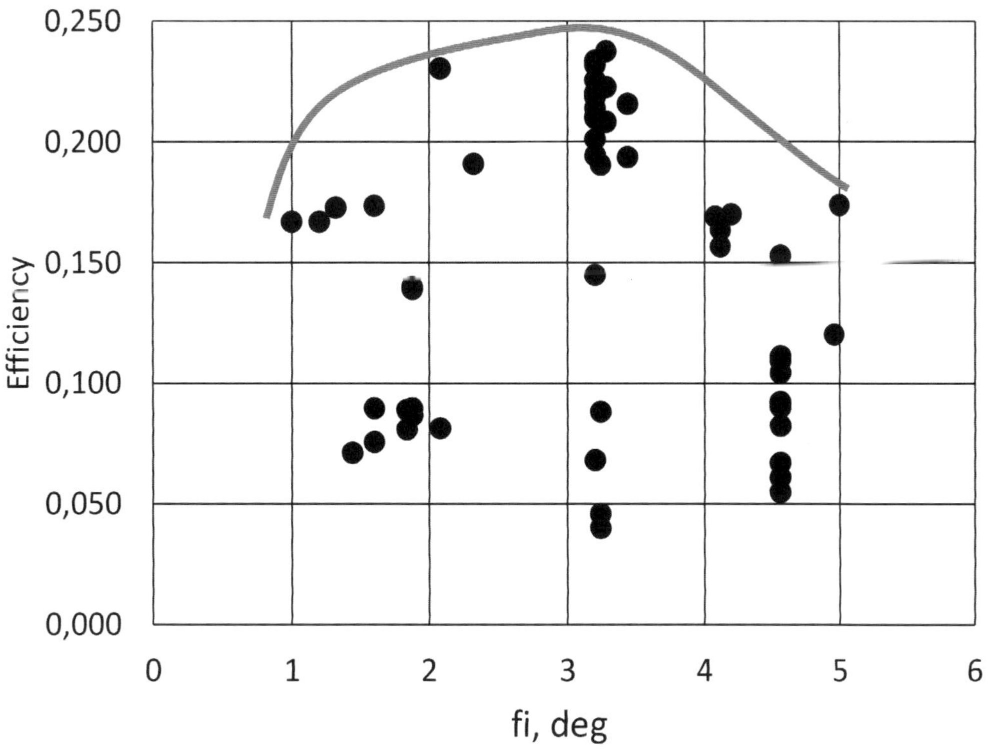

IOP Conf. Series: Materials Science and Engineering **779** (2020) 012056 doi:10.1088/1757-899X/779/1/012056

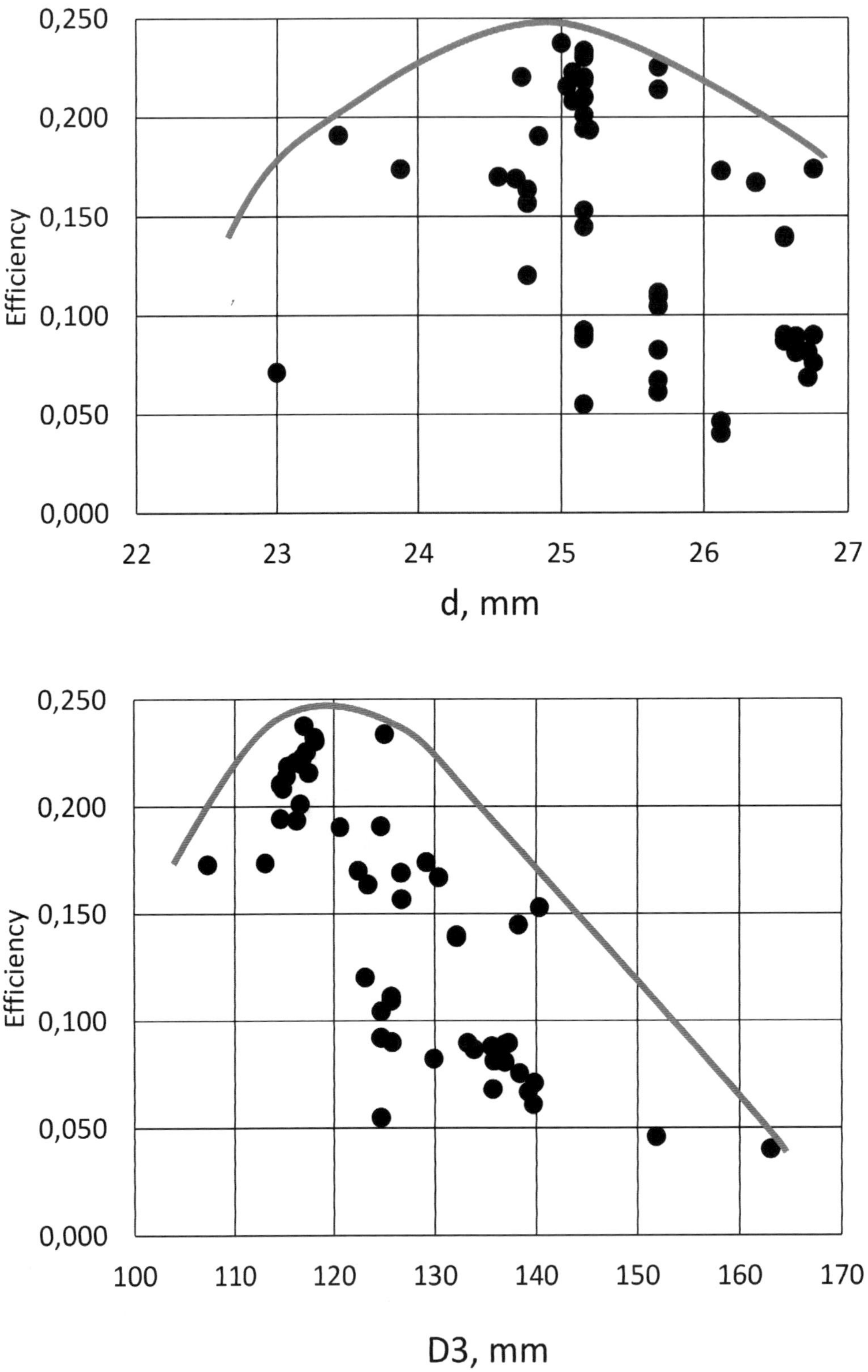

Figure 8. Efficiency dependencies on the geometric parameters

IOP Conf. Series: Materials Science and Engineering **779** (2020) 012056 doi:10.1088/1757-899X/779/1/012056

Conclusion

In the process of optimization the required speed value in the suction pipe was obtained, the power value was kept within permissible limits, and the highest value of the efficiency was achieved by changing the geometric dimensions of the elements of the flow part of the jet pump.

According to the optimization results, the dependences of efficiency, speed in the inlet pipe and power on the geometric parameters were obtained. The most demonstrative of them were given in this article.

It should be noted that the most effective diffuser opening angle turned out to be 3.2 °, which is less than the recommended values for the opening angle of classical stream (10-12 °). Most likely, this is due to the peripheral disposition of the nozzles of the active stream.

To reduce the time and volume of calculations, the inlet suction pipe was calculated by hydrodynamic modeling methods, its drag coefficient was determined and used to set the total pressure at the inlet.

Thus, in the course of the study, it was managed to obtain a flow part capable of performing the tasks facing the designed jet pump, and some interesting dependences were obtained.

References

[1] KalachevV. V. Strujnyenasosy. Teoriya, raschetiproektirovanie. 2017.

[2] Wang, X., Chen, Y.L., Zhao, Y., Xu, Y., Wang, B. Influence of suction chamber profile on flow field of annular jet pump. IOP Conference Series: Earth and Environmental Science. Vol 3.4, issue 3, 2019. DOI: 10.1088/1755-1315/304/3/032088.

[3] Panevnik, A.V.Choosing diameter of the operating nozzle of a hole jet pump. Neftyanoekhozyaystvo - Oil Industry, 2002.

[4] Darbandi, M., Sabzpoushan, S.,Schneider, G.Geometrical optimization of a steam jet-ejector using the computational fluid dynamics. Vol 2, 2018. DOI: 10.1115/FEDSM2018-83203.

[5] Cheremushkin V.A., Petrov A.I., Chaburko P.S. Using Stator Blades in Accessory Tracts of Hermetic Pumps // Mashiny I Ustanovki: Proektirovanie, Razrabotka I Ekspluataciya. 2017, no. 2, pp 1–12

[6] Lomakin V O 2015 Investigation of two-phase flow in axial-centrifugal impeller by hydrodynamic modeling methods Proceedings of 2015 International Conference on FluidPower and Mechatronics, FPM 2015 no 7337302 pp 1204–6

[7] Lomakin V. O. Complex optimization of the wet part of the hermetic pump by the LPTau search method // Pumps. Turbines. Systems. 2016, no. 1 (18), pp. 55–61

[8] Lomakin V, Cheremushkin V and Chaburko P 2018 Investigation of vortex and hysteresis effects in the inlet device of a centrifugal pump 2018 Global Fluid Power Society PhD Symposium, GFPS, no 8472374 DOI: 10.1109/GFPS.2018.8472374

[9] Tesch K, and Kaczorowska-Ditrich K 2018 Discrete-continuous optimization of an axial flow blood pump 23rd Fluid Mechanics Conference, FMC 2018 (Zawiercie, Poland) 1101 DOI: 10.1088/1742-6596/1101/1/012044

[10] V. Cheremushkin and A. Polyakov2019 Optimization of the output device of a disk pump for high viscous fluid. IOP Conf. Series: Materials Science and Engineering 589 (2019). doi:10.1088/1757-899X/589/1/012001

AUTHOR INDEX